Structure Determination of Organic Compounds

Ernö Pretsch · Philippe Bühlmann ·
Martin Badertscher

Structure Determination
of Organic Compounds

Tables of Spectral Data

Fourth, Revised and Enlarged Edition

 Springer

Prof. Dr. Ernö Pretsch
ETH Zürich
Institute of Biogeochemistry and
Pollutant Dynamics
Universitätsstr. 16
8092 Zürich
Switzerland
pretsche@ethz.ch

Prof. Dr. Philippe Bühlmann
University of Minnesota
Dept. of Chemistry
209 Pleasant Street SE.,
Minneapolis, MN 55455
USA
buhlmann@umn.edu

Dr. Martin Badertscher
ETH Zürich
Laboratory of Organic Chemistry
Wolfgang-Pauli-Str. 10
8093 Zürich
Switzerland
badertscher@org.chem.ethz.ch

ISBN 978-3-540-93809-5   e-ISBN 978-3-540-93810-1

DOI 10.1007/978-3-540-93810-1

Library of Congress Control Number: 2009920112

© Springer-Verlag Berlin Heidelberg 2009

Cover design: WMXDesign GmbH

Printed on acid-free paper

9 8 7 6 5 4 3 2 1

springer.com

Preface

The ongoing success of the earlier versions of this book motivated us to prepare a new edition. While modern techniques of nuclear magnetic resonance spectroscopy and mass spectrometry have changed the ways of data acquisition and greatly extended the capabilities of these methods, the basic parameters, such as chemical shifts, coupling constants, and fragmentation pathways remain the same. However, since the amount and quality of available data has considerably increased over the years, we decided to prepare a significantly revised manuscript. It follows the same basic concepts, i.e., it provides a representative, albeit limited set of reference data for the interpretation of ^{13}C NMR, ^{1}H NMR, IR, mass, and UV/Vis spectra. We also added a new chapter with reference data for ^{19}F and ^{31}P NMR spectroscopy and, in the chapter on infrared spectroscopy, we newly refer to important Raman bands.

Since operating systems of computers become outdated much faster than printed media, we decided against providing a compact disk with this new edition. The limited versions of the NMR spectra estimation programs can be downloaded from the home page of the developing company (www.upstream.ch/support/book_downloads.html).

We thank numerous colleagues who helped us in many different ways to complete the manuscript. We are particularly indebted to Dr. Dorothée Wegmann for her expertise with which she eliminated many errors and inconsistencies of the earlier versions. Special thanks are due to Prof. Wolfgang Robien for providing us with reference data from his outstanding ^{13}C NMR database, CSEARCH. Another high-quality source of information was the Spectral Database System of the National Institute of Advanced Industrial Science and Technology (http://riodb01.ibase.aist.go.jp/sdbs/), Tsukuba, Ibaraki (Japan).

In spite of great efforts and many checks to eliminate errors, it is likely that some mistakes or inconsistencies remain. We would like to encourage our readers to contact us with comments and suggestions under one of the following addresses: Prof. Ernö Pretsch, Institute of Biogeochemistry and Pollutant Dynamics, ETH Zürich, CH-8092 Zürich, Switzerland, e-mail: pretsche@ethz.ch, Prof. Philippe Bühlmann, Department of Chemistry, University of Minnesota, 207 Pleasant St. SE, Minneapolis, MN 55455, USA, e-mail: buhlmann@umn.edu, or Dr. Martin Badertscher, Laboratory of Organic Chemistry, ETH Zürich, CH-8093 Zürich, Switzerland, e-mail: badertscher@org.chem.ethz.ch.

Zürich and Minneapolis, November 2008

Contents

1 Introduction

1.1 Scope and Organization

The present data collection is intended to serve as an aid in the interpretation of molecular spectra for the elucidation and confirmation of the structure of organic compounds. It consists of reference data, spectra, and empirical correlations from ^{1}H, ^{13}C, ^{19}F, and ^{31}P nuclear magnetic resonance (NMR), infrared (IR), mass, and ultraviolet–visible (UV/Vis) spectroscopy. It is to be viewed as a supplement to textbooks and specific reference works dealing with these spectroscopic techniques. The use of this book to interpret spectra only requires the knowledge of basic principles of the techniques, but its content is structured in a way that it will serve as a reference book also to specialists.

Chapters 2 and 3 contain Summary Tables and Combined Tables of the most relevant spectral characteristics of structural elements. While Chapter 2 is organized according to the different spectroscopic methods, Chapter 3 for each class of structural elements supplies spectroscopic information obtained with various techniques. These two chapters should assist users less familiar with spectra interpretation to identify the classes of structural elements present in samples of their interest. The four chapters with data from ^{13}C NMR, ^{1}H NMR, IR spectroscopy, and mass spectrometry are ordered in the same manner by compound types. These cover the various carbon skeletons (alkyl, alkenyl, alkynyl, alicyclic, aromatic, and heteroaromatic), the most important substituents (halogen, single-bonded oxygen, nitrogen, sulfur, and carbonyl), and some specific compound classes (miscellaneous compounds and natural products). Finally, a spectra collection of common solvents, auxiliary compounds (such as matrix materials and references), and commonly found impurities is provided with each method. Not only the strictly analogous order of the data but also the optical marks on the edge of the pages help fast cross-referencing between the various spectroscopic techniques. Because their data sets are less comprehensive, the chapters on ^{19}F and ^{31}P NMR and UV/Vis are organized somewhat differently. Although currently UV/Vis spectroscopy is only marginally relevant to structure elucidation, its importance might increase by the advent of high-throughput analyses. Also, the reference data presented in the UV/Vis chapter are useful in connection with optical sensors and the widely applied UV/Vis detectors in chromatography and electrophoresis.

Since a great part of the tabulated data either comes from our own measurements or is based on a large body of literature data, comprehensive references to published sources are not included. Whenever possible, the data refer to conventional modes and conditions of measurement. For example, unless the solvent is indicated, the NMR chemical shifts were normally determined with deuterochloroform. Likewise, the IR spectra were measured using solvents of low polarity, such as chloroform or carbon disulfide. Mass spectral data were recorded with electron impact ionization at 70 eV.

While retaining the basic structure of the previous editions, numerous reference entries have been updated and new entries have been added. Altogether, about 20% of the data is new. The chapter on ^{19}F and ^{31}P NMR is entirely new, and the section on IR spectroscopy now includes references to important Raman bands.

1.2 Abbreviations and Symbols

al	aliphatic
alk	alkyl
alken	alkenyl
ar	aromatic
as	asymmetric
ax	axial
comb	combination vibration
d	doublet
δ	IR: deformation vibration
	NMR: chemical shift
DFTMP	1,1-difluoro-1-(trimethylsilyl)methylphosphonic acid
DMSO	dimethyl sulfoxide
eq	equatorial
ε	molar absorptivity
frag	fragment
γ	skeletal vibration
gem	geminal
hal	halogen
ip	in plane vibration
J	coupling constant
liq	liquid
$M^{+\cdot}$	molecular radical ion
m/z	mass to charge ratio
$\tilde{v}$	wavenumber
oop	out of plane vibration
sh	shoulder
st	stretching vibration
sy	symmetric
TFA	trifluoroacetic acid
THF	tetrahydrofuran
TMS	tetramethylsilane
vic	vicinal

2 Summary Tables

2.1 General Tables

2.1.1 Calculation of the Number of Double Bond Equivalents from the Molecular Formula

General Equation

$$\text{double bond equivalents} = 1 + \tfrac{1}{2} \sum_i n_i \, (v_i - 2)$$

n_i: number of atoms of element i in molecular formula
v_i: formal valence of element i

Short Cut

For compounds containing only C, H, O, N, S, and halogens, the following steps permit a quick and simple calculation of the number of double bond equivalents:

1. O and divalent S are deleted from the molecular formula
2. Halogens are replaced by hydrogen
3. Trivalent N is replaced by CH
4. The resulting hydrocarbon, C_nH_x, is compared with the saturated hydrocarbon, C_nH_{2n+2}. Each double bond equivalent reduces the number of hydrogen atoms by 2:

$$\text{double bond equivalents} = \tfrac{1}{2} \, (2\,n + 2 - x)$$

2.1.2 Properties of Selected Nuclei

Isotope	Natural abundance [%]	Spin quantum number, I	Frequency [MHz] at 2.35 Tesla	Relative sensitivity of nucleus	Relative sensitivity at natural abundance	Electric quadrupole moment [e × 10^{-24} cm^2]
^{1}H	99.985	1/2	100.0	1	1	
^{2}H	0.015	1	15.4	9.6×10^{-3}	1.5×10^{-6}	2.8×10^{-3}
^{3}H	0.000	1/2	106.7	1.2	0	
^{10}B	19.58	3	10.7	2.0×10^{-2}	3.9×10^{-3}	7.4×10^{-2}
^{11}B	80.42	3/2	32.1	1.6×10^{-1}	1.3×10^{-1}	3.6×10^{-2}
^{13}C	1.108	1/2	25.1	1.6×10^{-2}	1.8×10^{-4}	
^{14}N	99.635	1	7.3	1.0×10^{-3}	1.0×10^{-3}	1.9×10^{-2}
^{15}N	0.365	1/2	10.1	1.0×10^{-3}	3.8×10^{-6}	
^{17}O	0.037	5/2	13.6	2.9×10^{-2}	1.1×10^{-5}	-2.6×10^{-2}
^{19}F	100.000	1/2	94.1	8.3×10^{-1}	8.3×10^{-1}	
^{31}P	100.000	1/2	40.5	6.6×10^{-2}	6.6×10^{-2}	
^{33}S	0.76	3/2	7.6	2.3×10^{-3}	1.7×10^{-5}	-6.4×10^{-2}
^{117}Sn	7.61	1/2	35.6	4.5×10^{-2}	3.4×10^{-3}	
^{119}Sn	8.58	1/2	37.3	5.2×10^{-2}	4.4×10^{-3}	
^{195}Pt	33.8	1/2	21.5	9.9×10^{-3}	3.4×10^{-3}	
^{199}Hg	16.84	1/2	17.8	5.7×10^{-3}	9.5×10^{-4}	
^{207}Pb	22.6	1/2	20.9	9.2×10^{-3}	2.1×10^{-4}	

2.2 ^{13}C NMR Spectroscopy

Summary of the Regions of Chemical Shifts, δ (in ppm), for Carbon Atoms in Various Chemical Environments (carbon atoms are specified as follows: Q for CH$_3$, T for CH$_2$, D for CH, and S for C)

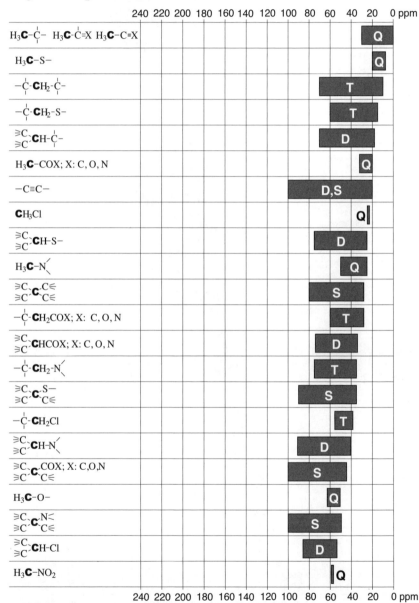

	240	220	200	180	160	140	120	100	80	60	40	20	0 ppm
$-\overset{\text{\textbar}}{\underset{\text{\textbar}}{C}}-\mathbf{CH_2}-O-$									T				
$\overset{\geqslant C}{\underset{\geqslant C}{}}\mathbf{CH}-O-$								D					
$-\overset{\text{\textbar}}{\underset{\text{\textbar}}{C}}-\mathbf{CH_2}-NO_2$									T				
$\overset{\geqslant C}{\underset{\geqslant C}{}}\mathbf{C}\overset{Cl}{\underset{C\leqslant}{}}$								S					
$\overset{\geqslant C}{\underset{\geqslant C}{}}\mathbf{CH}-NO_2$								D					
$\overset{\geqslant C}{\underset{\geqslant C}{}}\mathbf{C}\overset{O-}{\underset{C\leqslant}{}}$								S					
$\overset{\geqslant C}{\underset{\geqslant C}{}}\mathbf{C}\overset{NO_2}{\underset{C\leqslant}{}}$								S					
$\overset{H}{\underset{H}{}}\mathbf{C}=C\big<$							T						
$\overset{X}{\bigcirc}\mathbf{C}$-H; X: any substituent						D							
$\big>C=C\big<$						D, S							
$\big>\mathbf{C}\overset{O-}{\underset{O-}{}}$					**T, D, S**								
$\bigcirc\mathbf{C}$-X; X: any substituent					S								
N ring $\mathbf{C}$-X; X: any substituent					D, S								
$-\mathbf{C}\equiv N$							S						
N ring $\mathbf{C}=X$; X: any substituent				D, S									
$\big>\mathbf{C}=N-C\leqslant$ $\big>\mathbf{C}=N-O-$				D, S									
α,β-unsat. $\mathbf{C}$OX; X: O, N, Cl				S									
α,β-unsat. $\mathbf{C}$OOH				S									
$-\overset{\text{\textbar}}{\underset{\text{\textbar}}{C}}-\mathbf{C}OX$; X: O, N, Cl				S									
$-\overset{\text{\textbar}}{\underset{\text{\textbar}}{C}}-\mathbf{C}OOH$			S										
α,β-unsat. $\mathbf{C}$OH		S											
$-\overset{\text{\textbar}}{\underset{\text{\textbar}}{C}}-\mathbf{C}SX$; X: O, N		S											
α,β-unsat. $\big>\mathbf{C}=O$		S											
$-\overset{\text{\textbar}}{\underset{\text{\textbar}}{C}}-\mathbf{C}HO$	D												
$\overset{\geqslant C}{\underset{\geqslant C}{}}\mathbf{C}=O$		S											
$\overset{\geqslant C}{\underset{\geqslant C}{}}\mathbf{C}=S$	S												

| | 240 | 220 | 200 | 180 | 160 | 140 | 120 | 100 | 80 | 60 | 40 | 20 | 0 ppm |

^{13}C Chemical Shifts of Carbonyl Groups (δ in ppm)

R	R–CHO	R–COCH$_3$	R–COOH	R–COO$^-$
–H	197.0	200.5	166.3	171.3
–CH$_3$	200.5	206.7	176.9	182.6
–CH$_2$CH$_3$	202.7	207.6	180.4	185.1
–CH(CH$_3$)$_2$	204.6	211.8	184.1	
–C(CH$_3$)$_3$	205.6	213.5	185.9	188.6
–n-C$_8$H$_{17}$	202.6	207.9	180.7	183.1
–CH$_2$Cl	193.3	200.1	173.7	175.9
–CHCl$_2$		193.6	170.4	171.8
–CCl$_3$	176.9	186.3	167.1	167.6
–cyclohexyl	204.7	209.4	182.1	185.4
–CH=CH$_2$	194.4	197.5	171.7	174.5
–C≡CH	176.8	183.6	156.5	
–phenyl	192.0	196.9	172.6	177.6

R	R–CHO	R–COCH$_3$	R–COOH	R–COO$^-$
–H	161.6	167.6	158.5	
–CH$_3$	171.3	173.4	167.4	170.4
–CH$_2$CH$_3$	173.3	177.2	170.3	174.7
–CH(CH$_3$)$_2$	177.4		172.8	178.0
–C(CH$_3$)$_3$	178.8	180.9	173.9	180.3
–n-C$_8$H$_{17}$	174.4	176.3	169.4	173.8
–CH$_2$Cl	167.8	168.3	162.1	167.7
–CHCl$_2$	165.1		157.6	165.5
–CCl$_3$	162.5		154.1	
–cyclohexyl	175.3	177.3		176.3
–CH=CH$_2$	166.5	168.3		165.6
–C≡CH	153.4			
–phenyl	166.8	169.7	162.8	168.0

2.3 ^{1}H NMR Spectroscopy

Summary of the Regions of Chemical Shifts, δ (in ppm), for Hydrogen Atoms in Various Chemical Environments

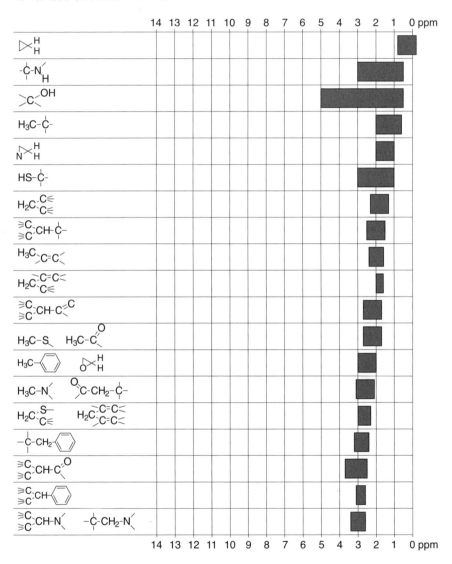

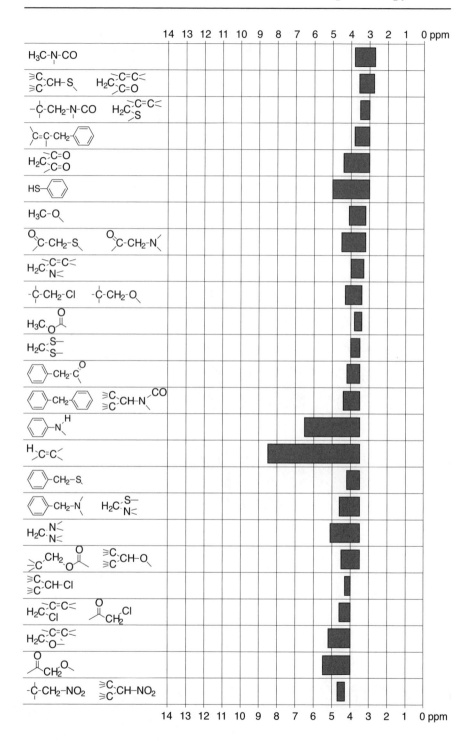

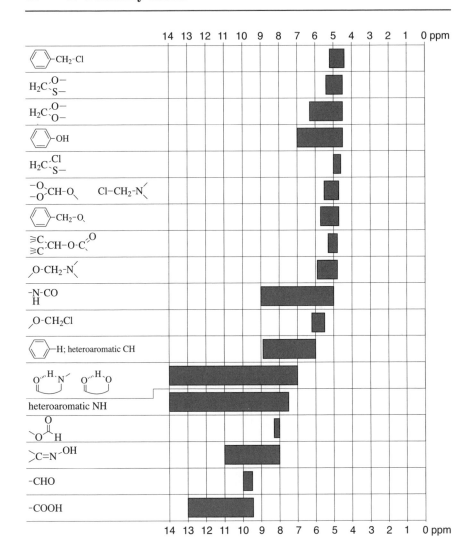

2.4 IR Spectroscopy

Summary of the Most Important IR Absorption Bands ($\tilde{\nu}$ in cm^{-1})

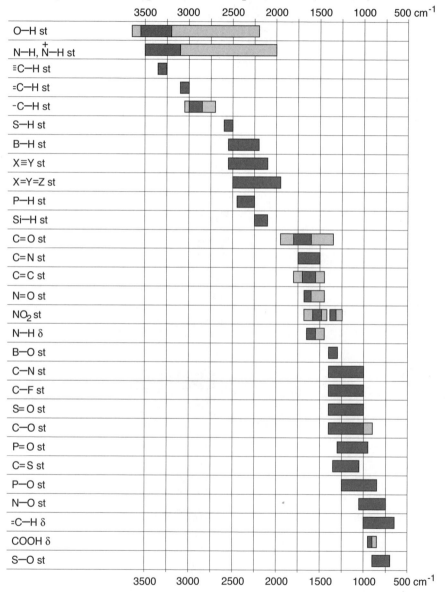

Summary of IR Absorption Bands of Carbonyl Groups ($\tilde{\nu}$ in cm^{-1})

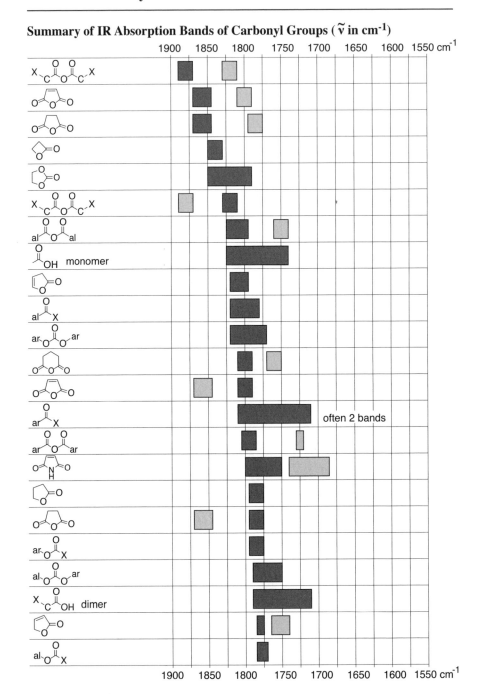

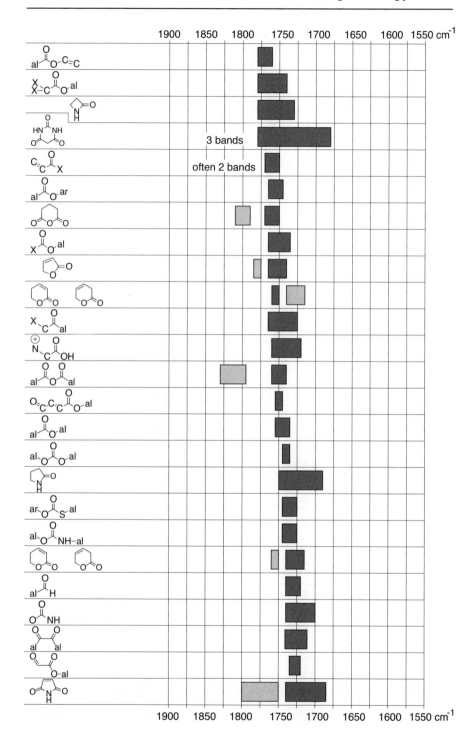

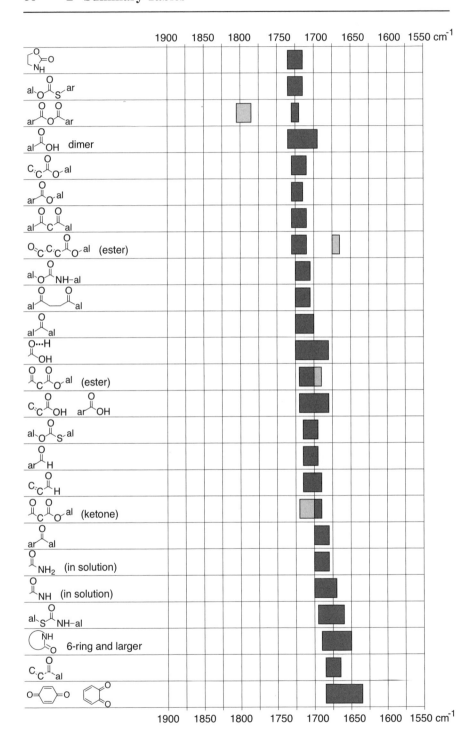

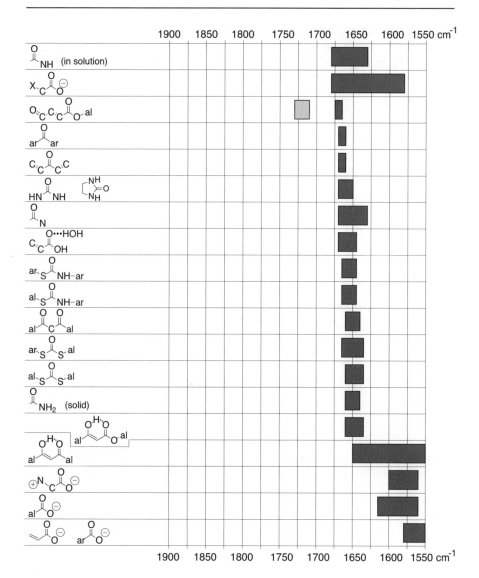

2.5 Mass Spectrometry

2.5.1 Average Masses of Naturally Occurring Elements with Masses and Representative Relative Abundances of Isotopes [1–3]

Element Isotope	Mass	Abundance	Element Isotope	Mass	Abundance
H	1.00794[a,b]	(in water)	**F**	18.998403	
^{1}H	1.007825	100[c]	^{19}F	18.998403	100
^{2}H	2.014102	0.0115			
			Ne	20.1797[a]	(in air)
He	4.002602[a]	(in air)	^{20}Ne	19.992440	100[c]
^{3}He	3.016029	0.000134	^{21}Ne	20.993847	0.38
^{4}He	4.002603	100	^{22}Ne	21.991385	10.22
Li	6.941[a]		**Na**	22.989769	
^{6}Li	6.015123	8.21[d]	^{23}Na	22.989769	100
^{7}Li	7.016005	100			
			Mg	24.3050	
Be	9.012182		^{24}Mg	23.985042	100
^{9}Be	9.012182	100	^{25}Mg	24.985837	12.66
			^{26}Mg	25.982593	13.94
B	10.811[a]				
^{10}B	10.012937	24.8[c]	**Al**	26.981538	
^{11}B	11.009305	100	^{27}Al	26.981538	100
C	12.0107[a]		**Si**	28.0855[a]	
^{12}C	12.000000	100	^{28}Si	27.976927	100
^{13}C	13.003355	1.08	^{29}Si	28.976495	5.080
			^{30}Si	29.973770	3.353
N	14.0067[a]				
^{14}N	14.003074	100	**P**	30.973762	
^{15}N	15.000109	0.365	^{31}P	30.973762	100
O	15.9994[a]		**S**	32.065[a]	
^{16}O	15.994915	100	^{32}S	31.972071	100[c]
^{17}O	16.999132	0.038	^{33}S	32.971459	0.79
^{18}O	17.999161	0.205	^{34}S	33.967867	4.47
			^{36}S	35.967081	0.01

Element Isotope	Mass	Abundance	Element Isotope	Mass	Abundance
Cl	35.453		**Cr**	51.9961	
^{35}Cl	34.968853	100^c	^{50}Cr	49.946044	5.186
^{37}Cl	36.965903	32.0	^{52}Cr	51.940508	100
			^{53}Cr	52.940649	11.339
Ar	39.948^a	(in air)	^{54}Cr	53.938880	2.823
^{36}Ar	35.967545	0.3379			
^{38}Ar	37.962732	0.0635	**Mn**	54.938045	
^{40}Ar	39.962383	100	^{55}Mn	54.938045	100
K	39.0983		**Fe**	55.845	
^{39}K	38.963707	100	^{54}Fe	53.939611	6.370
^{40}K	39.963998	0.0125	^{56}Fe	55.934938	100
^{41}K	40.961826	7.2167	^{57}Fe	56.935394	2.309
			^{58}Fe	57.933276	0.307
Ca	40.078				
^{40}Ca	39.962591	100	**Co**	58.933195	
^{42}Ca	41.958618	0.667	^{59}Co	58.933195	100
^{43}Ca	42.958767	0.139			
^{44}Ca	43.955482	2.152	**Ni**	58.6934	
^{46}Ca	45.953693	0.004	^{58}Ni	57.935343	100
^{48}Ca	47.952534	0.193	^{60}Ni	59.930786	38.5198
			^{61}Ni	60.931056	1.6744
Sc	44.955912		^{62}Ni	61.928345	5.3388
^{45}Sc	44.955912	100	^{64}Ni	63.927966	1.3596
Ti	47.867		**Cu**	63.546^a	
^{46}Ti	45.952632	11.19	^{63}Cu	62.929598	100
^{47}Ti	46.951763	10.09	^{65}Cu	64.927790	44.61
^{48}Ti	47.947946	100			
^{49}Ti	48.947870	7.34	**Zn**	65.409	
^{50}Ti	49.944791	7.03	^{64}Zn	63.929142	100
			^{66}Zn	65.926033	57.96
V	50.9415		^{67}Zn	66.927127	8.49
^{50}V	49.947159	0.251	^{68}Zn	67.924844	39.41
^{51}V	50.943960	100	^{70}Zn	69.925319	1.31

Element Isotope	Mass	Abundance	Element Isotope	Mass	Abundance
Ga	69.723		**Rb**	85.4678	
[69]Ga	68.925574	100[c]	[85]Rb	84.911790	100
[71]Ga	70.924701	66.36	[87]Rb	86.909181	38.56
Ge	72.64		**Sr**	87.62[a]	
[70]Ge	69.924247	55.50	[84]Sr	83.913425	0.68
[72]Ge	71.922076	74.37	[86]Sr	85.909260	11.94
[73]Ge	72.923459	21.13	[87]Sr	86.908877	8.48
[74]Ge	73.921178	100	[88]Sr	87.905612	100
[76]Ge	75.921403	21.32			
			Y	88.905848	
As	74.921597		[89]Y	88.905848	100
[75]As	74.921597	100			
			Zr	91.224	
Se	78.96		[90]Zr	89.904704	100
[74]Se	73.922476	1.79	[91]Zr	90.905646	21.81
[76]Se	75.919214	18.89	[92]Zr	91.905041	33.33
[77]Se	76.919914	15.38	[94]Zr	93.906315	33.78
[78]Se	77.917309	47.91	[96]Zr	95.908273	5.44
[80]Se	79.916521	100			
[82]Se	81.916699	17.60	**Nb**	92.906378	
			[93]Nb	92.906378	100
Br	79.904				
[79]Br	78.918337	100	**Mo**	95.94	
[81]Br	80.916291	97.28	[92]Mo	91.906811	61.06
			[94]Mo	93.905088	38.16
Kr	83.798	(in air)	[95]Mo	94.905842	65.72
[78]Kr	77.920382	0.623[c]	[96]Mo	95.904680	68.95
[80]Kr	79.916379	4.011	[97]Mo	96.906022	39.52
[82]Kr	81.913484	20.343	[98]Mo	97.905408	100
[83]Kr	82.914136	20.180	[100]Mo	99.907477	39.98
[84]Kr	83.911507	100			
[86]Kr	85.910611	30.321			

Element Isotope	Mass	Abundance	Element Isotope	Mass	Abundance
Ru	101.07		**In**	114.818	
^{96}Ru	95.907598	17.56	^{113}In	112.904058	4.48
^{98}Ru	97.905287	5.93	^{115}In	114.903878	100
^{99}Ru	98.905939	40.44			
^{100}Ru	99.904220	39.94	**Sn**	118.710	
^{101}Ru	100.905582	54.07	^{112}Sn	111.904818	2.98
^{102}Ru	101.904349	100	^{114}Sn	113.902779	2.03
^{104}Ru	103.905433	59.02	^{115}Sn	114.903342	1.04
			^{116}Sn	115.901741	44.63
Rh	102.905504		^{117}Sn	116.902952	23.57
^{103}Rh	102.905504	100	^{118}Sn	117.901603	74.34
			^{119}Sn	118.903309	26.37
Pd	106.42		^{120}Sn	119.902195	100
^{102}Pd	101.905609	3.73	^{122}Sn	121.903439	14.21
^{104}Pd	103.904036	40.76	^{124}Sn	123.905274	17.77
^{105}Pd	104.905085	81.71			
^{106}Pd	105.903486	100	**Sb**	121.760	
^{108}Pd	107.903892	96.82	^{121}Sb	120.903816	100
^{110}Pd	109.905153	42.88	^{123}Sb	122.904214	74.79
Ag	107.8682		**Te**	127.60	
^{107}Ag	106.905097	100	^{120}Te	119.904020	0.26
^{109}Ag	108.904752	92.90	^{122}Te	121.903044	7.48
			^{123}Te	122.904270	2.61
Cd	112.411		^{124}Te	123.902818	13.91
^{106}Cd	105.906459	4.35	^{125}Te	124.904431	20.75
^{108}Cd	107.904184	3.10	^{126}Te	125.903312	55.28
^{110}Cd	109.903002	43.47	^{128}Te	127.904463	93.13
^{111}Cd	110.904178	44.55	^{130}Te	129.906224	100
^{112}Cd	111.902758	83.99			
^{113}Cd	112.904402	42.53	**I**	126.904473	
^{114}Cd	113.903359	100	^{127}I	126.904473	100
^{116}Cd	115.904756	26.07			

Element Isotope	Mass	Abundance	Element Isotope	Mass	Abundance
Xe	131.293		**Nd**	144.242	
^{124}Xe	123.905893	0.354^c	^{142}Nd	141.907723	100
^{126}Xe	125.904274	0.330	^{143}Nd	142.909815	44.9
^{128}Xe	127.903531	7.099	^{144}Nd	143.910087	87.5
^{129}Xe	128.904779	98.112	^{145}Nd	144.912574	30.5
^{130}Xe	129.903508	15.129	^{146}Nd	145.913117	63.2
^{131}Xe	130.905082	78.906	^{148}Nd	147.916893	21.0
^{132}Xe	131.904154	100	^{150}Nd	149.920891	20.6
^{134}Xe	133.905395	38.782			
^{136}Xe	135.907219	32.916	**Sm**	150.36	
			^{144}Sm	143.911999	11.48
Cs	132.905452		^{147}Sm	146.914898	56.04
^{133}Cs	132.905452	100	^{148}Sm	147.914823	42.02
			^{149}Sm	148.917185	51.66
Ba	137.327		^{150}Sm	149.917276	27.59
^{130}Ba	129.906321	0.148	^{152}Sm	151.919732	100
^{132}Ba	131.905061	0.141	^{154}Sm	153.922209	85.05
^{134}Ba	133.904508	3.371			
^{135}Ba	134.905689	9.194	**Eu**	151.964	
^{136}Ba	135.904576	10.954	^{151}Eu	150.919850	91.61
^{137}Ba	136.905827	15.666	^{153}Eu	152.921230	100
^{138}Ba	137.905247	100			
			Gd	157.25	
La	138.90547		^{152}Gd	151.919791	0.81
^{138}La	137.907112	0.090	^{154}Gd	153.920866	8.78
^{139}La	138.906353	100	^{155}Gd	154.922622	59.58
			^{156}Gd	155.922123	82.41
Ce	140.116		^{157}Gd	156.923960	63.00
^{136}Ce	135.907172	0.209	^{158}Gd	157.924104	100
^{138}Ce	137.905991	0.284	^{160}Gd	159.927054	88.00
^{140}Ce	139.905439	100			
^{142}Ce	141.909244	12.565	**Tb**	158.925347	
			^{159}Tb	158.925347	100
Pr	140.907653				
^{141}Pr	140.907653	100			

Element Isotope	Mass	Abundance	Element Isotope	Mass	Abundance
Dy	162.500		**Hf**	178.49	
^{156}Dy	155.924283	0.20	^{174}Hf	173.940046	0.46
^{158}Dy	157.924409	0.34	^{176}Hf	175.941409	14.99
^{160}Dy	159.925198	8.24	^{177}Hf	176.943221	53.02
^{161}Dy	160.926933	66.84	^{178}Hf	177.943699	77.77
^{162}Dy	161.926798	90.15	^{179}Hf	178.944816	38.83
^{163}Dy	162.928731	88.10	^{180}Hf	179.946550	100
^{164}Dy	163.929175	100			
			Ta	180.94788	
Ho	164.930322		^{180}Ta	179.947465	0.012
^{165}Ho	164.930322	100	^{181}Ta	180.947996	100
Er	167.259		**W**	183.84	
^{162}Er	161.928778	0.41	^{180}W	179.946704	0.39
^{164}Er	163.929200	4.78	^{182}W	181.948204	86.49
^{166}Er	165.930293	100	^{183}W	182.950223	46.70
^{167}Er	166.932048	68.26	^{184}W	183.950931	100.0
^{168}Er	167.932370	80.52	^{186}W	185.954364	92.79
^{170}Er	169.935464	44.50			
			Re	186.207	
Tm	168.934213		^{185}Re	184.952955	59.74
^{169}Tm	168.934213	100	^{187}Re	186.955753	100
Yb	173.04		**Os**	190.23	
^{168}Yb	167.933897	0.41	^{184}Os	183.952489	0.05
^{170}Yb	169.934762	9.55	^{186}Os	185.953838	3.90
^{171}Yb	170.936326	44.86	^{187}Os	186.955751	4.81
^{172}Yb	171.936382	68.58	^{188}Os	187.955838	32.47
^{173}Yb	172.938211	50.68	^{189}Os	188.958148	39.60
^{174}Yb	173.938862	100	^{190}Os	189.958447	64.39
^{176}Yb	175.942572	40.09	^{192}Os	191.961481	100
Lu	174.967		**Ir**	192.217	
^{175}Lu	174.940772	100	^{191}Ir	190.960594	59.49
^{176}Lu	175.942686	2.66	^{193}Ir	192.962926	100.0

Element Isotope	Mass	Abundance	Element Isotope	Mass	Abundance
Pt	195.084		**Tl**	204.3833	
[190]Pt	189.959932	0.041	[203]Tl	202.972344	41.88
[192]Pt	191.961038	2.311	[205]Tl	204.974428	100
[194]Pt	193.962680	97.443			
[195]Pt	194.964791	100	**Pb**	207.2[a]	
[196]Pt	195.964952	74.610	[204]Pb	203.973044	2.7
[198]Pt	197.967893	21.172	[206]Pb	205.974465	46.0
			[207]Pb	206.975897	42.2
Au	196.966569		[208]Pb	207.976653	100
[197]Au	196.966569	100			
			Bi	208.980399	
Hg	200.59		[209]Bi	208.980399	100
[196]Hg	195.965833	0.50			
[198]Hg	197.966769	33.39	**Th**	232.038055	
[199]Hg	198.968280	56.50	[232]Th	232.038055	100
[200]Hg	199.968326	77.36			
[201]Hg	200.970302	44.14	**U**	238.02891	
[202]Hg	201.970643	100	[234]U	234.040952	0.0054[e]
[204]Hg	203.973494	23.01	[235]U	235.043930	0.7257
			[238]U	238.050788	100

[a] Natural variations in the isotopic composition of terrestrial materials do not allow to give a more precise value.

[b] The mole ratio of ^2H in hydrogen from gas cylinders was reported to be as low as 0.000032.

[c] Commercially available materials may have substantially different isotopic compositions if they were subjected to undisclosed or inadvertent isotopic fractionation.

[d] Materials depleted in ^6Li are commercial sources of laboratory shelf reagents and are known to have ^6Li abundances in the range of 2.0007–7.672 atom percent, with natural materials at the higher end of this range. Average atomic masses vary between 6.939 and 6.996; if a more accurate value is required, it must be determined for the specific material.

[e] Materials depleted in ^{235}U are commercial sources of laboratory shelf reagents.

2.5.2 Ranges of Natural Isotope Abundances of Selected Elements [3]

Element Isotope	Range [atom %]	Element Isotope	Range [atom %]	Element Isotope	Range [atom %]
H		**Si**		**Sr**	
^{1}H	99.9816–99.9974	^{28}Si	92.205–92.241	^{84}Sr	0.55–0.58
^{2}H	0.0026–0.0184	^{29}Si	4.678–4.692	^{86}Sr	9.75–9.99
		^{30}Si	3.082–3.102	^{87}Sr	6.94–7.14
He				^{88}Sr	82.29–82.75
^{3}He	4.6×10^{-8}–0.0041	**S**			
^{4}He	99.9959–100	^{32}S	94.454–95.281	**Ce**	
		^{33}S	0.730–0.793	^{136}Ce	0.185–0.186
Li		^{34}S	3.976–4.734	^{138}Ce	0.251–0.254
^{6}Li	7.225–7.714	^{36}S	0.013–0.019	^{140}Ce	88.446–88.449
^{7}Li	92.275–92.786			^{142}Ce	11.114–11.114
		Cl			
B		^{35}Cl	75.644–75.923	**Nd**	
^{10}B	18.929–20.386	^{37}Cl	24.077–24.356	^{142}Nd	26.80–27.30
^{11}B	79.614– 81.071			^{143}Nd	12.12–12.32
		Ca		^{144}Nd	23.79–23.97
C		^{40}Ca	96.933–96.947	^{145}Nd	8.23–8.35
^{12}C	98.853–99.037	^{42}Ca	0.646–0.648	^{146}Nd	17.06–17.35
^{13}C	0.963–1.147	^{43}Ca	0.135–0.135	^{148}Nd	5.66–5.78
		^{44}Ca	2.082–2.092	^{150}Nd	5.53–5.69
N		^{46}Ca	0.004–0.004		
^{14}N	99.579–99.654	^{48}Ca	0.186–0.188	**Pb**	
^{15}N	0.346–0.421			^{204}Pb	1.04–1.65
		V		^{206}Pb	20.84–27.48
O		^{50}V	0.2487–0.2502	^{207}Pb	17.62–23.65
^{16}O	99.738–99.776	^{51}V	99.7498–99.7513	^{208}Pb	51.28–56.21
^{17}O	0.037–0.040				
^{18}O	0.188–0.222	**Cu**		**U**	
		^{63}Cu	68.983–69.338	^{234}U	0.0050–0.0059
Ne		^{65}Cu	30.662–31.017	^{235}U	0.7198–0.7207
^{20}Ne	88.47–90.51			^{238}U	99.2739–99.2752
^{21}Ne	0.27–1.71				
^{22}Ne	9.20– 9.96				

2.5.3 Isotope Patterns of Naturally Occurring Elements

The mass of the most abundant isotope is given under the symbol of the element. The lightest isotope is shown at the left end of the x axis.

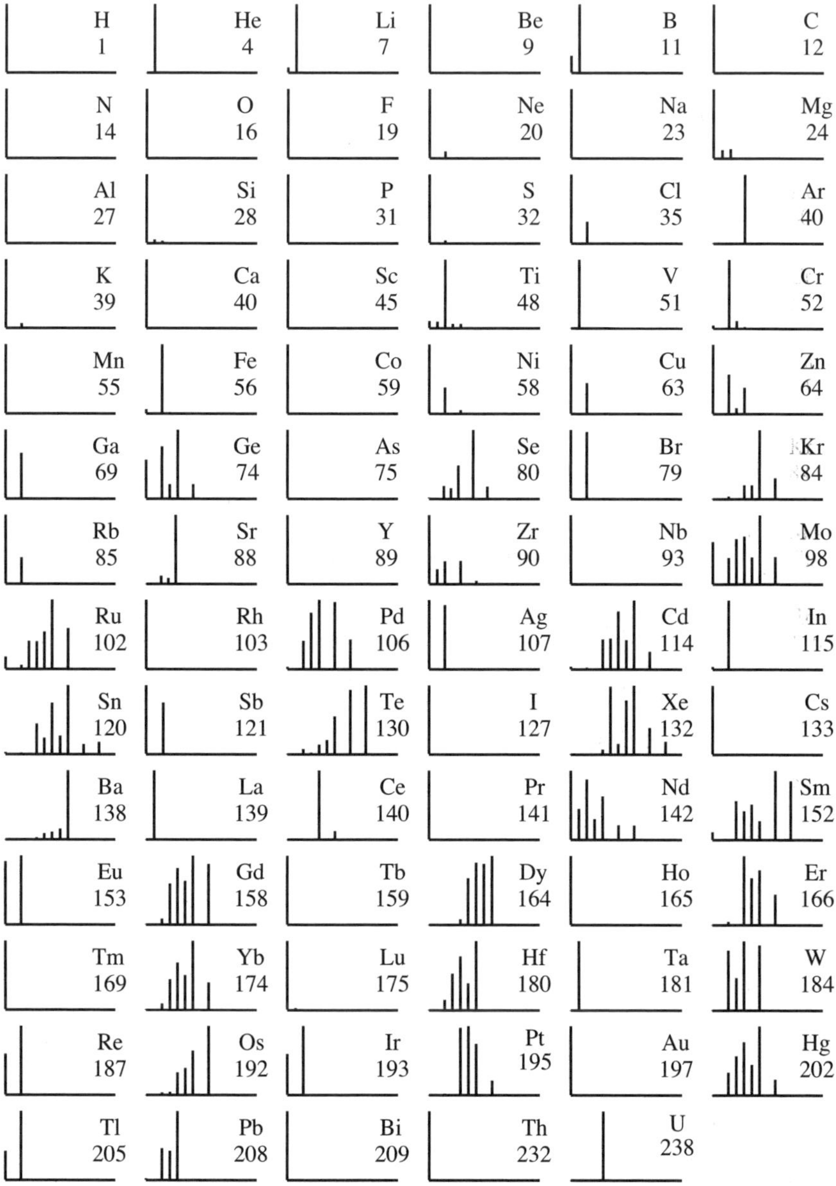

H	He	Li	Be	B	C
1	4	7	9	11	12
N	O	F	Ne	Na	Mg
14	16	19	20	23	24
Al	Si	P	S	Cl	Ar
27	28	31	32	35	40
K	Ca	Sc	Ti	V	Cr
39	40	45	48	51	52
Mn	Fe	Co	Ni	Cu	Zn
55	56	59	58	63	64
Ga	Ge	As	Se	Br	Kr
69	74	75	80	79	84
Rb	Sr	Y	Zr	Nb	Mo
85	88	89	90	93	98
Ru	Rh	Pd	Ag	Cd	In
102	103	106	107	114	115
Sn	Sb	Te	I	Xe	Cs
120	121	130	127	132	133
Ba	La	Ce	Pr	Nd	Sm
138	139	140	141	142	152
Eu	Gd	Tb	Dy	Ho	Er
153	158	159	164	165	166
Tm	Yb	Lu	Hf	Ta	W
169	174	175	180	181	184
Re	Os	Ir	Pt	Au	Hg
187	192	193	195	197	202
Tl	Pb	Bi	Th	U	
205	208	209	232	238	

2.5.4 Calculation of Isotope Distributions

The characteristic abundance patterns resulting from the combination of more than one polyisotopic element can be calculated from the relative abundances of the different isotopes. The following polynomial expression gives the isotope distribution of a polyisotopic molecule:

$$\{p_{i1}\,A^0 + p_{i2}\,A^{(m_{i2}\,-\,m_{i1})} + p_{i3}\,A^{(m_{i3}\,-\,m_{i1})} + \,...\}^{n_i} \times$$
$$\{p_{j1}\,A^0 + p_{j2}\,A^{(m_{j2}\,-\,m_{j1})} + p_{j3}\,A^{(m_{j3}\,-\,m_{j1})} + \,...\}^{n_j} \times \{...$$

where p_{ix} is the relative abundance of the xth isotope of element i, m_{ix} is the mass of the xth isotope of the element i, and the exponent n_i stands for the number of atoms of the element i in the molecule. The expansion of this polynomial expression after inserting the p_{ix} and m_{ix} values for all the isotopes 1, 2, 3, ... of the elements i, j, ... of a given molecule yields an expression that represents the isotope distribution:

$$w_0\,A^0 + w_r\,A^r + w_s\,A^s + w_t\,A^t + \,...$$

where the values of $w_0, w_r, w_s, w_t, ...$ are the relative abundances of $M^{+\cdot}$, $[M+r]^{+\cdot}$, $[M+s]^{+\cdot}$, $[M+t]^{+\cdot},...$, respectively. The use of $A^{(m_{ix}\,-\,m_{i1})}$ allows to determine the values of r, s, t,... simply by expanding the general polynomial. A numerical value for A, which has no intrinsic meaning, is never needed.

For example, for CBr_2Cl_2, the above equation gives rise to the following expression:

$$\{p_{12C}\,A^0 + p_{13C}\,A^{(m_{13C}\,-\,m_{12C})}\} \times$$
$$\{p_{79Br}\,A^0 + p_{81Br}\,A^{(m_{81Br}\,-\,m_{79Br})}\}^2 \times$$
$$\{p_{35Cl}\,A^0 + p_{37Cl}\,A^{(m_{37Cl}\,-\,m_{35Cl})}\}^2$$

For sufficient resolution, $(m_{ix} - m_{i1})$ and $(m_{jx} - m_{j1})$ differ from one another. This results in very complex isotope patterns even for very small molecules. Thus, owing to the occurrence of ^{12}C, ^{13}C, ^{79}Br, ^{81}Br, ^{35}Cl, and ^{37}Cl, there are 18 signals for CBr_2Cl_2. However, the limited resolution of many real life experiments can make many pairs of $(m_{ix} - m_{i1})$ and $(m_{jx} - m_{j1})$ indistinguishable within experimental error, thereby reducing the number of separate peaks. For example, at unit resolution, one obtains $(m_{13C} - m_{12C}) = 1$ and $(m_{81Br} - m_{79Br}) = (m_{37Cl} - m_{35Cl}) = 2$. Consequently, the expression for CBr_2Cl_2 becomes:

$$\{ p_{12_C} A^0 + p_{13_C} A^1 \} \times \{ p_{79_{Br}} A^0 + p_{81_{Br}} A^2 \}^2 \times \{ p_{35_{Cl}} A^0 + p_{37_{Cl}} A^2 \}^2 =$$

$$\{ p_{12_C} p_{79_{Br}}^2 p_{35_{Cl}}^2 \} A^0 +$$

$$\{ p_{13_C} p_{79_{Br}}^2 p_{35_{Cl}}^2 \} A^1 +$$

$$\{ p_{12_C} p_{79_{Br}} p_{81_{Br}} p_{35_{Cl}}^2 + p_{12_C} p_{79_{Br}}^2 p_{35_{Cl}} p_{37_{Cl}} \} A^2 +$$

$$\{ p_{13_C} p_{79_{Br}} p_{81_{Br}} p_{35_{Cl}}^2 + p_{13_C} p_{79_{Br}}^2 p_{35_{Cl}} p_{37_{Cl}} \} A^3 +$$

$$\{ p_{12_C} p_{81_{Br}}^2 p_{35_{Cl}}^2 + 4\, p_{12_C} p_{79_{Br}} p_{81_{Br}} p_{35_{Cl}} p_{37_{Cl}} + p_{12_C} p_{79_{Br}}^2 p_{37_{Cl}}^2 \} A^4 +$$

$$\{ p_{13_C} p_{81_{Br}}^2 p_{35_{Cl}}^2 + 4\, p_{13_C} p_{79_{Br}} p_{81_{Br}} p_{35_{Cl}} p_{37_{Cl}} + p_{13_C} p_{79_{Br}}^2 p_{37_{Cl}}^2 \} A^5 +$$

$$\{ p_{12_C} p_{79_{Br}} p_{81_{Br}} p_{37_{Cl}}^2 + p_{12_C} p_{81_{Br}}^2 p_{35_{Cl}} p_{37_{Cl}} \} A^6 +$$

$$\{ p_{13_C} p_{79_{Br}} p_{81_{Br}} p_{37_{Cl}}^2 + p_{13_C} p_{81_{Br}}^2 p_{35_{Cl}} p_{37_{Cl}} \} A^7 +$$

$$\{ p_{12_C} p_{81_{Br}}^2 p_{37_{Cl}}^2 \} A^8 +$$

$$\{ p_{13_C} p_{81_{Br}}^2 p_{37_{Cl}}^2 \} A^9$$

This shows that at unit resolution, CBr_2Cl_2 gives rise to only 10 peaks (M$^{+\cdot}$, [M+1]$^{+\cdot}$, [M+2]$^{+\cdot}$, ... [M+9]$^{+\cdot}$) rather than 18 peaks, as they would be expected for very high resolution. Moreover, the contribution of isotopes of low abundance can often be neglected without sacrificing much precision. For example, the effect of ^2H on isotope patterns is usually insignificant. Also, ^{13}C is often negligible when focussing on peaks of the series [M+2n]$^{+\cdot}$, which then results in patterns that are characteristic for halogens, sulfur, and silicon. In large molecules, however, isotopes of low abundance cannot be neglected. For example, in the case of buckminster fullerene (C_{60}), not only M$^{+\cdot}$ (relative intensity, 100%) and [M+1]$^{+\cdot}$ (64.80%), but also [M+2]$^{+\cdot}$ (20.65%), [M+3]$^{+\cdot}$ (4.31%), and even [M+4]$^{+\cdot}$ (0.66%) are quite significant ions.

With the above algorithm, typical isotope patterns can be readily calculated manually by applying the general equation and neglecting isotopes of low abundance. The outlined procedure can also be easily implemented and evaluated with generic computer software that allows simple calculations. Dedicated and user-friendly programs that already contain the necessary isotope abundances and masses are available. Incidentally, because the use of the above equation for systems with 1000 or more polyisotopic atoms results in excessive calculation times, more efficient but somewhat more complicated algorithms have been developed for implementation in dedicated programs [4]. Typical isotope patterns are given on the following pages.

2.5.5 Isotopic Abundances of Various Combinations of Chlorine, Bromine, Sulfur, and Silicon

Elements	Mass	Relative abundance	Elements	Mass	Relative abundance	Elements	Mass	Relative abundance
Cl_1	35	100	Br_1	79	100	S_1	32	100
	37	31.96		81	97.28		33	0.80
							34	4.52
Cl_2	70	100	Br_2	158	51.40			
	72	63.92		160	100	S_2	64	100
	74	10.21		162	48.64		65	1.60
							66	9.05
Cl_3	105	100	Br_3	237	34.27		68	0.20
	107	95.88		239	100			
	109	30.64		241	97.28	S_3	96	100
	111	3.26		243	31.54		97	2.40
							98	13.58
Cl_4	140	78.22	Br_4	316	17.61		99	0.22
	142	100		318	68.53		100	0.61
	144	47.94		320	100			
	146	10.21		322	64.85	S_4	128	100
	148	0.82		324	15.77		129	3.20
							130	18.12
Cl_5	175	62.53	Br_5	395	10.57		131	0.43
	177	100		397	51.40		132	1.23
	179	63.92		399	100			
	181	20.43		401	97.28	S_5	160	100
	183	3.26		403	47.32		161	4.00
	185	0.21		405	9.21		162	22.66
							163	0.72
Cl_6	210	52.15	Br_6	474	5.43		164	2.05
	212	100		476	31.70		166	0.09
	214	79.90		478	77.10			
	216	34.05		480	100			
	218	8.16		482	72.96			
	220	1.04		484	28.39			
	222	0.06		486	4.60			

Ele-ments	Mass	Relative abun-dance	Ele-ments	Mass	Relative abun-dance	Ele-ments	Mass	Relative abun-dance
Si_1	28	100	Si_2	56	100	Si_3	84	100
	29	5.08		57	10.15		85	15.23
	30	3.35		58	6.95		86	10.82
				59	0.34		87	1.03
				60	0.11		88	0.36
Cl_1Br_1	114	77.38	Cl_1Br_2	193	44.14	Cl_1Br_3	272	26.51
	116	100		195	100		274	85.85
	118	24.06		197	69.23		276	100
				199	13.35		278	48.46
							280	7.80
Cl_1Br_4	351	14.45	Cl_2Br_1	149	62.03	Cl_2Br_2	228	38.69
	353	60.84		151	100		230	100
	355	100		153	44.91		232	88.68
	357	79.42		155	6.16		234	31.09
	359	29.94					236	3.74
	361	4.14						
Cl_3Br_1	184	51.77	Cl_3Br_2	263	32.07	Cl_4Br_1	219	44.42
	186	100		265	93.14		221	100
	188	64.15		267	100		223	82.47
	190	17.12		269	49.27		225	32.28
	192	1.64		271	11.34		227	6.11
				273	0.99		229	0.45
Cl_1S_1	67	100	Cl_1S_2	99	100	Cl_2S_1	102	100
	68	0.80		100	1.60		103	0.80
	69	36.48		101	41.01		104	68.44
	70	0.26		102	0.58		105	0.51
	71	1.44		103	3.10		106	13.10
							108	0.46

Ele- ments	Mass	Relative abun- dance	Ele- ments	Mass	Relative abun- dance	Ele- ments	Mass	Relative abun- dance
Cl_1Si_1	63	100	Cl_2Si_1	98	100	Cl_3Si_1	133	100
	64	5.08		99	5.08		134	5.08
	65	35.31		100	67.27		135	99.23
	66	1.62		101	3.25		136	4.87
	67	1.07		102	12.35		137	33.85
				103	0.52		138	1.56
				104	0.34		139	4.29

2.5.6 Isotope Patterns of Combinations of Cl and Br

The signals are separated by 2 mass units. The mass for the most abundant signal is shown under the symbol of the element. The combination of the lightest isotopes is given on the left side of the x axis. See Chapter 2.5.5 for exact abundances of many of these combinations.

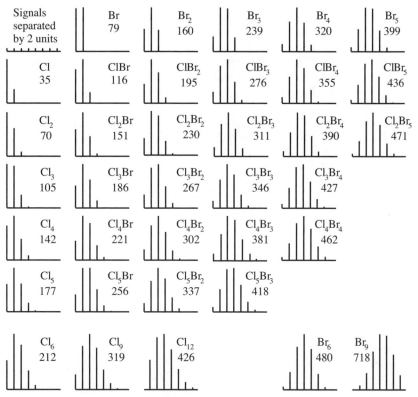

2.5.7 Indicators of the Presence of Heteroatoms

In low-resolution mass spectra, one often observes characteristic isotope patterns, specific masses of fragment ions, and characteristic mass differences (Δm) between the molecular ion ($M^{+\cdot}$) and fragment ions (frag$^+$) or between fragment ions. High resolution mass spectra can be used to confirm the elemental composition provided that the resolution is sufficient to discriminate alternative compositions. Moreover, tandem mass spectrometry (also called MS/MS) may be used to identify characteristic losses of heteroatoms from parent or fragment ions:

Indication of O: Δm 17 from $M^{+\cdot}$, in N-free compounds
Δm 18 from $M^{+\cdot}$
Δm 18 from frag$^+$, particularly in aliphatic compounds
Δm 28, 29 from $M^{+\cdot}$ for aromatic compounds
Δm 28 from frag$^+$ for aromatic compounds
m/z 15, relatively abundant
m/z 19
m/z 31, 45, 59, 73, ... $+ (14)_n$
m/z 32, 46, 60, 74, ... $+ (14)_n$
m/z 33, 47, 61, 75, ... $+ (14)_n$ for 2 × O, in absence of S
m/z 69 for aromatic compounds meta-disubstituted by O

Indication of N: $M^{+\cdot}$ odd-numbered (indicates odd number of N in $M^{+\cdot}$)
Large number of even-numbered fragment ions
Δm 17 from $M^{+\cdot}$ or frag$^+$, in O-free compounds
Δm 27 from $M^{+\cdot}$ or frag$^+$, for aromatic compounds or nitriles
Δm 30, 46 for nitro compounds
m/z 30, 44, 58, 72, ... $+ (14)_n$ for aliphatic compounds

Indication of S: Isotope peak $[M+2]^{+\cdot} \geq 5\%$ of $M^{+\cdot}$
Δm 33, 34, 47, 48, 64, 65 from $M^{+\cdot}$
Δm 34, 48, 64 from frag$^+$
m/z 33, 34, 35
m/z 45 in O-free compounds
m/z 47, 61, 75, 89, ... $+ (14)_n$ unless compound with 2 × O
m/z 48, 64 for *S*-oxides

Indication of F: Δm 19, 20, 50 from $M^{+\cdot}$
Δm 20 from frag$^+$
m/z 20
m/z 57 without m/z 55 in aromatics

Indication of Cl: Isotope peak $[M+2]^{+\cdot} \geq 33\%$ of $M^{+\cdot}$
Δm 35, 36 from $M^{+\cdot}$
Δm 36 from $frag^+$
m/z 35/37, 36/38, 49/51

Indication of Br: Isotope peak $[M+2]^{+\cdot} \geq 98\%$ of $M^{+\cdot}$
Δm 79, 80 from $M^{+\cdot}$
Δm 80 from $frag^+$
m/z 79/81, 80/82

Indication of I: Isotope peak $[M+1]^+$ of very low abundance at relatively high mass
Δm 127 from $M^{+\cdot}$
Δm 127, 128 from $frag^+$
m/z 127, 128, 254

Indication of P: m/z 47 in compounds without S or $2 \times$ O
m/z 99 without isotope peak at m/z 100 in alkyl phosphates

2.5.8 Rules for Determining the Relative Molecular Weight (M$_r$)

The molecular ion (M$^{+\cdot}$) is defined as the ion that comprises the most abundant isotopes of the elements in the molecule. Interestingly, the lightest isotopes of most elements frequently occurring in organic compounds and their common salts (H, C, N, O, F, Si, P, S, Cl, As, Br, I, Na, Mg, Al, K, Ca, Rb, Cs) are also the most abundant ones. Notable exceptions are B, Li, Se, Sr, and Ba.

M$^{+\cdot}$ is always accompanied by isotope peaks. Their relative abundance depends on the number and kind of the elements present and their natural isotopic distribution. The abundance of [M+1]$^{+\cdot}$ indicates the maximum number of carbon atoms (C$_{max}$) according to the following relationship:

$$C_{max} = 100 \times \text{intensity}([M+1]^{+\cdot}) \, / \, \{1.1 \times \text{intensity}(M^{+\cdot})\}$$

[M+2]$^{+\cdot}$ and higher masses indicate the number and kind of elements that have a relatively abundant heavier isotope (such as S, Si, Cl, Br). Note that, in analogy to the calculation of C$_{max}$, the ratio of the intensities of [M+2]$^{+\cdot}$ and M$^{+\cdot}$ for a compound with n silicon, o sulfur, p chlorine, or q bromine atoms can be approximated with quite high accuracy from $n \times 3.35\%$, $o \times 4.52\%$, $p \times 31.96\%$, or $q \times 97.28\%$, respectively (see also Chapters 2.5.4 to 2.5.6).

The mass of M$^{+\cdot}$ is always an even number if the molecule contains only elements for which the atomic mass and valence are both even- (C, O, S, Si) or both odd-numbered (H, P, F, Cl, Br, I). In the presence of other elements (e.g., ^{14}N) and isotope labels (e.g., ^{13}C, ^{2}H), M$^{+\cdot}$ becomes an odd number if they are present in an odd number.

The molecular ion can only form fragment ions of masses that differ from that of M$^{+\cdot}$ by chemically logical values (Δm). In this context, chemically illogical differences are $\Delta m = 3$ (in the absence of $\Delta m = 1$) to $\Delta m = 14$, $\Delta m = 21$ (in the absence of $\Delta m = 1$) to $\Delta m = 24$, $\Delta m = 37, 38$, and all Δm less than the mass of an element of characteristic isotope pattern in cases where the same isotope pattern is not retained in the fragment ion.

M$^{+\cdot}$ must contain all elements (and the maximum number of each) that are shown to be present in the fragment ions.

If ionization is performed by electron impact, M$^{+\cdot}$ is the ion with the lowest appearance potential.

If a pure sample flows into the ion source through a molecular leak, M$^{+\cdot}$ exhibits the same effusion rate as can be determined from the fragment ions. The abundance of M$^{+\cdot}$ is proportional to the sample pressure in the ion source.

For polar compounds, [M+H]$^{+}$ is often observed in mass spectra obtained not only with fast atom bombardment and atmospheric pressure chemical ionization but also with electron impact ionization. In this latter case, the abundance of [M+H]$^{+}$ changes in proportion to the square of the sample pressure in the ion source.

In the absence of a signal for M$^{+\cdot}$, the relative molecular weight must have a value that shows a logical and reasonable mass difference, Δm, to all the observed fragment ions.

2.5.9 Homologous Mass Series as Indications of Structural Type

Certain sequences of intensity maxima in the lower mass range and the masses of
unique signals are often characteristic of a particular compound type. The intensity
distribution of such ion series is in general smooth. Therefore, abrupt changes (maxima and minima) are of structural significance. The ion or ion series most indicative
of a particular compound type is set in italics.

Mass values, m/z	Elemental composition	Compound types
12 + 14n	C_nH_{2n-2}	alkenes, monocycloalkanes, alkynes, dienes, cycloalkenes, polycyclic alicyclics, cyclic alcohols
13 + 14n	C_nH_{2n-1}	alkanes, alkenes, *monocycloalkanes*, alkynes, dienes, cycloalkenes, polycyclic alicyclics, alcohols, alkyl ethers, cyclic alcohols, cyclo-alkanones, aliphatic acids, esters, lactones, thiols, sulfides, glycols, glycol ethers, alkyl chlorides
	$C_nH_{2n-3}O$	cycloalkanones
14 + 14n	C_nH_{2n}	alkanes, alkenes, monocycloalkanes, polycyclic alicyclics, alcohols, alkyl ethers, thiols, sulfides, alkyl chlorides
	$C_nH_{2n-2}O$	cycloalkanones
15 + 14n	C_nH_{2n+1}	*alkanes*, alkenes, monocycloalkanes, alkynes, dienes, cycloalkenes, polycyclic alicyclics, alkanones, alkanals, glycols, glycol ethers, alkyl chlorides, acid chlorides
	$C_nH_{2n-1}O$	alkanones, alkanals, *cyclic alcohols*, acid chlorides
16 + 14n	$C_nH_{2n}O$	*alkanones, alkanals*
	$C_nH_{2n+2}N$	*alkyl amines, aliphatic amides*
	$C_nH_{2n}NO$	aliphatic amides
17 + 14n	$C_nH_{2n+1}O$	*alcohols, alkyl ethers*, aliphatic acids, esters, lactones, glycols, glycol ethers
	$C_nH_{2n-1}O_2$	aliphatic acids, esters, lactones
18 + 14n	$C_nH_{2n}O_2$	*aliphatic acids, esters, lactones*
19 + 14n	$C_nH_{2n+3}O$	alcohols, alkyl ethers
	$C_nH_{2n+1}O_2$	aliphatic acids, esters, lactones
	$C_nH_{2n+1}O_2$	*glycols, glycol ethers*
	$C_nH_{2n+1}S$	*thiols, sulfides*
20 + 14n	$C_8H_8 + C_nH_{2n}$	alkylbenzenes
	$C_nH_{2n+2}O_2$	glycols, glycol ethers
	$C_nH_{2n+2}S$	thiols, sulfides

Mass values m/z	Elemental composition	Compound types
21 + 14n	$C_7H_7 + C_nH_{2n}$	alkylbenzenes
	C_7H_5O	aryl ketones
	$C_nH_{2n}Cl$	*alkyl chlorides*
	$C_nH_{2n}COCl$	acid chlorides
22 + 14n	$C_6H_6N + C_nH_{2n}$	alkylanilines
	C_nH_{2n-6}	polycyclic alicyclics
23 + 14n	C_nH_{2n-5}	polycyclic alicyclics
24 + 14n	C_nH_{2n-4}	polycyclic alicyclics
25 + 14n	C_nH_{2n-3}	*alkynes, dienes, cycloalkenes*, polycyclic alicyclics
39, 52±1, 64±1, 76±2, 91±1	$C_nH_{n\pm1}$	alkylbenzenes, aromatic hydrocarbons, phenols, aryl ethers, aryl ketones

2.5.10 Mass Correlation Table

Note: As long as it makes sense chemically, CH_2, CH_4, CH_3O, and O_2 in the formulae of the second column may be replaced by N, O, P, and S, respectively.

Mass	Ion	Product ion (and neutral particle lost)		Substructure or compound type
1		$[M+1]^+$, $[M-1]^-$		particularly in FAB spectra, in which M±1 occurs even for moderately basic and acidic compounds, but intensive $M^{+\cdot}$ without M±1 is unusual
7	$Li^{+\cdot}$	$[M+7]^+$		in FAB spectra in the presence of Li^+ (with isotope signal for ^6Li)
		$[M-7]^-$		in FAB spectra of organic Li^+ salts
12	$C^{+\cdot}$			
13	CH^+			
14	$CH_2^{+\cdot}$, N^+, N_2^{++}, CO^{++}			
15	CH_3^+	$[M-15]^{+\cdot}$	(CH_3)	nonspecific; *abundant:* methyl, N-ethylamines
16	$O^{+\cdot}$, NH_2^+, O_2^{++}	$[M-16]^{+\cdot}$	(CH_4)	methyl (rare)
			(O)	nitro compounds, sulfones, epoxides, N-oxides
			(NH_2)	primary amines
17	OH^+, $NH_3^{+\cdot}$	$[M-17]^{+\cdot}$	(OH)	acids (especially aromatic acids), hydroxylamines, N-oxides, nitro compounds, sulfoxides, tertiary alcohols
			(NH_3)	primary amines
18	$H_2O^{+\cdot}$, NH_4^+	$[M-18]^{+\cdot}$	(H_2O)	nonspecific; *abundant:* alcohols, some acids, aldehydes, ketones, lactones, cyclic ethers **O indicator**
19	H_3O^+, F^+	$[M-19]^+$	(F)	fluoro compounds **F indicator**
20	$HF^{+\cdot}$, Ar^{++}, CH_2CN^{++}	$[M-20]^{+\cdot}$	(HF)	fluoro compounds **F indicator**
21	$C_2H_2O^{++}$			
22	CO_2^{++}			

Mass	Ion	Product ion (and neutral particle lost)	Substructure or compound type
23	Na^+	$[M+23]^+$	in FAB spectra in the presence of Na^+; sometimes strong even if Na^+ is only an impurity
		$[M-23]^-$	in FAB spectra of organic Na^+ salts
24	$C_2^{+\cdot}$		
25	C_2H^+	$[M-25]^+$ (C_2H)	terminal acetylenyl
26	$C_2H_2^{+\cdot}$, CN^+	$[M-26]^{+\cdot}$ (C_2H_2)	aromatics
		(CN)	nitriles
27	$C_2H_3^+$, $HCN^{+\cdot}$	$[M-27]^+$ (C_2H_3)	terminal vinyl, some ethyl esters and N-ethylamides, ethyl phosphates
		$[M-27]^{+\cdot}$ (HCN)	aromatic N, nitriles
28	$C_2H_4^{+\cdot}$, $CO^{+\cdot}$, $N_2^{+\cdot}$, $HCNH^+$	$[M-28]^{+\cdot}$ (C_2H_4)	nonspecific; *abundant:* cyclohexenes, ethyl esters, propyl ketones, propyl-substituted aromatics
		(CO)	aromatic O, quinones, lactones, lactams, unsaturated cyclic ketones, allyl aldehydes
		(N_2)	diazo compounds; air (intensity 3.7 times larger than for $O_2^{+\cdot}$, m/z 32)
29	$C_2H_5^+$, CHO^+	$[M-29]^+$ (C_2H_5)	nonspecific; *abundant:* ethyl
		(CHO)	phenols, furans, aldehydes
30	$CH_2O^{+\cdot}$, $CH_2NH_2^+$, NO^+, $C_2H_6^{+\cdot}$, $BF^{+\cdot}$, $N_2H_2^{+\cdot}$	$[M-30]^{+\cdot}$ (C_2H_6) (CH_2O)	ethylalkanes, polymethyl compounds cyclic ethers, lactones, primary alcohols
		(NO)	nitro and nitroso compounds **N indicator**
31	CH_3O^+, $CH_3NH_2^{+\cdot}$, CF^+, $N_2H_3^+$	$[M-31]^+$ (CH_3O)	methyl esters, methyl ethers, primary alcohols **O indicator**
		(CH_3NH_2)	N-methylamines
		(N_2H_3)	hydrazides
32	$O_2^{+\cdot}$, $CH_3OH^{+\cdot}$, $N_2H_4^{+\cdot}$, $S^{+\cdot}$	$[M-32]^+$ (O_2)	cyclic peroxides; air (intensity 3.7 times smaller than for $N_2^{+\cdot}$, m/z 28)
		(CH_3OH)	methyl esters, methyl ethers
		(S)	sulfides (with ^{34}S isotope signal) **O indicator**

Mass	Ion	Product ion (and neutral particle lost)	Substructure or compound type
33	$CH_3OH_2^+$, SH^+, CH_2F^+	$[M-33]^+$ (SH)	nonspecific (with isotope signal for ^{34}S) **S indicator**
		$(CH_3 + H_2O)$	nonspecific **O indicator**
		(CH_2F)	fluoromethyl
34	$SH_2^{+\cdot}$	$[M-34]^{+\cdot}$ (SH_2)	nonspecific (with ^{34}S isotope signal) **S indicator**
		$(OH + OH)$	nitro compounds
35	SH_3^+, Cl^+	$[M-35]^+$ (Cl)	chloro compounds (with ^{37}Cl isotope signal)
		$(OH + H_2O)$	nitro compounds **2 × O indicator**
36	$HCl^{+\cdot}$, C_3^+	$[M-36]^{+\cdot}$ (HCl)	chloro compounds
		$(H_2O + H_2O)$	**2 × O indicator**
37	C_3H^+		
	$^{37}Cl^+$		chloro compounds (with isotope signal for ^{35}Cl)
38	$C_3H_2^{+\cdot}$		
39	$C_3H_3^+$	$[M-39]^+$ (C_3H_3)	aromatics
	K^+	$[M+39]^+$	in FAB spectra often strong even if K^+ is only an impurity (with isotope signal for ^{41}K)
		$[M-39]^-$	in FAB spectra of organic K^+ salts
40	$C_3H_4^{+\cdot}$, $Ar^{+\cdot}$, CH_2CN^+	$[M-40]^{+\cdot}$ (CH_2CN)	cyanomethyl
41	$C_3H_5^+$, $CH_3CN^{+\cdot}$	$[M-41]^+$ (C_3H_5)	alicyclics (*especially* polyalicyclics), alkenes
		(CH_3CN)	2-methyl-*N*-aromatics, *N*-methyl-anilines
42	$C_3H_6^{+\cdot}$, $C_2H_2O^{+\cdot}$, CON^+, $C_2H_4N^+$	$[M-42]^{+\cdot}$ (C_3H_6)	nonspecific; *abundant:* propyl esters, butyl ketones, butylaromatics, methylcyclohexenes
		(C_2H_2O)	acetates (*especially* enol acetates), acetamides, cyclohexenones, α,β-unsaturated ketones
43	$C_3H_7^+$, $C_2H_3O^+$, $CONH^{+\cdot}$	$[M-43]^+$ (C_3H_7)	nonspecific; *abundant:* propyl, alicyclics, cycloalkanones, cycloalkylamines, cycloalkanols, butylaromatics
		(CH_3CO)	methyl ketones, acetates, aromatic methyl ethers

Mass	Ion	Product ion (and neutral particle lost)	Substructure or compound type
44	$CO_2^{+\cdot}$, $C_2H_6N^+$, $C_2H_4O^{+\cdot}$, $CS^{+\cdot}$, $C_3H_8^{+\cdot}$, $CH_4Si^{+\cdot}$	$[M-44]^{+\cdot}$ (C_3H_8) (C_2H_6N) (C_2H_4O) (CO_2)	propylalkanes *N,N*-dimethylamines, *N*-ethylamines cycloalkanols, cyclic ethers, ethylene ketals, aliphatic aldehydes (McLafferty rearrangement) anhydrides, lactones, carboxylic acids
45	$C_2H_5O^+$, $C_2H_7N^{+\cdot}$, CHS^+ (with isotope signal for ^{34}S)	$[M-45]^+$ (C_2H_5O) (CHO_2) (C_2H_7N)	ethyl esters, ethyl ethers, lactones, ethyl sulfonates, ethyl sulfones carboxylic acids *N,N*-dimethylamines, *N*-ethylamines **O indicator** **S indicator**
46	$C_2H_5OH^{+\cdot}$, NO_2^+	$[M-46]^{+\cdot}$ (C_2H_6O) $(H_2O + C_2H_4)$ $(H_2O + CO)$ (NO_2)	ethyl esters, ethyl ethers, ethyl sulfonates primary alcohols carboxylic acids nitro compounds
47	CH_3S^+, CCl^+, $C_2H_5OH_2^+$, $CH(OH)_2^+$, PO^+	$[M-47]^+$ (CH_3S)	methyl sulfides (with isotope signal for ^{34}S) **2 × O indicator** **S indicator** **P indicator**
48	$CH_3SH^{+\cdot}$, $SO^{+\cdot}$, $CHCl^{+\cdot}$	$[M-48]^{+\cdot}$ (CH_4S) (SO)	methyl sulfides sulfoxides, sulfones, sulfonates (with isotope signal for ^{34}S)
49	CH_2Cl^+, $CH_3SH_2^+$ (with isotope signal for ^{34}S)	$[M-49]^+$ (CH_2Cl)	chloromethyl (with ^{37}Cl isotope signal)
50	$C_4H_2^{+\cdot}$, $CH_3Cl^{+\cdot}$, $CF_2^{+\cdot}$	$[M-50]^{+\cdot}$ (CF_2)	trifluoromethylaromatics, perfluoro-alicyclics
51	$C_4H_3^+$, CHF_2^+		

Mass	Ion	Product ion (and neutral particle lost)		Substructure or compound type
52	$C_4H_4^{+\cdot}$			
53	$C_4H_5^+$			
54	$C_4H_6^{+\cdot}$,	$[M-54]^{+\cdot}$	(C_4H_6)	cyclohexenes
	$C_2H_4CN^+$		(C_2H_4CN)	cyanoethyl
55	$C_4H_7^+$,	$[M-55]^+$	(C_4H_7)	nonspecific; *abundant:* alicyclics,
	$C_3H_3O^+$			butyl esters, N-butylamides
56	$C_4H_8^{+\cdot}$,	$[M-56]^{+\cdot}$	(C_4H_8)	butyl esters, N-butylamides, pentyl
	$C_3H_4O^{+\cdot}$			ketones, cyclohexenes, tetralins,
				pentylaromatics
			(C_3H_4O)	methylcyclohexenones, β-tetralones
57	$C_4H_9^+$,	$[M-57]^+$	(C_4H_9)	nonspecific
	$C_3H_5O^+$,		(C_3H_5O)	ethyl ketones
	$C_3H_2F^+$			
58	$C_3H_8N^+$,	$[M-58]^{+\cdot}$	(C_4H_{10})	alkanes
	$C_3H_6O^{+\cdot}$		(C_3H_6O)	α-methylalkanals, methyl ketones,
				isopropylidene glycols **N indicator**
				O indicator
59	$C_3H_7O^+$,	$[M-59]^+$	(C_3H_7O)	propyl esters, propyl ethers
	$C_2H_5NO^{+\cdot}$		$(C_2H_3O_2)$	methyl esters
			(C_3H_9N)	amines, amides **O indicator**
60	$C_2H_4O_2^{+\cdot}$,	$[M-60]^{+\cdot}$	(C_3H_8O)	propyl esters, propyl ethers
	$CH_2NO_2^+$,		$(C_2H_4O_2)$	acetates
	$C_2H_6NO^+$,		$(CH_3OH + CO)$	methyl esters **O indicator**
	$C_2H_4S^{+\cdot}$		(C_2H_4S)	
61	$C_2H_5O_2^+$,	$[M-61]^+$	$(C_2H_5O_2)$	glycols, ethylene ketals
	$C_2H_5S^+$			**2 × O indicator**
			(C_2H_5S)	ethyl sulfides (with ^{34}S isotope
				signal) **S indicator**
62	$C_2H_6O_2^{+\cdot}$,	$[M-62]^{+\cdot}$	$(C_2H_6O_2)$	methoxymethyl ethers, ethylene
	$C_2H_3Cl^{+\cdot}$			glycols, ethylene ketals
	$C_2H_6S^{+\cdot}$		(C_2H_6S)	ethyl sulfides (with ^{34}S isotope
				signal)
63	$C_5H_3^+$,	$[M-63]^+$	(C_2H_4Cl)	chloroethyl
	$C_2H_4Cl^+$,		$(CO + Cl)$	carboxylic acid chlorides (with ^{37}Cl
	$COCl^+$			isotope signal)
64	$C_5H_4^{+\cdot}$,	$[M-64]^{+\cdot}$	(SO_2)	sulfones, sulfonates
	$SO_2^{+\cdot}$, $S_2^{+\cdot}$		(S_2)	disulfides (with ^{34}S isotope signal)

Mass	Ion	Product ion (and neutral particle lost)		Substructure or compound type
65	$C_5H_5^+$, $H_2PO_2^+$	$[M-65]^+$	(S_2H) (SO_2H)	disulfides
66	$C_5H_6^{+\cdot}$ $S_2H_2^{+\cdot}$	$[M-66]^{+\cdot}$	(C_5H_6)	cyclopentenes disulfides (with ^{34}S isotope signal)
67	$C_5H_7^+$, $C_4H_3O^+$	$[M-67]^+$	(C_4H_3O)	furyl ketones
68	$C_5H_8^{+\cdot}$, $C_4H_4O^{+\cdot}$, $C_3H_6CN^+$	$[M-68]^{+\cdot}$	(C_5H_8) (C_4H_4O)	cyclohexenes, tetralins cyclohexenones, β-tetralones
69	$C_5H_9^+$, $C_4H_5O^+$, $C_3HO_2^+$	$[M-69]^+$	(C_5H_9)	alicyclics, alkenes
	CF_3^+		(CF_3)	trifluoromethyl
70	$C_5H_{10}^{+\cdot}$			alkanes, alkenes, alicyclics
	$C_4H_6O^{+\cdot}$			cycloalkanones
	$C_4H_8N^+$			pyrrolidines
71	$C_5H_{11}^+$			alkanes, larger alkyl groups
	$C_4H_7O^+$			alkanones, alkanals, tetrahydrofurans
72	$C_4H_8O^{+\cdot}$			alkanones, alkanals **O indicator**
	$C_4H_{10}N^+$			aliphatic amines **N indicator**
	$C_6^{+\cdot}$			perhalogenated benzenes
73	$C_4H_9O^+$			alcohols, ethers, esters **O indicator**
	$C_3H_5O_2^+$			carboxylic acids, esters, lactones
	$C_3H_9Si^+$			trimethylsilyl compounds
74	$C_4H_{10}O^{+\cdot}$			ethers
	$C_3H_6O_2^{+\cdot}$			methyl esters of carboxylic acids, α-methyl carboxylic acids
75	$C_3H_7O_2^+$			methyl acetals, glycols **2 × O indicator**
	$C_3H_7S^+$			sulfides, thiols (with ^{34}S isotope signal) **S indicator**
	$C_2H_7SiO^+$			trimethylsilyloxyl compounds
76	$C_6H_4^{+\cdot}$			aromatics
77	$C_6H_5^+$			aromatics
	$C_3H_6Cl^+$			chloro compounds (with ^{37}Cl isotope signal)

Mass	Ion	Compound type	
78	$C_6H_6^{+\cdot}$	aromatics	
	$C_5H_4N^+$	pyridines	
	$C_3H_7Cl^{+\cdot}$	chloro compounds (with ^{37}Cl isotope signal)	
79	$C_6H_7^+$	aromatics with H-containing substituents	
	$C_5H_5N^{+\cdot}$	pyridines, pyrroles	
	Br^+	bromo compounds (with ^{81}Br isotope signal)	
80	$C_6H_8^{+\cdot}$	cyclohexenes, polycyclic alicyclics	
	$C_5H_4O^{+\cdot}$	cyclopentenones	
	$HBr^{+\cdot}$	bromo compounds (with ^{81}Br isotope signal)	
	$C_5H_6N^+$	pyrroles, pyridines	
81	$C_6H_9^+$	cyclohexanes, cyclohexenyls, dienes	
	$C_5H_5O^+$	furans, pyrans	
	$^{81}Br^+$	bromo compounds (with ^{79}Br isotope signal)	
82	$C_6H_{10}^{+\cdot}$	cyclohexanes	
	$C_5H_6O^{+\cdot}$	cyclopentenones, dihydropyrans	
	$C_5H_8N^+$	tetrahydropyridines	
	$C_4H_6N_2^{+\cdot}$	pyrazoles, imidazoles	
	$CCl_2^{+\cdot}$	chloro compounds (with isotope signals at m/z 84 and 86)	
83	$C_6H_{11}^+$	alkenes, alicyclics, monosubstituted alkanes	
	$C_5H_7O^+$	cycloalkanones	
84	$C_5H_{10}N^+$	piperidines, *N*-methylpyrrolidines	
85	$C_6H_{13}^+$	alkanes	
	$C_5H_9O^+$	alkanones, alkanals, tetrahydropyrans, fatty acid derivatives	
	$CClF_2^+$	chlorofluoroalkanes (with ^{37}Cl isotope signal)	
86	$C_5H_{10}O^{+\cdot}$	alkanones, alkanals	
	$C_5H_{12}N^+$	aliphatic amines	**N indicator**
87	$C_5H_{11}O^+$	alcohols, ethers, esters	**O indicator**
	$C_4H_7O_2^+$	esters, carboxylic acids	
88	$C_4H_8O_2^{+\cdot}$	ethyl esters of carboxylic acids, α-methyl-methyl esters, α-C_2-carboxylic acids	
89	$C_4H_9O_2^+$	diols, glycol ethers	**2 × O indicator**
	$C_4H_9S^+$	sulfides (with ^{34}S isotope signal)	
90	$C_7H_6^{+\cdot}$	disubstituted aromatics	
91	$C_7H_7^+$	aromatics	
	$C_4H_8Cl^+$	alkyl chlorides (with ^{37}Cl isotope signal)	

Mass	Ion	Compound type
92	$C_7H_8^{+\cdot}$	alkylbenzenes
	$C_6H_6N^+$	alkylpyridines
93	$C_6H_5O^+$	phenols, phenol derivatives
	$C_6H_7N^{+\cdot}$	anilines
	CH_2Br^+	bromo compounds (with ^{81}Br isotope signal)
94	$C_6H_6O^{+\cdot}$	phenol esters, phenol ethers
	$C_5H_4NO^+$	pyrryl ketones, pyridone derivatives
95	$C_5H_3O_2^+$	furyl ketones
96	$C_7H_{12}^{+\cdot}$	alicyclics
97	$C_7H_{13}^+$	alicyclics, alkenes
	$C_6H_9O^+$	cycloalkanones
	$C_5H_5S^+$	alkylthiophenes (with ^{34}S isotope signal)
98	$C_6H_{12}N^+$	N-alkylpiperidines
99	$C_7H_{15}^+$	alkanes
	$C_6H_{11}O^+$	alkanones
	$C_5H_7O_2^+$	ethylene ketals
	$H_4PO_4^+$	alkyl phosphates
104	$C_8H_8^{+\cdot}$	tetralin derivatives, phenylethyl derivatives
	$C_7H_4O^{+\cdot}$	disubstituted α-ketobenzenes
105	$C_8H_9^+$	alkylaromatics
	$C_7H_5O^+$	benzoyl derivatives
	$C_6H_5N_2^+$	diazophenyl derivatives
106	$C_7H_8N^+$	alkylanilines
111	$C_5H_3OS^+$	thiophenoyl derivatives (with ^{34}S isotope signal)
115	$C_9H_7^+$	aromatics
	$C_6H_{11}O_2^+$	esters
	$C_5H_7O_3^+$	diesters
119	$C_9H_{11}^+$	alkylaromatics
	$C_8H_7O^+$	tolyl ketones
	$C_2F_5^+$	perfluoroethyl derivatives
	$C_7H_5NO^{+\cdot}$	phenyl carbamates
120	$C_7H_4O_2^{+\cdot}$	γ-benzopyrones, salicylic acid derivatives
	$C_8H_{10}N^+$	pyridines, anilines
121	$C_8H_9O^+$ and $C_7H_5O_2^+$	hydroxybenzene derivatives

Mass	Ion	Compound type
127	$C_{10}H_7^+$	naphthalenes
	$C_6H_7O_3^+$	unsaturated diesters
	$C_6H_6NCl^{+\cdot}$	chlorinated N-aromatics (with ^{37}Cl isotope signal)
	I^+	iodo compounds
128	$C_{10}H_8^{+\cdot}$	naphthalenes
	$C_6H_5OCl^{+\cdot}$	chlorinated hydroxybenzene derivatives (with ^{37}Cl isotope signal)
	$HI^{+\cdot}$	iodo compounds
130	$C_9H_8N^+$	quinolines, indoles
	$C_9H_6O^{+\cdot}$	naphthoquinones
131	$C_{10}H_{11}^+$	tetralins
	$C_5H_7S_2^+$	thioethylene ketals (with ^{34}S isotope signal)
	$C_3F_5^+$	perfluoroalkyl derivatives
135	$C_4H_8Br^+$	alkyl bromides (with ^{81}Br isotope signal at m/z 137)
141	$C_{11}H_9^+$	naphthalenes
142	$C_{10}H_8N^+$	quinolines
149	$C_8H_5O_3^+$	phthalates
152	$C_{12}H_8^{+\cdot}$	diphenyl aromatics
165	$C_{13}H_9^+$	diphenylmethane derivatives (fluorenyl cation)
167	$C_8H_7O_4^+$	phthalates
205	$C_{12}H_{13}O_3^+$	phthalates
223	$C_{12}H_{15}O_4^+$	phthalates

2.5.11 References

[1] M.E. Wieser, Atomic weights of the elements 2005, *J. Phys. Chem. Ref. Data* **2007**, *36*, 485.
[2] G. Audi, A.H. Wapstra, The AME2003 atomic mass evaluation, (II). Tables, graphs and references, *Nucl. Phys. A* **2003**, *729*, 337.
[3] J.K. Böhlke, J.R. de Laeter, P. De Bièvre, H. Hidaka, H.S. Peiser, K.J.R. Rosman, P.P.D. Taylor, Isotopic compositions of the elements, 2001, *J. Phys. Chem. Ref. Data* **2005**, *34*, 57.
[4] H. Kubinyi, Calculation of isotope distributions in mass spectrometry. A trivial solution for a non-trivial problem, *Anal. Chim. Acta* **1991**, *247*, 107.

2.6 UV/Vis Spectroscopy

UV/Vis Absorption Bands of Various Compound Types (A: alkyl or H; R: alkyl; sh: shoulder)

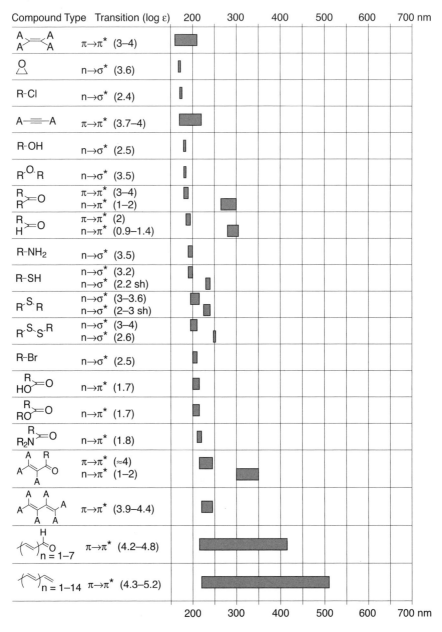

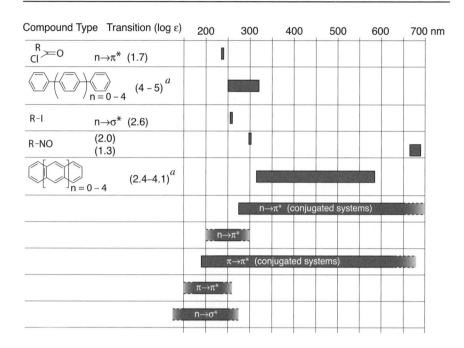

^{a} longest wavelength absorption maximum

3 Combination Tables

3.1 Alkanes, Cycloalkanes

	Assignment	Range	Comments
13C	CH_3	5–35 ppm	CH_3, CH_2, CH, and C can be differentiated
	CH_2	5–45 ppm	by multipulse experiments (DEPT, APT),
	CH	25–60 ppm	off-resonance decoupling, 2D CH correla-
	C	30–60 ppm	tion spectra, or based on relaxation times
			Lower shift values in three-membered rings
1H	CH_3	0.8–1.2 ppm	
	CH_2	1.1–1.8 ppm	Lower shift values in three-membered rings
	CH	1.1–1.8 ppm	
IR	CH st	3000–2840 cm^{-1}	Higher frequency in three-membered rings
	CH_3 δ as	≈1460 cm^{-1}	
	CH_2 δ	≈1460 cm^{-1}	
	CH_3 δ sy	≈1380 cm^{-1}	Doublet for *geminal* methyl groups
	CH_2 γ	770–720 cm^{-1}	In C–$(CH_2)_n$–C with n ≥ 4 at ca. 720 cm^{-1}
MS	Molecular ion	m/z 14n + 2	Weak in *n*-alkanes
			Very weak in isoalkanes
	Fragments		*n*-Alkanes: local maxima at 14n + 1, intensity variations: smooth, minimum at $[M-15]^+$
			Isoalkanes: local maxima at 14n + 1, intensity distribution: irregular (relative maxima due to fragmentation at branching points with charge retention at the most highly substituted C)
	Rearrange-		*n*-Alkanes: unspecific
	ments	m/z 14n	Isoalkanes: elimination of alkenes
		m/z 14n - 2	Monocycloalkanes: elimination of alkanes
UV			No absorption above 200 nm

3.2 Alkenes, Cycloalkenes

Assignment	Range	Comments
13C C=C	100–150 ppm	Considerable differences between Z and E:
C–(C=C)	10–60 ppm	

1H H–(C=C)	4.5–6.5 ppm	Coupling constants, $|J_{gem}|$ 0–3 Hz
		J_{cis} 5–12 Hz
		J_{trans} 12–18 Hz
CH$_3$–(C=C)	≈1.7 ppm	Coupling constants, $^3J_{CH_3–CH=C} \approx 7$ Hz
CH$_2$–(C=C)	≈2.0 ppm	$^3J_{CH_2–CH=C} \approx 7$ Hz

In rings, $|J|$ smaller: $n = 2, {}^3J \approx 0.5$ Hz
$n = 3, {}^3J \approx 1.5$ Hz
$n = 4, {}^3J \approx 4.0$ Hz

Long-range coupling, $^4J_{HC–C=CH}$ 0–2 Hz

IR H–C(=C) st	3100–3000 cm^{-1}	
C=C st	1690–1635 cm^{-1}	Of variable intensity
H–C(=C) δ oop	1000–675 cm^{-1}	
CH$_2$–(C=C) δ	1440 cm^{-1}	

MS Molecular ion	m/z 14n	Alkenes: moderate intensity
	m/z 14n - 2	Monocycloalkenes: medium intensity
Fragments	m/z 14n - 1	Local maxima for alkenes
	m/z 14n - 3	Local maxima for monocyclic alkenes
		Usually, double bonds cannot be localized
Rearrange-ments		n-Alkenes: unspecific except for:

Cyclohexenes: retro-Diels–Alder reaction:

UV C=C $\pi \to \pi^*$	< 210 nm	Isolated double bonds; for highly substituted
	(log ε 3–4)	double bonds often absorption tail
(C=C)$_2$ $\pi \to \pi^*$	215–280 nm	
	(log ε 3.5–4.5)	

3.3 Alkynes

	Assignment	Range	Comments
^{13}C	C≡C	65–85 ppm	Coupling constant $^2J_{HC≡^{13}C} ≈50$ Hz; often leading to unexpected signs of signals in DEPT spectra and unexpected signals in 2D heteronuclear correlation spectra
	C–(C≡C)	0–30 ppm	
^{1}H	H–(C≡C)	1.5–3.0 ppm	Coupling constants, $\|^4J_{CH-C≡CH}\| ≈3$ Hz $\|^5J_{CH-C≡C-CH}\| ≈3$ Hz
	CH$_3$–(C≡C)	≈1.8 ppm	
	CH$_2$–(C≡C)	≈2.2 ppm	
	CH–(C≡C)	≈2.6 ppm	
IR	H–C(≡C) st	3340–3250 cm^{-1}	Sharp, intensive
	C≡C st	2260–2100 cm^{-1}	Sometimes very week
MS	Molecular ion		Weak, in the case of 1-alkynes up to C$_7$ often absent
	Fragments		[M-1]$^+$ often significant
	Rearrangements		Extensive rearrangements, not very characteristic
UV	C≡C $\pi \rightarrow \pi^*$	< 210 nm (log ε 3.7–4.0)	Isolated double bonds; for highly substituted double bonds often absorption tail

3.4 Aromatic Hydrocarbons

	Assignment	Range	Comments
13C	ar C	120–150 ppm	Same ranges for polycyclic aromatic hydro-
	ar CH	110–130 ppm	carbons
	al C–C ar	10–60 ppm	
1H	H–C ar	6.5–7.5 ppm	In polycyclic aromatic hydrocarbons up to $\approx$9 ppm
			Coupling constants, $^{3}J_{ortho} \approx 7$ Hz
			$^{4}J_{meta} \approx 2$ Hz
			$^{5}J_{para} < 1$ Hz
	CH_3–C ar	$\approx$2.3 ppm	Often line broadening due to long-range cou-
	CH_2–C ar	$\approx$2.6 ppm	pling with aromatic protons
	CH–C ar	$\approx$2.9 ppm	
IR	ar C–H st	3080–3030 cm^{-1}	Often multiple bands, weak
	comb	2000–1650 cm^{-1}	Very weak
	ar C–C st	$\approx$1600 cm^{-1}	
		$\approx$1500 cm^{-1}	Of variable intensity, sometimes not all bands
		$\approx$1450 cm^{-1}	observable
	ar C–H δ oop	960–650 cm^{-1}	Strong, frequently multiple bands
MS	Molecular ion		Strong, often base peak
	Fragments	m/z 39, 50–53, 63–65, 75–78, $[M-26]^{+\cdot}$, $[M-39]^{+}$	Often doubly charged fragment ions
	Benzylic cleavage		m/z 91 (90, 92)
	Other typical frag- ments		m/z 127
			m/z 152
			m/z 165
	Rearrange- ments		

	Assignment	Range	Comments
UV		≈200–210 nm (log ε ≈4) ≈260 nm (log ε ≈2.4)	In benzene and alkylbenzenes

3.5 Heteroaromatic Compounds

	Assignment	Range	Comments
^{13}C	ar C–X	120–160 ppm	
	ar C–C	100–150 ppm	
^{1}H	H–C ar	6–9 ppm	Coupling constants in 6-membered rings similar to those in aromatic hydrocarbons; smaller in 5-membered rings
	H–N ar	7–14 ppm	Strongly solvent dependent, generally broad
IR	ar C–H st	3100–3000 cm^{-1}	Often multiple bands, weak
	ar N–H st	3500–2800 cm^{-1}	
	ar C–C st	≈1600 cm^{-1} ≈1500 cm^{-1} ≈1450 cm^{-1}	Often split, sometimes not all bands observable
	ar C–H δ oop	1000–650 cm^{-1}	Often strong, frequently multiple bands
MS	Molecular ion		Strong, often base peak
	Fragments	m/z 39, 50–53, 63–65, 75–78, [M-26]$^{+\cdot}$, [M-39]$^+$	Often doubly charged fragment ions
		m/z 45 [CHS]$^+$	S-Heteroaromatics
		Benzyl-analogous cleavage	
	Rearrangements		Loss of HCN (Δm 27, N-heteroaromatics) Loss of CO (Δm 28, O-heteroaromatics) Loss of CS (Δm 44, S-heteroaromatics)
UV			cf. UV/Vis Reference Spectra, Chapter 8.5.3

3.6 Halogen Compounds

	Assignment	Range	Comments
13C	al C–F	70–100 ppm	CF$_3$: ≈115 ppm
	(C)=C–F	125–175 ppm	Coupling, with ^{19}F (isotope abundance,
	C=(C–F)	65–115 ppm	100%; I = 1/2): $\mid^{1}$J$_{CF}\mid$ 100–300 Hz
	ar C–F	140–165 ppm	$\mid^{2}$J$_{CF}\mid$ 10–40 Hz
	ar C–(C–F)	105–135 ppm	$\mid^{3}$J$_{CF}\mid$ 5–10 Hz
			$\mid^{4}$J$_{CF}\mid$ 0–5 Hz
	al C–Cl	30–60 ppm	
	(C)=C–Cl	100–150 ppm	
	C=(C–Cl)	100–155 ppm	
	ar C–Cl	120–150 ppm	
	ar C–(C–Cl)	125–135 ppm	
	al C–Br	10–45 ppm	
	(C)=C–Br	90–140 ppm	
	C=(C–Br)	90–140 ppm	
	ar C–Br	110–140 ppm	
	ar C–(C–Br)	125–135 ppm	
	al C–I	-20 to +30 ppm	
	(C)=C–I	60–110 ppm	
	C=(C–I)	120–150 ppm	
	ar C–I	85–115 ppm	
	ar C–(C–I)	125–145 ppm	
1H	CH$_2$–F	≈4.3 ppm	Coupling, with ^{19}F (isotope abundance,
			100%; I = 1/2): $\mid^{2}$J$_{HF}\mid$ 40–80 Hz
			$\mid^{3}$J$_{HF}\mid$ 0–50 Hz
			$\mid^{4}$J$_{CF}\mid$ 0–5 Hz
	CH$_2$–Cl	≈3.5 ppm	
	CH$_2$–Br	≈3.4 ppm	
	CH$_2$–I	≈3.1 ppm	
	H–CX=C	5.5–8.0 ppm	Similar shifts for all halogens
	H–C=CF	4.0–6.0 ppm	
	H–C=CCl	4.5–6.5 ppm	
	H–C=CBr	5.0–7.0 ppm	
	H–C=CI	5.5–7.5 ppm	
	H–phenyl–hal	7.0–7.6 ppm	Shielding by F in *ortho* and *para* positions; small effects for Cl and Br; deshielding by I in *ortho*, and shielding in *meta* position
IR	C–F st	1400–1000 cm^{-1}	Strong
	C–Cl st	850–600 cm^{-1}	Strong
	C–Br st	700–500 cm^{-1}	Strong
	C–I st	650–450 cm^{-1}	Strong

	Assignment	Range	Comments
MS	Molecular ion		Often weak for saturated aliphatic halogen compounds, often absent from spectra of aliphatic polyhalogenated compounds
			Characteristic isotope pattern for Cl and Br
	Fragments	m/z 69, 50–53	CF_3
			Upon fragmentation of the C–hal bond, the positive charge preferably remains on the alkyl side, and on the halogen side upon fragmentation of the neighboring bond:

$$R-C-\!\!-hal \quad > \quad R-\!\!-C-hal$$

	Assignment	Range	Comments
	Rearrange-ments	$[M-20]^{+\cdot}$	HF elimination
		$[M-50]^{+\cdot}$ or $[frag-50]^+$	CF_2 elimination
		$[M-36]^{+\cdot}$	HCl elimination
UV	hal $n \rightarrow \pi^*$	≤280 nm (log ε ≈2.5)	For C–I; for C–Br and C–Cl in general only absorption tail, for C–F no absorption

3.7 Oxygen Compounds

3.7.1 Alcohols and Phenols

Assignment	Range	Comments
13C al C–OH	50–80 ppm	Shift with respect to C–H ≈50 ppm
al C–(C–OH)	10–60 ppm	Hardly any shift with respect to C–(C–CH$_3$)
al C–(C–C–OH)	10–60 ppm	Shift with respect to C–(C–C–CH$_3$) ≈-5 ppm
ar C–OH	140–155 ppm	Shift with respect to C–H ≈+25 ppm
ar C–(C–OH)	100–130 ppm	Shift with respect to C–(C–H): *ortho* ≈-13 ppm, *meta* ≈+1 ppm, *para* ≈-8 ppm
1H HO–C al	0.5–6 ppm	Often broad; position and shape strongly
HO–C ar	4–12 ppm	depend on experimental conditions
CH$_2$–(OH)	3.5–4.0 ppm	
CH–(OH)	3.8–4.2 ppm	
ar CH–(C–OH)	6.5–7.0 ppm	Shift with respect to CH–(C–H): *ortho* ≈-0.6 ppm, *meta* ≈-0.1 ppm, *para* ≈-0.5 ppm
IR O–H st	3650–3200 cm^{-1}	Position and shape depend on the degree of association. Often different bands for H-bonded and free OH
C–O(H) st	1260–970 cm^{-1}	Strong
MS Molecular ion		Aliphatic: weak, often missing in the case of primary and highly branched alcohols; in this case, peaks at highest mass are often due to [M-18]$^{+\cdot}$ or [M-15]$^{+}$ Aromatic: strong
Fragments	Aliphatic: m/z 31, 45, 59, ... [M-33]$^{+}$	Primary: m/z 31 > m/z 45 ≈ m/z 59 Secondary, tertiary: local maxima due to α-cleavage:

$$R\!-\!\overset{\displaystyle R}{\underset{\displaystyle H}{C}}\!-\!OH \; \xrightarrow{\;-\,R^{\bullet}\;} \; R\!-\!\overset{+\cdot}{C}\!=\!OH$$

	Aromatic: [M-28]$^{+\cdot}$ (CO) [M-29]$^{+}$ (CHO)	CO and CHO elimination also from fragments. H$_2$O elimination ([M-18]$^{+\cdot}$) only with alkyl substituent in *ortho* position
Rearrangements	Aliphatic: [M-18]$^{+\cdot}$ [M-46]$^{+\cdot}$	Elimination of H$_2$O from M$^{+\cdot}$ followed by alkene elimination; elimination of H$_2$O from products of α-cleavage
	Unsaturated	Vinylcarbinols: spectra similar to those of ketones Allyl alcohols: specific aldehyde elimination:

$$\left[R_1\diagup\!\!\diagdown\!\!\diagup\!\!\overset{OH}{\diagdown}\!\!R_2 \right]^{+\cdot} \xrightarrow{\;-\,R_2CHO\;} \left[R_1\diagup\!\!\diagdown\!\!\diagup \right]^{+\cdot}$$

	Assignment	Range	Comments
MS		Aromatic:	*Ortho* effect with appropriate substituents:

with Y–Z as –CO–OR, C–hal, –O–R, and similar

	Assignment	Range	Comments
UV	Aliphatic		No absorption above 200 nm
	Aromatic	≈200–210 nm	In alkaline solution, shift to longer wave-
		(log ε ≈3.8)	length and increase in intensity due to
		≈270 nm	deprotonation
		(log ε ≈2.4)	

3.7.2 Ethers

	Assignment	Range	Comments
13C	al C–O	50–90 ppm	Oxiranes: outside the normal range
	al C–(C–O)	10–60 ppm	Hardly any shift with respect to C–(C–CH$_3$)
	al C–(C–C–O)	10–60 ppm	Shift with respect to C–(C–C–CH$_3$) ≈-5 ppm
	O–C–O	85–110 ppm	
	(C)=C–O	115–165 ppm	Shift with respect to (C)=C–C ≈+15 ppm
	C=(C–O)	70–120 ppm	Shift with respect to C=(C–C) ≈-30 ppm
	ar C–O	140–155 ppm	Shift with respect to ar C–H ≈+25 ppm
	ar C–(C–O)	100–130 ppm	Shift with respect to ar C–(C–H):
			ortho ≈-15 ppm
			meta ≈+1 ppm
			para ≈-8 ppm
1H	CH$_3$–O	3.3–4.0 ppm	Singlet
	CH$_2$–O	3.4–4.2 ppm	
	O–CH$_2$–O	4.5–6.0 ppm	
	CH–O	3.5–4.3 ppm	
	CH(O)$_3$	≈ 5–6 ppm	
	H–C(O)=C	5.7–7.5 ppm	Shift with respect to H–C(H)=C ≈+1.2 ppm
	H–C=C–O	3.5–5.0 ppm	Shift with respect to H–C(=C–H) ≈-1 ppm
	ar CH–C–O	6.6–7.6 ppm	
IR	H–C(–O) st	2880–2815 cm^{-1}	For CH$_3$–O and CH$_2$–O; similar range for corresponding amines
	H–CH(O)$_2$ st	2880–2750 cm^{-1}	Two bands
	C–O–C st	1310–1000 cm^{-1}	Strong, sometimes two bands

	Assignment	Range	Comments
MS	Molecular ion		Aliphatic: weak, tendency to protonate
			Aromatic: strong
	Fragments	Aliphatic:	Base peak of aliphatic ethers generally due to
		m/z 31, 45, 59, ...	fragmentation of the bond next to the ether
		[M-33]$^+$	bond: $R_1-C-O-R_2$ $]^{+\cdot}$ $\xrightarrow{-R_1{}^{\cdot}}$ $C=\overset{+}{O}-R_2$
			or due to heterolytic cleavage of the C–O bond (especially for polyethers):
			R_1-O-R_2 $]^{+\cdot}$ $\xrightarrow{-R_1-O^{\cdot}}$ $R_2{}^+$
		Alkyl aryl ethers	Preferential loss of the alkyl chain
		Diaryl ethers	Preferential loss of CO (Δm 28) from M$^{+\cdot}$
			and/or [M-H]$^+$ as well as: $ar_1-O-\overset{\curvearrowright}{|}-ar_2$
	Rearrange-	Aliphatic:	Elimination of water or alcohol
	ments	[M-18]$^{+\cdot}$	
		[M-46]$^{+\cdot}$	
		Aromatic	Ethyl and higher alkyl ethers: alkene elimination to the phenol:

	Assignment		Comments
UV	Aliphatic		No absorption above 200 nm
	Aromatic		Shift to higher wavelength and increase in intensity due to the ether group

3.8 Nitrogen Compounds

3.8.1 Amines

	Assignment	Range	Comments
13C	al C–N	25–70 ppm	Shift with respect to C–H ≈+20 ppm
	al C–(C–N)	10–60 ppm	Shift with respect to C–(C–CH$_3$) ≈+2 ppm
	al C–(C–C–N)	10–60 ppm	Shift with respect to C–(C–C–CH$_3$) ≈-2 ppm
	(C)=C–N	120–170 ppm	Shift with respect to (C)=C–C ≈+20 ppm
	C=(C–N)	75–125 ppm	Shift with respect to C=(C–C) ≈-25 ppm
	ar C–N	130–150 ppm	Shift with respect to C–H ≈+20 ppm
	ar C–(C–N)	100–130 ppm	Shift with respect to C–(C–H):
			ortho ≈-15 ppm
			meta ≈+1 ppm
			para ≈-10 ppm
1H	HN–C al	0.5–4.0 ppm	
	HN–C ar	2.5–5.0 ppm	
	HN$^+$–C al or ar	6.0–9.0 ppm	Often broad
	CH$_3$–N	2.3–3.1 ppm	Singlet
	CH$_2$–N	2.5–3.5 ppm	
	CH–N	3.0–3.7 ppm	
	CH–N$^+$	3.2–4.0 ppm	
	ar CH–C–N	6.0–7.5 ppm	Shift with respect to CH–(C–H):
			ortho ≈-0.8 ppm
			meta ≈-0.2 ppm
			para ≈-0.7 ppm
	ar CH–C–N$^+$	7.5–8.0 ppm	Shift with respect to CH–(C–H):
			ortho ≈+0.7 ppm
			meta ≈+0.4 ppm
			para ≈+0.3 ppm
IR	N–H st	3500–3200 cm^{-1}	Position and shape depend on the degree of association. Often different bands for H-bonded and free NH. For NH$_2$, always at least two bands
	N$^+$–H st	3000–2000 cm^{-1}	Broad, similar to COOH but more structured
	N–H δ	1650–1550 cm^{-1}	Weak or absent
	N$^+$–H δ	1600–1460 cm^{-1}	Often weak
	H–C(–N) st	2850–2750 cm^{-1}	For CH$_3$–N and CH$_2$–N in amines; similar range for corresponding ethers

	Assignment	Range	Comments
MS	Molecular ion		Odd nominal mass number for odd number of N atoms
			Aliphatic: weak, tendency to protonate, $[M+H]^+$ is often important
			Aromatic: strong, no tendency to protonate
	Fragments	Aliphatic: m/z 30, 44, 58, ...	Base peak of aliphatic amines generally due to fragmentation of the bond next to the amine bond:
	Rearrange-ments		Elimination of alkenes following amine cleavage:
UV	Aliphatic		No absorption above 200 nm
	Aromatic		In acidic solutions, shift to lower wavelength and decrease in intensity

3.8.2 Nitro Compounds

	Assignment	Range	Comments
^{13}C	al C–NO$_2$	55–110 ppm	Shift with respect to C–H $\approx$+50 ppm
	al C–(C–NO$_2$)	10–50 ppm	Shift with respect to C–(C–C) $\approx$-6 ppm
	al C–(C–CNO$_2$)	10–60 ppm	Shift with respect to C–(C–C–C) $\approx$-2 ppm
	ar C–NO$_2$	130–150 ppm	Shift with respect to C–H $\approx$+20 ppm
	ar C–(C–NO$_2$)	120–140 ppm	Shift with respect to C–(C–H): *ortho* $\approx$-5 ppm, *meta* $\approx$+1 ppm, *para* $\approx$+6 ppm
^{1}H	al CH–NO$_2$	4.2–4.6 ppm	
	ar CH–C–NO$_2$	7.5–8.5 ppm	Shift with respect to CH–(C–H): *ortho* $\approx$+1 ppm, *meta* $\approx$+0.3 ppm, *para* $\approx$+0.4 ppm
IR	NO$_2$ st as	1660–1490 cm^{-1}	Strong to very strong
	NO$_2$ st sy	1390–1260 cm^{-1}	Strong to very strong
MS	Molecular ion		Odd nominal mass number for odd number of N atoms
			Aliphatic: weak or absent
			Aromatic: strong
	Fragments	$[M-16]^{+\cdot}$, $[M-46]^+$	
	Rearrange-ments	m/z 30, $[M-17]^+$, $[M-30]^+$, $[M-47]^{+\cdot}$	
UV	Aliphatic	$\approx$275 nm (log ε <2)	
	Aromatic	$\approx$350 nm (log ε $\approx$2)	

3.9 Thiols and Sulfides

	Assignment	Range	Comments
^{13}C	al C–S	5–60 ppm	No significant shift with respect to C–C
	ar C–S	120–140 ppm	
^{1}H	HS–C al	1.0–2.0 ppm	*Vicinal* coupling constant, J, 5–9 Hz
	HS–C ar	2.0–4.0 ppm	
	al CH–S	2.0–3.2 ppm	
	ar CH–S	7.0–7.5 ppm	
IR	S–H st	2600–2540 cm^{-1}	Frequently weak
MS	Molecular ion		^{34}S-isotope peak at [M+2]$^{+\cdot}$ ≈4.5%
			Aliphatic: intensity higher than for corresponding alcohols and ethers
	Fragments	m/z 47, 61, 75, ...	Sulfide cleavage:

$$R_1-S-CH_2-R_2 \Big]^{+\cdot} \xrightarrow{-R_2^\cdot} R_1-\overset{+}{S}=CH_2$$

	Rearrangements	m/z 34, 35, 48	
		[M-33]$^+$	Alkene elimination after sulfide cleavage
		[M-34]$^{+\cdot}$	
UV	Aliphatic	<225 nm (log ε 3–4)	
		220–250 nm (log ε 2–3)	

3.10 Carbonyl Compounds

3.10.1 Aldehydes

	Assignment	Range	Comments
13C	CHO	190–205 ppm	Coupling constant $^1J_{CH}$ 172 Hz
	al C–(CHO)	30–70 ppm	Coupling constant $|^2J_{CH}|$ 20–50 Hz
	al C–(C–CHO)	5–50 ppm	Shift with respect to C–(C–CH$_3$) ≈-10 ppm
	(C)=C–(CHO)	110–160 ppm	
	C=(C–CHO)	110–160 ppm	
	ar C–(CHO)	120–150 ppm	
1H	H–(C=O)	9.0–10.5 ppm	
	al CH–(CHO)	2.0–2.5 ppm	$^3J_{HH}$ 0–3 Hz
	(CH)=CH(CHO)	5.5–7.0 ppm	$^3J_{HH}$ ≈8 Hz
	CH=(CH–CHO)	5.5–7.0 ppm	
	ar CH–(C–CHO)	7.2–8.0 ppm	Shift with respect to CH–(C–H):
			ortho ≈+0.6 ppm
			meta ≈+0.2 ppm
			para ≈+0.3 ppm
IR	comb	2900–2700 cm^{-1}	Two weak bands
	C=O	1765–1645 cm^{-1}	Aliphatic: ≈1730 cm^{-1}
			Conjugated: ≈1690 cm^{-1}
MS	Molecular ion		Aliphatic: moderate
			Aromatic: strong
	Fragments	[M-1]$^+$	For aliphatic aldehydes, only significant up to C$_7$
		[M-29]$^+$	
	Rearrange-	m/z 44	Aliphatic aldehydes
	ments	[M-44]$^{+\cdot}$	

UV	n → π*	270–310 nm (log ε ≈1)	Saturated aldehydes
		≥207 nm (log ε ≈4)	α,β-Unsaturated aldehydes
		≥250 nm (log ε >3)	Aromatic aldehydes

3.10.2 Ketones

	Assignment	Range	Comments
¹³C	C=O	195–220 ppm	
	al C–(C=O)	25–70 ppm	
	al C–(C–C=O)	5–50 ppm	Shift with respect to C–(C–CH$_3$) ≈-6 ppm
	(C)=C–(C=O)	105–160 ppm	
	C=(C–C=O)	105–160 ppm	
	ar C–(C=O)	120–150 ppm	
¹H	al CH–(C=O)	2.0–3.6 ppm	al CH–C(=O)–C al 2.0–2.6 ppm
			al CH–C(=O)–C ar 2.5–3.6 ppm
	CH=CH–(C=O)	5.5–7.0 ppm	
	ar CH–(C–C=O)	7.2–8.0 ppm	Shift with respect to CH–(C–H):
			ortho ≈+0.6 ppm
			meta ≈+0.1 ppm
			para ≈+0.2 ppm
IR	C=O st	1775–1650 cm^{-1}	Aliphatic: ≈1715 cm^{-1}
			Cyclic: ring size ≥6: ≈1715 cm^{-1}
			ring size <6: ≥1750 cm^{-1}
			Conjugated: ≈1690–1665 cm^{-1}
MS	Molecular ion		Aliphatic: moderate
			Aromatic: strong
	Fragments		Ketone cleavages:

$$\left[\ \overset{O}{\underset{R_1\quad R_2}{C}}\ \right]^{+\cdot} \longrightarrow R_1CO^+,\ R_1^+,\ R_2CO^+,\ R_2^+$$

| | Rearrange-ments | m/z 44 [M-44]$^{+\cdot}$ | Aliphatic ketones |

	Assignment	Range	Comments
UV	π → π*	<200 nm (log ε 3–4)	Saturated ketones
	n → π*	250–300 nm (log ε 1–2)	Saturated ketones
		≥215 nm (log ε ≈4)	α,β-Unsaturated ketones
		≥245 nm (log ε >3)	Aromatic ketones

3.10.3 Carboxylic Acids

	Assignment	Range	Comments
13C	COOH	170–185 ppm	For COO⁻, shift with respect to COOH: 0 to +8 ppm
	al C–(COOH)	25–70 ppm	
	al C–(C–COOH)	5–50 ppm	Shift with respect to C–(C–CH$_3$) ≈-6 ppm
	(C)=C–(COOH)	105–160 ppm	
	C=(C–COOH)	105–160 ppm	
	ar C–(COOH)	120–150 ppm	
1H	COOH	10.0–13.0 ppm	Position and shape strongly depend on experimental conditions
	al CH–(COOH)	2.0–2.6 ppm	
	CH=CH–(COOH)	5.2–7.5 ppm	
	ar CH–(C–COOH)	7.2–8.0 ppm	Shift with respect to CH–(C–H): *ortho* ≈+0.8 ppm, *meta* ≈+0.2 ppm, *para* ≈+0.3 ppm
IR	COO–H st	3550–2500 cm⁻¹	Broad
	C=O st	1800–1650 cm⁻¹	Aliphatic: ≈1715 cm⁻¹ Conjugated: ≈1695 cm⁻¹ For COO⁻, two bands: 1580 and 1420 cm⁻¹
	COO–H δ oop	≈920 cm⁻¹	For dimers
MS	Molecular ion		Aliphatic: moderate, strong for long chains, tendency to protonate Aromatic: strong
	Fragments	[M-17]⁺ [M-45]⁺	Strong for aromatic acids
	Rearrange-ments	m/z 60, 61 [M-18]⁺˙	Aliphatic acids Aromatic acids *Ortho* effect with aromatic acids:

			for X: CH$_2$, O, N, S, etc.

	Assignment	Range	Comments
UV	n → π*	<220 nm (log ε 1–2)	Saturated acids
		≥193 nm (log ε ≈4)	α,β-Unsaturated acids
		≥230 nm (log ε >3)	Aromatic acids

3.10.4 Esters and Lactones

	Assignment	Range	Comments
13C	COOR	165–180 ppm	Shift with respect to COOH -5 to -10 ppm
	al C–(COOR)	20–70 ppm	
	al C–(OCOR)	50–100 ppm	Shift with respect to C–(OH) +2 to +10 ppm
	(C)=C–(COOR)	105–160 ppm	
	C=(C–COOR)	105–160 ppm	
	(C)=C–(OCOR)	100–150 ppm	
	C=(C–OCOR)	80–130 ppm	
	ar C–(COOR)	120–150 ppm	
	ar C–(OCOR)	130–160 ppm	
	ar C=(C–OCOR)	105–130 ppm	
1H	al CH–COOR	2.0–2.5 ppm	$CH_3COOR \approx 2.0$ ppm
			$CH_2COOR \approx 2.3$ ppm
			$CHCOOR \approx 2.5$ ppm
	al CH–OCOR	3.5–5.3 ppm	$CH_3OCOR \approx 3.5–3.9$ ppm
			$CH_2COOR \approx 4.0–4.5$ ppm
			$CHCOOR \approx 4.8–5.3$ ppm
	CH=CH–COOR	5.5–8.0 ppm	Shift with respect to CH=CH–H:
			gem $\approx +0.8$ ppm, *cis* $\approx +1.1$ ppm
			trans $\approx +0.5$ ppm
	CH=CH–OCOR	6.0–8.0 ppm	Shift with respect to CH=CH–H:
			gem $\approx +2.1$ ppm, *cis* ≈ -0.4 ppm
			trans ≈ -0.6 ppm
	ar CH–C–COOR	7.0–8.0 ppm	Shift with respect to CH–(C–H):
			ortho $\approx +0.7$ ppm, *meta* $\approx +0.1$ ppm,
			para $\approx +0.2$ ppm
	ar CH–C–OCOR	6.8–7.5 ppm	Shift with respect to CH–(C–H):
			ortho ≈ -0.2 ppm, *meta* ≈ 0 ppm,
			para ≈ -0.1 ppm
IR	C=O st	1745–1730 cm⁻¹	Strong; range for aliphatic esters
			Higher wavenumbers for hal–C–COOR, COO–C=C, COO–C ar, and for small ring lactones
			Lower wavenumbers for C=C–COOR and ar C–COOR
	C–O st	1330–1050 cm⁻¹	Mostly two bands, at least one of them strong
			For COO⁻, two bands: 1580 and 1420 cm⁻¹

Assignment	Range	Comments
MS Molecular ion		Aliphatic esters: weak, tendency to protonate Aliphatic lactones: medium to weak, tendency to protonate Aromatic esters and lactones: strong
Fragments	$[M - RO]^+$ $[M - ROCO]^+$	Esters Esters Lactones: loss of α-substituents (attached to ether carbon), decarbonylation, for aromatic lactones also double decarbonylation
Rearrangements		Alkene elimination from the alcohol moiety:

$$\left[R_1\!-\!CH\!-\!H \cdots O\!-\!C(=O)\!-\!R_2 \right]^{+\cdot} \longrightarrow \left[R_1\!-\!CH=CH_2 \ + \ R_2\!-\!COOH \right]^{+\cdot}$$

Elimination of the alcohol side chain with double H transfer (for alcohols with $C_{n>2}$):

$$\left[R_1\!-\!COOR_2 \right]^{+\cdot} \longrightarrow R_1\!-\!C(OH)_2^+ \ \ \overset{+}{OH}$$

Alcohol elimination from *ortho*-substituted aromatic esters:

$$\left[\text{aromatic ester with } COOR, \ X\!-\!H \right]^{+\cdot} \ \xrightarrow{\ -\,ROH\ } \ \left[\text{product } C{=}O, \ X \right]^{+\cdot} \quad \text{for X: } CH_2, \ O, \ N, \ S, \text{ etc.}$$

| | $[M\text{-}18]^{+\cdot}$ | Lactones |
| **UV** $n \rightarrow \pi^*$ | <220 nm (log ε 1–2) ≥193 nm (log $\varepsilon \approx$4) ≥230 nm (log ε >3) | Aliphatic esters α,β-Unsaturated esters Aromatic esters |

3.10.5 Amides and Lactams

	Assignment	Range	Comments
¹³C	CONR$_2$	165–180 ppm	
	al C–(ĊONR$_2$)	20–70 ppm	
	al C–(C–CONR$_2$)	5–50 ppm	Shift with respect to C–(C–CH$_3$) ≈-6 ppm
	al C–(NCOR)	25–80 ppm	Shift with respect to C–(NH) ≈-1 to -2 ppm
	C=C–(CONR$_2$)	105–160 ppm	
	ar C–(CONR$_2$)	120–150 ppm	
	ar C–(NCOR)	110–150 ppm	
¹H	CONH	5–10 ppm	Frequently broad to very broad; splitting due to H–N–C–H coupling often recognizable only in the CH signal
	al CH–CONR$_2$	2.0–2.5 ppm	
	al CH–NCOR	2.7–4.8 ppm	CH$_3$NCOR ≈2.7–3.0 ppm CH$_2$NCOR ≈3.1–3.5 ppm CHNCOR ≈3.8–4.8 ppm
	CH=CH–CONR$_2$	5.2–7.5 ppm	Shift with respect to CH=CH–(H): *gem* ≈+1.4 ppm, *cis* ≈+1.0 ppm *trans* ≈+0.5 ppm
	C=CH–NCOR	6.0–8.0 ppm	Shift with respect to CH=CH–(H):
	CH=C–NCOR	4.5–6.0 ppm	*gem* ≈+2.1 ppm, *cis* ≈-0.6 ppm *trans* ≈-0.7 ppm
	ar CH–C(CONR$_2$)	7.5–8.5 ppm	Shift with respect to CH–C–(H): *ortho* ≈+0.6 ppm, *meta* ≈+0.1 ppm, *para* ≈+0.2 ppm
	ar CH–C(NCOR)	6.8–7.5 ppm	Shift with respect to CH–C–(H): *ortho* ≈0 ppm, *meta* ≈0 ppm, *para* ≈-0.2 ppm
IR	N–H st	3500–3100 cm⁻¹	Position and shape depend on the extent of association, often different bands for H-bonded and free NH, always at least two bands for NH$_2$
	C=O st (amide I)	1700–1650 cm⁻¹	Strong; range for amides as well as for δ- and larger lactams, higher wavenumbers for β- and γ-lactams
	N–H δ and N–C=O st sy (amide II)	1630–1510 cm⁻¹	Often strong, missing in the case of tertiary amides and lactams

	Assignment	Range	Comments
MS	Molecular ion		Aliphatic amides: moderate, tendency to protonate Aromatic amides: strong
	Fragments		Amides: cleavage on both sides of the carbonyl group followed by loss of CO; large number of fragments of even mass Lactams: loss of α-substituent, loss of CO
	Rearrange-ments		Amides: elimination of the amine moiety, elimination of alkene from the amine or acid moiety in analogy to esters
		$[M\text{-}18]^{+\cdot}$	Lactams
UV	$n \rightarrow \pi^*$	<220 nm (log ε 1–2)	Aliphatic amides and lactams

4 ^{13}C NMR Spectroscopy

4.1 Alkanes

4.1.1 Chemical Shifts

^{13}C Chemical Shifts (δ in ppm)

-2.3
CH$_4$

7.3
CH$_3$
|
CH$_3$

15.9

15.4

13.0

24.8

24.1

25.0

22.8

34.8   14.2

32.0   22.3

11.8   30.1

31.3

27.7

32.2   14.2

23.1

29.5   23.1

32.4   14.1

32.1   14.1

29.5   22.8

^{13}C Chemical Shifts of Methyl Groups (δ in ppm)

	Substituent R	δ_{CH_3-R}		Substituent R	δ_{CH_3-R}
	–H	-2.3	**C**	–2-pyridyl	24.2
C	–CH$_3$	7.3		–3-pyridyl	18.0
	–CH$_2$CH$_3$	15.4		–4-pyridyl	20.6
	–CH(CH$_3$)$_2$	24.1		–2-furyl	13.7
	–C(CH$_3$)$_3$	31.3		–2-thienyl	14.7
	–(CH$_2$)$_6$CH$_3$	14.1		–2-pyrrolyl	11.8
	–CH$_2$–phenyl	15.7		–2-indolyl	13.4
	–CH$_2$F	15.8		–3-indolyl	9.8
	–CH$_2$Cl	18.7		–4-indolyl	21.6
	–CH$_2$Br	19.1		–5-indolyl	21.5
	–CH$_2$I	20.4		–6-indolyl	21.7
	–CHCl$_2$	31.6		–7-indolyl	16.6
	–CHBr$_2$	31.8	**X**	–F	71.6
	–CCl$_3$	46.3		–Cl	25.6
	–CBr$_3$	49.4		–Br	9.6
	–CH$_2$OH	18.2		–I	-24.0
	–CH$_2$OCH$_3$	14.7	**O**	–OH	50.2
	–CH$_2$OCH$_2$CH$_3$	15.4		–OCH$_3$	60.9
	–CH$_2$OCH=CH$_2$	14.6		–OCH$_2$CH$_3$	57.6
	–CH$_2$O–phenyl	14.9		–OCH(CH$_3$)$_2$	54.9
	–CH$_2$OCOCH$_3$	14.4		–OC(CH$_3$)$_3$	49.4
	–CH$_2$NH$_2$	19.0		–OCH$_2$CH=CH$_2$	57.4
	–CH$_2$NHCH$_3$	14.3		–O–cyclohexyl	55.1
	–CH$_2$N(CH$_3$)$_2$	12.8		–OCH=CH$_2$	52.5
	–CH$_2$NO$_2$	12.3		–O–phenyl	54.8
	–CH$_2$SH	19.7		–OCOCH$_3$	51.5
	–CH$_2$S(O)$_2$CH$_3$	6.7		–OCO–cyclohexyl	51.2
	–CH$_2$S(O)$_2$OH	8.0		–OCOCH=CH$_2$	51.5
	–CH$_2$CHO	5.2		–OCO–phenyl	51.8
	–CH$_2$COCH$_3$	7.0		–OCOOCH$_3$	54.9
	–CH$_2$COOH	9.6		–OS(O)$_2$–4-tolyl	56.3
	–cyclopentyl	20.5		–OS(O)$_2$OCH$_3$	59.1
	–cyclohexyl	23.1		–OP(OCH$_3$)$_2$	48.8
	–CH=CH$_2$	18.7	**N**	–NH$_2$	28.3
	–C≡CH	3.7		–NH$_3$$^+$	26.5
	–phenyl	21.4		–NHCH$_3$	38.2
	–1-naphthyl	19.1		–NH–cyclohexyl	33.5
	–2-naphthyl	21.5		–NH–phenyl	30.2

Substituent R	δ_{CH_3-R}		Substituent R	δ_{CH_3-R}
N –N(CH$_3$)$_2$	47.5	**O**	–COCH=CH$_2$	25.7
–*N*-pyrrolidinyl	42.7	‖	–CO–cyclohexyl	27.6
–*N*-piperidinyl	47.7	**C**	–CO–phenyl	25.7
–N(CH$_3$)phenyl	39.9		–COOH	21.7
–*N*-pyrrolyl	35.9		–COO$^-$	24.4
–*N*-imidazolyl	32.2		–COOCH$_3$	20.6
–*N*-pyrazolyl	38.4		–COOCOCH$_3$	21.8
–*N*-indolyl	32.1		–CONH$_2$	22.3
–NHCOCH$_3$	26.1		–CON(CH$_3$)$_2$	21.5
–N(CH$_3$)CHO	31.5, 36.5		–COSH	32.6
			–COSCH$_3$	30.2
–N(CH$_3$)COCH$_3$	35.0, 38.0		–COCOCH$_3$	23.2
			–COCl	33.6
–N(CH$_3$)P[N(CH$_3$)$_2$]$_2$	33.9		–COBr	39.1
–NO$_2$	61.2		–COSi(CH$_3$)$_3$	35.7
–C≡N	1.7	**M**	–Li	-16.6
–NC	26.8		–B(CH$_3$)$_2$	14.8
–NCS	29.1		–B$^-$(CH$_3$)$_3$ Li$^+$	6.2
S –SH	6.5		–Si(CH$_3$)$_2$CH=CH$_2$	-2.0
–SCH$_3$	19.3		–SiCl$_3$	9.8
–S–*n*-C$_8$H$_{17}$	15.5		–Ge(CH$_3$)$_3$	-3.6
–S–phenyl	15.6		–Sn(CH$_3$)$_3$	-9.3
–SSCH$_3$	22.0		–Pb(CH$_3$)$_3$	-4.2
–S(O)CH$_3$	40.1		–P(CH$_3$)(*n*-C$_4$H$_9$)	14.4
–S(O)$_2$CH$_3$	42.6		–P$^+$(CH$_3$)$_3$ I$^-$	10.7
–S(O)$_2$CH$_2$CH$_3$	39.3		–As(CH$_3$)$_2$	11.2
–S(O)$_2$Cl	52.6		–As$^+$(CH$_3$)$_3$ I$^-$	8.4
–S(O)$_2$OH	39.6		–In(CH$_3$)$_2$	-6.3
–S(O)$_2$ONa	41.1			
O –CHO	31.2			
‖ –COCH$_3$	30.7			
–COCH$_2$CH$_3$	27.5			
C –COCCl$_3$	21.1			

^{13}C Chemical Shifts of Monosubstituted Alkanes (δ in ppm)

Substituent	Methyl $-CH_3$	Ethyl $-CH_2$	$-CH_3$	1-Propyl $-CH_2$	$-CH_2$	$-CH_3$
X –H	-2.3	7.3	7.3	15.4	15.9	15.4
–CH=CH$_2$	18.7	27.4	13.4	36.2	22.4	13.6
–C≡CH	3.7	12.3	13.8	20.6	22.2	13.4
–phenyl	21.4	29.1	15.8	38.3	24.8	13.8
–F	71.6	80.1	15.8	85.2	23.6	9.2
–Cl	25.6	39.9	18.9	46.8	26.3	11.6
–Br	9.6	27.6	19.4	35.6	26.4	13.0
–I	-24.0	-1.6	20.6	9.1	27.0	15.3
O –OH	50.2	57.8	18.2	64.2	25.9	10.3
–OCH$_3$	60.9	67.7	14.7	74.5	23.2	10.5
–OCH$_2$CH$_3$	57.6	66.0	15.4	72.5	23.2	10.7
–OCH(CH$_3$)$_2$	54.9					
–OC(CH$_3$)$_3$	49.4	56.8	16.4			
–O–phenyl	54.8	63.2	14.9	69.4	22.8	10.6
–OCOCH$_3$	51.5	60.4	14.4	66.2	22.4	10.5
–OCO–phenyl	51.8	60.8	14.4	66.4	22.2	10.5
–OS(O)$_2$–4-tolyl	56.3	66.9	14.7	72.2	22.3	10.0
N –NH$_2$	28.3	36.9	19.0	44.6	27.4	11.5
–NHCH$_3$	38.2	45.9	14.3	54.0	23.2	12.5
–N(CH$_3$)$_2$	47.6	53.6	12.8	61.8	20.6	11.9
–NHCOCH$_3$	26.1	34.4	14.6	40.7	22.5	11.1
–NO$_2$	61.2	70.8	12.3	77.4	21.2	10.8
–C≡N	1.7	10.8	10.6	19.3	19.0	13.3
–NC	26.8	36.4	15.3	43.4	22.9	11.0
S –SH	6.5	19.1	19.7	26.4	27.6	12.6
–SCH$_3$	19.3					
–SSCH$_3$	22.0	31.8	14.7			
–S(O)CH$_3$	40.1					
–S(O)$_2$CH$_3$	42.6	48.2	6.7	56.3	16.3	13.0
–S(O)$_2$Cl	52.6	60.2	9.1	67.1	18.4	12.1
–S(O)$_2$OH	39.6	46.7	8.0	53.7	18.8	13.7
O ‖ C –CHO	31.3	36.7	5.2	45.7	15.7	13.3
–COCH$_3$	30.7	35.2	7.0	45.2	17.5	13.5
–CO–phenyl	25.7	31.7	8.3	40.4	17.7	13.8
–COOH	21.7	28.5	9.6	36.2	18.7	13.7
–COOCH$_3$	20.6	27.2	9.2	35.6	18.9	13.8
–CONH$_2$	22.3	29.0	9.7			
–COCl	33.6	41.0	9.3	48.9	18.8	13.0

^{13}C Chemical Shifts of Monosubstituted Alkanes (δ in ppm, contd.)

		2-Propyl		tert-Butyl	
		–CH	–CH$_3$	–C	–CH$_3$
	–H	15.9	15.4	25.0	24.1
C	–CH=CH$_2$	32.3	22.1	33.8	29.4
	–C≡CH	20.3	22.8	27.4	31.1
	–phenyl	34.3	24.0	34.6	31.4
X	–F	87.3	22.6	93.5	28.3
	–Cl	53.7	27.3	66.7	34.6
	–Br	44.8	28.5	62.1	36.4
	–I	20.9	31.2	43.0	40.4
O	–OH	64.0	25.3	68.9	31.2
	–OCH$_3$	72.6	21.4	72.7	27.0
	–OCH$_2$CH$_3$			72.6	27.7
	–OCH(CH$_3$)$_2$	68.5	23.0	73.0	28.5
	–OC(CH$_3$)$_3$	63.5	25.2	76.3	33.8
	–O–phenyl	69.3	22.0		
	–OCOCH$_3$	67.5	21.9	79.9	28.1
	–OCO–phenyl	68.2	21.9	80.7	28.2
N	–NH$_2$	43.0	26.5	47.2	32.9
	–NHCH$_3$	50.5	22.5	50.4	28.2
	–N(CH$_3$)$_2$	55.5	18.7	53.6	25.4
	–NHCOCH$_3$	40.5	22.3	49.9	28.6
	–NO$_2$	78.8	20.8	85.2	26.9
	–C≡N	19.8	19.9	28.1	28.5
	–NC	45.5	23.4	54.0	30.7
S	–SH	29.9	27.4	41.1	35.0
	–SCH$_2$CH$_3$	34.4	23.4		
	–S(O)$_2$CH$_3$	53.5	15.2	57.6	22.7
	–S(O)$_2$Cl	67.6	17.1	74.2	24.5
	–S(O)$_2$OH	52.9	16.8	55.9	25.0
O	–CHO	41.1	15.5	42.4	23.4
‖	–COCH$_3$	41.6	18.2	44.3	26.5
C	–CO–phenyl	35.2	19.1	43.5	27.9
	–COOH	34.1	18.8	38.7	27.1
	–COOCH$_3$	34.1	19.1	38.7	27.3
	–CONH$_2$	34.9	19.5	38.6	27.6
	–COCl	46.5	19.0	49.4	27.1

^{13}C Chemical Shifts of 1-Substituted *n*-Octanes (δ in ppm)

Substituent	1 –CH$_2$	2 –CH$_2$	3 –CH$_2$	4 –CH$_2$	5 –CH$_2$	6 –CH$_2$	7 –CH$_2$	8 –CH$_3$
–H	14.1	22.8	32.1	29.5	29.5	32.1	22.8	14.1
–CH=CH$_2$	34.5	~29.6	~29.6	~29.6	~29.6	32.2	23.0	13.9
–phenyl	36.2	31.7	~29.6	~29.6	~29.6	32.1	22.8	14.1
–F	84.2	30.6	25.3	29.3	29.3	31.9	22.7	14.1
–Cl	45.1	32.8	27.0	29.0	29.2	31.9	22.8	14.1
–Br	33.8	33.0	28.3	28.8	29.2	31.8	22.7	14.1
–I	6.9	33.7	30.6	28.6	29.1	31.8	22.6	14.1
–OH	63.1	32.9	25.9	29.5	29.4	31.9	22.8	14.1
–O–*n*-C$_8$H$_{17}$	71.1	30.0	26.3	29.6	29.4	32.0	22.8	14.1
–O–phenyl	68.0	26.2	29.3	29.4	29.4	31.9	22.7	14.1
–OCO–*n*-propyl	64.4	28.8	26.1	29.3	29.3	31.9	22.8	14.1
–OCO–phenyl	65.1	28.8	26.1	29.3	29.3	31.9	22.7	14.1
–ONO	68.3	29.2	26.0	29.3	29.3	31.9	22.7	14.0
–NH$_2$	42.4	34.1	27.0	29.6	29.4	31.9	22.7	14.1
–N(CH$_3$)$_2$	60.1	29.5*	≈27.9*	≈27.7*	29.7*	32.0	22.8	14.4
–N$^+$(CH$_3$)$_3$ Cl$^-$	66.6	26.2	23.2	29.1*	29.0*	31.6	22.5	14.0
–NO$_2$	75.8	26.2	27.9	≈29.6	≈29.6	31.4	22.6	14.0
–C≡N	17.2	25.5	≈29.9	≈29.9	≈29.9	31.8	22.7	14.0
–SH	24.7	34.2	28.5	29.2	29.1	31.9	22.7	14.1
–SCH$_3$	34.5	29.0	29.4	29.4	29.4	31.9	22.8	14.1
–S(O)–*n*-C$_8$H$_{17}$	52.6	≈29.1	≈29.1	≈29.1	≈29.1	31.8	22.7	14.1
–CHO	44.0	22.2	≈29.3	≈29.3	≈29.3	31.9	22.7	14.1
–COCH$_3$	43.7	24.1	≈29.5	≈29.5	≈29.5	32.0	22.8	14.1
–CO–phenyl	38.6	24.4	29.5	29.5	29.5	31.9	22.7	14.0
–COOH	34.2	24.8	≈29.3	≈29.3	≈29.3	31.9	22.7	14.1
–COOCH$_3$	34.2	25.1	29.3	29.3	29.3	31.9	22.8	14.1
–CONH$_2$	35.5	25.4	29.1	29.1	29.1	31.6	22.3	14.0
–COCl	47.2	25.1	28.5	29.1	29.1	31.8	22.7	14.1
–Si(OCH$_3$)$_3$	9.2	22.7	33.2	29.3	29.3	32.0	22.7	14.1

* Assignment uncertain

Estimation of ¹³C Chemical Shifts of Aliphatic Compounds (δ in ppm)

The chemical shifts of sp^3-hybridized carbon atoms can be estimated with the help of an additivity rule using the shift value of methane (-2.3 ppm) and increments (Z) for substituents in α, β, γ, and δ position (see next pages). Some substituents occupy two positions. Thus, the quaternary carbon atom **c** in the example given below is in δ position relative to the carbon atom **a** since the sp^3-hybridized oxygen of the β-COO group occupies the γ position. This simple linear model needs corrections in case of strong branching of the observed C atom and/or its neighbors (steric corrections, S). Substituents for which such corrections are necessary are those with varying branching, i.e., a varying number of directly bonded H atoms. They are marked with an asterisk (*) in the Table of Increments (next page). Further correction terms are needed if γ substituents are in a sterically fixed position (conformational corrections, K).

The chemical shifts estimated with this additivity rule, in general, differ by less than ca. 4 ppm from the experimental values. Larger discrepancies may be expected for highly branched systems (particularly for quaternary carbon atoms). For carbon atoms bearing several halogen, oxygen, and/or other strongly deshielding substituents, additional correction terms are needed [1]. Without such corrections, deviations can be so large as to render the rule useless.

Example: Estimation of chemical shifts for *N*-(*tert*-butoxycarbonyl)alanine

a			b		
	base value	-2.3		base value	-2.3
	1 α-C	9.1		1 α-C	9.1
	1 α-COOH	20.1		1 β-COOH	2.0
	1 α-NH	28.3		1 β-NH	11.3
	1 β-COO	2.0		1 γ-COO	-2.8
	1 δ-C	0.3		1 S(prim,3)	-1.1
	1 S(tert,2)	-3.7		estimated	16.2
	estimated	53.8		exp	17.3
	exp	49.0			
c			d		
	base value	-2.3		base value	-2.3
	3 α-C	27.3		1 α-C	9.1
	1 α-OCO	56.5		2 β-C	18.8
	1 γ-NH	-5.1		1 β-OCO	6.5
	1 δ-C	0.3		1 δ-NH	0.0
	3 S(quat,1)	-4.5		1 S(prim,4)	-3.4
	estimated	72.2		estimated	28.7
	exp	78.1		exp	28.1

Estimation of ^{13}C Chemical Shifts of Aliphatic Compounds (δ in ppm)

$$\delta = -2.3 + \sum_i Z_i + \sum_j S_j + \sum_k K_k$$

Substituent		Increment Z_i for substituents in position			
		α	β	γ	δ
	–H	0.0	0.0	0.0	0.0
C	–C*≤	9.1	9.4	-2.5	0.3
	–C*=C<	19.5	6.9	-2.1	0.4
	–C≡C–	4.4	5.6	-3.4	-0.6
	–phenyl	22.1	9.3	-2.6	0.3
X	–F	70.1	7.8	-6.8	0.0
	–Cl	31.0	10.0	-5.1	-0.5
	–Br	18.9	11.0	-3.8	-0.7
	–I	–7.2	10.9	-1.5	-0.9
O	–O–*	49.0	10.1	-6.2	0.3
	–OCO–	56.5	6.5	-6.0	0.0
	–ONO	54.3	6.1	-6.5	-0.5
N	–N*<	28.3	11.3	-5.1	0.0
	–N^{+},*≤	30.7	5.4	-7.2	-1.4
	–NH$_3^+$	26.0	7.5	-4.6	0.0
	–NO$_2$	61.6	3.1	-4.6	-1.0
	–C≡N	3.1	2.4	-3.3	-0.5
	–NC	31.5	7.6	-3.0	0.0
S	–S*–	10.6	11.4	-3.6	-0.4
	–SCO–	17.0	6.5	-3.1	0.0
	–S*(O)–	31.1	7.0	-3.5	0.5
	–S*(O)$_2$–	30.3	7.0	-3.7	0.3
	–S(O)$_2$Cl	54.5	3.4	-3.0	0.0
	–SCN	23.0	9.7	-3.0	0.0
O‖C	–CHO	29.9	-0.6	-2.7	0.0
	–CO–	22.5	3.0	-3.0	0.0
	–COOH	20.1	2.0	-2.8	0.0
	–COO$^-$	24.5	3.5	-2.5	0.0
	–COO–	22.6	2.0	-2.8	0.0
	–CO–N<	22.0	2.6	-3.2	-0.4
	–COCl	33.1	2.3	-3.6	0.0
	–C=NOH syn	11.7	0.6	-1.8	0.0
	–C=NOH anti	16.1	4.3	-1.5	0.0
	–CS–N<	33.1	7.7	-2.5	0.6
	–Sn	-5.2	4.0	-0.3	0.0

Steric Corrections, S_j

Observed ^{13}C center	S for number of substituents at the α atom[a]			
	1	2	3	4
primary (CH$_3$)	0.0	0.0	-1.1	-3.4
secondary (CH$_2$)	0.0	0.0	-2.5	-6.0
tertiary (CH)	0.0	-3.7	-8.5	-10.0
quaternary (C)	-1.5	-8.0	-10.0	-12.5

[a] To be applied to each of the neighboring atoms that has an unspecified number of non-hydrogen substituents (marked with an asterisk (*) in the Table of Increments, Z_i).

Conformational Corrections, K_k, for γ Substituents

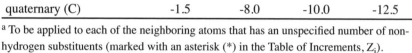

Conformation		K
synperiplanar (eclipsed)	C X	-4.0
synclinal (gauche)	C X	-1.0
anticlinal	C X	0.0
antiperiplanar (anti)	C X	2.0
not fixed		0.0

One can also use the chemical shifts of a reference compound as the base value if its structure is closely related to that assumed for the unknown. The increments corresponding to the structural elements missing in the reference compound are then added to the base value, while those of structural elements present in the reference but absent in the unknown are subtracted (see example on next page).

Example: Estimation of the chemical shifts for the carbon atoms **a** and **b** in *N-(tert-butoxycarbonyl)alanine* using the chemical shifts of valine as base values (**a'**, **b'**):

Target: Reference:

a	base value (**a'**)	61.9	**b**	base value (**b'**)	30.3
	1 β-COO	2.0		1 γ-COO	-2.8
	1 δ-C	0.3		1 S(prim,3)	-1.1
	1 S(tert,2)	-3.7		- 2 α-C	-18.2
	- 2 β-C	-18.8		- 1 S(tert,3)	8.5
	- 1 S(tert,3)	8.5		estimated	16.6
	estimated	50.2		exp	17.3
	exp	49.0			

4.1.2 Coupling Constants

^{13}C-^{1}H Coupling Constants

Coupling through one bond ($^1J_{CH}$ in Hz)

The ^{13}C-^{1}H coupling constant of 125 Hz in methane increases in the presence of electronegative substituents and can be estimated by using the following additivity rule:

$$J_{CHZ_1Z_2Z_3} = 125.0 + \sum_i Z_i$$

Substituent	Increment Z_i	Substituent	Increment Z_i
–H	0.0	–Br	27.0
–CH$_3$	1.0	–I	26.0
–C(CH$_3$)$_3$	-3.0	–OH	18.0
–CH$_2$Cl	3.0	–O–phenyl	18.0
–CH$_2$Br	3.0	–NH$_2$	8.0
–CH$_2$I	7.0	–NHCH$_3$	7.0
–CHCl$_2$	6.0	–N(CH$_3$)$_2$	6.0
–CCl$_3$	9.0	–C≡N	11.0
–C≡C	7.0	–S(O)CH$_3$	13.0
–phenyl	1.0	–CHO	2.0
–F	24.0	–COCH$_3$	-1.0
–Cl	27.0	–COOH	5.5

Example: Estimation of ^{13}C-^{1}H coupling constant of CHCl$_3$:
J = 125.0 + 3 × 27.0 = 206.0 Hz (exp: 209.0 Hz).

Coupling through more than one bond ($|J_{CH}|$ in Hz)

The coupling constants can be estimated from the corresponding ^1H-^1H coupling constants [2]: $J_{CH} \approx 0.62\, J_{HH}$

Typical values:	Examples:	
$^2J_{CH}$ 1–6	^1H–CH_2–$^{13}CH_3$	4.5
$^3J_{CH}$ 0–10	^1H–CH_2–CH_2–$^{13}CH_3$	5.8

The ^{13}C-^1H coupling constants for coupling across three bonds depend on the dihedral angle in the same way as the vicinal ^1H-^1H coupling constants (see Chapter 5.1.2):

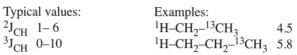

^{13}C-^{13}C Coupling Constants ($|J_{CC}|$ in Hz)

H_3C–CH_3 ^1J 34.6

$^1J_{ab}$ 34.6

$^2J_{ac}$ 4.6
$^3J_{ad}$ 4.6
$^2J_{bd}$ <1

H_3C–O–CH_3 $^2J_{ab}$ 2.4

H_3C–C(=O)–CH_3 $^2J_{ab}$ 16.1

H–C(=O)–N(CH$_3$)–CH_3 $^2J_{ab}$ 0.5
$^2J_{ac}$ 4.9

The ^{13}C-^{13}C coupling constants for coupling over three bonds depend on the dihedral angle in the same way as the vicinal ^1H-^1H (see Chapter 5.1.2) and ^{13}C-^1H coupling constants. Maximum values of ca. 4–6 Hz are observed for dihedral angles of 0° and 180° and minimal values around 0 Hz at 90°.

4.1.3 References

[1] A. Fürst, E. Pretsch, W. Robien, A comprehensive parameter set for the prediction of the ^{13}C NMR chemical shifts of sp^3-hybridized carbon atoms in organic compounds, *Anal. Chim. Acta* **1990**, *233*, 213.
[2] J.L. Marshall, Carbon-Carbon and Carbon-Proton NMR couplings, Verlag Chemie International, Deerfield Beach, FL, 1983.

4.2 Alkenes

4.2.1 Chemical Shifts

^{13}C Chemical Shifts (δ in ppm)

C = C   The ^{13}C chemical shifts of the carbons of C=C double bonds typically range from ca. 80–160 ppm; a wider range of 40–210 ppm is observed with O and N substituents. In unsaturated *acyclic hydrocarbons*, they can be predicted with high accuracy (see below). To estimate the ^{13}C chemical shifts in all other *substituted alkenes*, one can use the substituent effects listed for chemical shifts in vinyl groups. However, since no configuration-dependent parameters are available, the values thus estimated are less accurate than those for unsaturated acyclic hydrocarbons.

The ^{13}C chemical shifts of sp^3-hybridized carbon atoms in the vicinity of double bonds can be estimated using the additivity rule given in Chapter 4.1.1. The conformational correction factors, K, for γ substituents of *cis*- vs. *trans*-disubstituted alkenes differ by 6 ppm because the relative position of these substituents is fixed by the double bond.

Estimation of the ^{13}C Chemical Shifts of sp^2-Hybridized Carbon Atoms in Unsaturated Acyclic Hydrocarbons (δ in ppm)

$$C–C–C'=C–C–C–C$$
$$\gamma' \ \beta' \ \alpha' \ \uparrow \ \alpha \ \beta \ \gamma$$

Base value: 123.3

Increments for C substituents:

at C atom under consideration (**C**)		at neighboring C atom (**C'**)	
α	10.6	α'	-7.9
β	4.9	β'	-1.8
γ	-1.5	γ'	1.5

Steric corrections:

• for each pair of *cis*-α,α'-substituents	-1.1
• for a pair of geminal α,α-substituents	-4.8
• for a pair of geminal α',α'-substituents	2.5
• if one or more β-substituents are present	2.3

Example: Estimation of chemical shifts of *cis*-4-methyl-2-pentene

a   b

C = C

a				b		
	base value	123.3			base value	123.3
	1 α-C	10.6			1 α-C	10.6
	1 α'-C	-7.9			2 β-C	9.8
	2 β'-C	-3.6			1 α'-C	-7.9
	cis-α,α'	-1.1			*cis*-α,α'	-1.1
	estimated	121.3			1 β-substituent	2.3
	exp	121.8			estimated	137.0
					exp	138.8

Effect of Substituents on the ^{13}C Chemical Shifts of Vinyl Compounds (δ in ppm)

$$\overset{1}{R}-\overset{2}{CH}=CH_2 \qquad \delta_{C_i} = 123.3 + Z_i$$

	Substituent R	Z_1	Z_2		Substituent R	Z_1	Z_2
	–H	0.0	0.0	**O**	–OH	25.7	-35.3
C	–CH$_3$	12.9	-7.4		–OCH$_3$	29.4	-38.9
	–CH$_2$CH$_3$	17.2	-9.8		–OCH$_2$CH$_3$	28.8	-37.1
	–CH$_2$CH$_2$CH$_3$	15.7	-8.8		–O(CH$_2$)$_3$CH$_3$	28.1	-40.4
	–CH(CH$_3$)$_2$	22.7	-12.0		–OCOCH$_3$	18.4	-26.7
	–(CH$_2$)$_3$–	14.6	-8.9	**N**	–N(CH$_3$)$_2$	28.0*	-32.0*
	–C(CH$_3$)$_3$	26.0	-14.8		–N$^+$(CH$_3$)$_3$	19.8	-10.6
	–CH$_2$Cl	10.2	-6.0		–*N*-pyrrolidonyl	6.5	-29.2
	–CH$_2$Br	10.9	-4.5		–NO$_2$	22.3	-0.9
	–CH$_2$I	14.2	-4.0		–C≡N	-15.1	14.2
	–CH$_2$OH	14.2	-8.4		–NC	-3.9	-2.7
	–CH$_2$OCH$_2$CH$_3$	12.3	-8.8	**S**	–SCH$_2$CH$_3$	9.0	-12.8
	–CH=CH$_2$	13.6	-7.0		–S(O)$_2$CH=CH$_2$	14.3	7.9
	–C≡CH	-6.0	5.9	**O**	–CHO	15.3	14.5
	–phenyl	12.5	-11.0	‖	–COCH$_3$	13.8	4.7
X	–F	24.9	-34.3		–COOH	5.0	9.8
	–Cl	2.8	-6.1	**C**	–COOCH$_2$CH$_3$	6.3	7.0
	–Br	-8.6	-0.9		–COCl	8.1	14.0
	–I	-38.1	7.0	**Si**	–Si(CH$_3$)$_3$	16.9	6.7
					–SiCl$_3$	8.7	16.1

* Estimated values

The values listed on the preceding page can also be used to estimate the ^{13}C chemical shifts of sp^2-hybridized carbon atoms in alkenes with more than one substituent (note that the *cis/trans* configuration is not taken into account):

$$\delta_{C_i} = 123.3 + \Sigma\, Z_i$$

C=C

Example: Estimation of chemical shifts of 1-bromo-1-propene

$$\overset{a}{}\quad\overset{b}{}$$
$$Br-CH=CH-CH_3$$

a	base value	123.3	**b**	base value	123.3
	$Z_1(Br)$	-8.6		$Z_2(Br)$	-0.9
	$Z_2(CH_3)$	-7.4		$Z_1(CH_3)$	12.9
	estimated	107.3		estimated	135.3
	exp	108.9 *(cis)*		exp	129.4 *(cis)*
		104.7 *(trans)*			132.7 *(trans)*

The following examples show some larger deviations between measured and estimated (in parentheses) chemical shifts. This is usually to be expected when several substituents are present that strongly interact with the π electrons of the double bond:

NC, a, b N(CH₃)₂ / C=C / NC, N(CH₃)₂

a 39.1 (29.1)
b 171.0 (207.7)

H, a, b N(CH₃)₂ / C=C / H, N(CH₃)₂

a 69.2 (59.3)
b 163.0 (179.3)

H, a, b NO₂ / C=C / (CH₃)₂N, H

a 151.0 (150.4)
b 111.4 (113.6)

H, a, b OCH₃ / C=C / H, OCH₃

a 54.7 (45.5)
b 167.9 (182.1)

^{13}C Chemical Shifts of *cis*- and *trans*-1,2-Disubstituted Alkenes (δ in ppm)

Substituent R	R R >=< H H	R H >=< H R
–CH$_3$	123.3	124.5
–CH$_2$CH$_3$	131.2	131.3
–Cl	118.1	119.9
–Br	116.4	109.4
–I	96.5	79.4
–C≡N	120.8	120.2
–OCH$_3$	130.3	135.2
–COOH	130.4	134.2
–COOCH$_3$	130.1	133.5

^{13}C Chemical Shifts of Enols (δ in ppm)

The carbon atom bonded to the enolic OH group is strongly deshielded so that its shift is close to that of a carbonyl carbon. The other carbon atom of the double bond is strongly shielded.

Enol:

a	22.5
b	190.5
c	99.0

Ketone:

a	28.5
b	201.1
c	56.6

C=C

a	28.3
b	32.8
c	46.2
d	191.1
e	103.3

a	28.3
b	31.0
c	54.2
d	203.6
e	57.3

^{13}C Chemical Shifts of Allenes (δ in ppm)

R_1	R_2	R_3	a	b	c
–H	–H	–H	74.8	213.5	74.8
–CH$_3$	–H	–H	84.4	210.4	74.1
–CH$_3$	–CH$_3$	–H	93.4	207.3	72.1
–CH$_3$	–H	–CH$_3$	85.4	207.1	85.4
–CH$_2$CH$_3$	–H	–H	91.7	208.9	75.3
–C(CH$_3$)$_3$	–C(CH$_3$)$_3$	–H	119.6	207.0	75.8
–CH=CH$_2$	–H	–H	93.9	211.4	75.1
–C≡CH	–H	–H	74.8	217.7	77.3
–phenyl	–H	–H	94.4	210.0	78.8
–F	–H	–H	129.8	200.2	93.9
–Cl	–H	–H	88.8	207.9	84.5
–Br	–H	–H	72.7	207.6	83.8
–I	–H	–H	35.3	208.0	78.3
–OCH$_3$	–H	–H	123.1	202.0	90.3
–N(CH$_3$)$_2$	–H	–H	113.1	204.2	85.5
–C≡N	–H	–H	67.4	218.7	80.7
–SCH$_3$	–H	–H	90.0	206.1	81.3
–COOH	–H	–H	88.1	217.7	80.0

4.2.2 Coupling Constants

^{13}C-^{1}H Coupling Constants ($|J_{CH}|$ in Hz)

Coupling through one bond

$C=C$ $CH_2=CH_2$ $^1J_{CH}$ 156.4 $CH_2=C=CH_2$ $^1J_{CH}$ 167.8

Coupling through two bonds

$^2J_{CH}$ -2.4 $^2J_{CH}$ 6.9

Additivity rule for the estimation of $^2J_{CH}$ of alkenes: see [2].

Coupling through three bonds

The *trans*-^{1}H–C=C–^{13}C coupling constant of alkenes is always larger than the corresponding *cis* coupling constant so that an assignment is possible if both isomers are available: see [3].

$$\begin{array}{ll} \text{H a} \quad ^{13}\text{CH}_3 \\ \quad \backslash \quad / \\ \quad C=C \\ \quad / \quad \backslash \\ \text{H b} \quad \text{H} \end{array} \quad \begin{array}{l} ^3J_{ac} \ 7.6 \\ ^3J_{bc} \ 12.6 \end{array}$$

$$\begin{array}{ll} \text{H a} \quad ^{13}\text{CH}_3 \\ \quad \backslash \quad / \\ \quad C=C \\ \quad / \quad \backslash \\ \text{H b} \quad \text{Cl} \end{array} \quad \begin{array}{l} ^3J_{ac} \ 4.1 \\ ^3J_{bc} \ 8.1 \end{array}$$

$$\begin{array}{ll} \text{H a} \quad ^{13}\text{COOH} \\ \quad \backslash \quad / \\ \quad C=C \\ \quad / \quad \backslash \\ \text{H b} \quad \text{H} \end{array} \quad \begin{array}{l} ^3J_{ac} \ 7.6 \\ ^3J_{bc} \ 14.1 \end{array}$$

$$\begin{array}{ll} \text{H a} \quad ^{13}\text{COOH} \\ \quad \backslash \quad / \\ \quad C=C \\ \quad / \quad \backslash \\ \text{H b} \quad \text{CH}_3 \end{array} \quad \begin{array}{l} ^3J_{ac} \ 7.6 \\ ^3J_{bc} \ 14.1 \end{array}$$

$$\begin{array}{ll} \text{H} \quad ^{13}\text{COOH} \\ \quad \backslash \quad / \\ \quad C=C \\ \quad / \quad \backslash \\ \text{H}_3\text{C} \quad ^{13}\text{CH}_3 \ \text{b} \end{array} \quad \begin{array}{l} ^3J_{ab} \ 7.7 \\ ^3J_{ac} \ 7.4 \end{array}$$

$$\begin{array}{ll} \text{H}_3\text{C} \quad ^{13}\text{COOH} \\ \quad \backslash \quad / \\ \quad C=C \\ \quad / \quad \backslash \\ \text{H} \quad ^{13}\text{CH}_3 \\ \text{a} \quad\quad \text{b} \end{array} \quad \begin{array}{l} ^3J_{ab} \ 6.9 \\ ^3J_{ac} \ 13.2 \end{array}$$

^{13}C-^{13}C Coupling Constants ($|J_{CC}|$ in Hz)

$CH_2=CH_2$ $^1J_{CC}$ 67.6

$$\overset{a}{C}H_2=\overset{b}{C}H-\overset{c}{C}H_3 \quad \begin{array}{l} ^1J_{ab} \ 70.0 \\ ^1J_{bc} \ 41.9 \end{array}$$

$CH_2=C=CH_2$ $^1J_{CC}$ 98.7

$\begin{array}{ll} ^1J_{ab} \ 68.8 & ^2J_{ac} \ <1 \\ ^1J_{bc} \ 53.7 & ^3J_{ad} \ 9.0 \end{array}$

4.2.3 References

[1] R.H.A.M. Janssen, R.J.J.Ch. Lousberg, M.J.A. de Bie, An additivity relation for carbon-13 chemical shifts in substituted allenes, *Recl. Trav. Chim. Pays-Bas* **1981**, *100*, 85.

[2] U. Vögeli, D. Herz, W. von Philipsborn, Geminal C,H spin coupling in substituted alkenes, *Org. Magn. Reson.* **1980**, *13*, 200.

[3] U. Vögeli, W. von Philipsborn, Vicinal C,H spin coupling in substituted alkenes. *Org. Magn. Reson.* **1975**, *7*, 617.

4.3 Alkynes

4.3.1 Chemical Shifts

^{13}C Chemical Shifts of Alkynes (δ in ppm)

$$\overset{\text{a}\quad\text{b}}{R-C\equiv C-H}$$

Substituent R	a	b
–H	71.9	71.9
–CH$_3$	80.4	68.3
–CH$_2$CH$_3$	85.5	67.1
–CH$_2$CH$_2$CH$_3$	84.0	68.7
–CH$_2$CH$_2$CH$_2$CH$_3$	83.0	66.0
–CH(CH$_3$)$_2$	89.2	67.6
–C(CH$_3$)$_3$	92.6	66.8
–cyclohexyl	88.7	68.3
–CH$_2$OH	83.0	73.8
–CH=CH$_2$	82.8	80.0
–C≡C–CH$_3$	68.8	64.7
–phenyl	84.6	78.3
–OCH$_2$CH$_3$	90.9	26.5
–SCH$_2$CH$_3$	72.6	81.4
–CHO	81.8	83.1
–COCH$_3$	81.9	78.1
–COOH	74.0	78.6
–COOCH$_3$	74.8	75.6

C ≡ C

Additivity rule for estimating the chemical shifts of *sp*-hybridized carbon atoms in alkynes: see [1].

4.3.2 Coupling Constants

^{13}C-^{1}H Coupling Constants ($|J_{CH}|$ in Hz) [2]

$$\overset{\text{a}\quad\text{b}\quad\text{c}}{H-C\equiv C-H}$$
$^1J_{ab}$ 249.0
$^2J_{ac}$ 49.3 (in substituted acetylenes: 40–60)

$$\overset{\text{a}\quad\text{b}\quad\text{c}\quad\text{d e}}{H-C\equiv C-CH_3}$$
$^2J_{ac}$ 50.1 $^3J_{ad}$ 3.4
$^2J_{ce}$ -10.4 $^3J_{be}$ 4.7

$$\overset{\text{a}\qquad\text{b}\quad\text{c}}{H_3C-C\equiv C-CH_3}$$
$^2J_{ab}$ -10.3 $^3J_{ac}$ 4.3

With acetylenes, the results of multipulse experiments (such as DEPT, INEPT, SEFT, or APT) to determine the number of protons attached to the carbon atoms must be interpreted with care. As a consequence of the unusually large ^{13}C-^{1}H coupling constants through one and two bonds, the sign of the signals may be opposite to the expected one. For the same reasons, unexpected signals may occur in two-dimensional heteronuclear correlation spectra (HSQC, HMBC).

^{13}C-^{13}C Coupling Constants ($|^1J_{CC}|$ in Hz)

$C\equiv C$

$$H-C\equiv C-H \quad ^1J_{CC} \quad 171.5$$

$$\overset{a \quad\; b \quad\; c}{H-C\equiv C-C\equiv C-H} \quad \begin{array}{l} ^1J_{ab} \quad 190.3 \\ ^1J_{bc} \quad 153.4 \end{array}$$

4.3.3 References

[1]   W. Höbold, R. Radeglia, D. Klose, Inkrementen-Berechnung von ^{13}C-chemischen Verschiebungen in *n*-Alkinen, *J. Prakt. Chem.* **1976**, *318*, 519.
[2]   K. Hayamizu, O. Yamamoto, ^{13}C,^{1}H Spin coupling constants of dimethylacetylene, *Org. Magn. Reson.* **1980**, *13*, 460.

4.4 Alicyclics

4.4.1 Chemical Shifts

Saturated Monocyclic Alicyclics (δ in ppm)

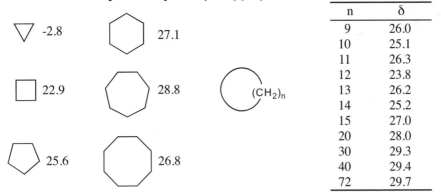

n	δ
9	26.0
10	25.1
11	26.3
12	23.8
13	26.2
14	25.2
15	27.0
20	28.0
30	29.3
40	29.4
72	29.7

^{13}C Chemical Shifts of Monosubstituted Cyclopropanes (δ in ppm)

$$R\!-\!\overset{a}{\triangleleft}\,^{b}$$

	Substituent R	a	b	other
	–H	-2.8	-2.8	
C	–CH$_3$	4.9	5.6	CH$_3$ 19.4
	–CH$_2$CH$_3$	12.8	4.1	CH$_2$ 27.8, CH$_3$ 13.6
	–CH$_2$CH$_2$CH$_2$CH$_3$	10.9	4.4	1-CH$_2$ 34.7, 2-CH$_2$ 32.0
	–C(CH$_3$)$_3$	22.7	0.3	C 29.3, CH$_3$ 28.2
	–CH$_2$Cl	13.6	5.5	CH$_2$ 50.3
	–CH$_2$OH	12.7	2.2	CH$_2$ 66.5
	–CH=CH$_2$	14.7	6.6	CH 142.4, CH$_2$ 111.5
	–phenyl	15.3	9.2	C 143.9, CH 125.3–128.2
X	–Cl	27.3	8.9	
	–Br	14.2	9.1	
	–I	-20.1	10.4	
O	–OH	45.7	6.8	
N	–NH$_2$	24.0	7.4	
	–NO$_2$	54.3	11.7	
	–C≡N	-4.5	6.2	CN 121.5
O‖C	–CHO	22.7	7.4	CO 202.1
	–COCH$_3$	20.1	9.6	CO 207.3, CH$_3$ 29.1
	–CO–phenyl	17.1	11.5	
	–COOH	12.7	8.9	CO 181.6
	–COOCH$_3$	12.2	7.7	CO 174.7, CH$_3$ 51.1

^{13}C Chemical Shifts of Monosubstituted Cyclopentanes (δ in ppm)

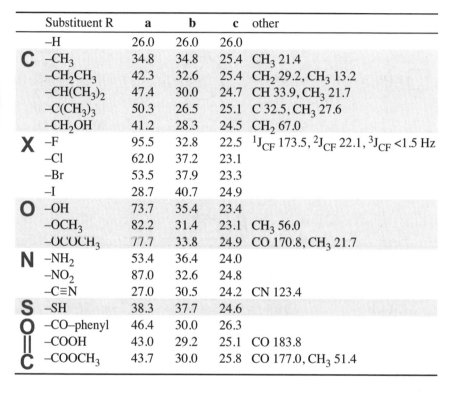

	Substituent R	a	b	c	other
	–H	26.0	26.0	26.0	
C	–CH$_3$	34.8	34.8	25.4	CH$_3$ 21.4
	–CH$_2$CH$_3$	42.3	32.6	25.4	CH$_2$ 29.2, CH$_3$ 13.2
	–CH(CH$_3$)$_2$	47.4	30.0	24.7	CH 33.9, CH$_3$ 21.7
	–C(CH$_3$)$_3$	50.3	26.5	25.1	C 32.5, CH$_3$ 27.6
	–CH$_2$OH	41.2	28.3	24.5	CH$_2$ 67.0
X	–F	95.5	32.8	22.5	^{1}J$_{CF}$ 173.5, ^{2}J$_{CF}$ 22.1, ^{3}J$_{CF}$ <1.5 Hz
	–Cl	62.0	37.2	23.1	
	–Br	53.5	37.9	23.3	
	–I	28.7	40.7	24.9	
O	–OH	73.7	35.4	23.4	
	–OCH$_3$	82.2	31.4	23.1	CH$_3$ 56.0
	–OCOCH$_3$	77.7	33.8	24.9	CO 170.8, CH$_3$ 21.7
N	–NH$_2$	53.4	36.4	24.0	
	–NO$_2$	87.0	32.6	24.8	
	–C≡N	27.0	30.5	24.2	CN 123.4
S	–SH	38.3	37.7	24.6	
O	–CO–phenyl	46.4	30.0	26.3	
‖	–COOH	43.0	29.2	25.1	CO 183.8
C	–COOCH$_3$	43.7	30.0	25.8	CO 177.0, CH$_3$ 51.4

^{13}C Chemical Shifts of Equatorially and Axially Monosubstituted Cyclohexanes (δ in ppm)

Substituent R	a	b	c	d	a	b	c	d
–H	27.1	27.1	27.1	27.1	27.1	27.1	27.1	27.1
C –CH$_3$	33.2	36.0	27.1	27.0	28.4	32.4	20.6	26.9
–CH$_2$CH$_3$	40.1	33.4	26.9	27.2	35.5	30.0	21.4	27.1
–CH$_2$CH$_2$CH$_3$	40.0	33.6	26.6	26.9				
–CH(CH$_3$)$_2$	44.6	30.0	26.8	27.3	41.1	30.2	21.6	27.1
–CH$_2$CH$_2$CH$_2$CH$_3$	38.4	34.1	27.1	27.3				
–C(CH$_3$)$_3$	48.8	28.1	27.7	27.1				
–cyclohexyl	44.3	30.8	27.4	27.4				
–CH=CH$_2$	42.1	32.3	26.0	27.1	37.0	30.0	21.2	27.1
–C≡CH	28.7	32.1	25.2	24.4	28.0	30.0	21.2	25.7
–phenyl	45.1	34.9	27.4	26.7	35.2	30.1	21.9	27.7
X –F	91.0	32.8	23.6	25.3	88.1	30.1	19.8	25.0
–Cl	59.8	37.4	26.1	25.4	60.1	33.9	20.4	26.0
–Br	52.4	38.3	27.3	25.6	55.4	34.9	21.5	26.4
–I	31.2	40.1	28.3	25.4	38.3	36.0	22.8	26.1
O –OH	70.4	35.8	25.1	26.3	65.5	33.2	20.5	27.1
–OCH$_3$	79.2	32.2	24.5	26.4	74.9	30.0	21.1	26.6
–OCOCH$_3$	72.3	32.2	24.4	26.1				
–OCO–phenyl	72.8	31.5	24.1	24.7	69.0	29.3	20.3	24.7
–OSi(CH$_3$)$_3$	70.5	36.0	24.7	25.0	66.1	33.1	19.8	25.0
N –NH$_2$	51.1	37.6	25.8	26.3	47.4	33.8	20.0	27.1
–NHCH$_3$	58.7	32.7	25.7	26.8				
–N(CH$_3$)$_2$	64.3	29.2	26.5	26.9				
–NH$_3^+$Cl$^-$	51.8	32.2	24.8	25.2				
–N=C=N–cyclohexyl	55.7	35.0	24.8	25.5				
–NO$_2$	84.6	31.4	24.7	25.5				
–N$_3$	59.5	31.5	24.5	24.5	56.8	29.0	20.1	25.2
–C≡N	28.0	29.6	24.6	25.1	26.4	27.4	21.9	25.0
–NC	51.9	33.7	24.4	25.2	50.3	30.5	20.1	25.2
–NCS	55.3	33.9	24.5	24.8	52.8	31.3	20.4	24.8
S –SH	38.3	38.1	26.6	25.3	35.9	33.1	19.4	25.7
O –CHO	50.1	26.0	25.2	26.1	46.4	24.7	22.7	27.1
‖ –COCH$_3$	51.5	29.0	26.6	26.3				
C –COOH	43.7	29.6	26.2	26.6				
–COO$^-$	47.2	30.9	26.9	26.9				
–COOCH$_3$	43.4	29.6	26.0	26.4	39.1	27.7	24.1	26.7
–COCl	55.4	29.7	25.5	25.9				

Estimation of ^{13}C Chemical Shifts of Alicyclic Compounds (δ in ppm)

The ^{13}C chemical shift of the parent compound (e.g., 22.9 for cyclobutane, 26.0 for cyclopentane, and 27.1 ppm for cyclohexane) and the same increments as for alkanes (see Chapter 4.1) can be used to estimate the chemical shifts of sp^3-hybridized carbon atoms of alicyclic compounds. Appropriate use of the conformational correction terms, K, is especially important with axial and equatorial substituents in cyclohexanes. The additivity rule is, however, not suitable for estimating chemical shifts of substituted cyclopropanes.

^{13}C Chemical Shifts of Unsaturated Alicyclics (δ in ppm)

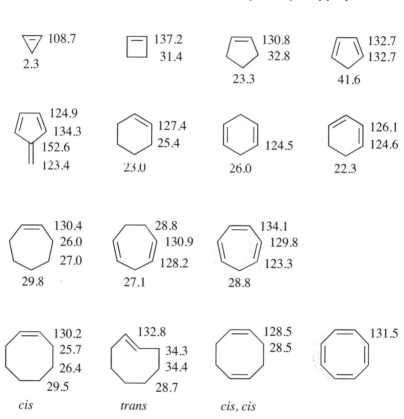

108.7
2.3

137.2
31.4

130.8
32.8
23.3

132.7
132.7
41.6

124.9
134.3
152.6
123.4

127.4
25.4
23.0

124.5
26.0

126.1
124.6
22.3

130.4
26.0
27.0
29.8

28.8
130.9
128.2
27.1

134.1
129.8
123.3
28.8

130.2
25.7
26.4
29.5
cis

132.8
34.3
34.4
28.7
trans

128.5
28.5
cis, cis

131.5

^{13}C Chemical Shifts of Condensed Alicyclics (δ in ppm)

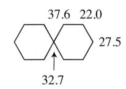

27.6 16.7
20.2   5.8

23.9 9.4
21.5   10.3

28.1 H 33.3
22.9   24.6
H

31.8 H 45.4
26.5   29.4
H

28.0 H 39.9
23.8   29.9
22.6
H

32.4 H 47.3
27.1   31.7
22.1
H

43.3
H 34.3
26.4
H

36.8
H 29.7
24.5
H

44.0
H 34.6
27.1
H

24.5 38.7 38.7
26.8   24.1
42.6

37.6 22.0
27.5
32.7

38.5
36.5
29.8

33.2
29.7
9.9

32.2
23.2
15.0

48.8
42.0
24.8 135.8

75.2
50.4
143.2

24.6
26.7

37.9
28.5

47.3

24.4
28.4
28.8

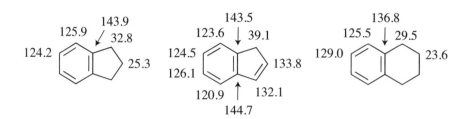

4.4.2 Coupling Constants

^{13}C-^{1}H Coupling Constants

Coupling through one bond ($|^1J_{CH}|$ in Hz)

Coupling through two bonds ($|^2J_{CH}|$ in Hz)

Coupling through three bonds ($|^3J_{CH}|$ in Hz)

H 2.1

^{13}C ——— H 8.1

^{13}C-^{13}C Coupling Constants ($|^1J_{CC}|$ in Hz)

12.4	
a b c CH₃	$^1J_{ab}$ 13.4 $^1J_{bc}$ 44.0
	32.7

4.5 Aromatic Hydrocarbons

4.5.1 Chemical Shifts

13C Chemical Shifts of Aromatic Hydrocarbons (δ in ppm) [1]

128.5

133.7   128.0
126.0

131.8
126.2 | 128.1
125.3

126.3
122.4
131.9
126.3
128.3
130.1
126.6

125.5
124.6
130.9
127.0
124.6

135.2
119.7
140.1
137.4
123.9
137.4

143.9
125.9 / 32.8
124.2
25.3

143.5
123.6 / 39.1
124.5
126.1
133.8
120.9 \ 132.1
144.7

141.6   119.7
126.5
126.5
36.8 | 124.8
143.2

136.8
125.5 / 29.5
129.0
23.6

137.3
128.0
37.7   128.0

29.2   134.7
127.5
127.5
137.3 123.9 127.5

139.7   30.3
145.9
119.5
128.2
132.1   122.7

128.7   129.7
140.0
124.3
127.9
128.4   127.4

143.2

Effect of Substituents on ^{13}C Chemical Shifts of Monosubstituted Benzenes (δ in ppm)

$$\delta_{C_i} = 128.5 + Z_i$$

Substituent R	Z_1	Z_2	Z_3	Z_4
–CH$_3$	9.2	0.7	-0.1	-3.0
–CH$_2$CH$_3$	11.7	-0.6	-0.1	-2.8
–CH$_2$CH$_2$CH$_3$	10.3	-0.2	0.1	-2.7
–CH(CH$_3$)$_2$	20.2	-2.2	-0.3	-2.8
–CH$_2$CH$_2$CH$_2$CH$_3$	10.9	-0.2	-0.2	-2.8
–C(CH$_3$)$_3$	18.6	-3.3	-0.4	-3.1
–cyclopropyl	15.1	-3.3	-0.6	-3.6
–cyclopentyl	17.8	-1.5	-0.4	-2.9
–cyclohexyl	16.3	-1.8	-0.3	-2.8
–1-adamantyl	22.2	-2.9	-0.5	-3.1
–CH$_2$F	8.5	-0.7	0.4	0.5
–CF$_3$	2.5	-3.2	0.3	3.3
–CH$_2$Cl	9.3	0.3	0.2	0.0
–CHCl$_2$	11.9	-2.4	0.1	1.2
–CCl$_3$	16.3	-1.7	-0.1	1.8
–CH$_2$Br	9.5	0.7	0.3	0.2
–CH$_2$I	10.5	0.0	0.0	-0.9
–CH$_2$OH	12.4	-1.2	0.2	-1.1
–CH$_2$OCH$_3$	8.7	-0.9	-0.1	-0.9
–CH$_2$NH$_2$	14.9	-1.4	-0.2	-2.0
–CH$_2$NHCH$_3$	12.6	-0.3	-0.3	-1.8
–CH$_2$N(CH$_3$)$_2$	7.8	0.5	-0.3	-1.5
–CH$_2$NO$_2$	2.2	2.2	2.2	1.2
–CH$_2$CN	1.6	0.5	-0.8	-0.7
–CH$_2$SH	12.5	-0.6	0.0	-1.6
–CH$_2$SCH$_3$	9.8	0.4	-0.1	-1.6
–CH$_2$S(O)CH$_3$	0.8	1.5	0.4	-0.2
–CH$_2$S(O)$_2$CH$_3$	-0.1	2.1	0.6	0.6
–CH$_2$CHO	7.4	1.3	0.5	-1.1
–CH$_2$COCH$_3$	5.8	0.8	0.1	-1.6
–CH$_2$COOH	6.5	1.4	0.4	-1.2
–CH$_2$Li	32.2	-22.0	-0.4	-24.3
–CH=CH$_2$	8.9	-2.3	-0.1	-0.8
–C(CH$_3$)=CH$_2$	12.6	-3.1	-0.4	-1.2
–C≡CH	-6.2	3.6	-0.4	-0.3
–phenyl	8.1	-1.1	0.5	-1.1
–2-pyridyl	11.2	-1.4	0.5	-1.4
–4-pyridyl	9.6	-1.6	0.5	0.5

	Substituent R	Z_1	Z_2	Z_3	Z_4
X	–F	33.6	-13.0	1.6	-4.4
	–Cl	5.3	0.4	1.4	-1.9
	–Br	-5.4	3.3	2.2	-1.0
	–I	-31.2	8.9	1.6	-1.1
O	–OH	28.8	-12.8	1.4	-7.4
	–ONa	39.6	-8.2	1.9	-13.6
	–OCH$_3$	33.5	-14.4	1.0	-7.7
	–OCH=CH$_2$	28.2	-11.5	0.7	-5.8
	–O–phenyl	27.6	-11.2	-0.3	-6.9
	–OCOCH$_3$	22.4	-7.1	0.4	-3.2
	–OSi(CH$_3$)$_3$	26.8	-8.4	0.9	-7.1
	–OPO(O–phenyl)$_2$	21.9	-8.4	1.2	-3.0
	–OCN	25.0	-12.7	2.6	-1.0
N	–NH$_2$	18.2	-13.4	0.8	-10.0
	–NHCH$_3$	15.0	-16.2	0.8	-11.6
	–N(CH$_3$)$_2$	16.0	-15.4	0.9	-10.5
	–NH–phenyl	14.7	-10.6	0.9	-10.5
	–N(phenyl)$_2$	13.1	-7.0	0.9	-5.6
	–NH$_3^+$	0.1	-5.8	2.2	2.2
	–NH$_2^+$CH(CH$_3$)$_2$	5.5	-4.1	1.1	0.7
	–N$^+$(CH$_3$)$_3$	19.5	-7.3	2.5	2.4
	–N(O)(CH$_3$)$_2$	26.2	-8.4	0.8	0.6
	–NHCOCH$_3$	9.7	-8.1	0.2	-4.4
	–NHOH	21.5	-13.1	-2.2	-5.3
	–NHNH$_2$	22.8	-16.5	0.5	-9.6
	–N=CH–phenyl	24.7	-6.5	1.3	-1.5
	–N=NCH$_3$	22.2	-6.2	0.5	-3.0
	–NO	37.4	-7.6	0.8	7.1
	–NO$_2$	19.9	-4.9	0.9	6.1
	–C≡N	-16.0	3.5	0.7	4.3
	–NC	-1.8	-2.2	1.4	0.9
	–NCO	5.1	-3.7	1.1	-2.8
	–NCS	3.0	-2.7	1.3	-1.0
	–N$^+$≡N	-12.7	6.0	5.7	16.0
S	–SH	4.0	0.7	0.3	-3.2
	–SCH$_3$	10.0	-1.9	0.2	-3.6
	–SC(CH$_3$)$_3$	4.5	9.0	-0.3	0.0
	–S(CH$_3$)$_2^+$	-1.0	3.1	2.2	6.3
	–SCH=CH$_2$	5.8	2.0	0.2	-1.8
	–S–phenyl	7.3	2.5	0.6	-1.5
	–S–S–phenyl	7.5	-1.3	0.8	-1.1
	–S(O)CH$_3$	17.6	-5.0	1.1	2.4
	–S(O)$_2$CH$_3$	12.3	-1.4	0.8	5.1
	–S(O)$_2$OH	15.0	-2.2	1.3	3.8
	–S(O)$_2$OCH$_3$	6.4	-0.6	1.5	5.9

	Substituent R	Z_1	Z_2	Z_3	Z_4
S	–S(O)$_2$F	4.6	0.0	1.5	7.5
	–S(O)$_2$Cl	15.6	-1.7	1.2	6.8
	–S(O)$_2$NH$_2$	10.8	-3.0	0.3	3.2
	–SCN	-3.7	2.5	2.2	2.2
O	–CHO	8.2	1.2	0.5	5.8
‖	–COCH$_3$	8.9	0.1	-0.1	4.4
C	–COCF$_3$	-5.6	1.8	0.7	6.7
	–COC≡CH	7.4	1.0	0.0	5.9
	–CO–phenyl	9.3	1.6	-0.3	3.7
	–COOH	2.1	1.6	-0.1	5.2
	–COONa	9.7	4.6	2.2	4.6
	–COOCH$_3$	2.0	1.2	-0.1	4.3
	–CONH$_2$	5.0	-1.2	0.1	3.4
	–CON(CH$_3$)$_2$	6.0	-1.5	-0.2	1.0
	–COCl	4.7	2.7	0.3	6.6
	–COSH	6.2	-0.6	0.2	5.4
	–CH=NCH$_3$	8.8	0.5	0.1	2.3
	–CS–phenyl	18.7	1.0	-0.6	2.4
P	–P(CH$_3$)$_2$	13.6	1.6	-0.6	-1.0
	–P(phenyl)$_2$	8.9	5.2	0.0	0.1
	–P$^+$(phenyl)$_2$CH$_3$	-9.7	5.2	2.0	6.7
	–PO(CH$_3$)$_2$	2.5	1.1	0.1	3.0
	–PO(phenyl)$_2$	5.8	3.9	-0.1	3.0
	–PO(OH)$_2$	-1.9	3.6	1.5	5.6
	–PO(OCH$_2$CH$_3$)$_2$	1.6	3.6	-0.2	3.4
	–PS(CH$_3$)$_2$	6.7	2.0	0.2	2.9
	–PS(OCH$_2$CH$_3$)$_2$	6.1	2.8	-0.4	3.4
M	–Li	-43.2	-12.7	2.4	3.1
	–MgBr	-35.8	-11.4	2.7	4.0
	–SiH$_3$	-0.5	7.3	-0.4	1.3
	–SiH$_2$CH$_3$	4.8	6.3	-0.5	1.0
	–Si(CH$_3$)$_3$	11.6	4.9	-0.7	0.4
	–Si(phenyl)$_3$	5.8	7.9	-0.6	1.1
	–SiCl$_3$	3.0	4.6	0.1	4.2
	–Ge(CH$_3$)$_3$	13.7	4.5	-0.5	-0.2
	–Sn(CH$_3$)$_3$	13.2	7.2	-0.4	-0.4
	–Pb(CH$_3$)$_3$	20.1	8.0	-0.1	-1.0
	–AsH$_2$	1.7	7.9	0.8	0.0
	–As(phenyl)$_2$	11.1	5.0	0.1	-0.1
	–As(O)(OH)$_2$	3.8	1.6	0.8	4.5
	–SeCH=CH$_2$	0.7	4.7	0.4	-1.4
	–SeCN	-5.3	5.1	2.9	2.1
	–Sb(phenyl)$_2$	9.8	7.7	0.3	0.0
	–Hg–phenyl	41.6	9.3	-0.9	-1.6
	–HgCl	22.5	8.0	-0.6	-0.9

Estimation of ^{13}C Chemical Shifts of Multiply Substituted Benzenes and Naphthalenes (δ in ppm)

The ^{13}C chemical shifts of multiply substituted benzenes and naphthalenes (see next pages) can be estimated using the substituent effects in the corresponding monosubstituted hydrocarbons.

Example: Estimation of the chemical shifts for 3,5-dimethylnitrobenzene

C-1	base value	128.5		**C-2**	base value	128.5
	$Z_1(NO_2)$	19.9			$Z_2(NO_2)$	-4.9
	$2\,Z_3(CH_3)$	-0.2			$Z_2(CH_3)$	0.7
	estimated	148.2			$Z_4(CH_3)$	-3.0
	exp	148.5			estimated	121.3
					exp	121.7

C-3	base value	128.5		**C-4**	base value	128.5
	$Z_1(CH_3)$	9.2			$2\,Z_2(CH_3)$	1.4
	$Z_3(CH_3)$	-0.1			$Z_4(NO_2)$	6.1
	$Z_3(NO_2)$	0.9			estimated	136.0
	estimated	138.5			exp	136.2
	exp	139.6				

Larger discrepancies between estimated and experimental values are to be expected if the substituents are *ortho* to each other or if strongly electron-donating and electron-accepting groups occur simultaneously.

Effect of Substituents in Position 1 on ^{13}C Chemical Shifts of Monosubstituted Naphthalenes (δ in ppm)

for R: H $\delta_{C_1} = 128.0$

$\delta_{C_2} = 125.9$

$\delta_{C_9} = 133.6$

	Substituent R	C-1	C-2	C-3	C-4	C-5	C-6	C-7	C-8	C-9	C-10
C	–CH$_3$	6.0	0.5	0.6	-1.8	0.3	-0.7	-0.5	-4.1	-1.1	-0.2
	–C(CH$_3$)$_3$	17.9	-2.8	-0.9	-0.6	1.6	-1.4	-1.4	-1.2	-1.6	2.2
	–CH$_2$Br	4.0	1.1	-0.9	1.3	0.5	-0.1	0.3	-4.6	-2.8	0.1
	–CH$_2$OH	8.2	-0.9	-0.6	0.1	0.5	-0.3	0.1	-4.5	-2.6	0.0
	–CF$_3$	-1.9	-1.3	-1.8	5.0	1.0	0.8	2.0	-3.4	1.0	-3.9
X	–F	31.5	-16.1	0.1	-3.8	0.1	1.4	0.7	-7.1	-9.3	2.1
	–Cl	3.9	0.2	-0.2	-0.9	0.2	3.1	0.8	-3.6	-2.8	1.0
	–Br	-5.4	3.6	-0.2	-0.5	-0.1	0.4	1.0	-1.3	-2.0	0.6
	–I	-28.4	12.3	1.7	1.7	1.4	1.6	2.6	4.4	1.3	1.3
O	–OH	23.5	-17.2	-0.1	-7.3	-0.4	0.5	0.3	-6.6	-9.3	1.0
	–OCH$_3$	27.3	-22.3	-0.2	-7.9	-0.7	0.3	-0.9	-6.1	-8.1	0.8
	–OCOCH$_3$	18.6	-7.9	-0.6	-2.1	0.0	0.4	0.4	-6.9	-6.9	0.9
N	–NH$_2$	14.0	-16.5	0.3	-9.3	0.3	-0.3	-1.3	-7.3	-10.2	0.6
	–N(CH$_3$)$_2$	23.7	-11.2	0.6	-4.6	1.0	0.4	-0.3	-3.2	-3.9	2.1
	–NH$_3$$^+$	-3.8	-4.6	-0.9	3.4	1.4	2.1	2.8	-9.0	-7.4	1.2
	–NHCOCH$_3$	5.7	-4.4	-0.5	-3.0	0.0	-0.1	-0.3	-5.3	-5.9	0.1
	–NO$_2$	18.5	-2.1	-2.0	6.5	0.5	1.3	3.4	-5.1	-8.7	0.6
	–C≡N	-19.2	5.1	-2.4	3.8	-0.7	0.2	1.2	-4.5	-2.8	-2.2
O ‖ C	–CHO	2.9	10.8	-1.4	6.7	0.2	0.6	2.7	-3.5	-3.6	-0.3
	–COCH$_3$	6.9	2.9	-1.7	4.9	0.3	0.4	2.0	-2.0	-3.5	0.2
	–COOH	-1.5	3.6	-2.4	4.3	-0.6	-0.9	0.6	-3.2	-3.2	-0.8
	–COOCH$_3$	-0.9	4.5	-1.2	5.4	0.7	0.5	1.9	-1.8	-1.9	0.5
	–CON(CH$_3$)$_2$	6.8	-2.1	-0.8	0.9	0.4	0.4	1.0	0.1	-4.1	-0.2
	–COCl	1.2	10.6	-0.5	9.3	1.9	2.1	4.5	-2.1	-2.1	1.0
	–Si(CH$_3$)$_3$	9.8	5.1	-0.4	1.7	1.2	-0.8	-0.7	0.1	3.8	0.2

Effect of Substituents in Position 2 on ^{13}C Chemical Shifts of Monosubstituted Naphthalenes (δ in ppm)

for R: H $\quad \delta_{C_1} = 128.0$

$\delta_{C_2} = 125.9$

$\delta_{C_9} = 133.6$

	Substituent R	C-1	C-2	C-3	C-4	C-5	C-6	C-7	C-8	C-9	C-10
C	–CH₃	-1.3	9.3	2.0	-0.8	-0.5	-1.1	-0.2	-0.6	-0.1	-2.0
	–C(CH₃)₃	-3.3	22.5	-3.0	-0.4	0.0	-0.7	-0.2	-0.6	0.4	-1.3
	–CH₂Br	-1.7	9.0	1.9	-0.4	-0.5	0.7	0.3	0.6	-0.6	-0.7
	–CH₂OH	-2.7	12.3	-4.4	-0.1	-0.4*	-0.2*	0.1*	-0.2*	-0.3	-0.8
	–CF₃	-2.0	1.9	-4.2	1.1*	0.1*	2.4*	1.5	1.1	-1.1	1.3
X	–F	-17.0	34.9	-9.6	2.4	0.0	-0.7	1.1	-0.6	0.7	-3.0
	–Cl	-1.4	5.7	0.8	1.5	-0.2	0.2	1.1	-1.1	0.7	-1.9
	–Br	1.8	-6.2	3.1	1.5	-0.3	0.2	0.8	-1.1	-2.0	0.7
	–I	9.2	-34.1	9.0	2.3	0.5	1.3	1.5	-0.6	2.1	-0.8
O	–OH	-18.6	27.3	-8.3	1.8	-0.3	-2.4	0.5	-1.7	0.9	-4.7
	–OCH₃	-22.2	31.8	-7.1	1.5	-0.3	-2.2	0.5	-1.2	1.0	-4.3
	–OCOCH₃	-9.5	22.5	-4.8	1.3	-0.4	-0.3	0.6	-0.4	0.1	-2.2
N	–NH₂	-20.6	16.7	-8.9	-0.2	-1.6	-4.8	-0.9	-3.5	-0.1	-7.0
	–N(CH₃)₂	-21.1	23.6	-8.8	1.2	0.0	-3.4	0.7	-1.1	2.4	-5.9
	–NH₃⁺	-5.9	-0.3	-6.5	3.2	0.2	2.3	2.0	0.2	0.1	-0.3
	–NHCOCH₃	-11.0	9.6	-5.7	0.6	-0.4*	-0.9	1.6*	-1.6	0.2	-3.0
	–NO₂	-3.4	20.0	-6.7	1.7	0.1	4.0	2.2	2.1	-1.1	2.4
	–C≡N	5.8	-16.7	0.1	1.0	-0.2	3.0	1.6	0.2	-1.6	0.7
O ‖ C	–CHO	6.2	7.9	-3.6	0.8	-0.3	2.9	0.9	1.8	2.4	-1.4
	–COCH₃	1.9	8.3	-2.2	0.2	-0.4	2.3	0.7	1.4	1.8	-1.3
	–COOH	2.7	2.4	-0.6	0.2	-0.3	2.4	0.9	1.3	-1.3	1.5
	–COOCH₃	3.0	1.8	-0.5	0.2	-0.1	2.4	0.9	1.4	-1.0	1.9
	–COCl	2.5	9.1	-0.7	0.2*	-0.4	2.2*	0.8	1.2		-1.4
	–Si(CH₃)₃	5.8	11.9	3.9	-1.0	0.1	0.3	-0.2	0.1	-0.5	0.2

* Assignment uncertain

4.5.2 Coupling Constants

^{13}C-^{1}H Coupling Constants (|J| in Hz)

	In benzene:	In derivatives:
$^1J_{^{13}CH_a}$	159.0	
$^2J_{^{13}CH_b}$	1.0	1–4
$^3J_{^{13}CH_c}$	7.6	7–10
$^4J_{^{13}CH_d}$	-1.3	

^{13}C-^{13}C Coupling Constants (|J| in Hz)

$^1J_{ab}$	57.0
$^2J_{ac}$	2.5
$^3J_{ad}$	10.0

$^1J_{ab}$	44.2
$^2J_{ac}$	3.1
$^3J_{ad}$	3.8
$^4J_{ae}$	0.9

4.5.3 References

[1] P.E. Hansen, ^{13}C NMR of polycyclic aromatic hydrocarbons. A review, *Org. Magn. Reson.* **1979**, *12*, 109.

4.6 Heteroaromatic Compounds

4.6.1 Chemical Shifts

¹³C Chemical Shifts of Monocyclic Heteroaromatics (δ in ppm)

furan: 109.9 / 143.0 (O)	pyrrole: 107.7 / 118.0 (N–H)	thiophene: 126.4 / 124.9 (S)	selenophene: 128.8 / 131.0 (Se)
oxazole: 150.6 / 125.4 / 138.1 (N, O)	imidazole: 136.2 / 122.3 / 122.3 (N, N–H)	thiazole: 152.7 / 143.2 / 118.6 (N, S)	tellurophene: 137.6 / 126.2 (Te)
isoxazole: 150.0 / 100.5 / 158.9 (N, O)	pyrazole: 133.3 / 104.7 / 133.3 (N, N–H)	isothiazole: 157.0 / 123.4 / 147.8 (N, S)	1,2,3-thiadiazole: 147.3 / 135.8 (N, N, S)
1,2,3-triazole: 147.9 (N–N, N–H)	1,2,4-triazole: 147.4 / 147.4 (N, N, N–H)	130.4 (N, N, N–H)	1,2,3,4-tetrazole: 143.3 (N, N, N, N–H)
120.1 / 134.6 (HN, N–H, +)	109.0 / 135.0 (HN, N, +)	126.8 / 145.1 (N, N, –)	103.4 / 138.5 (N, N, –)

135.9	148.4	146.0	135.6	125.7
pyridine: 123.7 / 149.8 (N)	129.0 / 142.5 (N–H, +)	128.6 / 146.0 (N, +)	128.0 / 139.6 (N, +)	127.2 / 139.4 (N)
		I⁻ CH₃ 49.8 (in ethanol)	Cl⁻ OH (in DMSO)	N→O

pyridazine: 126.5 / 151.4 (N–N)	pyrimidine: 158.0 / 121.4 / 156.4 (N, N)	pyrazine: 144.9 (N, N)	1,3,5-triazine: 166.5 (N, N, N)	1,2,4,5-tetrazine: 160.9 (N, N, N, N)

Effect of Substituents in Position 2 on ^{13}C Chemical Shifts of Monosubstituted Pyridines (δ in ppm)

for R: H $\delta_{C_{2,6}} = 149.8$

$\delta_{C_{3,5}} = 123.7$

$\delta_{C_4} = 135.9$

	Substituent R	C-2	C-3	C-4	C-5	C-6
C	–CH$_3$	8.6	-0.5	0.3	-3.0	-0.7
	–CH$_2$CH$_3$	13.7	-1.7	0.4	-2.8	-0.6
	–CH=CH$_2$	5.9	-1.3	1.1	-2.5	-0.3
	–phenyl	7.7	-1.6	0.8	-3.2	0.2
X	–F	13.9	-14.0	5.4	-2.5	-2.0
	–Cl	1.8	0.8	2.8	-1.4	0.0
	–Br	-7.5	4.6	2.6	-1.1	0.5
	–I	-31.6	11.3	1.7	-0.8	1.0
O	–OH*	15.5	3.6	-1.1	-17.0	-8.2
	–OCH$_3$	14.3	-12.7	2.6	-7.1	-2.9
	–O–phenyl	13.9	-12.2	3.5	-5.3	-2.0
	–OCOCH$_3$	7.6	-7.3	3.4	-1.8	-1.6
N	–NH$_2$	8.4	-15.1	1.8	-9.7	-1.6
	–NHCH$_3$	10.9	-16.2	1.5	-11.3	-1.3
	–N(CH$_3$)$_2$	9.6	-17.9	1.2	-12.3	-1.9
	–NHCOCH$_3$	1.4	-9.8	2.6	-3.9	-2.1
	–NO$_2$	6.9	-5.7	3.9	5.4	-0.8
	–C≡N	-15.8	4.8	1.1	3.2	1.4
S	–SH	30.4	10.7	2.1	-10.6	-12.1
	–SCH$_3$	10.2	-4.6	0.0	-2.2	-0.5
	–S(O)CH$_3$	16.2	-4.4	2.2	0.9	-0.2
	–S(O)$_2$CH$_3$	8.5	-2.6	2.4	3.7	0.3
O‖C	–CHO	3.0	-2.0	1.2	4.2	0.4
	–COCH$_3$	3.8	-2.1	0.9	3.4	-0.8
	–COOH	-3.7	0.0	2.5	4.2	-1.7
	–COOCH$_3$	-1.7	1.5	1.1	3.3	0.0
	–CONH$_2$	-0.3	-1.2	1.4	2.8	-1.5
M	–Si(CH$_3$)$_3$	18.6	5.0	-2.0	-1.1	0.3
	–Sn(CH$_3$)$_3$	23.3	7.6	-2.7	-1.7	0.6
	–Pb(CH$_3$)$_3$	33.4	9.2	-2.6	-2.3	1.1

* Keto form (2-pyridone)

Effect of Substituents in Position 3 on ^{13}C Chemical Shifts of Monosubstituted Pyridines (δ in ppm)

for R: H $\delta_{C2,6} = 149.8$
           $\delta_{C3,5} = 123.7$
           $\delta_{C4} = 135.9$

Substituent R	C-2	C-3	C-4	C-5	C-6
C $-CH_3$	1.3	8.9	0.0	-0.9	-2.3
$-CH_2CH_3$	-0.4	15.4	-0.8	-0.5	-2.7
$-$phenyl	-1.4	12.8	-1.8	-0.3	-1.3
X $-F$	-11.5	36.1	-13.2	0.8	-3.9
$-Cl$	-0.3	8.1	-0.4	0.6	-1.4
$-Br$	2.1	-2.7	2.7	1.1	-0.9
$-I$	7.1	-28.5	8.9	2.3	0.3
O $-OH$	-10.7	31.3	-12.4	1.2	-8.6
$-OCH_3$	-12.5	31.5	-15.9	0.1	-8.4
$-OCOCH_3$	-6.5	23.4	-7.0	-0.1	-3.2
N $-NH_2$	-11.9	21.4	-14.4	0.8	-10.8
$-NHCH_3$	-13.6	23.1	-18.2	0.6	-11.9
$-N(CH_3)_2$	-14.0	23.3	-17.1	0.1	-11.6
$-C\equiv N$	3.6	-13.8	4.2	0.5	4.2
S $-SH$	-12.8	26.1	-11.3	7.3	-2.8
$-SCH_3$	-13.6	24.6	-11.7	10.6	-3.0
O $-CHO$	2.4	7.8	-0.2	0.5	5.4
‖ $-COCH_3$	3.5	8.5	-0.7	-0.2	0.0
C $-COOH$	-6.4	13.0	11.1	4.3	-6.0
$-COOCH_3$	-0.6	1.0	-0.5	-1.8	1.8
$-CONH_2$	2.7	5.9	1.1	1.2	-1.5
M $-Si(CH_3)_3$	2.7	9.1	3.0	-2.3	-1.2
$-Ge(CH_3)_3$	3.9	12.8	4.2	-0.4	-0.1
$-Sn(CH_3)_3$	5.9	13.0	7.1	0.1	-0.3
$-Sn(n\text{-}C_4H_9)_3$	6.6	12.6	7.7	0.0	-0.4
$-Pb(n\text{-}C_4H_9)_3$	7.1	21.7	8.5	0.9	-1.8

Effect of Substituents in Position 4 on ^{13}C Chemical Shifts of Monosubstituted Pyridines (δ in ppm)

R
|
4
5   3
6   2
N

for R: H $\delta_{C_{2,6}}$ = 149.8

$\delta_{C_{3,5}}$ = 123.7

δ_{C_4} = 135.9

Substituent R	C-2	C-3	C-4
C –CH$_3$	0.5	0.7	10.6
–CH$_2$CH$_3$	-0.1	-0.5	16.8
–CH(CH$_3$)$_2$	0.4	-1.9	21.2
–C(CH$_3$)$_3$	0.9	-2.6	23.9
–CH=CH$_2$	0.3	-3.0	8.4
–phenyl	0.4	-2.2	12.2
X –F	2.7	-11.9	32.8
–Br	3.0	3.3	-3.2
–I	0.2	9.1	-30.8
O –OH*	-9.8	-6.2	45.4
–OCH$_3$	0.9	-13.9	29.0
–OCOCH$_3$	1.7	-6.7	23.9
N –NH$_2$	0.7	-13.8	19.3
–NHCH$_3$	0.5	-15.9	19.8
–N(CH$_3$)$_2$	0.6	-16.3	19.2
–C≡N	2.1	2.1	-15.9
S –SH	-16.9	5.9	54.3
–SCH$_3$	0.1	-3.3	14.6
O‖C –CHO	1.7	-0.7	5.3
–COCH$_3$	1.6	-2.7	6.6
–COOCH$_3$	1.0	-0.8	1.4
–CONH$_2$	0.4	-0.9	6.2
M –Si(CH$_3$)$_3$	-2.8	2.4	11.9
–Ge(CH$_3$)$_3$	-1.1	4.4	16.8
–Sn(CH$_3$)$_3$	-1.1	7.3	16.2
–Pb(CH$_3$)$_3$	-0.5	9.1	24.6

* Keto form (4-pyridone)

Estimation of ^{13}C Chemical Shifts of Multiply Substituted Pyridines (δ in ppm)

The ^{13}C chemical shifts in multiply substituted pyridines can be estimated using the substituent effects in the monosubstituted parent compound.

Example: Estimation of the chemical shifts for 2-amino-5-methylpyridine

C-2	base value	149.8		**C-3**	base value	123.7
	2-NH$_2$	8.4			2-NH$_2$	-15.1
	5-CH$_3$	-2.3			5-CH$_3$	-0.9
	estimated	155.9			estimated	107.7
	exp	156.9			exp	108.4
C-4	base value	135.9		**C-5**	base value	123.7
	2-NH$_2$	1.8			2-NH$_2$	-9.7
	5-CH$_3$	0.0			5-CH$_3$	8.9
	estimated	137.7			estimated	122.9
	exp	138.6			exp	122.5
C-6	base value	149.8				
	2-NH$_2$	-1.6				
	5-CH$_3$	1.3				
	estimated	149.5				
	exp	147.6				

Larger discrepancies between estimated and experimental values are to be expected if the substituents are *ortho* to each other and if strongly electron-donating and -accepting groups occur simultaneously. Also, tautomerization and zwitterion formation have large effects on ^{13}C chemical shifts.

^{13}C Chemical Shifts of Condensed Heteroaromatics (δ in ppm)

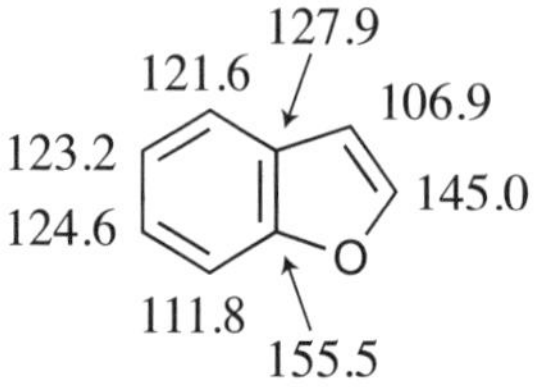

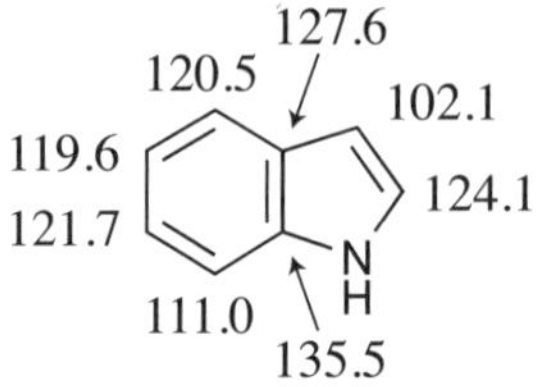

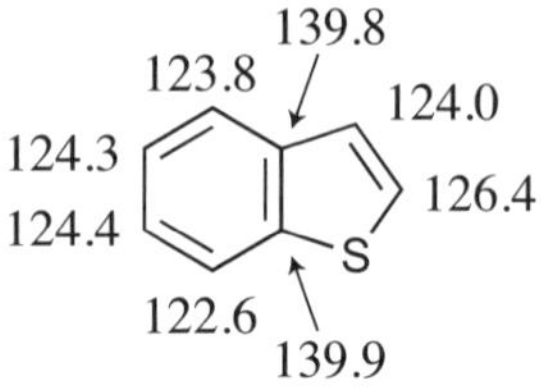

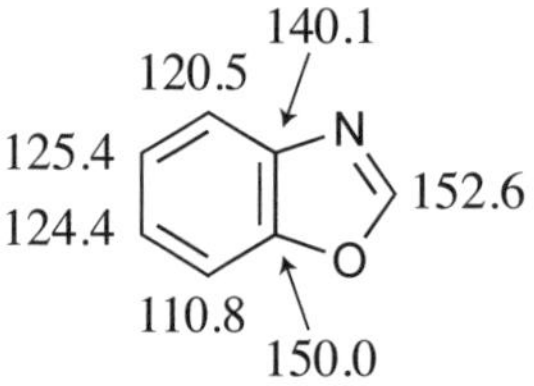

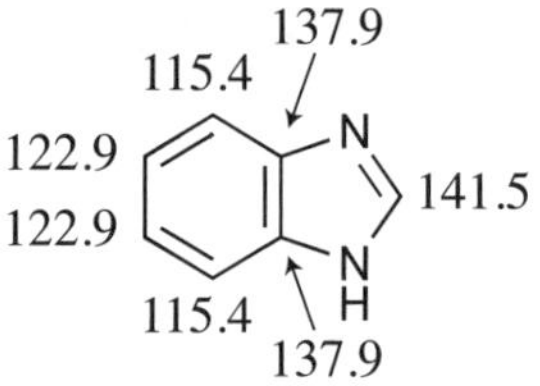

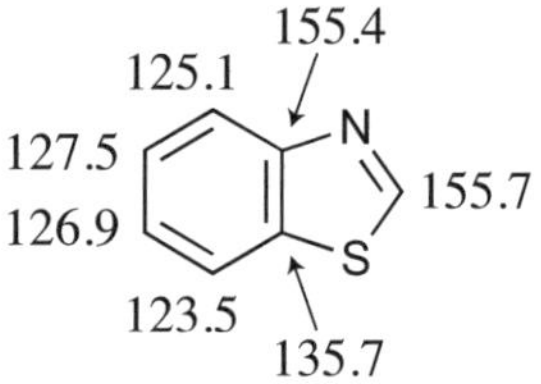

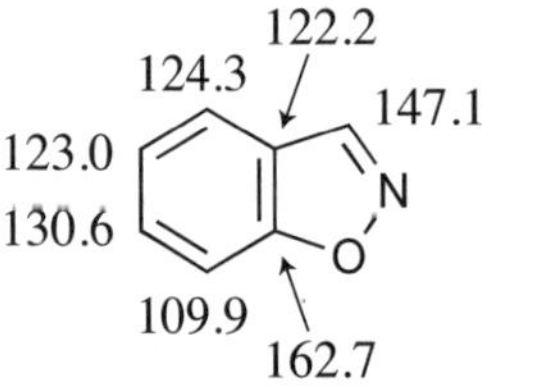

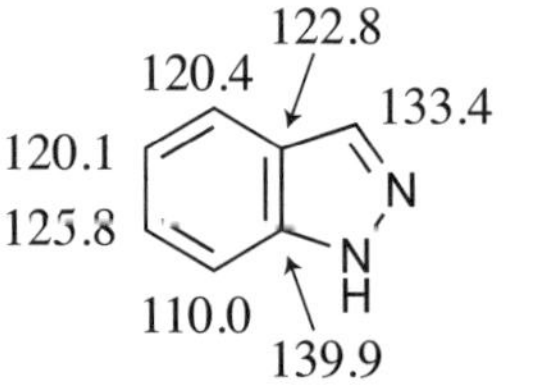

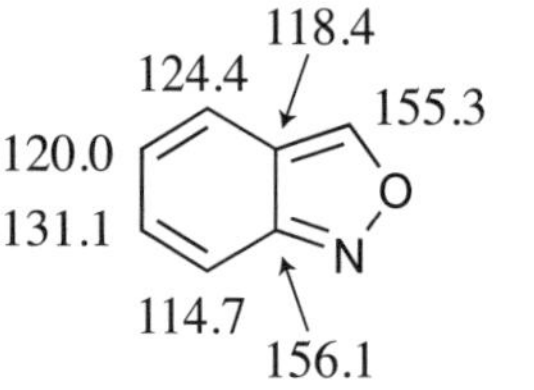

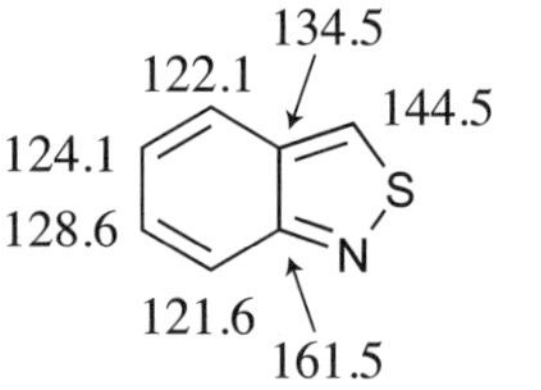

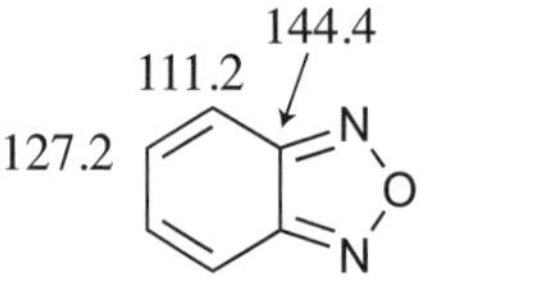

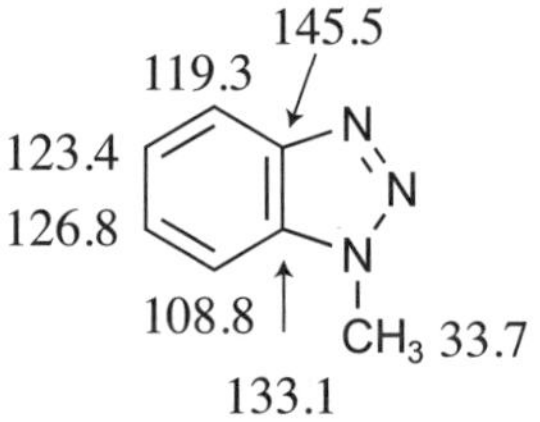

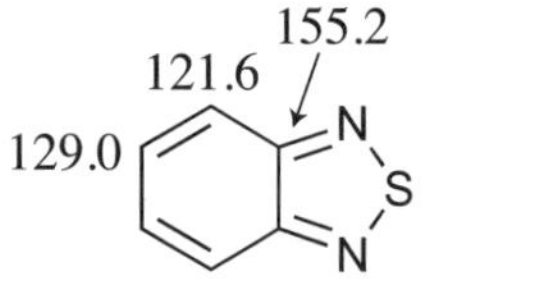

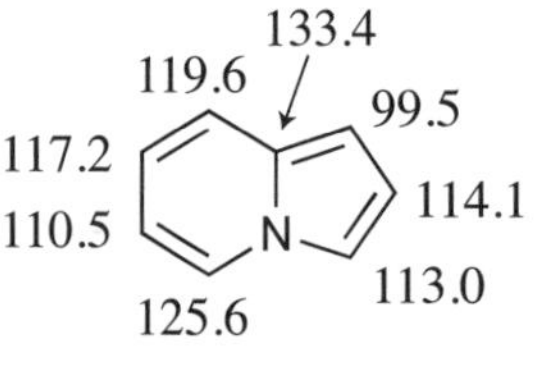

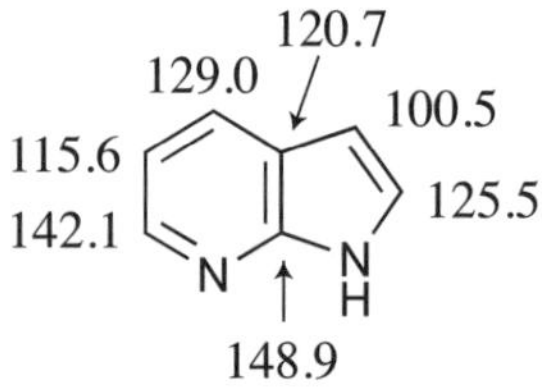

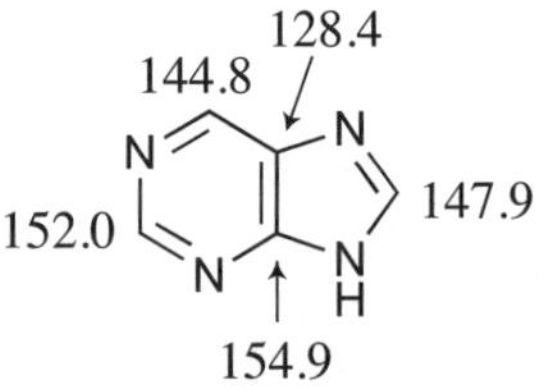

127.9
121.6
123.2
124.6
106.9
145.0
111.8
155.5

127.6
120.5
119.6
121.7
102.1
124.1
111.0
N
H
135.5

139.8
123.8
124.3
124.4
124.0
126.4
122.6
S
139.9

140.1
120.5
125.4
124.4
N
152.6
O
110.8
150.0

137.9
115.4
122.9
122.9
N
141.5
N
H
115.4
137.9

155.4
125.1
127.5
126.9
N
155.7
S
123.5
135.7

122.2
124.3
123.0
130.6
147.1
N
O
109.9
162.7

122.8
120.4
120.1
125.8
133.4
N
N
H
110.0
139.9

118.4
124.4
120.0
131.1
155.3
O
N
114.7
156.1

134.5
122.1
124.1
128.6
144.5
S
N
121.6
161.5

144.4
111.2
127.2
N
O
N

145.5
119.3
123.4
126.8
N
N
N
108.8
CH$_3$ 33.7
133.1

155.2
121.6
129.0
N
S
N

133.4
119.6
117.2
110.5
99.5
N
114.1
113.0
125.6

120.7
129.0
115.6
142.1
100.5
N
N
H
125.5
148.9

128.4
144.8
152.0
N
N
N
H
147.9
154.9

128.0
127.6 ↓ 135.7
126.3 [quinoline] 120.8
129.2 150.0
129.2 ↑
148.1

135.7
126.4 ↓ 120.4
130.2 [isoquinoline] 142.9
127.2
127.5 ↑ 152.4
128.6

126.9
128.0 ↓ 124.7
132.2 [cinnoline] 146.1
132.1
129.5 ↑
151.0

125.2
127.4 ↓ 155.9
127.9 [quinazoline] 160.7
134.1
128.6 ↑
150.1

142.8
129.6 ↓
129.4 [quinoxaline] 144.8

126.7
126.7 ↓ 152.0
133.1 [phthalazine]

124.2
↓ 120.6
[dibenzofuran] 122.6
127.0
156.2 ↑ 111.6

122.6
↓ 120.0
[carbazole] 118.4
125.4
139.6 ↑ 110.8

134.9
↓ 121.9
[dibenzothiophene] 124.6
127.0
138.5 ↑ 122.9

142.7
↓ 114.5
[phenoxazine] 120.0
123.0
131.8 ↑ 112.8

126.6
135.8 ↓ 129.5
[acridine] 128.3
125.5
149.1 ↑ 130.3

116.8
↓ 126.7
[phenothiazine] 121.3
125.6
145.7 ↑ 113.8

142.2
↓ 116.2
[dibenzodioxin] 123.6

144.0
↓ 130.9
[phenazine] 130.2

119.9
↓ 127.4*
[phenoxathiin] 124.2
126.5*
151.9 ↑ 117.5

* Assignment uncertain

4.6.2 Coupling Constants

^{13}C-^{1}H Coupling Constants (|^{1}J| in Hz)

| | $^{1}J_{C_aH_a}$ | 175 |
| | $^{1}J_{C_bH_b}$ | 202 |

| | $^{1}J_{C_aH_a}$ | 169 |
| | $^{1}J_{C_bH_b}$ | 183 |

| | $^{1}J_{C_aH_a}$ | 206 |
| | $^{1}J_{C_bH_b}$ | 189 |

| | $^{1}J_{C_aH_a}$ | 186 |
| | $^{1}J_{C_bH_b}$ | 177 |

| | $^{1}J_{C_aH_a}$ | 194 |

| | $^{1}J_{C_aH_a}$ | 209 |

	$^{1}J_{C_aH_a}$	161
	$^{1}J_{C_bH_b}$	163
	$^{1}J_{C_cH_c}$	178

^{13}C-^{13}C Coupling Constants (|^{1}J| in Hz)

| | $^{1}J_{C_aC_b}$ | 69.1 |

| | $^{1}J_{C_aC_b}$ | 65.6 |

| | $^{1}J_{C_aC_b}$ | 64.2 |

| | $^{1}J_{C_aC_b}$ | 53.7 |
| | $^{1}J_{C_bC_c}$ | 54.3 |

4.7 Halogen Compounds

The additivity rules for estimating the ^{13}C chemical shifts of various skeletons can be applied to those haloalkanes that do not have more than one halogen atom at a given carbon atom. In all other cases, the simple linear models fail but correction terms for non-additivity are available for halomethanes and derivatives (see [1, 2]).

4.7.1 Fluoro Compounds

^{19}F (natural abundance 100%) has a spin quantum number I of 1/2. The signals of carbon atoms up to a distance of about four bonds are split by coupling to ^{19}F.

^{13}C Chemical Shifts and ^{19}F-^{13}C Coupling Constants (δ in ppm, $|J|$ in Hz)

CH$_3$F 71.6   CH$_2$F$_2$ 109.0   CHF$_3$ 116.4   CF$_4$ 118.5
$\quad$ $^1J_{CF}$ 161.9 $\quad$ $^1J_{CF}$ 234.8 $\quad$ $^1J_{CF}$ 274.3 $\quad$ $^1J_{CF}$ 259.2

Hal

$^2J_{CF}$ 21.1 $^2J_{CF}$ 19.5 $^2J_{CF}$ 22.4   28.3
$\quad$ 16.4 $\quad$ 23.6 $\quad$ 23.0

83.7   9.2   85.2   87.8   93.5
$^1J_{CF}$ 160.1 $^3J_{CF}$ 6.7 $^1J_{CF}$ 163.3 $^1J_{CF}$ 162.1

$^4J_{CF} \approx 0$ $^2J_{CF}$ 18.3   116.2 $^2J_{CF}$ 24.8
14.1   31.9   29.3   30.6   F$_3$C$-$CF$_3$ 88.5

22.7   29.3   25.3   84.2 $^1J_{CF}$ 271
$\quad$ $^3J_{CF}$ 6.2 $^1J_{CF}$ 164.8 $^2J_{CF}$ 48.1   147.7
                                                                            $^1J_{CF}$ 267.2

$^1J_{CF}$ 177 $^1J_{CF}$ 239 $^1J_{CF}$ 283.2
$\quad$ 78.9 $\quad$ 108.1 $\quad$ 115.0

FH$_2$C$\diagdown$$\diagup$OH   F$_2HC\diagdown$$\diagup$OH   F$_3C\diagdown$$\diagup$OH
$\quad\quad$ 173.5 $\quad\quad$ 167.2 $\quad\quad$ 163.0
O $^2J_{CF}$ 22   O $^2J_{CF}$ 28   O $^2J_{CF}$ 43.6

F
91.0; $^{1}J_{CF}$ 170.6
32.8; $^{2}J_{CF}$ 19.0
23.6; $^{3}J_{CF}$ 7.6
25.3; $^{4}J_{CF}$ 1.5

F
163.3; $^{1}J_{CF}$ 245.1
115.5; $^{2}J_{CF}$ 21.0
130.1; $^{3}J_{CF}$ 7.8
124.1; $^{4}J_{CF}$ 3.2

CH$_2$F 84.9; $^{1}J_{CF}$ 166.0
137.0; $^{2}J_{CF}$ 16.5
127.8; $^{3}J_{CF}$ 6.1
128.9; $^{4}J_{CF}$ 1
129.0; $^{5}J_{CF}$ 3

CHF$_2$ 114.8; $^{1}J_{CF}$ 238.6
134.4; $^{2}J_{CF}$ 22.2
125.5; $^{3}J_{CF}$ 5.6
128.6
130.7

CF$_3$ 124.5; $^{1}J_{CF}$ 272.2
131.0; $^{2}J_{CF}$ 36.6
125.3; $^{3}J_{CF}$ 3.7
128.8; $^{4}J_{CF}$ 1
131.8; $^{5}J_{CF} \approx 0$

F
168.7; $^{1}J_{CF}$ 261.8
111.8; $^{2}J_{CF}$ 16.1
152.5; $^{3}J_{CF}$ 6.4

122.7; $^{2}J_{CF}$ 17.7
124.5; $^{3}J_{CF}$ 4.3
145.9; $^{4}J_{CF}$ 3.7
F
159.8; $^{1}J_{CF}$ 255.1
138.3; $^{2}J_{CF}$ 22.5

141.3; $^{3}J_{CF}$ 7.5
121.2; $^{4}J_{CF}$ 4.2
147.8; $^{3}J_{CF}$ 14.9
109.7; $^{2}J_{CF}$ 37.6
F
163.7; $^{1}J_{CF}$ 236.3

Hal

Estimation of ^{13}C Chemical Shifts of Linear Perfluoroalkanes (δ in ppm)

$$\delta = 124.8 + \sum Z_i$$

Increments Z_i for the CF$_2$ or CF$_3$ substituent in position:		
α	β	γ
-8.6	1.8	0.5

Example: Estimation of the chemical shifts in perfluorobutane

F$_3$C—CF$_2$—CF$_2$—CF$_3$

CF$_3$			CF$_2$		
base value		124.8	base value		124.8
1 α-CF$_2$		-8.6	1 α-CF$_3$		-8.6
1 β-CF$_2$		1.8	1 α-CF$_2$		-8.6
1 γ-CF$_3$		0.5	1 β-CF$_3$		1.8
estimated		118.5	estimated		109.4
exp		118.5	exp		109.3

4.7.2 Chloro Compounds

^{13}C Chemical Shifts of Chloro Compounds (δ in ppm)

25.6
CH$_3$Cl

54.0
CH$_2$Cl$_2$

77.2
CHCl$_3$

96.1
CCl$_4$

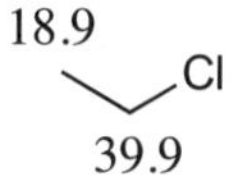

18.9
 Cl
39.9

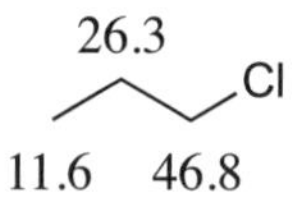

26.3
 Cl
11.6   46.8

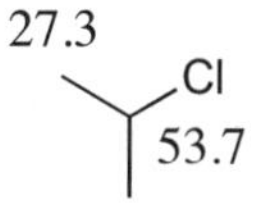

27.3
 Cl
53.7

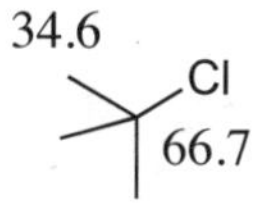

34.6
 Cl
66.7

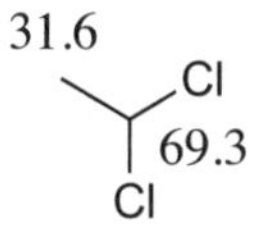

31.6
 Cl
69.3
Cl

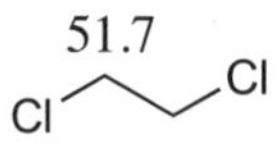

51.7
Cl   Cl

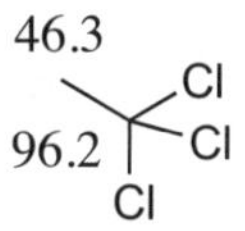

46.3
 Cl
96.2   Cl
Cl

105.3
Cl$_3$C—CCl$_3$

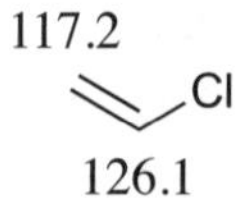

117.2
 Cl
126.1

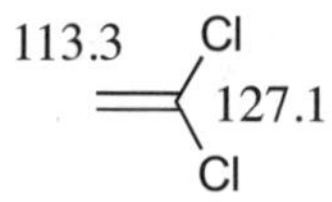

113.3   Cl
        127.1
        Cl

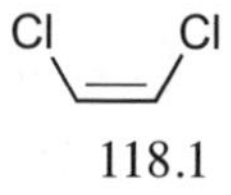

Cl   Cl
118.1

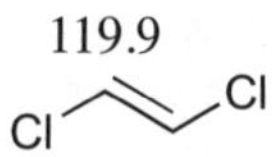

119.9
 Cl
Cl

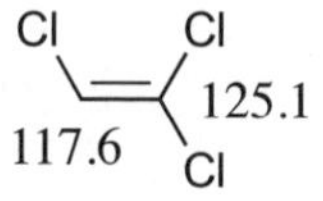

Hal

Cl   Cl
       125.1
117.6 Cl

Cl   Cl
       121.3
Cl   Cl

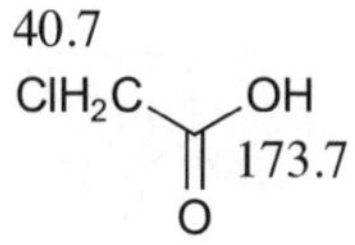

40.7
ClH$_2$C   OH
        173.7
       O

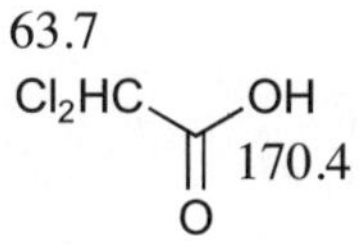

63.7
Cl$_2$HC   OH
        170.4
       O

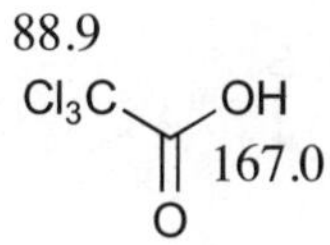

88.9
Cl$_3$C   OH
       167.0
      O

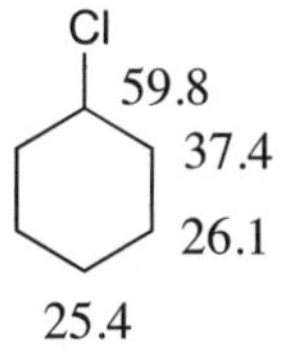

Cl
 59.8
 37.4
 26.1
25.4

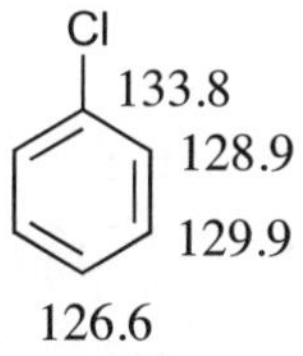

Cl
 133.8
 128.9
 129.9
126.6

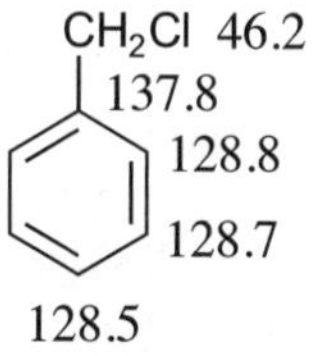

CH$_2$Cl 46.2
 137.8
 128.8
 128.7
128.5

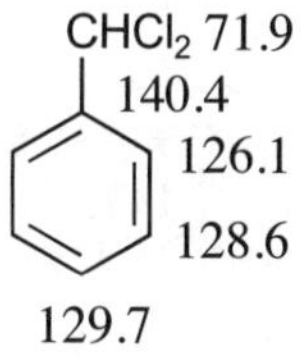

CHCl$_2$ 71.9
 140.4
 126.1
 128.6
129.7

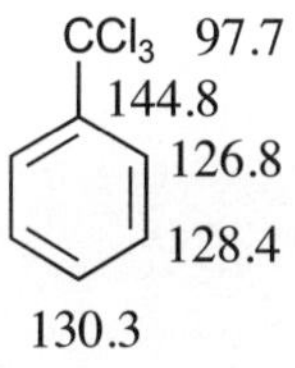

CCl$_3$ 97.7
 144.8
 126.8
 128.4
130.3

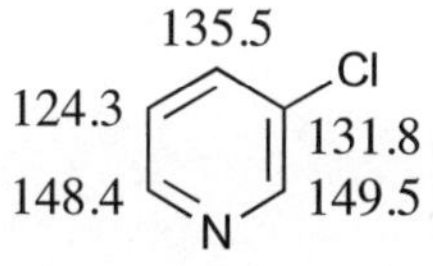

135.5
      Cl
124.3   131.8
148.4 N 149.5

138.7
122.3   124.5
          151.6
149.8 N Cl

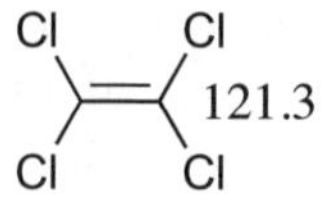

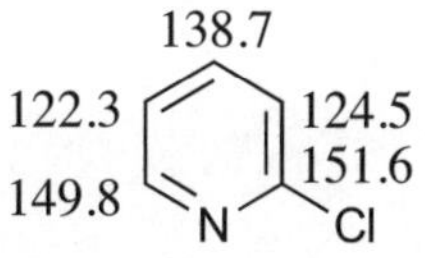

4.7.3 Bromo Compounds

^{13}C Chemical Shifts of Bromo Compounds (δ in ppm)

9.6
CH$_3$Br

21.4
CH$_2$Br$_2$

12.1
CHBr$_3$

-28.7
CBr$_4$

19.4
Br
27.6

26.4
Br
13.0   35.6

28.5
Br
44.8

36.4
Br
62.1

31.8
Br
40.1
Br

32.4
Br
Br

49.4
Br
31.5   Br
Br

53.4
Br$_3$C−CBr$_3$

122.4
Br
114.7

127.2   Br
97.0
Br

Br   Br
116.4

109.4
Br
Br

Br   Br
95.0
112.4   Br

Br   Br
93.7
Br   Br

25.9
BrH$_2$C   OH
172.0
O

31.3
Br$_2$HC   OH
169.7
O

Br
52.4
38.3
27.3
25.6

Br
123.1
131.8
130.7
127.5

CH$_2$Br 33.4
138.0
129.2
128.8
128.7

CHBr$_2$ 41.2
141.9
126.5
128.6
129.8

CBr$_3$ 36.5
147.0
126.5
128.1
130.1

Br
132.7
127.0
N   152.8

138.6
Br
124.8   121.0
148.9   N   151.9

138.5
122.6   128.3
142.3
150.3   N   Br

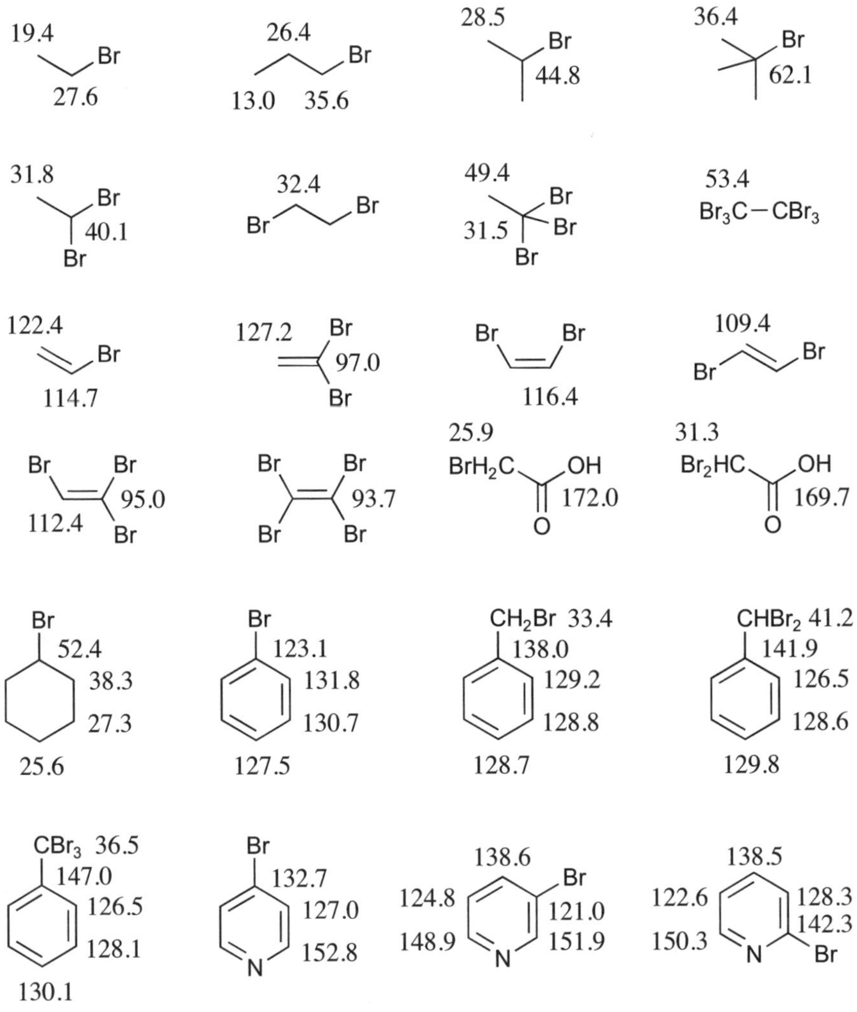

4.7.4 Iodo Compounds

^{13}C Chemical Shifts of Iodo Compounds (δ in ppm)

-24.0
CH$_3$I

-54.0
CH$_2$I$_2$

-139.9
CHI$_3$

-292.5
CI$_4$

20.6
$\diagdown\diagup$ I
-1.6

27.0
$\diagup\diagdown\diagup$ I
15.3   9.1

31.2
$\diagdown$ I
20.9

40.4
$\diagdown$ I
43.0

3.0
I $\diagup\diagdown\diagup$ I

130.3
$\diagdown$ I
85.2

I $\diagdown\diagup$ I
96.5

79.4
I $\diagup\diagdown$ I

I
31.2
40.1
28.3
25.4

I
97.3
137.4
130.1
127.4

CH$_2$I 5.8
139.2
128.7
128.7
127.8

CH I$_2$ -0.6
145.2
126.3
128.2
129.0

Hal

I
105.1
132.8
150.0
N

144.8
126.0
95.2
150.1   156.9
N

137.6
122.9
135.0
118.2
150.8
N   I

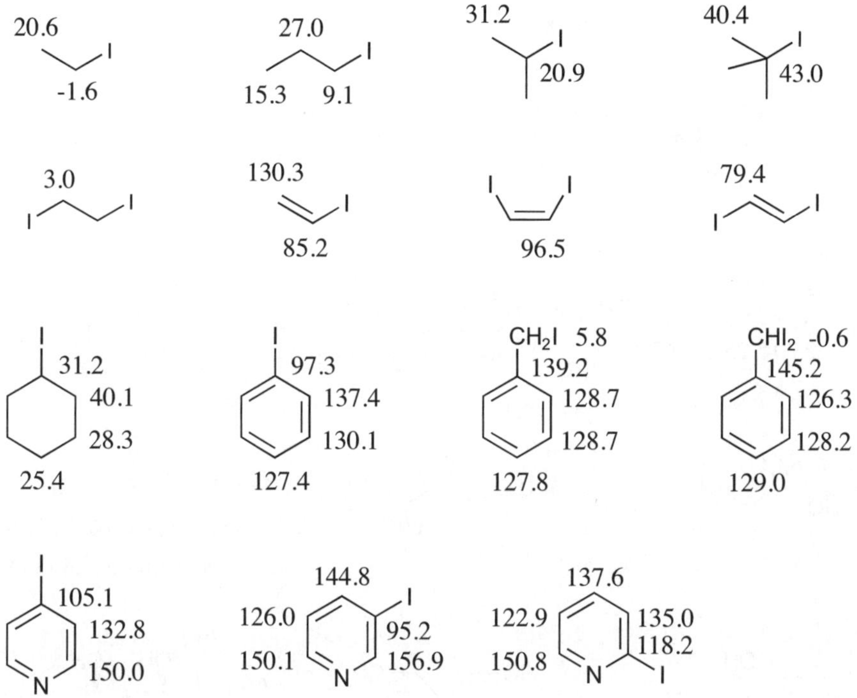

4.7.5 References

[1] G.R. Somayajulu, J.R. Kennedy, T.M. Vickrey, B.J. Zwolinski, Carbon-13 chemical shifts for 70 halomethanes, *J. Magn. Reson.* **1979**, *33*, 559.
[2] A. Fürst, W. Robien, E. Pretsch, A comprehensive parameter set for the prediction of the ^{13}C NMR chemical shifts of *sp^3*-hybridized carbon atoms in organic compounds, *Anal. Chim. Acta* **1990**, *233*, 213.
[3] D.W. Ovenall, J.J. Chang, Carbon-13 NMR of fluorinated compounds using wide-band fluorine decoupling, *J. Magn. Reson.* **1977**, *25*, 361.

4.8 Alcohols, Ethers, and Related Compounds

4.8.1 Alcohols

^{13}C Chemical Shifts of Alcohols (δ in ppm)

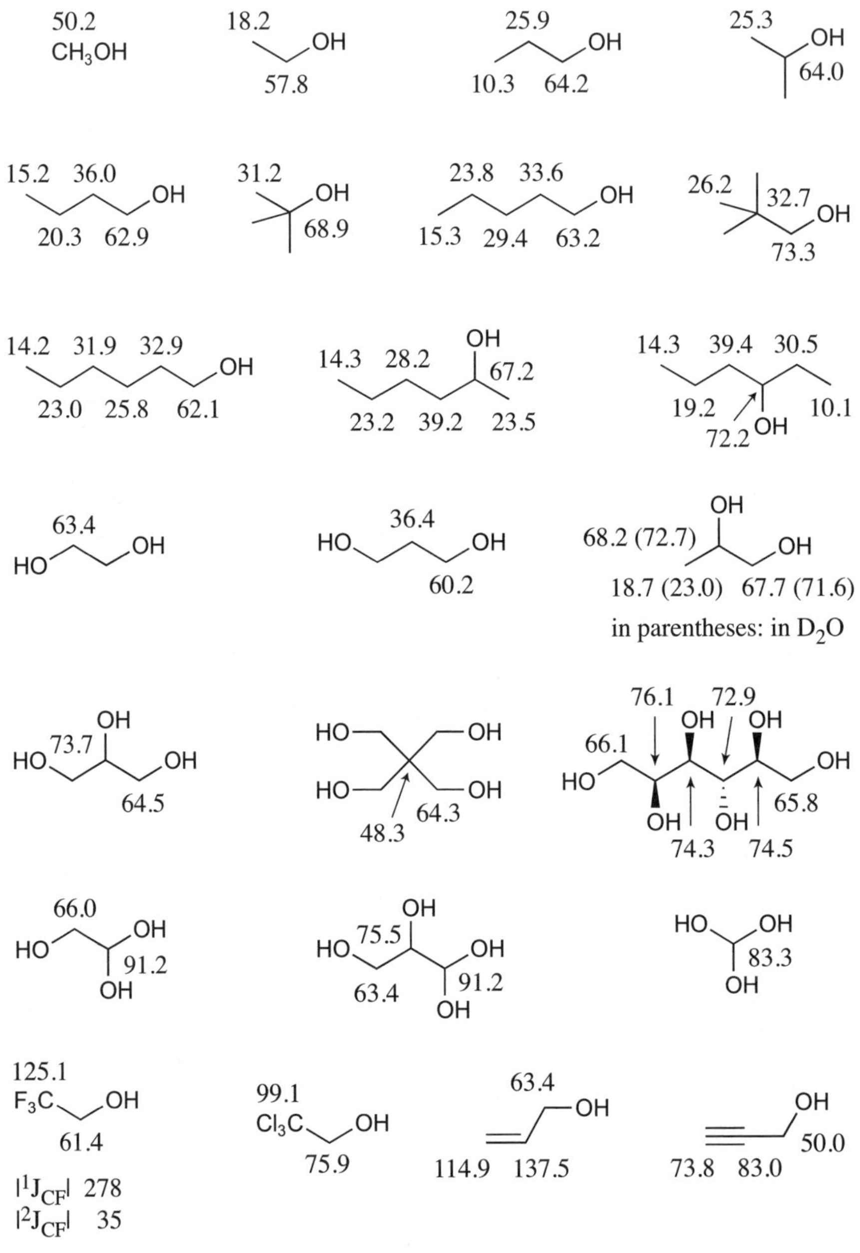

50.2
CH$_3$OH

18.2
OH
57.8

25.9
OH
10.3   64.2

25.3
OH
64.0

15.2 36.0
OH
20.3 62.9

31.2
OH
68.9

23.8 33.6
OH
15.3 29.4 63.2

26.2 | 32.7
OH
73.3

14.2 31.9 32.9
OH
23.0 25.8 62.1

14.3 28.2
OH
67.2
23.2 39.2 23.5

14.3 39.4 30.5
19.2 10.1
72.2 OH

63.4
HO OH

36.4
HO OH
60.2

OH
68.2 (72.7) OH
18.7 (23.0) 67.7 (71.6)
in parentheses: in D$_2$O

OH
HO 73.7 OH
64.5

HO OH
HO OH
48.3 64.3

76.1   72.9
OH OH
66.1
HO OH
65.8
OH OH
74.3   74.5

66.0
HO OH
91.2
OH

OH
75.5
HO OH
63.4 91.2
OH

HO OH
83.3
OH

125.1
F$_3$C OH
61.4
$|^1J_{CF}|$ 278
$|^2J_{CF}|$ 35

99.1
Cl$_3$C OH
75.9

63.4
OH
114.9 137.5

OH
50.0
73.8 83.0

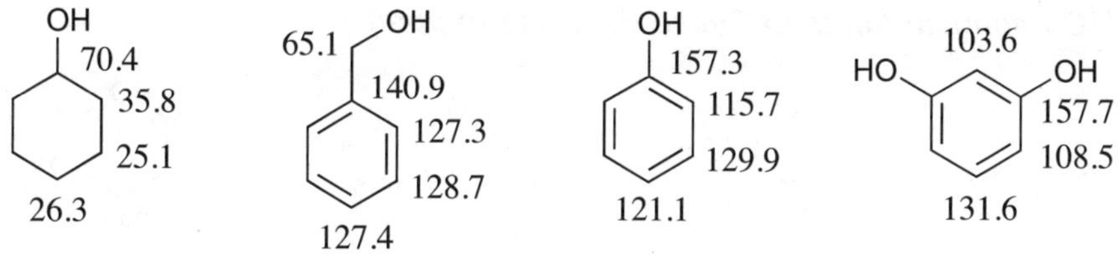

OH
70.4
35.8
25.1
26.3

65.1   OH
140.9
127.3
128.7
127.4

OH
157.3
115.7
129.9
121.1

103.6
HO   OH
157.7
108.5
131.6

^{13}C Chemical Shifts of Enols (δ in ppm)

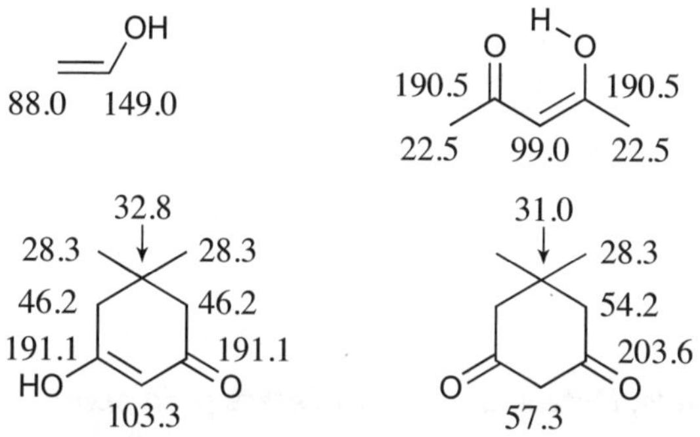

OH
88.0   149.0

190.5   H O   190.5
22.5   99.0   22.5

O   O
201.1
56.6   28.5

32.8
28.3   28.3
46.2   46.2
191.1   191.1
HO   O
103.3

31.0
28.3
54.2
203.6
O   O
57.3

4.8.2 Ethers

^{13}C Chemical Shifts of Ethers (δ in ppm)

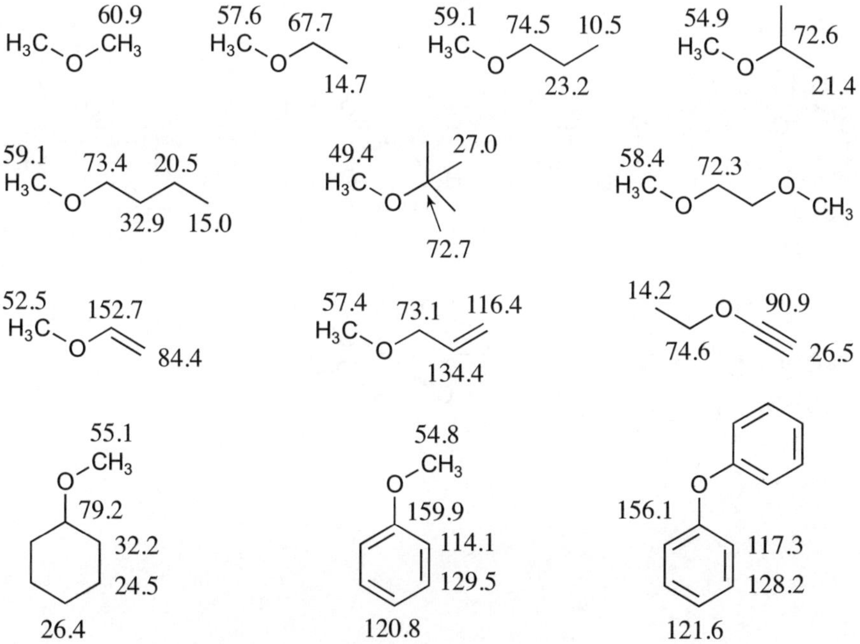

60.9
H_3C O CH_3

57.6 67.7
H_3C O
14.7

59.1 74.5 10.5
H_3C O
23.2

54.9
H_3C O
72.6
21.4

59.1 73.4 20.5
H_3C O
32.9 15.0

49.4   27.0
H_3C O
72.7

58.4 72.3
H_3C O   O CH_3

52.5 152.7
H_3C O
84.4

57.4 73.1 116.4
H_3C O
134.4

14.2   90.9
O
74.6   26.5

55.1
O CH_3
79.2
32.2
24.5
26.4

54.8
O CH_3
159.9
114.1
129.5
120.8

156.1
O
117.3
128.2
121.6

^{13}C Chemical Shifts of Cyclic Ethers (δ in ppm)

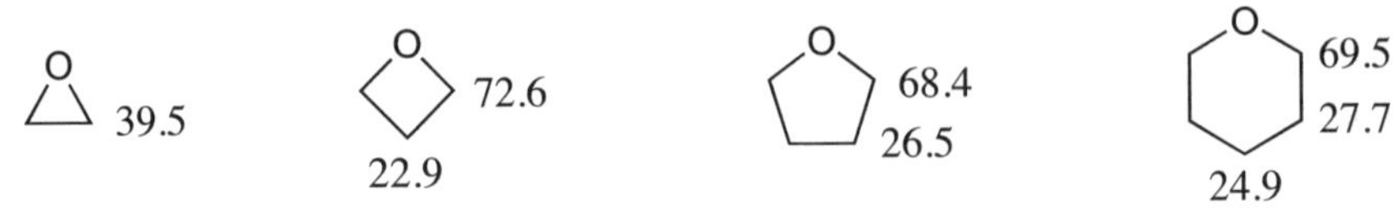

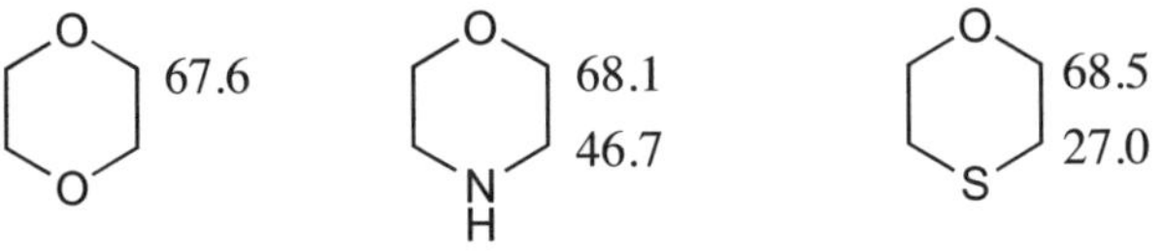

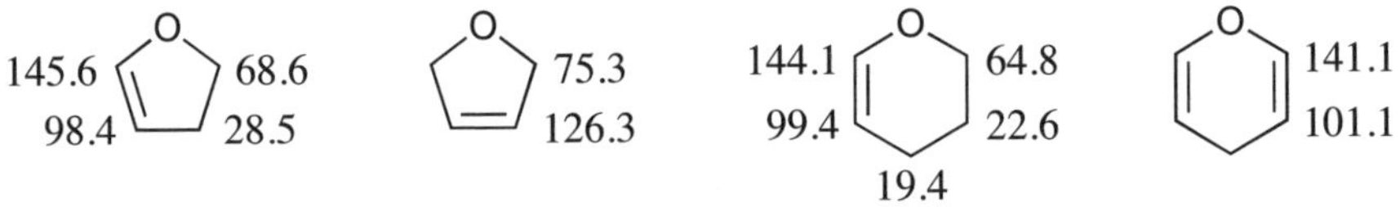

39.5   72.6   22.9   68.4   26.5   69.5   27.7   24.9

67.6   68.1   46.7   68.5   27.0

145.6   68.6   98.4   28.5   75.3   126.3   144.1   64.8   99.4   22.6   19.4   141.1   101.1

^{13}C Chemical Shifts of Acetals, Ketals, and Ortho Esters (δ in ppm)

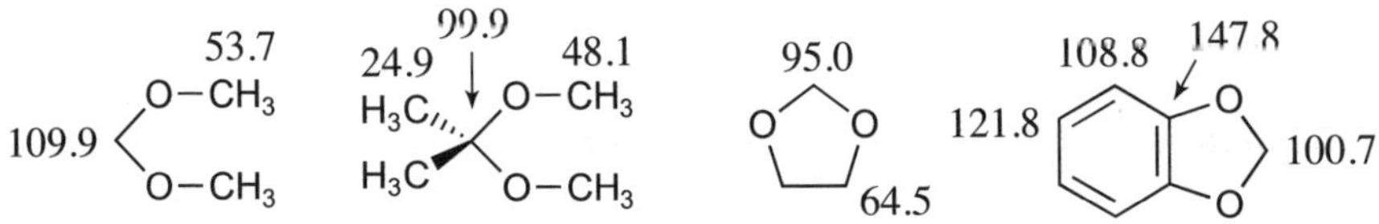

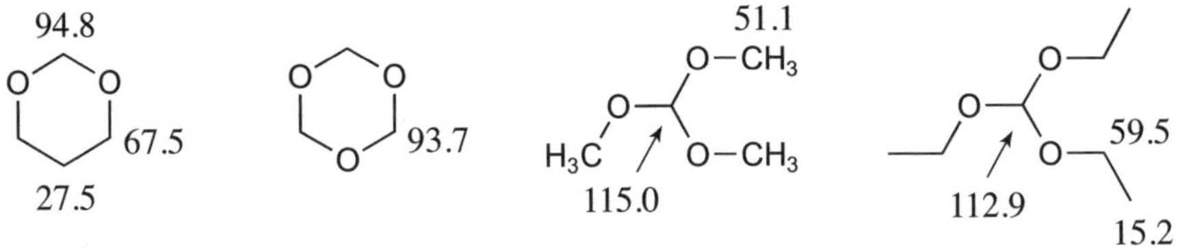

O

53.7   O–CH$_3$ 109.9   O–CH$_3$

24.9   99.9   48.1   H$_3$C$_{\prime\prime\prime}$ O–CH$_3$ H$_3$C   O–CH$_3$

95.0   O   O   121.8   64.5

108.8   147 8   O   100.7   O

94.8   O   O   67.5   27.5

O   O   O   93.7

51.1   O–CH$_3$ O   H$_3$C   O–CH$_3$ 115.0

O   O   O   59.5   112.9   15.2

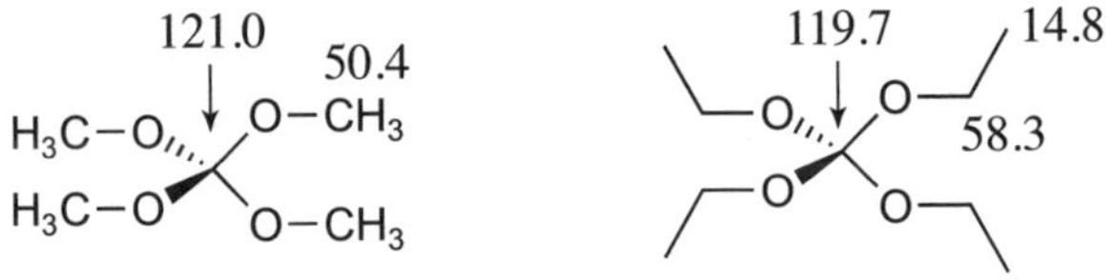

121.0   50.4   H$_3$C–O$_{\prime\prime\prime}$ O–CH$_3$ H$_3$C–O   O–CH$_3$

119.7   14.8   O$_{\prime\prime\prime}$ O   58.3   O   O

4.9 Nitrogen Compounds

4.9.1 Amines

^{13}C Chemical Shifts of Amines and Ammonium Salts (δ in ppm)

The protonation of amines causes a shielding of the carbon atoms in the vicinity of the nitrogen. This shielding amounts to -2 ppm for an α carbon atom, -3 to -4 for a β carbon, and -0.5 to -1.0 ppm for a γ carbon. The most frequent exceptions occur in branched systems: Tertiary and quaternary carbon atoms in the α-position are generally deshielded by protonation of the nitrogen ($\Delta\delta = +0.5$ to +9 ppm) [1]. In the following, shifts induced by protonation ($\delta_{amine\ hydrochloride} - \delta_{amine}$, measured in D_2O) are given in parentheses.

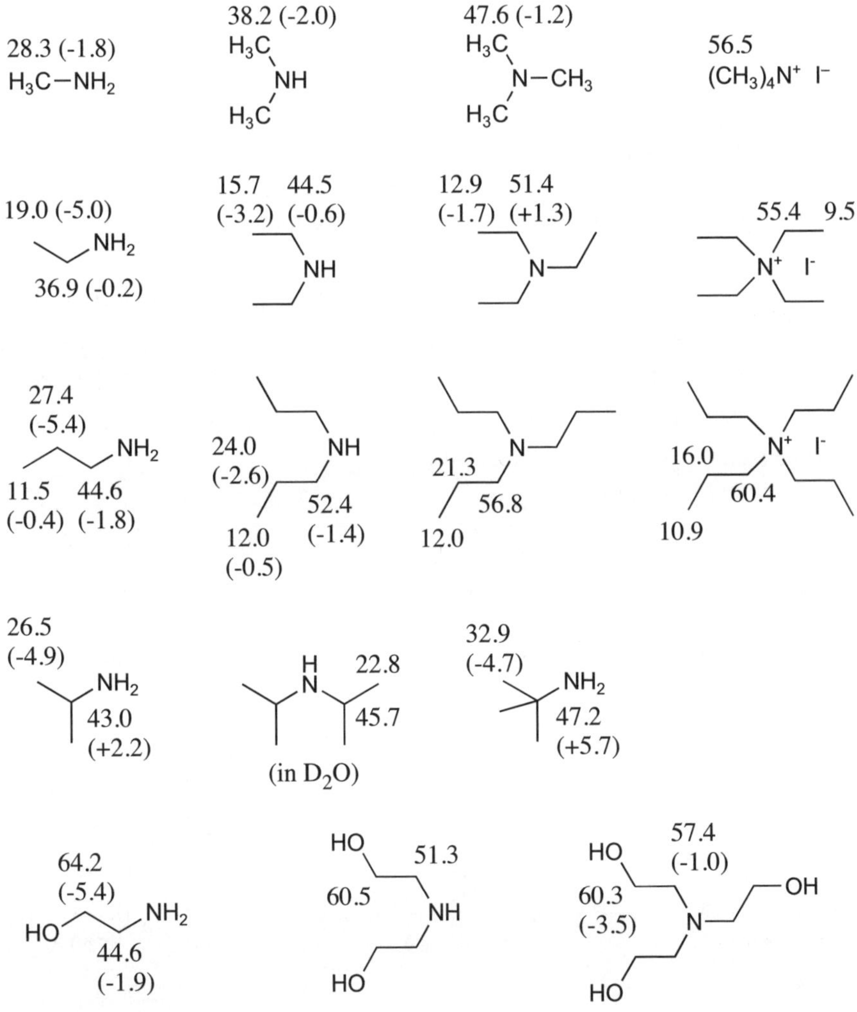

28.3 (-1.8)
H_3C-NH_2

38.2 (-2.0)
H_3C
NH
H_3C

47.6 (-1.2)
H_3C
N—CH_3
H_3C

56.5
$(CH_3)_4N^+\ I^-$

19.0 (-5.0)
NH$_2$
36.9 (-0.2)

15.7 44.5
(-3.2) (-0.6)
NH

12.9 51.4
(-1.7) (+1.3)
N

55.4 9.5
N$^+$ I$^-$

27.4
(-5.4)
NH$_2$
11.5 44.6
(-0.4) (-1.8)

24.0
(-2.6)
NH
52.4
12.0 (-1.4)
(-0.5)

21.3
N
56.8
12.0

16.0
N$^+$ I$^-$
60.4
10.9

N

26.5
(-4.9)
NH$_2$
43.0
(+2.2)

H 22.8
N
45.7
(in D_2O)

32.9
(-4.7)
NH$_2$
47.2
(+5.7)

64.2
(-5.4)
NH$_2$
HO
44.6
(-1.9)

HO 51.3
60.5
NH
HO

57.4
HO (-1.0)
—OH
60.3
(-3.5) N
HO

14.3
(-2.6)

H_3C—CH_2—$\overset{H}{N}$—CH_3

45.9 35.2
(-0.4) (-1.8)

23.2
(-2.9)

—$\overset{H}{N}$—CH_3

12.5 54.0 36.1
(-0.9) (-2.1) (-2.0)

22.5
(-3.1)

—$\overset{H}{N}$—CH_3

50.5   33.9
(+1.9)   (-2.5)

28.2
(-1.2)

—$\overset{H}{N}$—CH_3

50.4   28.5
(+6.6)   (-2.7)

12.8
(-2.1)

$\overset{CH_3}{N}$—CH_3

53.6 44.6
(+0.5) (-1.3)

20.6
(-2.0)

$\overset{CH_3}{N}$—CH_3

11.9 61.8 45.2
(-0.8) (-1.6) (-1.2)

18.7
(-1.3)

$\overset{CH_3}{N}$—CH_3

55.5   40.9
(+3.8)   (-0.8)

25.4
(-0.8)

$\overset{CH_3}{N}$—CH_3

53.6   38.7
(+8.9)   (+0.2)

H_3C—$\overset{H}{N}$—CH_2—CH_2—$\overset{H}{N}$—CH_3

51.2   36.6
(-3.0;   (-1.3;
-2.7*)   -0.5*)

* doubly protonated form

H_3C—$\overset{CH_3}{N}$—CH_2—CH_2—$\overset{CH_3}{N}$—CH_3

57.2   46.1
(-2.6;   (-1.0;
-2.3*)   -0.2*)

* doubly protonated form

44.8
CH_2=CH—CH_2—NH_2

113.6 139.9

NH_2
51.1 (+0.7)
37.6 (-5.4)
25.8 (-1.0)
26.3 (-1.1)

HN—CH_3 33.5 (-1.5)
58.7 (+0.6)
32.7 (-2.7)
25.7 (-0.3)
26.8 (-0.7)

H_3C—N—CH_3 41.1 (-0.7)
64.3 (+2.4)
29.2 (-1.6)
26.5 (-0.9)
26.9 (-1.2)

NH_2
146.7
115.1
129.3
118.5

HN—CH_3 30.2
149.9
112.3
129.3
116.9

H_3C—N—CH_3 39.9
151.0
113.1
129.4
117.0

46.3 —NH_2
143.4
127.1
128.3
126.5

HN—
143.2
117.9
129.4
118.0

N
141.6
121.5
129.4
122.9

^{13}C Chemical Shifts of Cyclic Amines (δ in ppm)

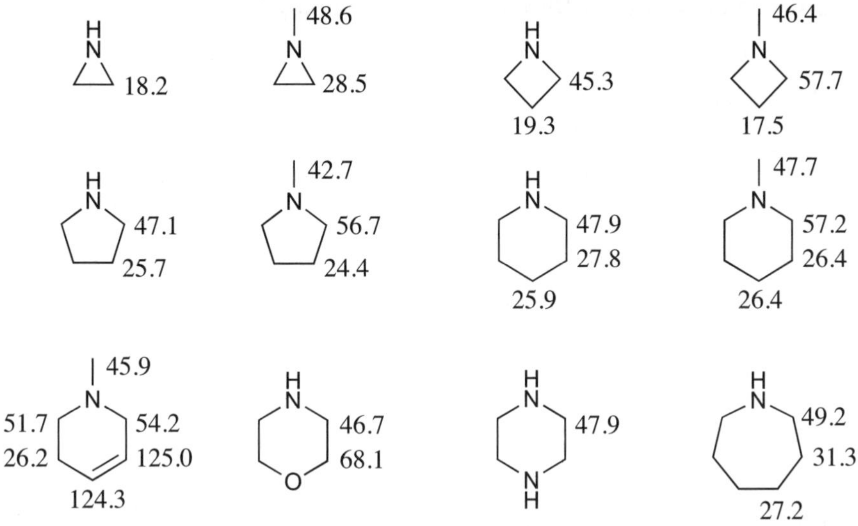

H
N
18.2

| 48.6
N
28.5

H
N
45.3
19.3

| 46.4
N
57.7
17.5

H
N
47.1
25.7

| 42.7
N
56.7
24.4

H
N
47.9
27.8
25.9

| 47.7
N
57.2
26.4
26.4

| 45.9
N
51.7   54.2
26.2   125.0
124.3

H
N
46.7
68.1
O

H
N
47.9
N
H

H
N
49.2
31.3
27.2

4.9.2 Nitro and Nitroso Compounds

^{13}C Chemical Shifts of Nitro and Nitroso Compounds (δ in ppm)

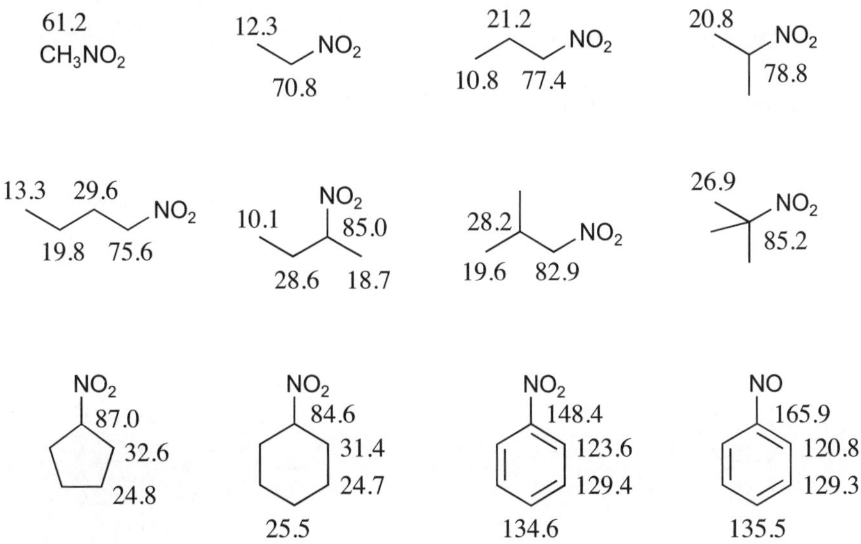

N

61.2
CH_3NO_2

12.3
NO_2
70.8

21.2
NO_2
10.8 77.4

20.8
NO_2
78.8

13.3 29.6
NO_2
19.8 75.6

NO_2
10.1   85.0
28.6 18.7

28.2
NO_2
19.6 82.9

26.9
NO_2
85.2

NO_2
87.0
32.6
24.8

NO_2
84.6
31.4
24.7
25.5

NO_2
148.4
123.6
129.4
134.6

NO
165.9
120.8
129.3
135.5

4.9.3 Nitrosamines and Nitramines

^{13}C Chemical Shifts of Nitrosamines (δ in ppm)

^{13}C Chemical Shifts of Nitramines (δ in ppm)

4.9.4 Azo and Azoxy Compounds

^{13}C Chemical Shifts of Azo and Azoxy Compounds (δ in ppm)

4.9.5 Imines and Oximes

^{13}C Chemical Shifts of Imines and Oximes (δ in ppm)

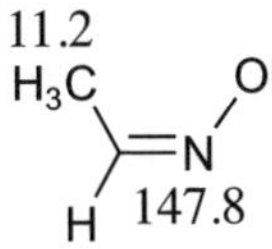

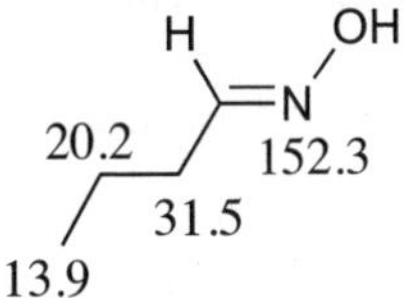

11.2
H$_3$C   OH
=N
H 147.8

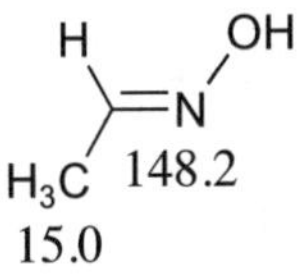

H   OH
=N
H$_3$C 148.2
15.0

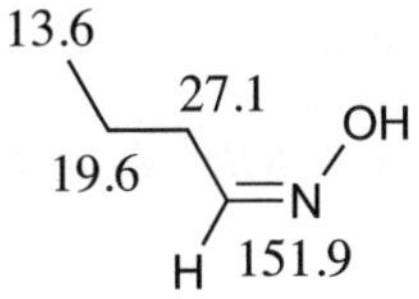

13.6
27.1
19.6   OH
=N
H 151.9

H   OH
=N
20.2   152.3
31.5
13.9

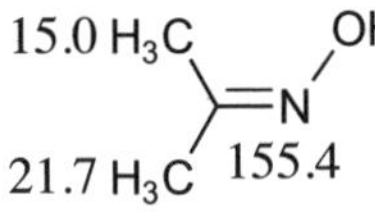

15.0 H$_3$C   OH
=N
21.7 H$_3$C 155.4

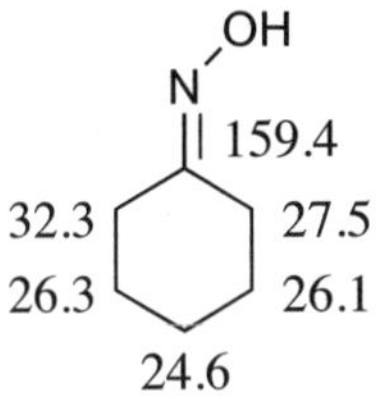

OH
N
159.4
32.3   27.5
26.3   26.1
24.6

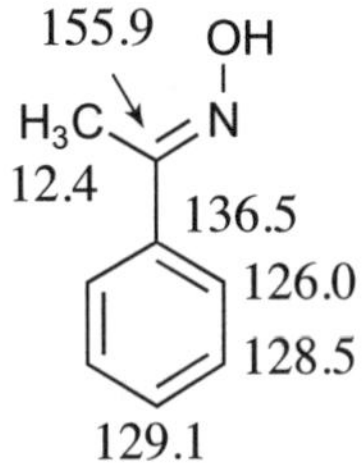

155.9 OH
H$_3$C   N
12.4   136.5
126.0
128.5
129.1

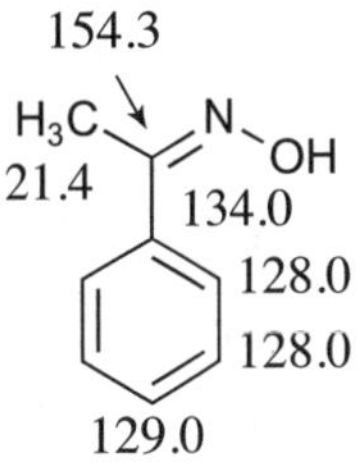

154.3
H$_3$C   N
21.4   OH
134.0
128.0
128.0
129.0

4.9.6 Hydrazones and Carbodiimides

^{13}C Chemical Shifts of Hydrazones and Carbodiimides (δ in ppm)

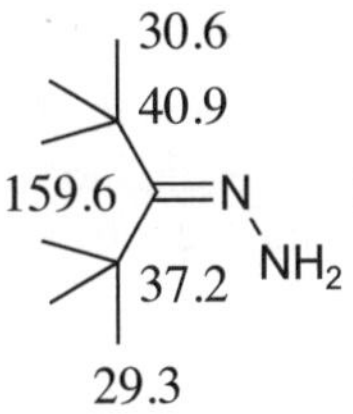

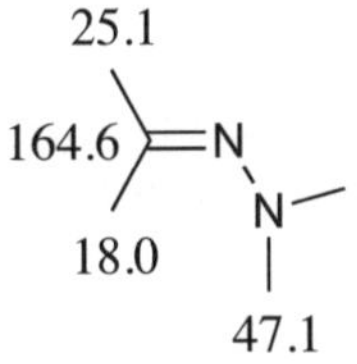

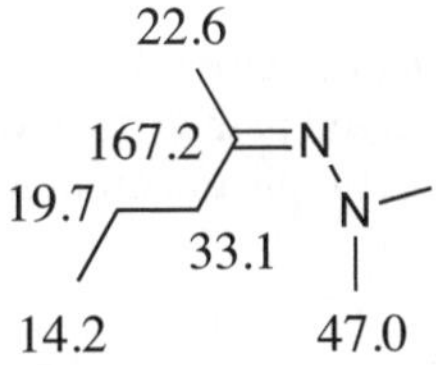

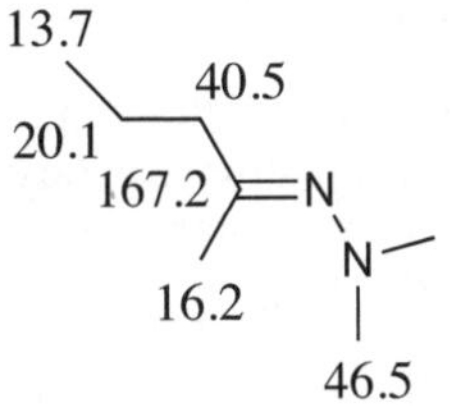

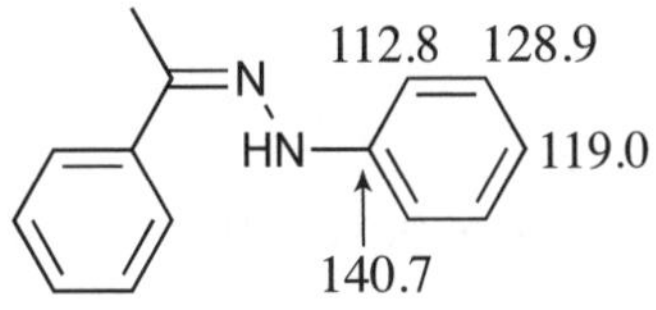

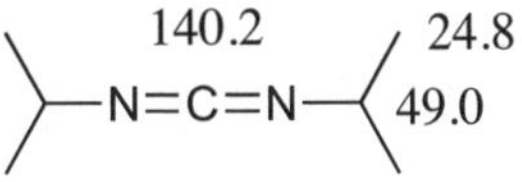

30.6
40.9
159.6 =N
37.2 NH$_2$
29.3

25.1
164.6 =N
N
18.0
47.1

22.6
167.2 =N
19.7   N
14.2   33.1
47.0

13.7
40.5
20.1
167.2 =N
N
16.2
46.5

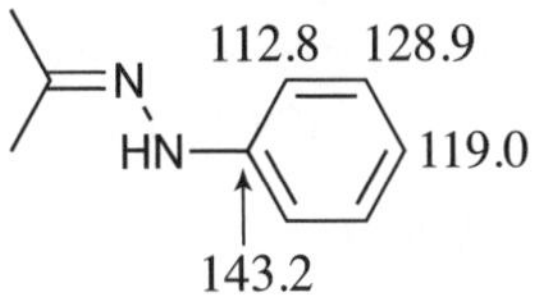

=N
HN   112.8 128.9
119.0
143.2

=N
HN   112.8 128.9
119.0
140.7

140.2   24.8
N=C=N   49.0

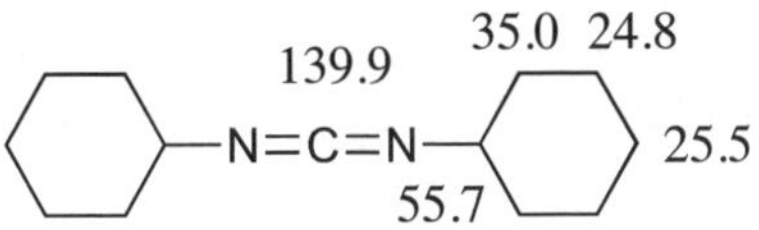

35.0 24.8
139.9
N=C=N   25.5
55.7

N

4.9.7 Nitriles and Isonitriles

^{13}C Chemical Shifts of Nitriles (δ in ppm)

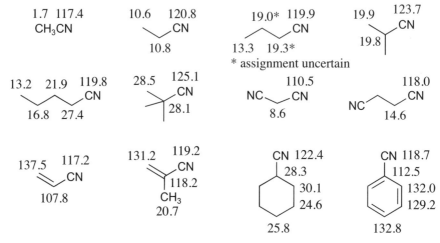

1.7 117.4
CH$_3$CN

10.6   120.8
　　　CN
　　10.8

19.0* 119.9
　　　　CN
13.3   19.3*
* assignment uncertain

19.9   123.7
　　　　CN
19.8

13.2   21.9 119.8
　　　　　　CN
16.8   27.4

28.5   125.1
　　　　CN
　　　28.1

110.5
NC   CN
　　8.6

118.0
NC　　　CN
　　　14.6

137.5   117.2
　　　　CN
　107.8

131.2   119.2
　　　　CN
　　　　118.2
　　　　CH$_3$
　　　20.7

CN 122.4
　28.3
　　30.1
　　24.6
25.8

CN 118.7
　112.5
　　132.0
　　129.2
132.8

^{13}C Chemical Shifts and ^{13}C–^{14}N Couplings of Isonitriles (δ in ppm, $|J|$ in Hz)

Because of the symmetrical electron distribution around the nitrogen atom, the ^{13}C-^{14}N coupling can be observed in the ^{13}C NMR spectra of isonitriles, leading to triplets with intensities of 1:1:1 (spin quantum number of ^{14}N: I = 1, natural abundance, 99.6%).

^{2}J 7.5 ^{1}J 5.8
26.8   158.2
CH$_3$NC

^{3}J $\approx$ 0 ^{1}J 5.3
15.3   156.8
　　　　　NC
　36.4
　^{2}J 6.5

^{3}J $\approx$ 0 ^{1}J 5.0
120.6   165.7
　　　　　NC
　119.4
　^{2}J 11.7

NC 165.7; ^{1}J 5.2
　126.7; ^{1}J 13.2
　126.3; ^{2}J $\approx$ 0
　129.9; ^{3}J $\approx$ 0
129.4; ^{4}J $\approx$ 0

4.9.8 Isocyanates, Thiocyanates, and Isothiocyanates

^{13}C Chemical Shifts (δ in ppm)

26.3   121.5
CH$_3$NCO

13.6   34.2
　　　　NCO
20.4   43.3

125 (broad)
　　NCO

110.7   124.2
　　　NCO
　124.7

NCO 124.9
　133.6
　　124.8
　　129.6
125.8

15.4   111.8
　　　SCN
　28.7

130.4
SCN$^-$ NH$_4^+$

29.3 128.7
CH$_3$NCS

13.3   32.3
　　　　NCS
20.0   45.0

131 (broad)
　NCS

N

4.10 Sulfur Compounds

4.10.1 Thiols

^{13}C Chemical Shifts (δ in ppm)

6.5
CH₃SH

19.7
SH
19.1

27.6
SH
12.6   26.4

27.4
SH
29.9

12.0   35.7
SH
21.0   23.7

35.0
SH
41.1

22.2   33.9
SH
14.0   30.6   24.6

31.8 SH
28.1   38.8

14.0   31.4   34.1
SH
22.6   28.1   24.7

28.7
HS   SH

64.2
HO   SH
27.3

SH
38.3
37.7
24.6

SH
38.5
38.5
26.8
25.9

28.8   SH
141.0
127.9
128.5
126.8

SH
130.6
129.2
128.8
125.3

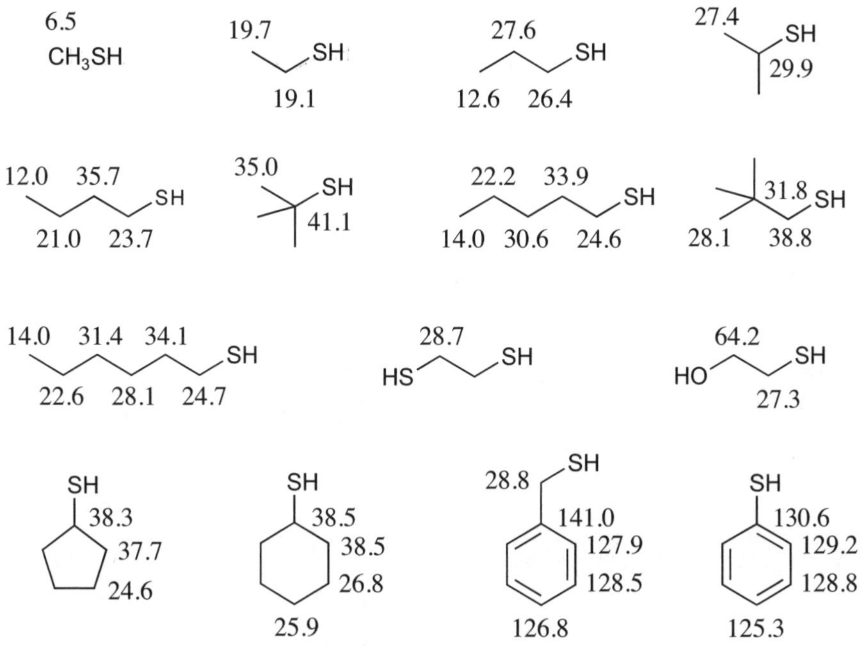

4.10.2 Sulfides

^{13}C Chemical Shifts (δ in ppm)

S

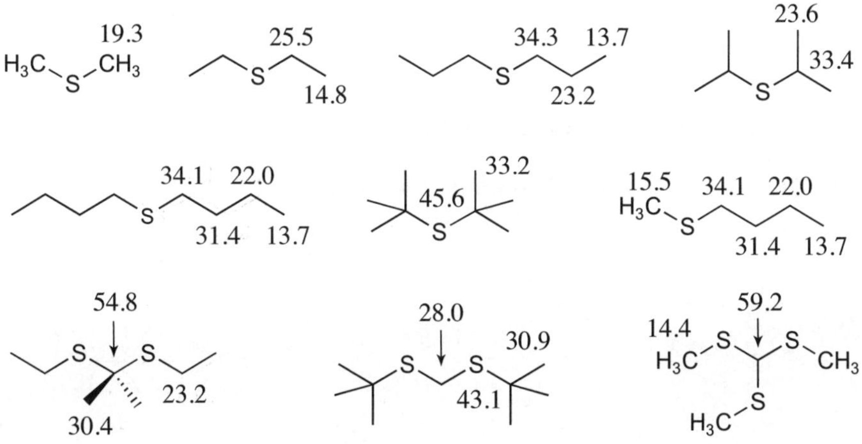

19.3
H₃C   S   CH₃

25.5
S
14.8

34.3   13.7
S
23.2

23.6
33.4
S

34.1   22.0
S
31.4   13.7

33.2
45.6
S

15.5   34.1   22.0
H₃C   S
31.4   13.7

54.8
S   S
23.2
30.4

28.0
S   S
30.9
43.1

59.2
14.4   S   S
H₃C   CH₃
S
H₃C

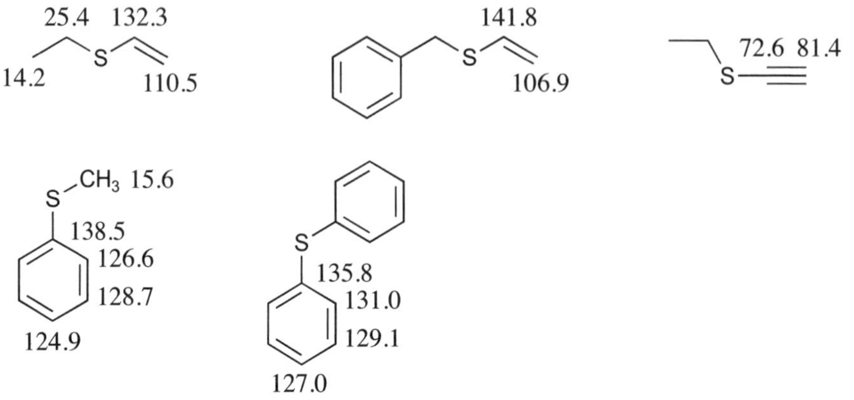

25.4 132.3

14.2   110.5

141.8

106.9

72.6 81.4

—CH₃ 15.6

138.5
126.6
128.7

124.9

135.8
131.0
129.1

127.0

¹³C Chemical Shifts of Cyclic Sulfides (δ in ppm)

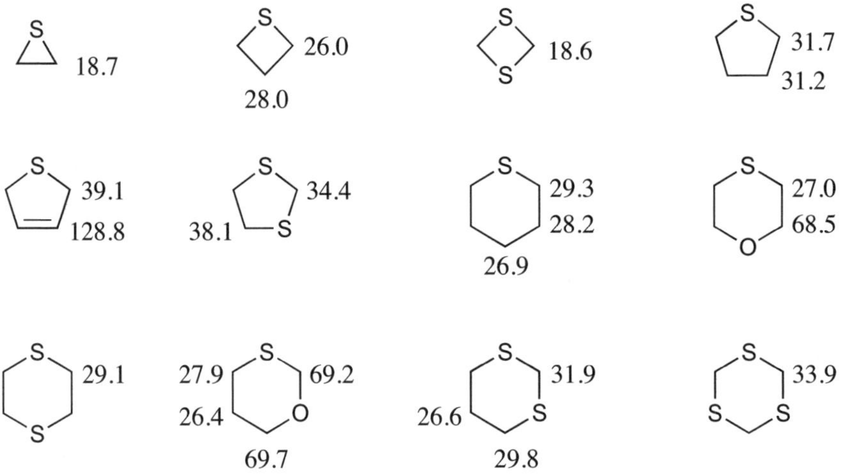

18.7

26.0

28.0

18.6

31.7
31.2

39.1
128.8

34.4

38.1

29.3
28.2

26.9

27.0
68.5

29.1

27.9 69.2
26.4

69.7

31.9

26.6

29.8

33.9

4.10.3 Disulfides and Sulfonium Salts

¹³C Chemical Shifts of Disulfides (δ in ppm)

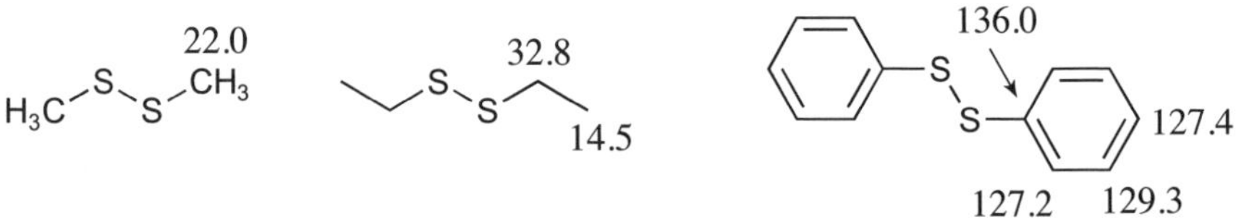

22.0
CH₃

H₃C

32.8

14.5

136.0

127.4

127.2   129.3

S

^{13}C Chemical Shifts of Sulfonium Salts (δ in ppm)

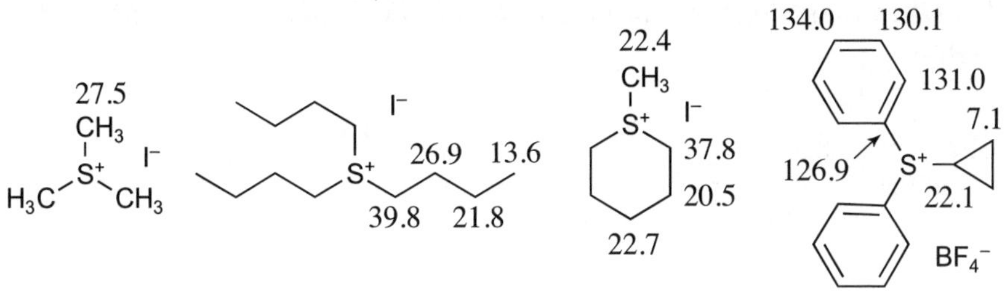

27.5
CH₃
S⁺ I⁻
H₃C   CH₃

26.9 13.6
S⁺ I⁻
39.8 21.8

22.4
CH₃
S⁺ I⁻
37.8
20.5
22.7

134.0 130.1
131.0
7.1
126.9 S⁺
22.1
BF₄⁻

4.10.4 Sulfoxides and Sulfones

^{13}C Chemical Shifts of Sulfoxides and Sulfones (δ in ppm)

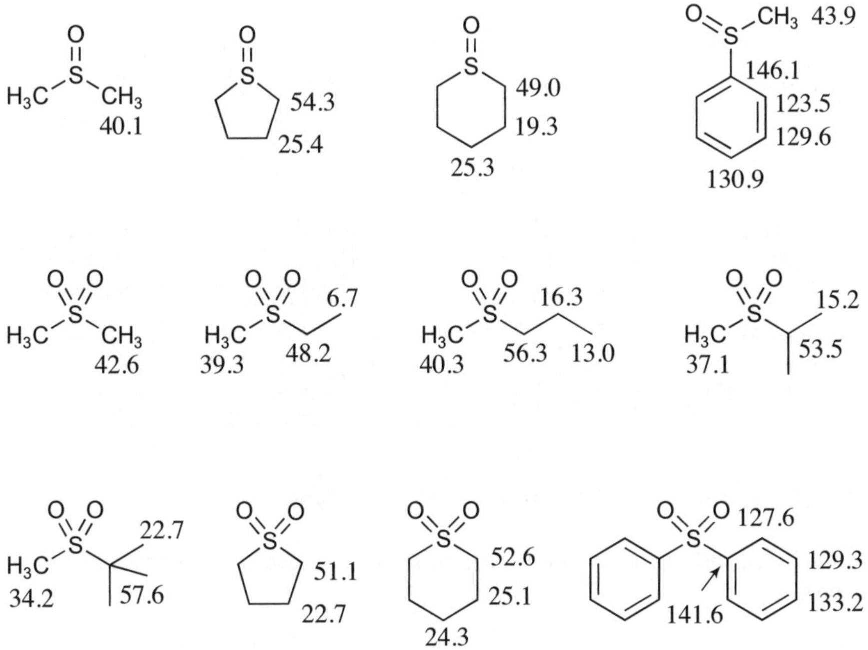

O
‖
H₃C   S   CH₃
40.1

O
‖
S
54.3
25.4

O
‖
S
49.0
19.3
25.3

O=S   CH₃ 43.9
146.1
123.5
129.6
130.9

O O
\‖/
H₃C   S   CH₃
42.6

O O
\‖/
H₃C   S   6.7
39.3   48.2

O O
\‖/
H₃C   S   16.3
40.3   56.3 13.0

O O
\‖/
H₃C   S   15.2
37.1   53.5

S

O O
\‖/   22.7
H₃C   S
34.2   57.6

O O
\‖/
S
51.1
22.7

O O
\‖/
S   52.6
25.1
24.3

O O 127.6
\‖/
S   129.3
141.6   133.2

4.10.5 Sulfonic and Sulfinic Acids and Derivatives

^{13}C Chemical Shifts of Sulfonic and Sulfinic Acids and Derivatives (δ in ppm)

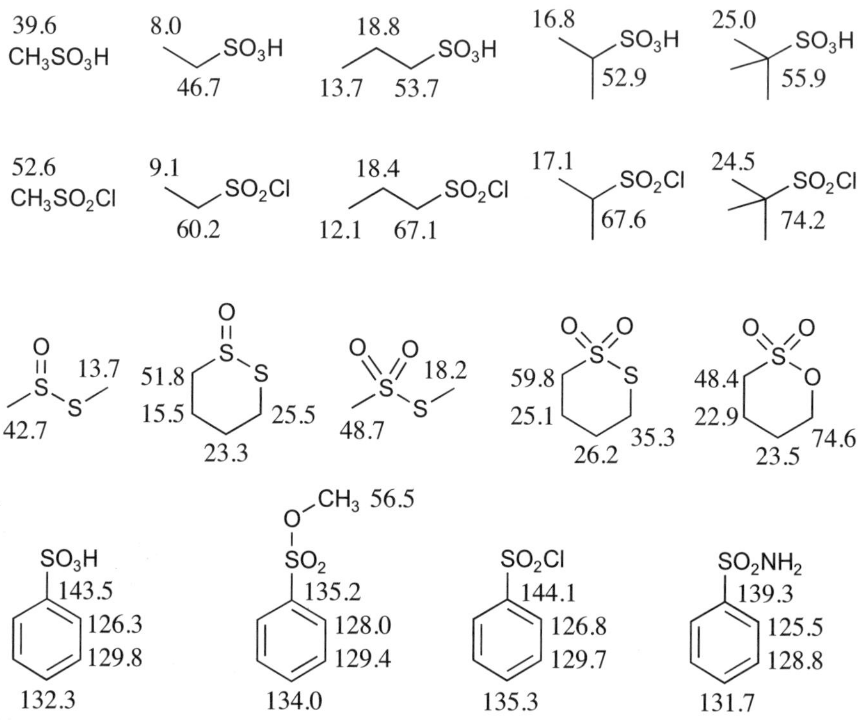

39.6
CH$_3$SO$_3$H

8.0
SO$_3$H
46.7

18.8
SO$_3$H
13.7   53.7

16.8
SO$_3$H
52.9

25.0
SO$_3$H
55.9

52.6
CH$_3$SO$_2$Cl

9.1
SO$_2$Cl
60.2

18.4
SO$_2$Cl
12.1   67.1

17.1
SO$_2$Cl
67.6

24.5
SO$_2$Cl
74.2

13.7   51.8
15.5
42.7

23.3   25.5

18.2
48.7

59.8
25.1
26.2   35.3

48.4
22.9
23.5   74.6

SO$_3$H
143.5
126.3
129.8
132.3

O—CH$_3$ 56.5
SO$_2$
135.2
128.0
129.4
134.0

SO$_2$Cl
144.1
126.8
129.7
135.3

SO$_2$NH$_2$
139.3
125.5
128.8
131.7

4.10.6 Sulfurous and Sulfuric Acid Derivatives

^{13}C Chemical Shifts of Sulfurous and Sulfuric Acid Derivatives (δ in ppm)

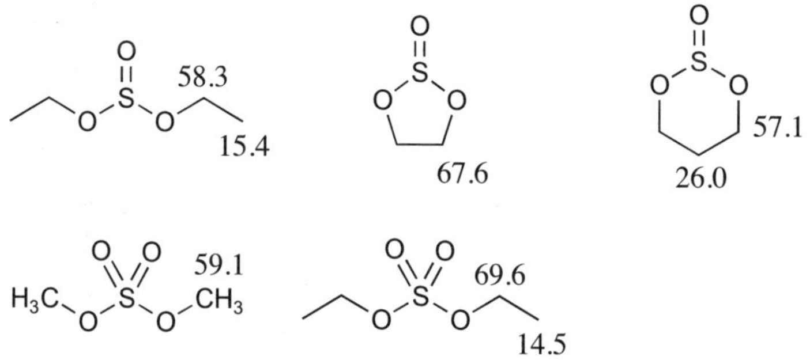

58.3
15.4

67.6

57.1
26.0

59.1
H$_3$C—O—S—O—CH$_3$

69.6
14.5

4.10.7 Sulfur-Containing Carbonyl Derivatives

^{13}C Chemical Shifts (δ in ppm)

The ^{13}C chemical shifts of thiocarbonyl groups are higher by about 30 ppm than those of the corresponding carbonyl groups:

$$\delta_{C=S} \approx 1.5 \times \delta_{C=O} - 57.5$$

Carbonyl groups of thiocarboxylic acids and their esters are deshielded by about 20 ppm with respect to the corresponding oxygen compounds.

4.11 Carbonyl Compounds

4.11.1 Aldehydes

Additivity Rule for Estimating the ^{13}C Chemical Shifts of Aldehyde Carbon Atoms (δ in ppm)

$$\delta_{C=O} = 193.0 + \Sigma\ Z_i$$

$$-C_\beta-C_\alpha-CHO$$

Substituent i	Z_α	Z_β
$-C\!\!<$	6.5	2.6
$-CH=CH_2$	-0.8	0.0
$-CH=CH-CH_3$	0.2	0.0
$-$phenyl	-1.2	0.0

^{13}C Chemical Shifts of Aldehydes (δ in ppm)

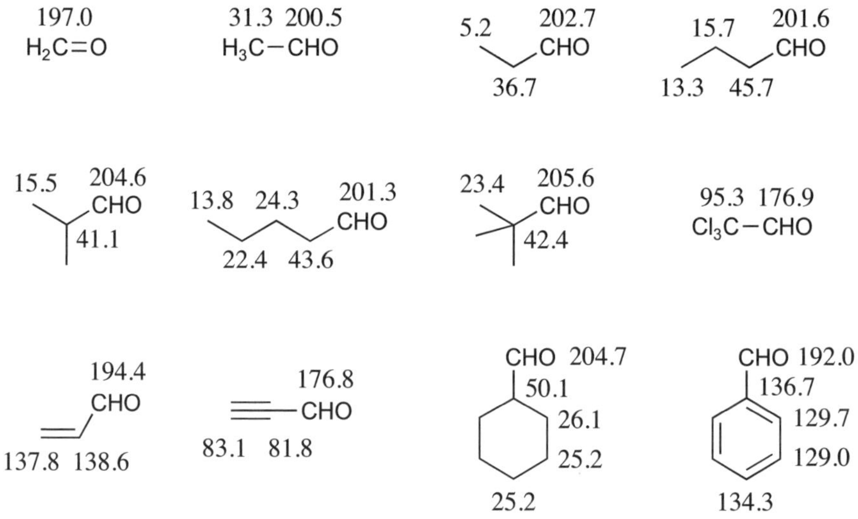

C = X

197.0
$H_2C=O$

31.3 200.5
H_3C-CHO

5.2   202.7
      CHO
36.7

15.7   201.6
       CHO
13.3 45.7

15.5 204.6
      CHO
      41.1

13.8 24.3   201.3
             CHO
22.4 43.6

23.4 205.6
      CHO
      42.4

95.3 176.9
Cl_3C-CHO

194.4
CHO
137.8 138.6

176.8
═══CHO
83.1 81.8

CHO 204.7
     50.1
     26.1
     25.2
25.2

CHO 192.0
    136.7
    129.7
    129.0
134.3

4.11.2 Ketones

Additivity Rule for Estimating the ^{13}C Chemical Shifts of Ketone Carbon Atoms (δ in ppm)

$$\delta_{C=O} = 193.0 + \Sigma\, Z_i$$

$$-C_\beta-C_\alpha-\overset{\overset{\displaystyle O}{\|}}{C}-C_\alpha-C_\beta-$$

Substituent i	Z_α	Z_β
$-C\hspace{-0.3em}\leqslant$	6.5	2.6
$-CH=CH_2$	-0.8	0.0
$-CH=CH-CH_3$	0.2	0.0
$-phenyl$	-1.2	0.0

^{13}C Chemical Shifts of Ketones (δ in ppm)

C = X

^{13}C Chemical Shifts of Halogenated Aliphatic Ketones (δ in ppm)

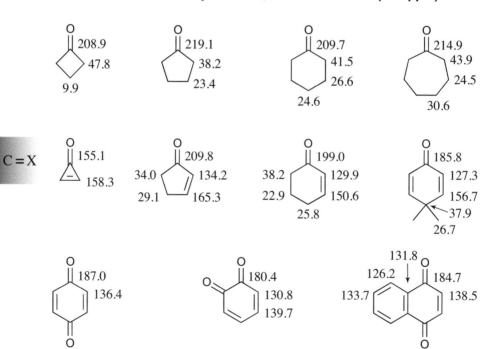

^{13}C Chemical Shifts of Diketones (δ in ppm)

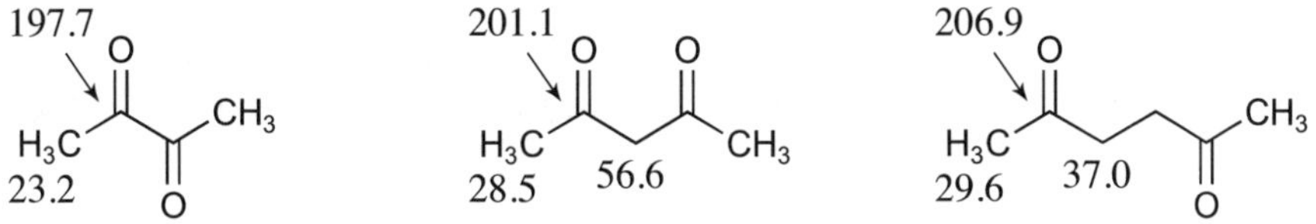

197.7

H$_3$C — 23.2 ... CH$_3$ (with O and O)

201.1

H$_3$C 28.5 ... 56.6 ... CH$_3$ (with O and O)

206.9

H$_3$C 29.6 ... 37.0 ... CH$_3$ (with O and O)

Enol form: see Chapter 4.8.1

4.11.3 Carboxylic Acids

Additivity Rule for Estimating the ^{13}C Chemical Shifts of Carboxyl Carbon Atoms (δ in ppm)

$$\delta_{C=O} = 166.0 + \Sigma\ Z_i$$

$$-C_\gamma-C_\beta-C_\alpha-COOH$$

Substituent i	Z_α	Z_β	Z_γ
$-C\lessgtr$	12.0	3.0	-1.0
$-CH=CH_2$	5.0	0.5	-1.5
-phenyl	6.0	1.0	-2.0

^{13}C Chemical Shifts of Carboxylic Acids (δ in ppm)

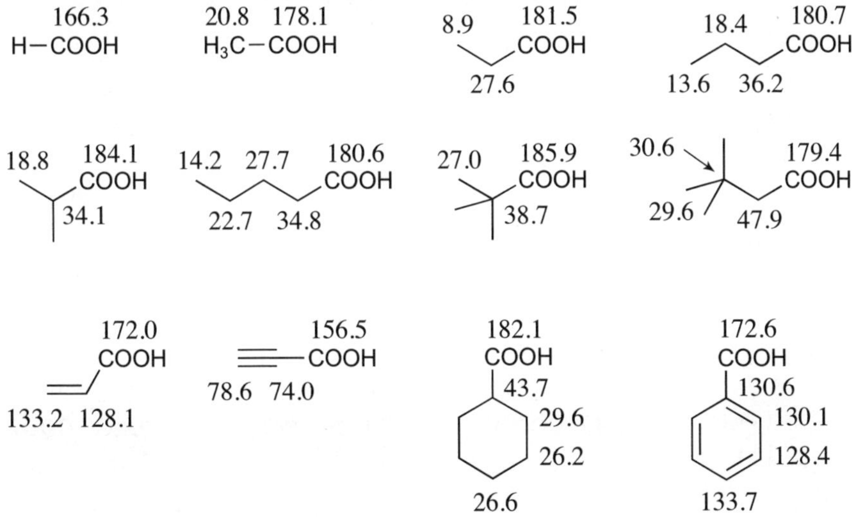

166.3
H—COOH

20.8 178.1
H$_3$C—COOH

8.9 181.5
COOH
27.6

18.4 180.7
COOH
13.6 36.2

C = X

18.8 184.1
COOH
34.1

14.2 27.7 180.6
COOH
22.7 34.8

27.0 185.9
COOH
38.7

30.6 179.4
COOH
29.6 47.9

172.0
COOH
133.2 128.1

156.5
≡—COOH
78.6 74.0

182.1
COOH
43.7
29.6
26.2
26.6

172.6
COOH
130.6
130.1
128.4
133.7

^{13}C Chemical Shifts of Halogenated Carboxylic Acids (δ in ppm)

115.0 163.0	40.7 173.7	63.7 170.4	88.9 167.1
F₃C—COOH	ClH₂C—COOH	Cl₂HC—COOH	Cl₃C—COOH

^{13}C Chemical Shifts of Dicarboxylic Acids (δ in ppm)

160.1
COOH
|
COOH

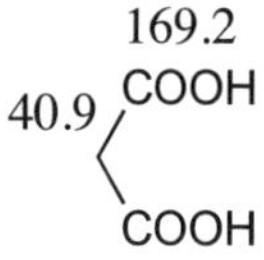

169.2
40.9 COOH

COOH

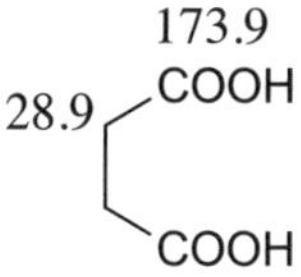

173.9
28.9 COOH

COOH

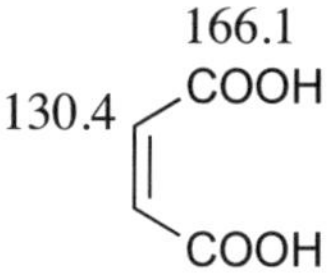

166.1
130.4 COOH

COOH

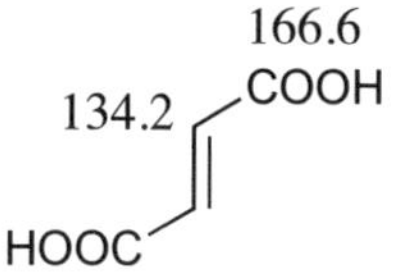

166.6
134.2 COOH

HOOC

^{13}C Chemical Shifts of Carboxylate Anions (δ in ppm)

Measured in water unless indicated otherwise.

171.3
H—COO⁻

24.4   182.6
20.8* 177.6*
H₃C—COO⁻

* in CDCl₃

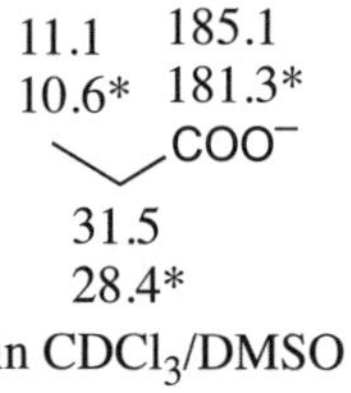

11.1   185.1
10.6* 181.3*
COO⁻

31.5
28.4*
* in CDCl₃/DMSO

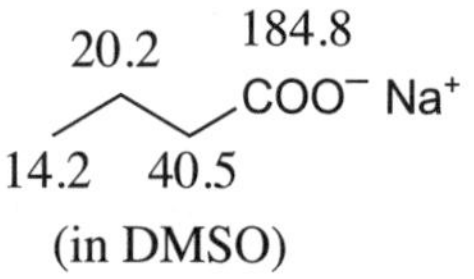

20.2   184.8
COO⁻ Na⁺

14.2   40.5
(in DMSO)

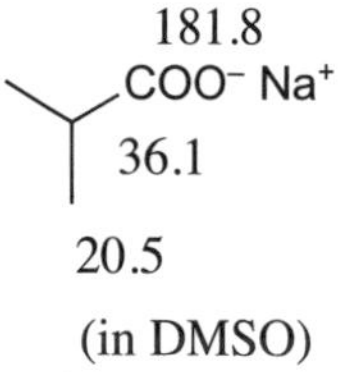

C = X

181.8
COO⁻ Na⁺
36.1

20.5
(in DMSO)

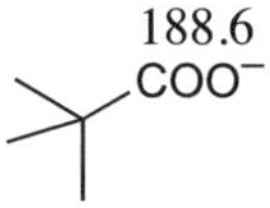

188.6
COO⁻

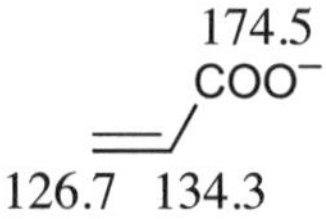

174.5
COO⁻

126.7 134.3

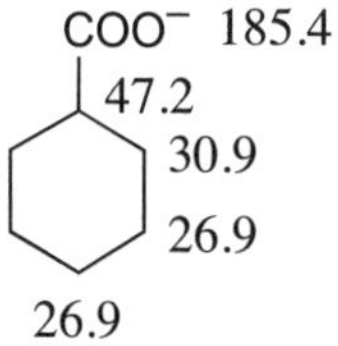

COO⁻ 185.4
47.2
30.9
26.9

26.9

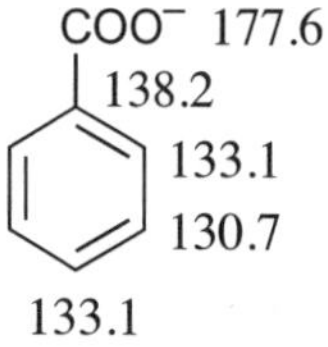

COO⁻ 177.6
138.2
133.1
130.7

133.1

45.0   175.9
ClH₂C—COO⁻

65.6   171.8
Cl₂HC—COO⁻

96.2   167.6
Cl₃C—COO⁻

4.11.4 Esters and Lactones

Additivity Rule for Estimating the ^{13}C Chemical Shifts of Ester Carbon Atoms (δ in ppm)

$$\delta_{C=O} = 166.0 + \Sigma\, Z_i$$

$$-C_\gamma-C_\beta-C_\alpha-COO-C_{\alpha'}-$$

Substituent i	Z_α	Z_β	Z_γ	$Z_{\alpha'}$
$-C\leqslant$	12.0	3.0	-1.0	-5.0
$-CH=CH_2$	5.0			-9.0
$-phenyl$	6.0	1.0		-8.0

^{13}C Chemical Shifts of Acetic Acid Esters (δ in ppm)

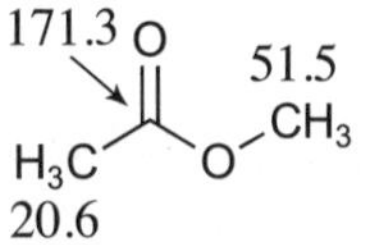

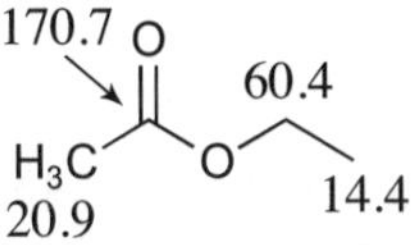

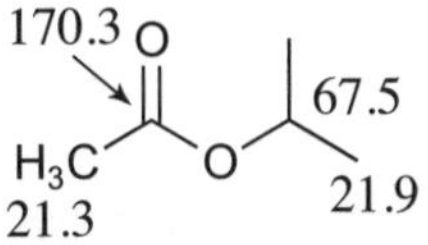

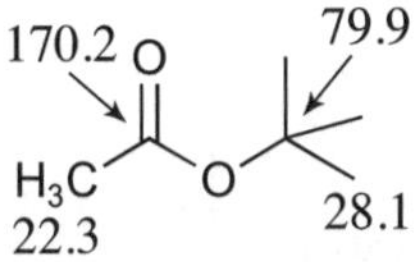

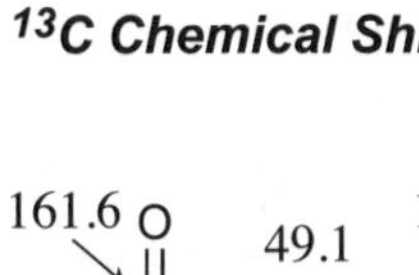

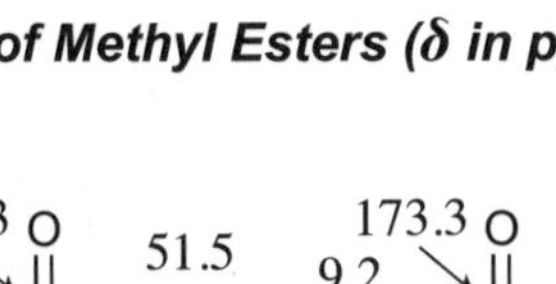

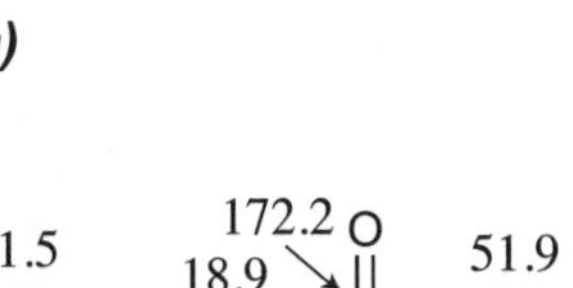

171.3 O 51.5 CH₃ H₃C 20.6

170.7 O 60.4 H₃C 20.9 14.4

170.3 O 67.5 H₃C 21.3 21.9

170.2 O 79.9 H₃C 22.3 28.1

168.0 O 141.4 H₃C 20.6 97.5

169.2 O 72.3 26.1 24.4 H₃C 21.0 32.2

169.2 O 150.9 125.3 128.9 H₃C 20.8 121.4

^{13}C Chemical Shifts of Methyl Esters (δ in ppm)

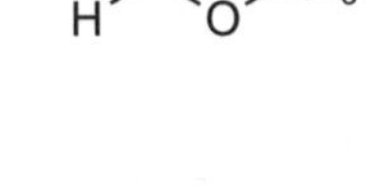

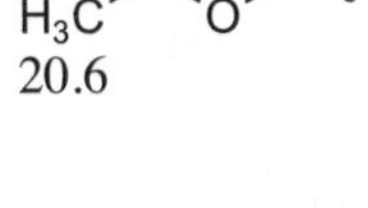

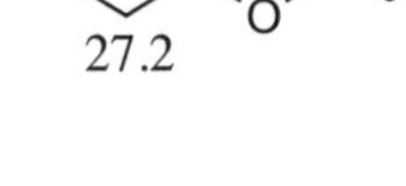

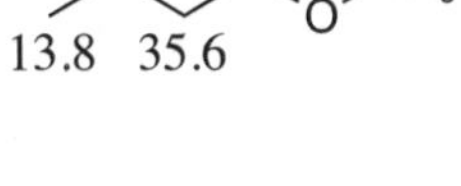

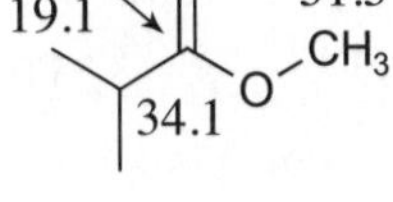

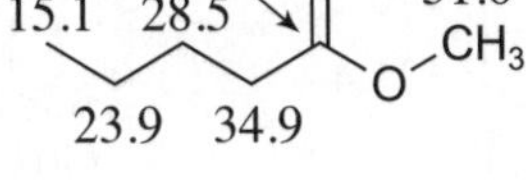

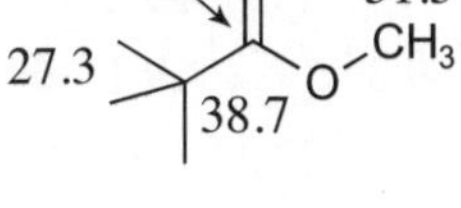

161.6 O 49.1 CH₃ H O

171.3 O 51.5 CH₃ H₃C 20.6

173.3 O 51.5 9.2 CH₃ 27.2

172.2 O 51.9 18.9 CH₃ 13.8 35.6

177.4 O 51.5 19.1 CH₃ 34.1

173.4 O 51.6 15.1 28.5 CH₃ 23.9 34.9

178.8 O 51.5 27.3 CH₃ 38.7

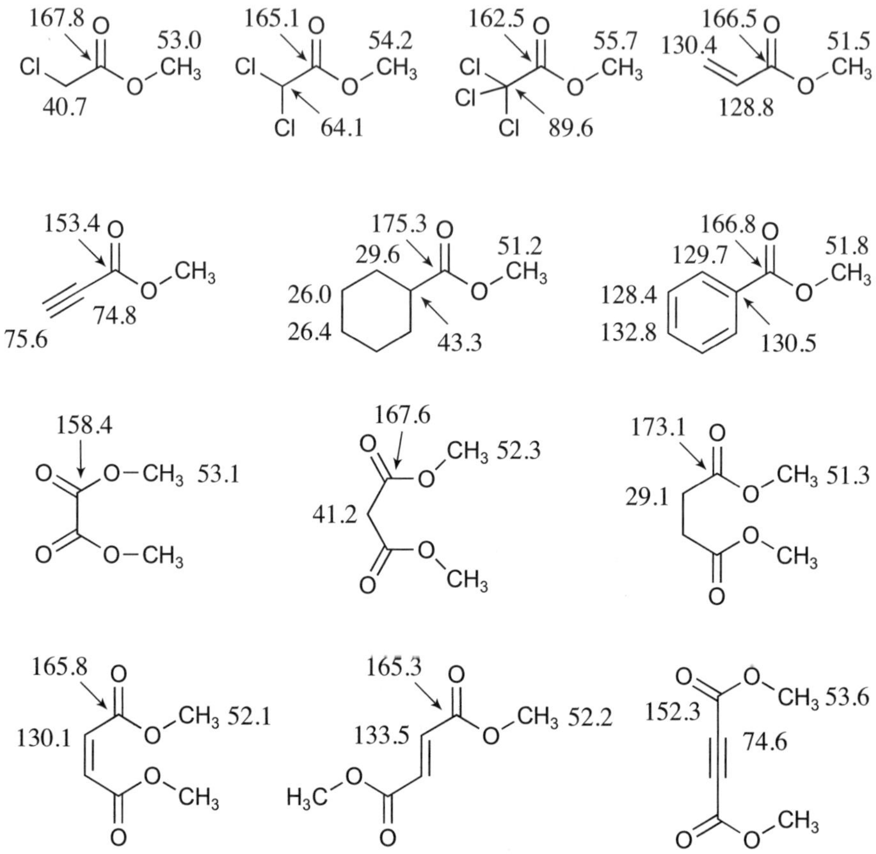

167.8 O   53.0
Cl   O–CH$_3$
40.7

165.1 O   54.2
Cl   O–CH$_3$
Cl 64.1

162.5 O   55.7
Cl   O–CH$_3$
Cl
Cl 89.6

166.5 O 130.4   51.5
    O–CH$_3$
128.8

153.4 O
    O–CH$_3$
74.8
75.6

175.3 O 29.6   51.2
    O–CH$_3$
26.0
26.4   43.3

166.8 O 129.7   51.8
    O–CH$_3$
128.4
132.8   130.5

158.4
O   O–CH$_3$ 53.1
O   O–CH$_3$

167.6
O   CH$_3$ 52.3
    O
41.2
O   O   CH$_3$

173.1 O
    O–CH$_3$ 51.3
29.1
O   O–CH$_3$

165.8 O
130.1   O–CH$_3$ 52.1
    O–CH$_3$
O

165.3 O
133.5   O–CH$_3$ 52.2
H$_3$C–O   O
O

O   O–CH$_3$ 53.6
152.3   74.6
O   O–CH$_3$

^{13}C Chemical Shifts of Lactones (δ in ppm)

C = X

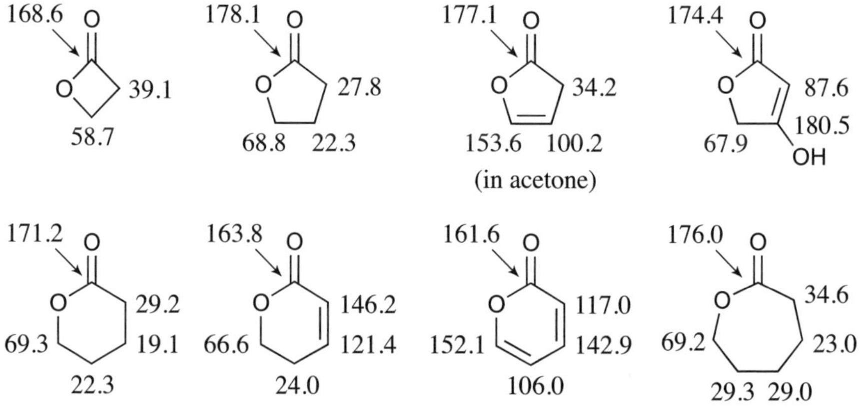

168.6 O
O   39.1
58.7

178.1 O
O   27.8
68.8   22.3

177.1 O
O   34.2
153.6 100.2
(in acetone)

174.4 O
O   87.6
67.9   180.5
    OH

171.2 O
O   29.2
69.3   19.1
22.3

163.8 O
O   146.2
66.6   121.4
24.0

161.6 O
O   117.0
152.1   142.9
106.0

176.0 O
O   34.6
69.2   23.0
29.3 29.0

4.11.5 Amides and Lactams

Additivity Rule for Estimating the ^{13}C Chemical Shifts of Amide Carbon Atoms (δ in ppm)

$$\delta_{C=O} = 166.0 + \Sigma\, Z_i$$

$$-C_\gamma-C_\beta-C_\alpha-CO-N \begin{cases} C_{\alpha'}-C_{\beta'} \\ C_{\alpha'}-C_{\beta'} \end{cases}$$

Substituent i	Z_α	Z_β	Z_γ	$Z_{\alpha'}$	$Z_{\beta'}$
$-C\leqslant$	7.7	4.5	-0.7	-1.5	-0.3
$-CH=CH_2$	3.3				
$-$phenyl	4.7			-4.5	

^{13}C Chemical Shifts of Amides (δ in ppm)

Formamides:

Primary and Secondary Acetamides:

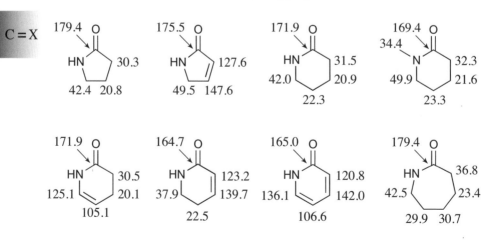

Tertiary Amides:

^{13}C Chemical Shifts of Lactams (δ in ppm)

C = X

4.11.6 Miscellaneous Carbonyl Derivatives

^{13}C Chemical Shifts of Carboxylic Acid Halides (δ in ppm)

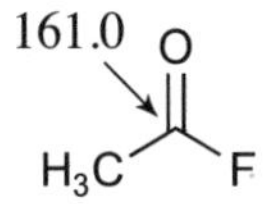

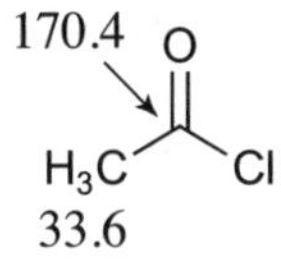

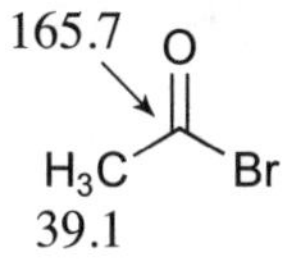

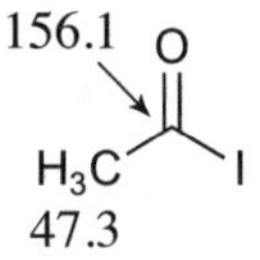

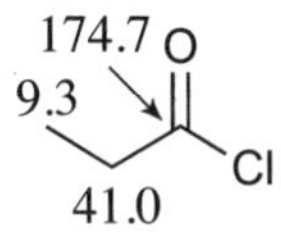

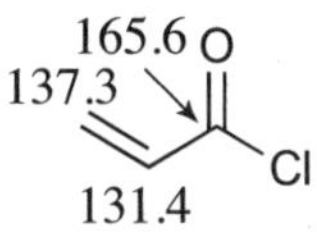

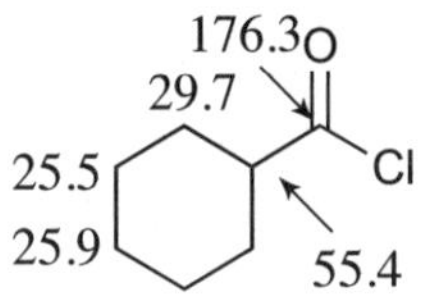

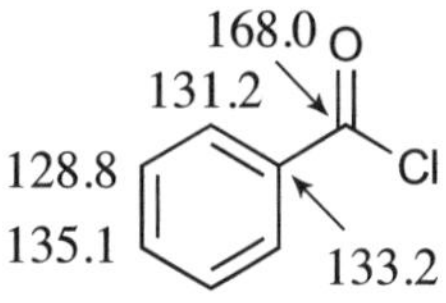

161.0
H$_3$C—C(=O)—F

170.4
H$_3$C—C(=O)—Cl
33.6

165.7
H$_3$C—C(=O)—Br
39.1

156.1
H$_3$C—C(=O)—I
47.3

174.7
9.3
—CH$_2$—C(=O)—Cl
41.0

165.6
137.3
CH$_2$=CH—C(=O)—Cl
131.4

176.3
29.7
cyclohexyl—C(=O)—Cl
25.5
25.9
55.4

168.0
131.2
C$_6$H$_4$—C(=O)—Cl
128.8
135.1
133.2

^{13}C Chemical Shifts of Carboxylic Acid Anhydrides (δ in ppm)

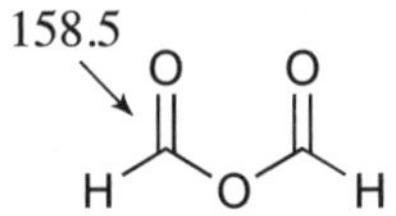

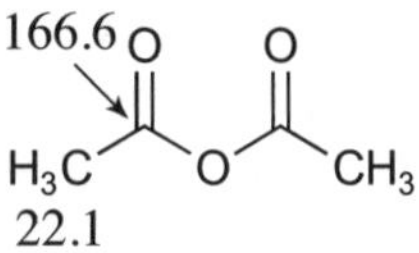

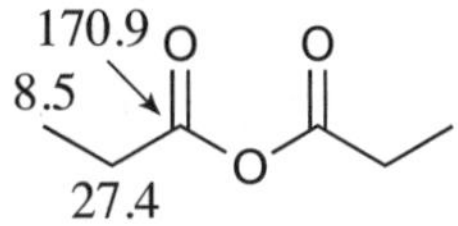

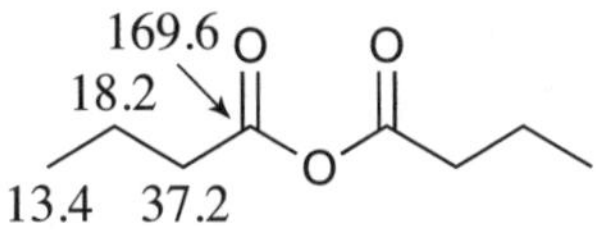

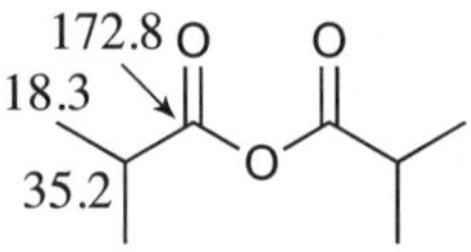

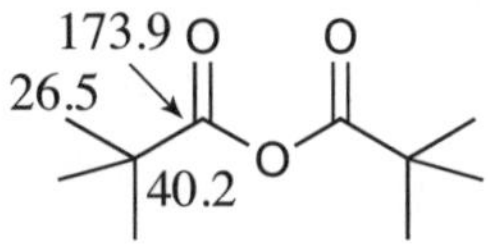

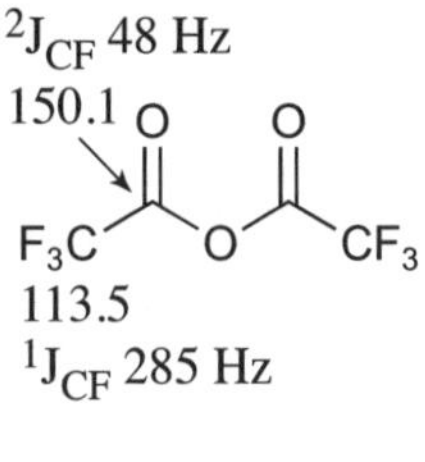

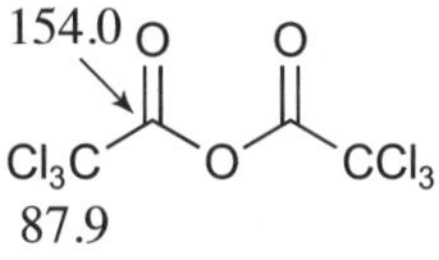

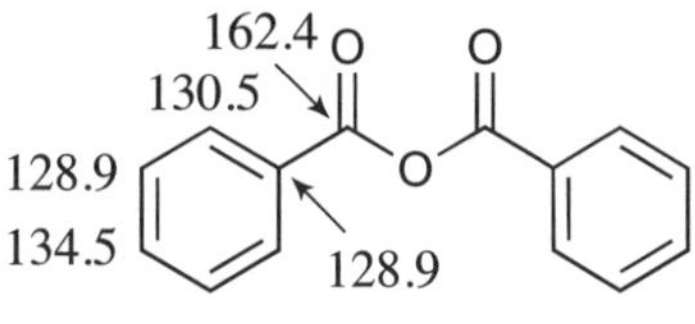

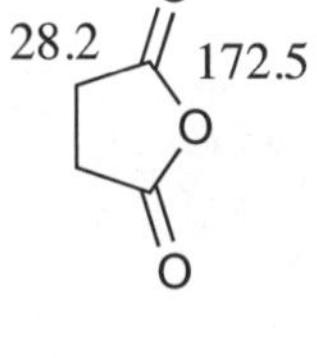

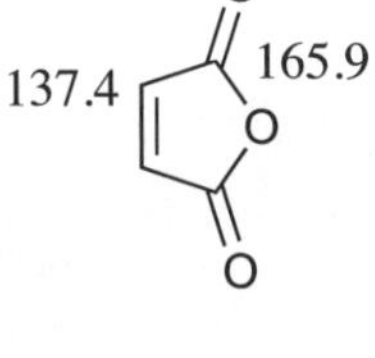

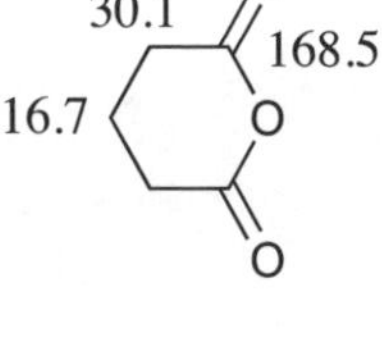

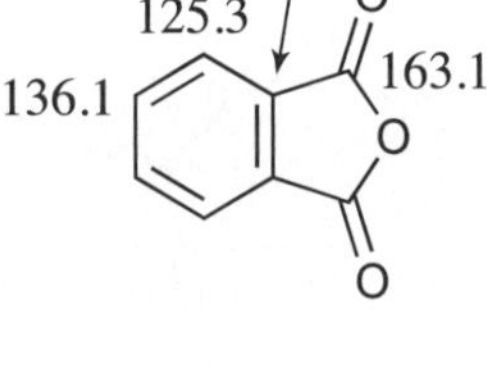

158.5
H—C(=O)—O—C(=O)—H

166.6
H$_3$C—C(=O)—O—C(=O)—CH$_3$
22.1

170.9
8.5
—C(=O)—O—C(=O)—
27.4

169.6
18.2
13.4   37.2
—C(=O)—O—C(=O)—

172.8
18.3
35.2
—C(=O)—O—C(=O)—

173.9
26.5
40.2
—C(=O)—O—C(=O)—

$^2J_{CF}$ 48 Hz
150.1
F$_3$C—C(=O)—O—C(=O)—CF$_3$
113.5
$^1J_{CF}$ 285 Hz

154.0
Cl$_3$C—C(=O)—O—C(=O)—CCl$_3$
87.9

162.4
130.5
128.9
134.5
C$_6$H$_5$—C(=O)—O—C(=O)—C$_6$H$_5$
128.9

C = X

28.2
172.5

137.4
165.9

30.1
16.7
168.5

131.1
125.3
136.1
163.1

^{13}C Chemical Shifts of Carboxylic Acid Imides (δ in ppm)

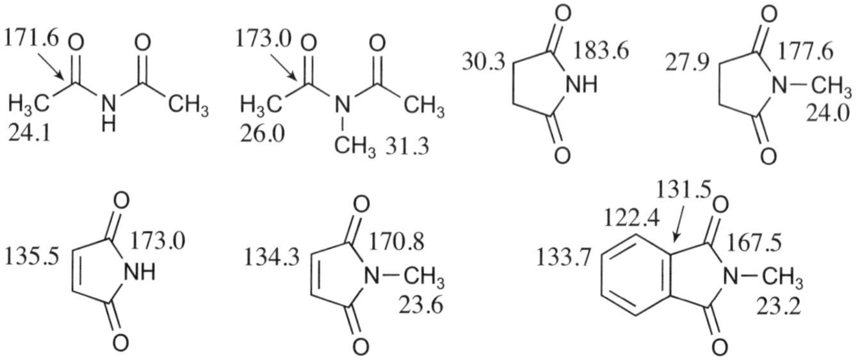

171.6 O   O
H₃C   N   CH₃
24.1   H

173.0 O   O
H₃C   N   CH₃
26.0   CH₃ 31.3

30.3   183.6
        NH
O

27.9   177.6
        N–CH₃
        24.0
O

135.5   173.0
         NH
O
O

134.3   170.8
         N–CH₃
         23.6
O
O

131.5
122.4   167.5
133.7   N–CH₃
         23.2
O

^{13}C Chemical Shifts of Carbonic Acid Derivatives (δ in ppm)

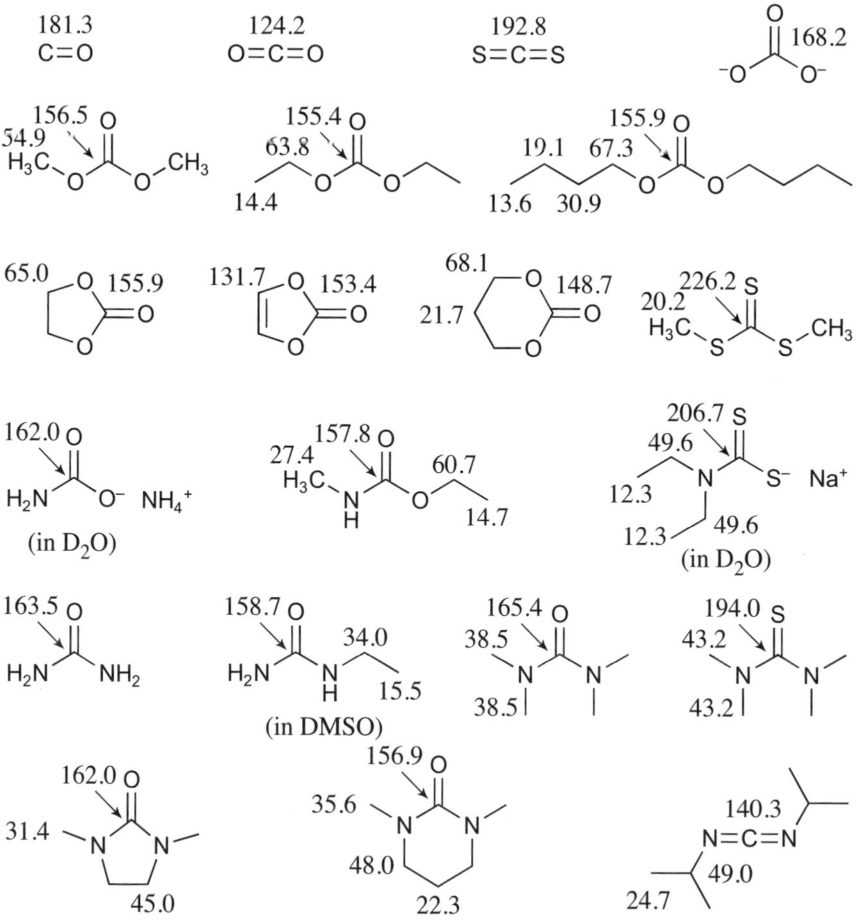

181.3
C=O

124.2
O=C=O

192.8
S=C=S

O
‖   168.2
⁻O   O⁻

156.5 O
54.9
H₃C O   O–CH₃

155.4 O
63.8
O   O
14.4

155.9 O
19.1 67.3
O   O
13.6   30.9

65.0 O 155.9
=O
O

131.7 O 153.4
=O
O

68.1
O 148.7
21.7   =O
O

226.2 S
20.2
H₃C S   S–CH₃

C = X

162.0 O
H₂N O⁻ NH₄⁺
(in D₂O)

157.8 O
27.4   60.7
H₃C N   O
     H   14.7

206.7 S
49.6
N   S⁻ Na⁺
12.3
12.3   49.6
(in D₂O)

163.5 O
H₂N NH₂

158.7 O
          34.0
H₂N N
     H   15.5
(in DMSO)

165.4 O
38.5
N   N
38.5

194.0 S
43.2
N   N
43.2

162.0 O
31.4 N   N
       N
45.0

156.9 O
35.6 N   N
48.0
22.3

140.3
N=C=N
49.0
24.7

4.12 Miscellaneous Compounds

4.12.1 Compounds with Group IV Elements

^{13}C Chemical Shifts of Silicon Compounds (δ in ppm)

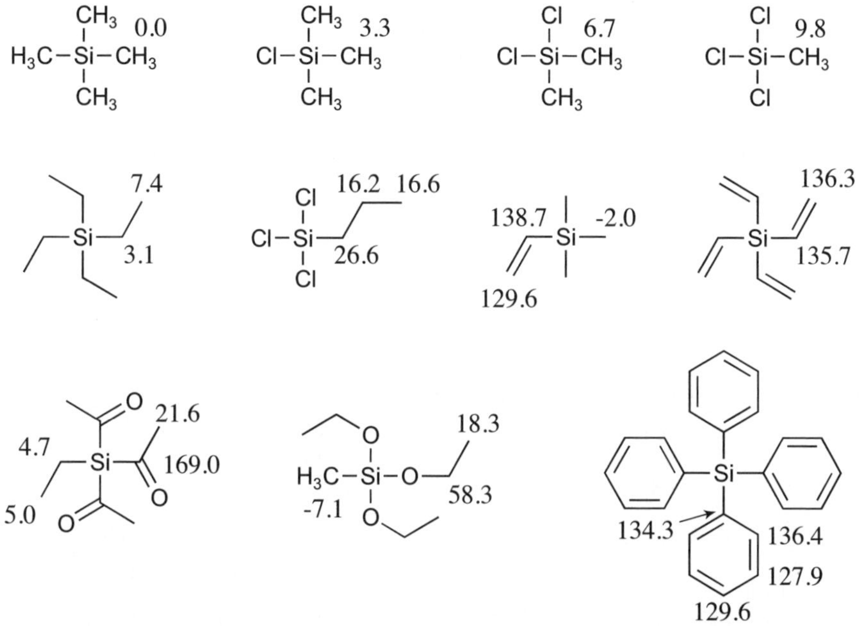

CH$_3$ 0.0
H$_3$C–Si–CH$_3$
CH$_3$

CH$_3$ 3.3
Cl–Si–CH$_3$
CH$_3$

Cl 6.7
Cl–Si–CH$_3$
CH$_3$

Cl 9.8
Cl–Si–CH$_3$
Cl

7.4 / Si / 3.1

Cl 16.2 16.6
Cl–Si–
Cl 26.6

138.7 | –2.0
Si–
129.6

136.3
Si
135.7

4.7 / Si / 5.0
O 21.6
169.0
O

18.3
H$_3$C–Si–O–
–7.1 O 58.3

Si
134.3   136.4
127.9
129.6

^{13}C Chemical Shifts and Coupling Constants of Germanium and Lead Compounds (δ in ppm, |J| in Hz)

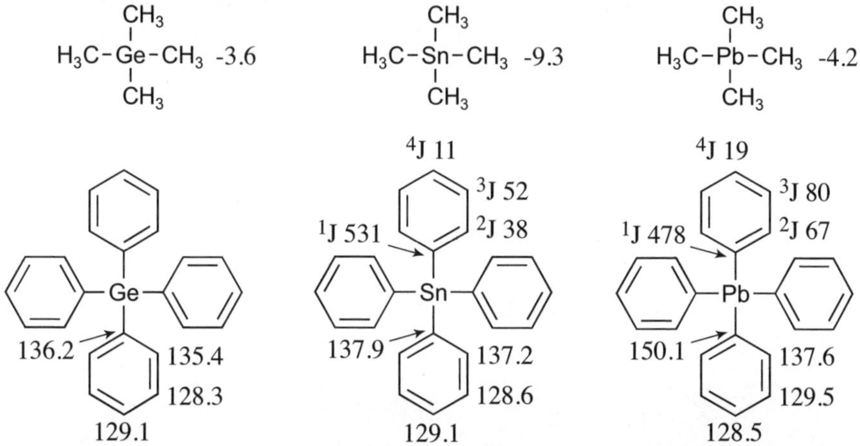

CH$_3$
H$_3$C–Ge–CH$_3$ –3.6
CH$_3$

CH$_3$
H$_3$C–Sn–CH$_3$ –9.3
CH$_3$

CH$_3$
H$_3$C–Pb–CH$_3$ –4.2
CH$_3$

Ge
136.2   135.4
128.3
129.1

^{4}J 11
^{3}J 52
^{1}J 531 ^{2}J 38
Sn
137.9   137.2
128.6
129.1
Couplings with ^{119}Sn
(8.58%, I = 1/2)

^{4}J 19
^{3}J 80
^{1}J 478 ^{2}J 67
Pb
150.1   137.6
129.5
128.5
Couplings with ^{207}Pb
(22.6%, I = 1/2)

P Si

4.12.2 Phosphorus Compounds

³¹P (natural abundance, 100%) has a spin quantum number I of 1/2. Couplings to protons through up to 3–4 bonds are usually observed.

Phosphines and Phosphonium Compounds (δ in ppm, |J₃₁ₚ₁₃c| in Hz)

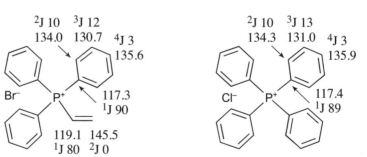

Phosphine Oxides and Sulfides (δ in ppm, |J_{31P13C}| in Hz)

Phosphinic and Phosphorous Acid Derivatives (δ in ppm, |J_{31P13C}| in Hz)

P Si

Phosphoric Acid Derivatives (δ in ppm, |J₃₁ₚ₁₃c| in Hz)

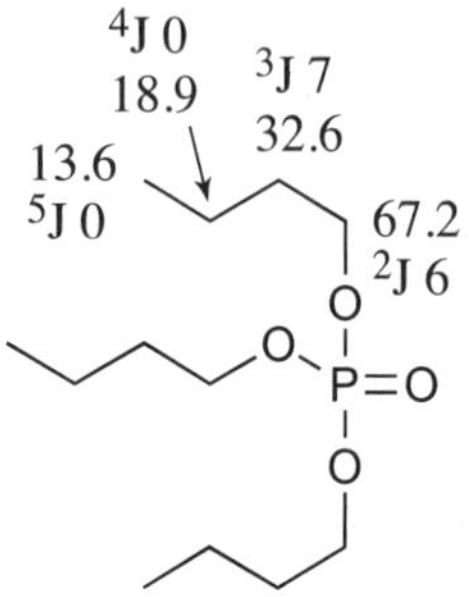

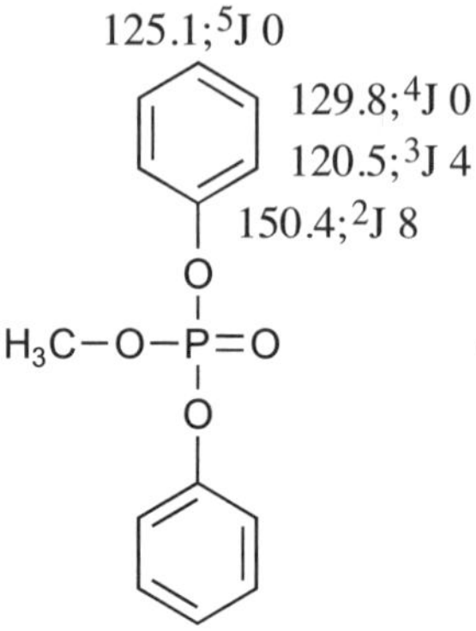

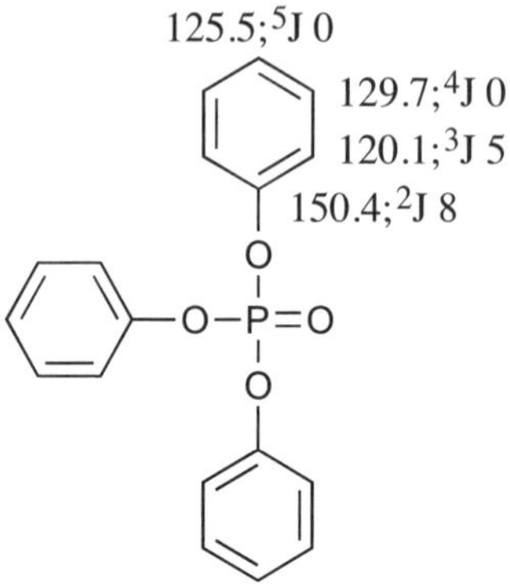

^{4}J 0
18.9 ^{3}J 7
32.6
13.6
^{5}J 0
67.2
^{2}J 6
P=O

125.1;^{5}J 0
129.8;^{4}J 0
120.5;^{3}J 4
150.4;^{2}J 8
H₃C−O−P=O

125.5;^{5}J 0
129.7;^{4}J 0
120.1;^{3}J 5
150.4;^{2}J 8
O−P=O

Phosphoranes and Phosphorus Ylides (δ in ppm, |J₃₁ₚ₁₃c| in Hz)

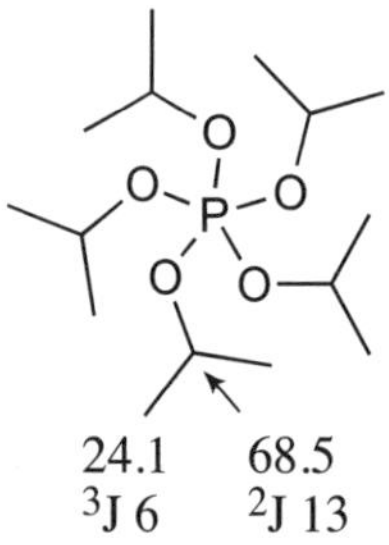

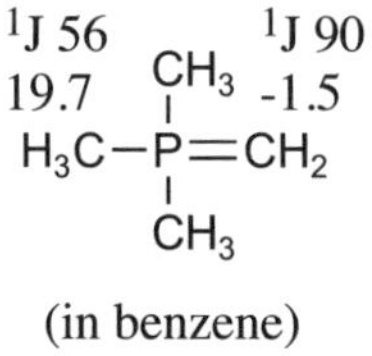

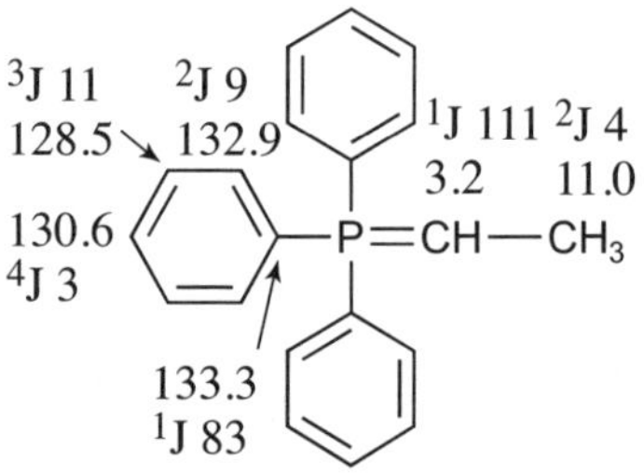

24.1   68.5
^{3}J 6 ^{2}J 13

^{1}J 56   CH₃ ^{1}J 90
19.7   -1.5
H₃C−P=CH₂
CH₃

(in benzene)

^{3}J 11 ^{2}J 9
128.5   132.9 ^{1}J 111 ^{2}J 4
3.2   11.0
130.6   P=CH−CH₃
^{4}J 3

133.3
^{1}J 83

4.12.3 Miscellaneous Organometallic Compounds

Lithium Compounds (δ in ppm)

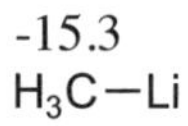

-15.3
H₃C−Li

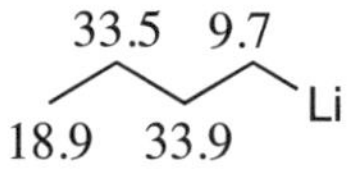

33.5   9.7
Li
18.9   33.9

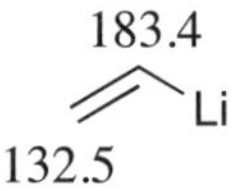

183.4
Li
132.5

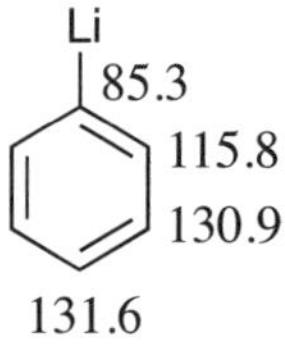

Li
85.3
115.8
130.9
131.6

Magnesium Compounds (δ in ppm)

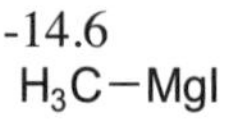

-14.6
H₃C−MgI

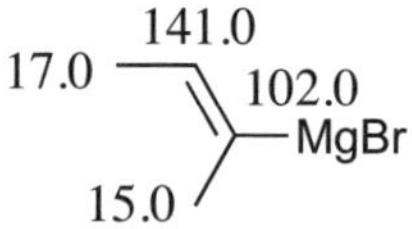

141.0
17.0
102.0
MgBr
15.0

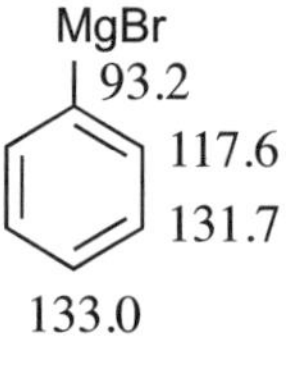

MgBr
93.2
117.6
131.7
133.0

Boron Compounds (δ in ppm, |J| in Hz)

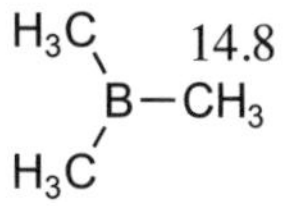

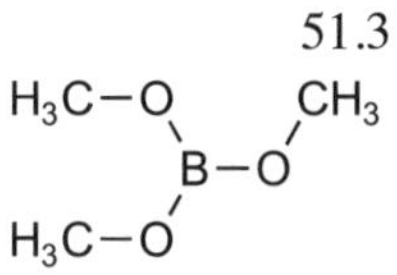

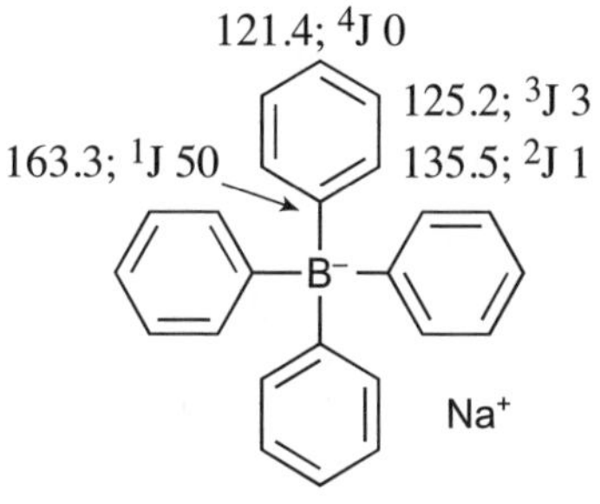

H₃C 14.8
B—CH₃
H₃C

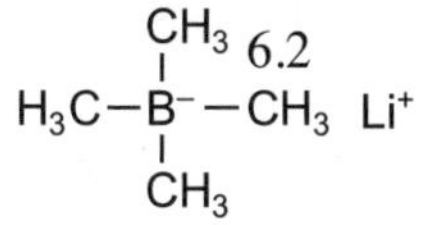

CH₃ 6.2
H₃C—B⁻—CH₃ Li⁺
CH₃

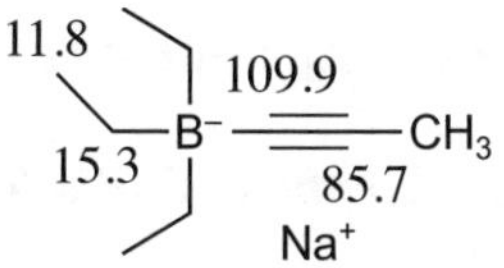

11.8
109.9
B⁻ ≡≡ —CH₃
15.3   85.7
Na⁺

51.3
H₃C—O   CH₃
B—O
H₃C—O

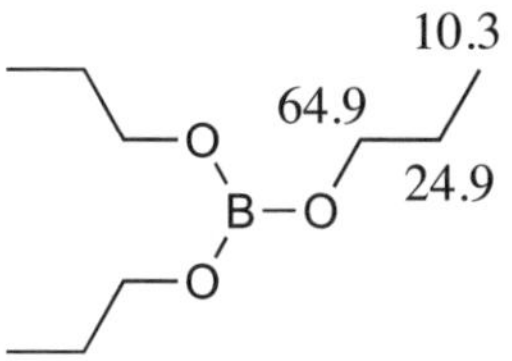

10.3
O   64.9
B—O   24.9
O

163.3; ¹J 50

121.4; ⁴J 0
125.2; ³J 3
135.5; ²J 1
B⁻

Na⁺

Couplings with ¹¹B
(80.4%, I = 3/2)

Arsenic, Antimony, and Bismuth Compounds (δ in ppm)

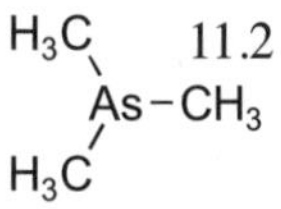

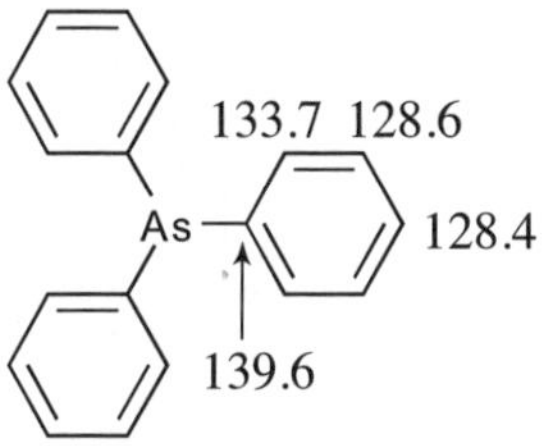

H₃C 11.2
As—CH₃
H₃C

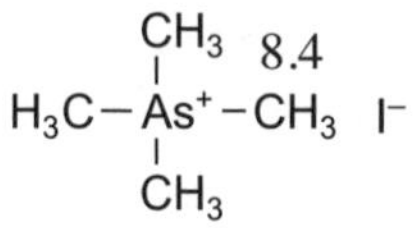

CH₃ 8.4
H₃C—As⁺—CH₃ I⁻
CH₃

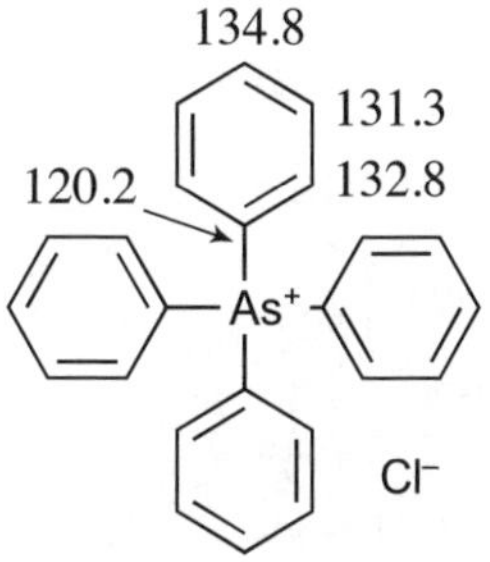

134.8
131.3
120.2   132.8
As⁺

Cl⁻

133.7 128.6
As   128.4
139.6

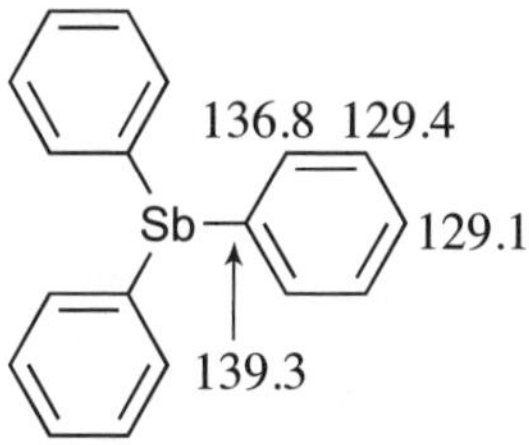

136.8 129.4
Sb   129.1
139.3

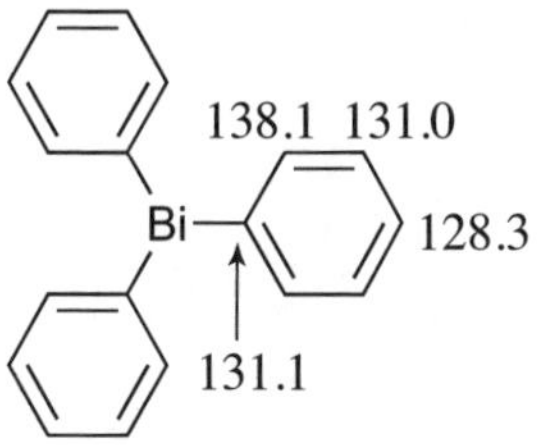

138.1 131.0
Bi   128.3
131.1

P Si

Mercury Compounds (δ in ppm, |J| in Hz)

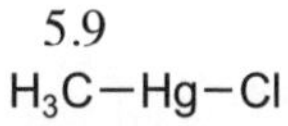

5.9
H₃C—Hg—Cl

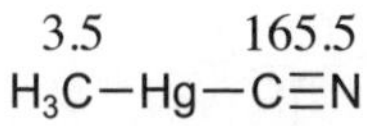

3.5   165.5
H₃C—Hg—C≡N

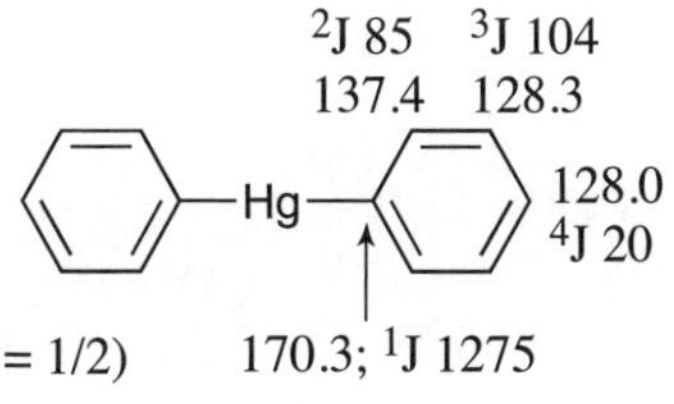

²J 85   ³J 104
137.4   128.3
Hg   128.0
⁴J 20

170.3; ¹J 1275

Couplings with ¹⁹⁹Hg (16.8%, I = 1/2)

4.13 Natural Products

4.13.1 Amino Acids

^{13}C Chemical Shifts (δ in ppm; solvent: water)

^+H_3N—CH$_2$(41.5)—C(=O)(171.2)—OH
(pH 0.45)

^+H_3N—CH$_2$(42.8)—C(=O)(173.6)—O$^-$
(pH 4.53)

H_2N—CH$_2$(46.0)—C(=O)(182.7)—O$^-$
(pH 12.01)

^+H_3N—(36.6)—(32.2)—C(=O)(175.7)—OH
(pH 0.49)

^+H_3N—(38.8)—(34.8)—C(=O)(179.4)—O$^-$
(pH 5.03)

H_2N—(39.3)—(41.6)—C(=O)(182.7)—O$^-$
(pH 12.56)

^+H_3N—CH(50.1)(CH$_3$ 16.5)—C(=O)(174.0)—OH
(pH 0.43)

^+H_3N—CH(51.9)(CH$_3$ 17.5)—C(=O)(177.0)—O$^-$
(pH 4.96)

H_2N—CH(52.7)(CH$_3$ 21.7)—C(=O)(185.7)—O$^-$
(pH 12.52)

^+H_3N—CH(59.6)(CH(30.0)(17.9)(18.5))—C(=O)(172.7)—OH
(pH 1.34)

^+H_3N—CH(61.9)(CH(30.3)(17.8)(19.2))—C(=O)(175.4)—O$^-$
(pH 5.64)

H_2N—CH(63.2)(CH(32.9)(17.9)(20.3))—C(=O)(184.1)—O$^-$
(pH 12.60)

^+H_3N—CH(52.8)(CH$_2$(40.1)—CH(25.1)(22.1)(22.7))—C(=O)(174.0)—OH
(pH 0.37)

^+H_3N—CH(54.4)(CH$_2$(40.7)—CH(25.1)(21.8)(22.9))—C(=O)(176.3)—O$^-$
(pH 7.00)

H_2N—CH(55.9)(CH$_2$(45.5)—CH(25.6)(22.5)(23.7))—C(=O)(185.4)—O$^-$
(pH 13.00)

^+H_3N—CH(58.7)(CH(37.1)(CH$_3$ 15.3)(CH$_2$(25.9)—CH$_3$ 12.1))—C(=O)(172.8)—OH
(pH 0.28)

^+H_3N—CH(60.9)(CH(37.1)(CH$_3$ 15.9)(CH$_2$(25.6)—CH$_3$ 12.4))—C(=O)(175.0)—O$^-$
(pH 6.04)

H_2N—CH(62.3)(CH(39.8)(CH$_3$ 16.7)(CH$_2$(25.2)—CH$_3$ 12.3))—C(=O)(184.1)—O$^-$
(pH 12.84)

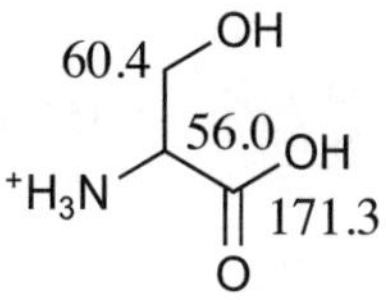

60.4 — OH
56.0 OH
^+H_3N 171.3
O
(pH 1.12)

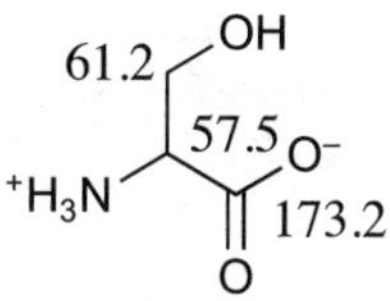

61.2 — OH
57.5 O$^-$
^+H_3N 173.2
O
(pH 6.05)

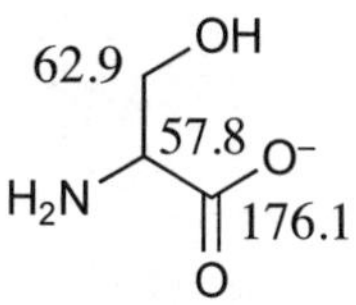

62.9 — OH
57.8 O$^-$
H_2N 176.1
O
(pH 9.28)

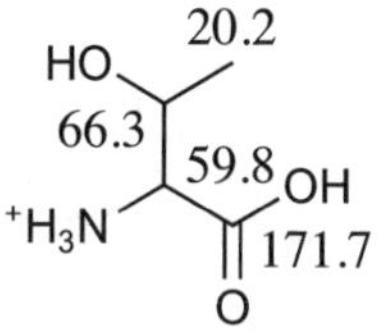

20.2
HO
66.3
59.8 OH
^+H_3N 171.7
O
(pH 1.36)

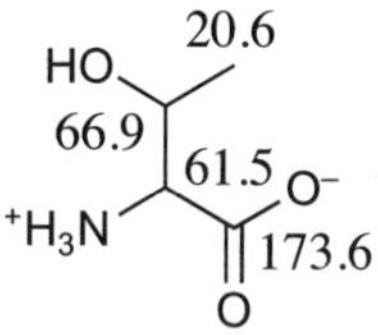

20.6
HO
66.9
61.5 O$^-$
^+H_3N 173.6
O
(pH 5.87)

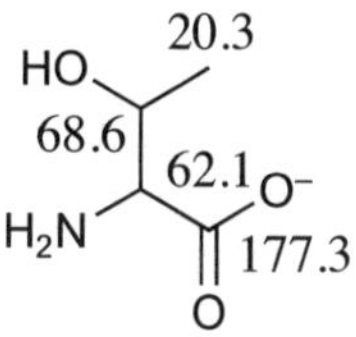

20.3
HO
68.6
62.1 O$^-$
H_2N 177.3
O
(pH 9.27)

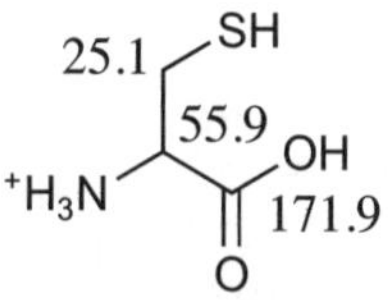

25.1 — SH
55.9 OH
^+H_3N 171.9
O
(pH 1.75)

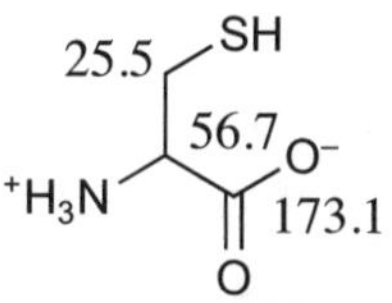

25.5 — SH
56.7 O$^-$
^+H_3N 173.1
O
(pH 5.14)

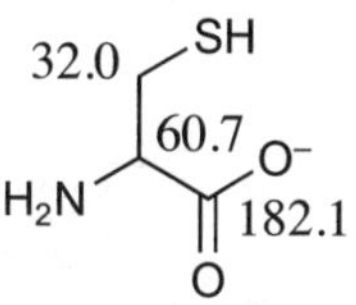

32.0 — SH
60.7 O$^-$
H_2N 182.1
O
(pH 11.02)

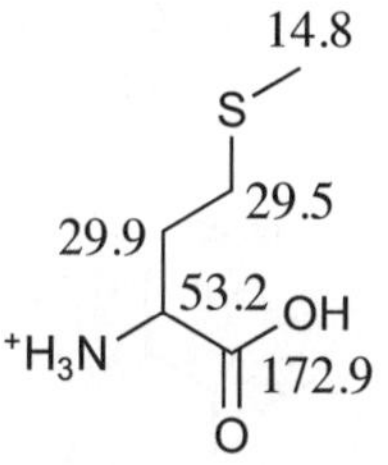

14.8
S
29.5
29.9
53.2 OH
^+H_3N 172.9
O
(pH 1.55)

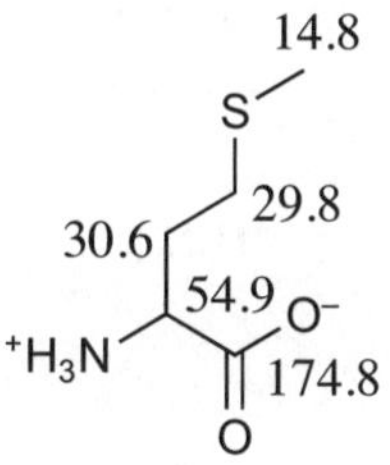

14.8
S
29.8
30.6
54.9 O$^-$
^+H_3N 174.8
O
(pH 5.83)

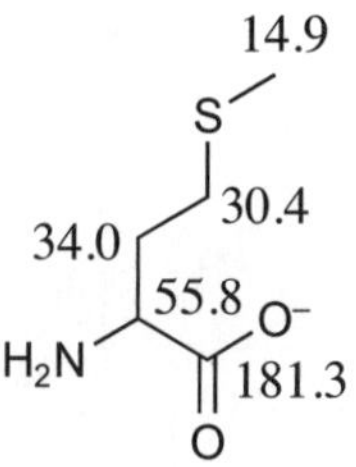

14.9
S
30.4
34.0
55.8 O$^-$
H_2N 181.3
O
(pH 9.80)

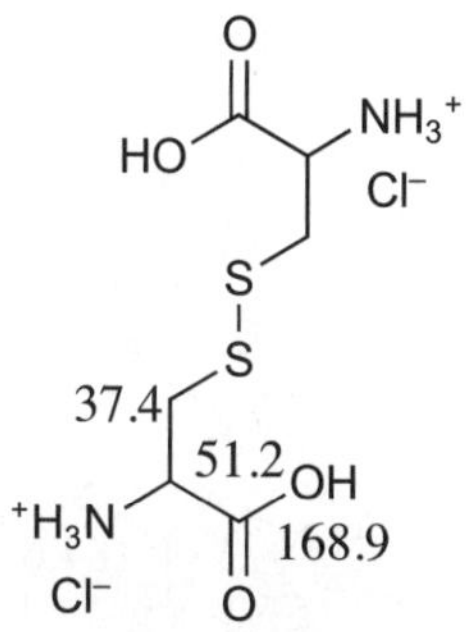

O
HO
NH$_3$$^+$
Cl$^-$
S
S
37.4
51.2 OH
^+H_3N 168.9
Cl$^-$ O
(in DMSO)

O
$^-$O
NH$_3$$^+$
S
S
44.1
55.8 O$^-$
^+H_3N 180.7
O
(in D$_2$O)

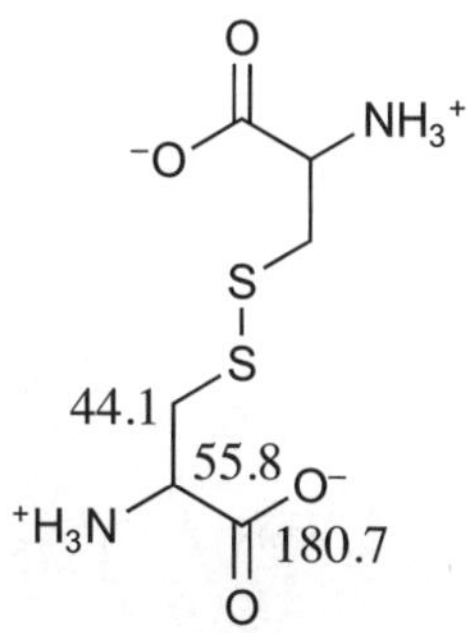

Natural
Products

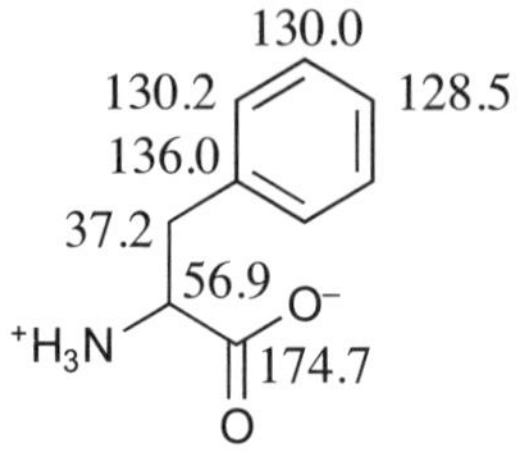

130.0
130.2   128.5
136.0
37.2
56.9 O$^-$
$^+$H$_3$N   174.7
O

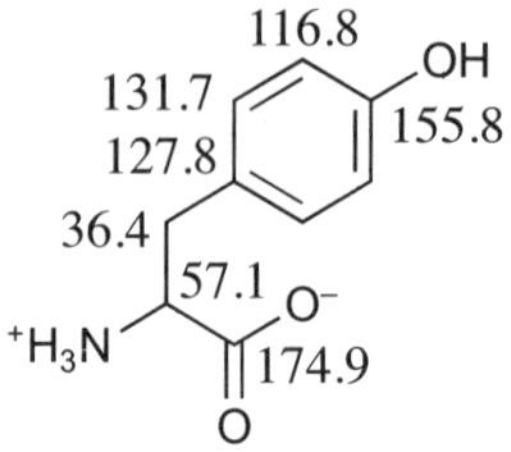

116.8 OH
131.7   155.8
127.8
36.4
57.1 O$^-$
$^+$H$_3$N   174.9
O

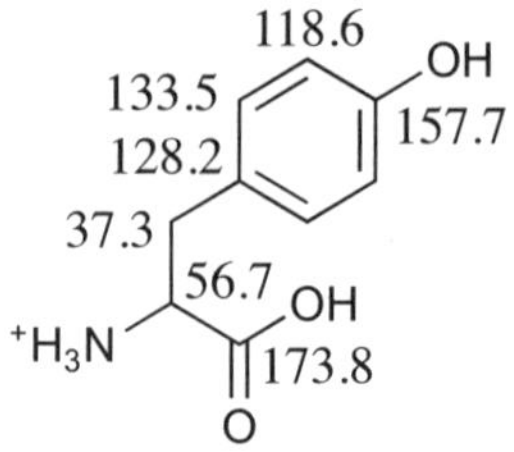

118.6 OH
133.5   157.7
128.2
37.3
56.7 OH
$^+$H$_3$N   173.8
O
(in D$_2$O/DCl)

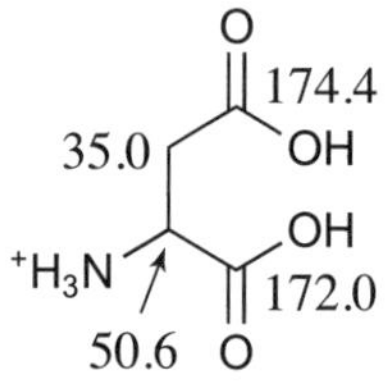

O 174.4
35.0   OH
OH
$^+$H$_3$N   172.0
50.6 O
(pH 0.41)

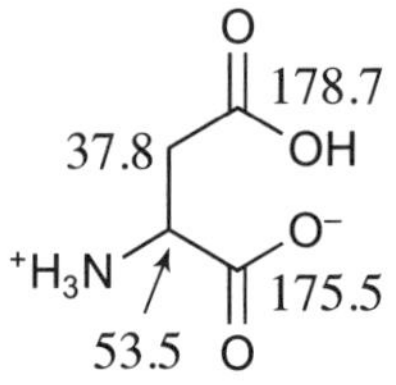

O 178.7
37.8   OH
O$^-$
$^+$H$_3$N   175.5
53.5 O
(pH 6.73)

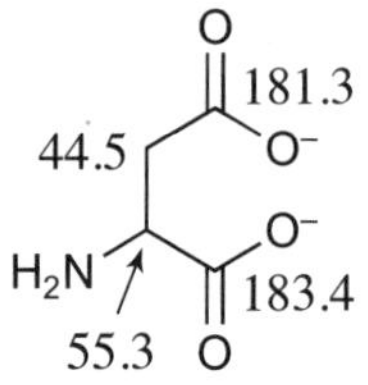

O 181.3
44.5   O$^-$
O$^-$
H$_2$N   183.4
55.3 O
(pH 12.73)

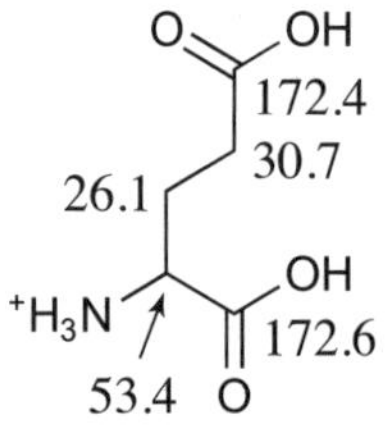

O OH
172.4
26.1   30.7
OH
$^+$H$_3$N   172.6
53.4 O
(pH 0.32)

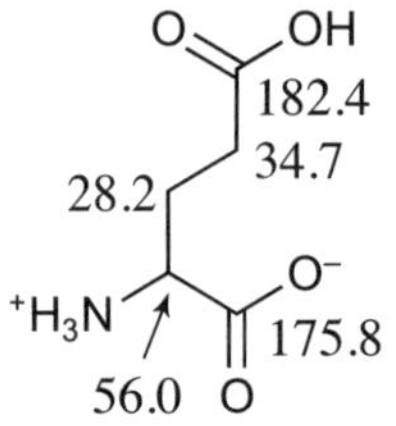

O OH
182.4
28.2   34.7
O$^-$
$^+$H$_3$N   175.8
56.0 O
(pH 6.95)

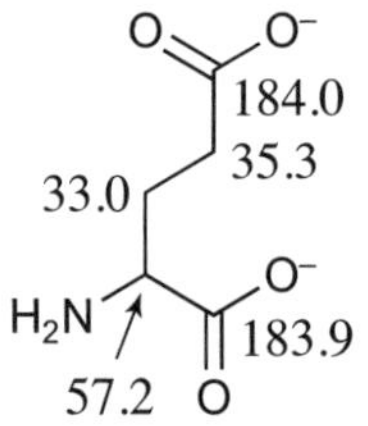

O O$^-$
184.0
33.0   35.3
O$^-$
H$_2$N   183.9
57.2 O
(pH 12.51)

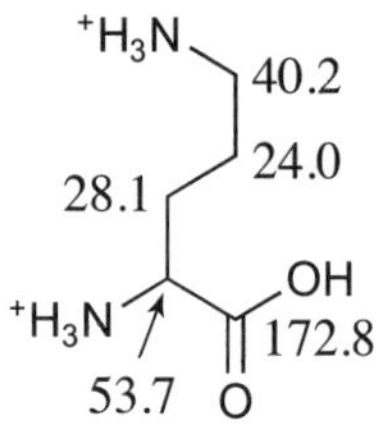

$^+$H$_3$N
40.2
28.1   24.0
OH
$^+$H$_3$N   172.8
53.7 O
(pH 0.46)

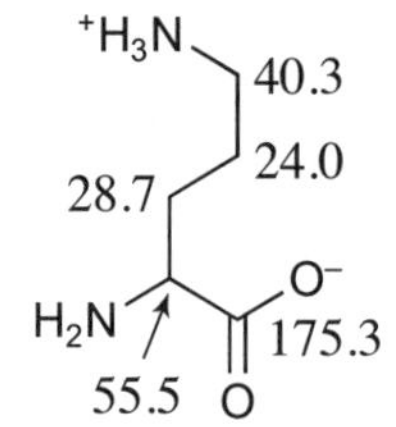

$^+$H$_3$N
40.3
28.7   24.0
O$^-$
H$_2$N   175.3
55.5 O
(pH 5.02)

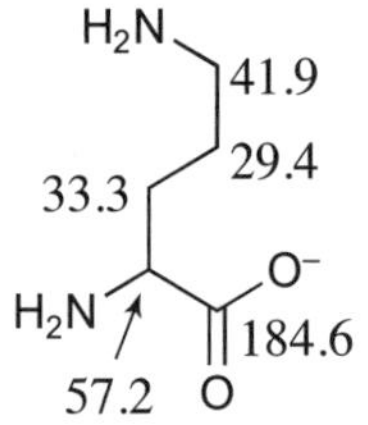

H$_2$N
41.9
33.3   29.4
O$^-$
H$_2$N   184.6
57.2 O
(pH 13.53)

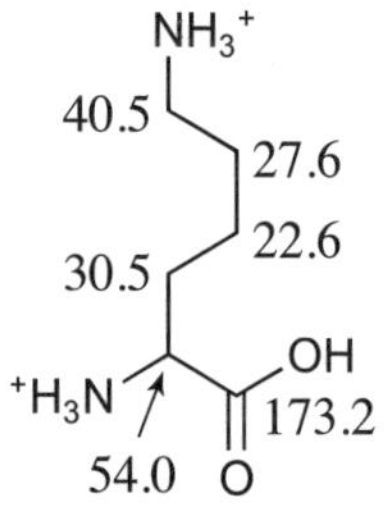

NH$_3$$^+$
40.5   27.6
30.5   22.6
OH
$^+$H$_3$N   173.2
54.0 O
(pH 0.50)

NH$_3$$^+$
40.5   27.7
31.2   22.6
O$^-$
H$_2$N   175.8
55.9 O
(pH 6.03)

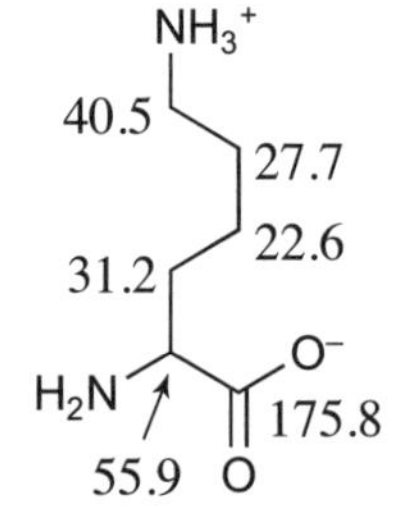

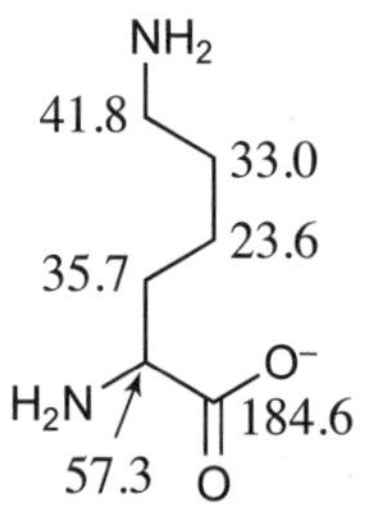

NH$_2$
41.8   33.0
35.7   23.6
O$^-$
H$_2$N   184.6
57.3 O
(pH 13.85)

Natural
Products

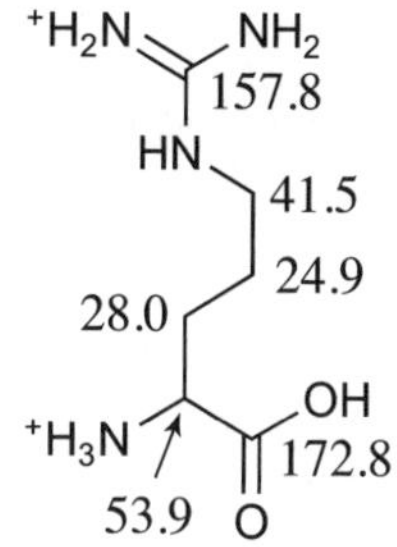

(pH 1.33)

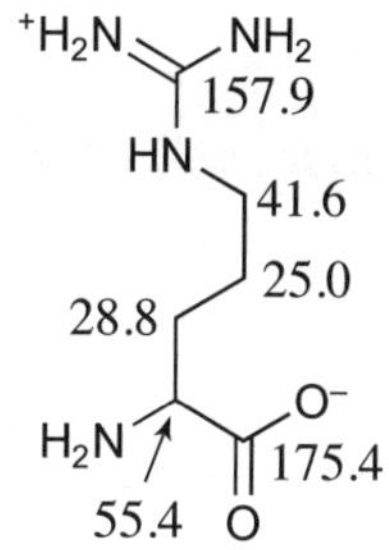

(pH 7.87)

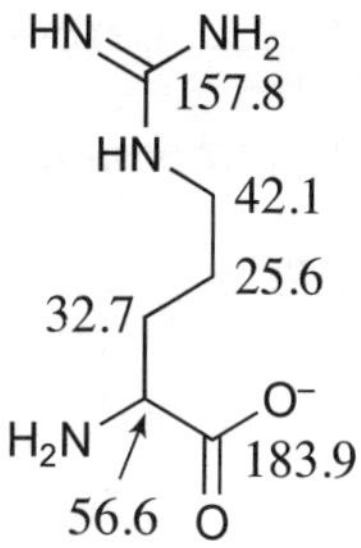

(pH 11.52)

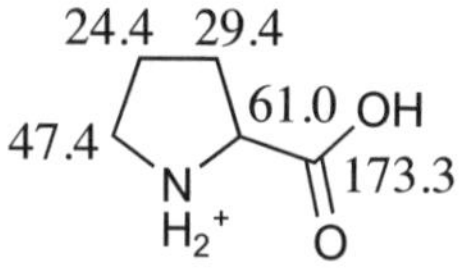

(pH 1.27)

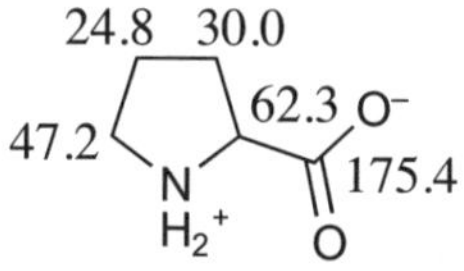

(pH 7.26)

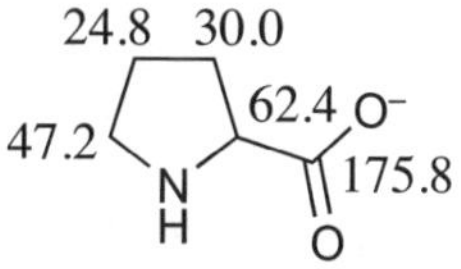

(pH 9.80)

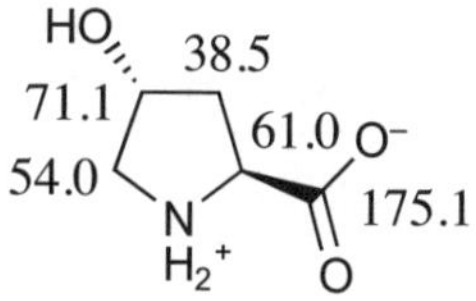

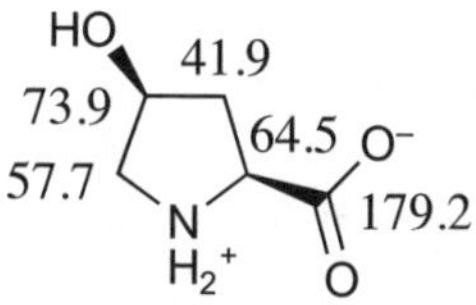

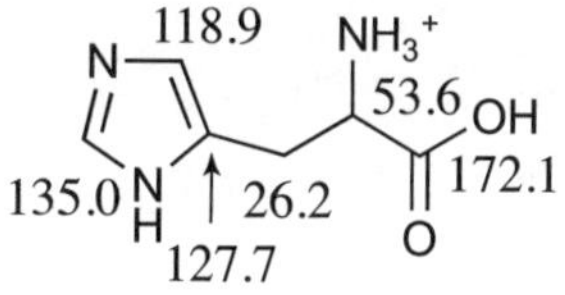

(pH 1.74)

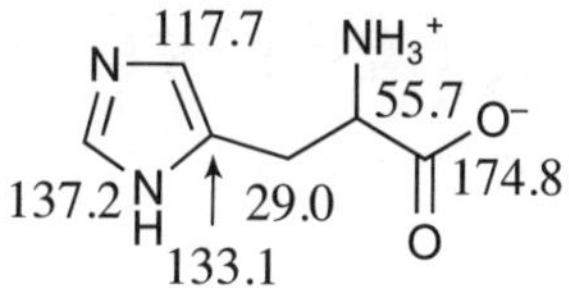

(pH 7.82)

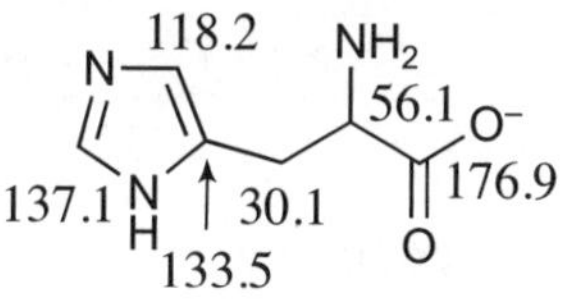

(pH 9.21)

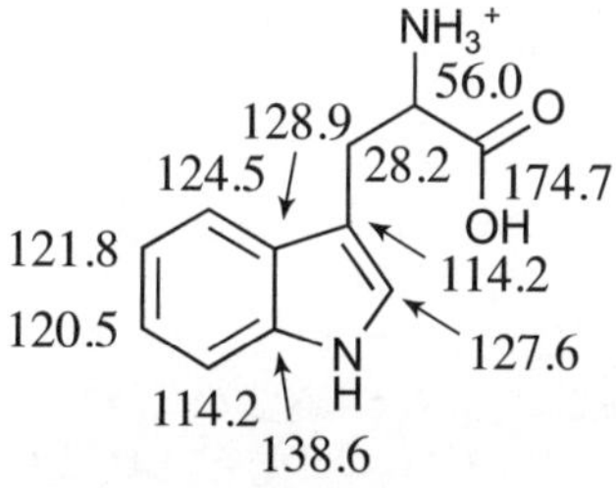

(in D$_2$O/DCl)

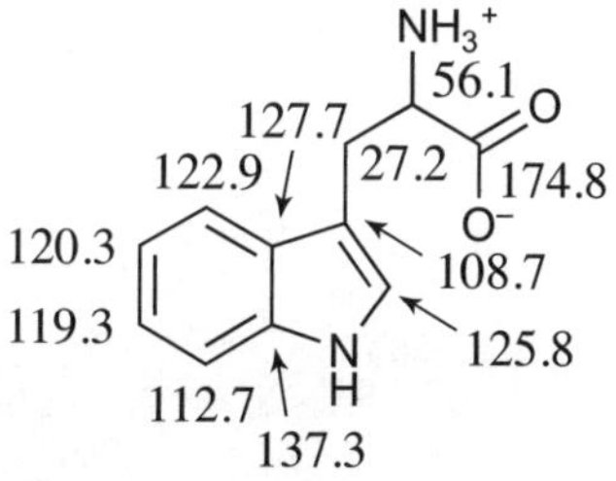

(80 °C)

Natural
Products

4.13.2 Carbohydrates

^{13}C Chemical Shifts of Monosaccharides (δ in ppm)

Ribose

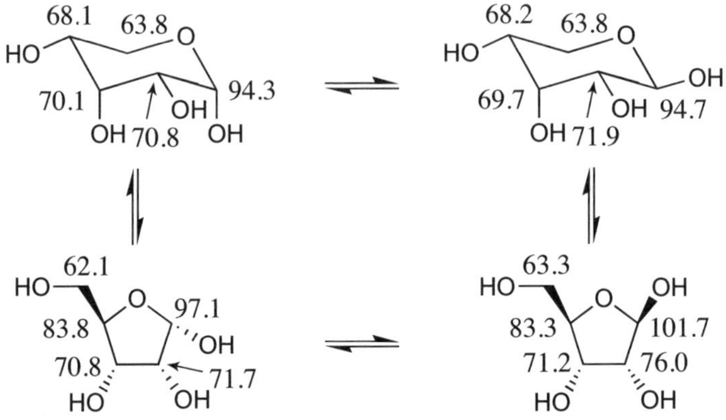

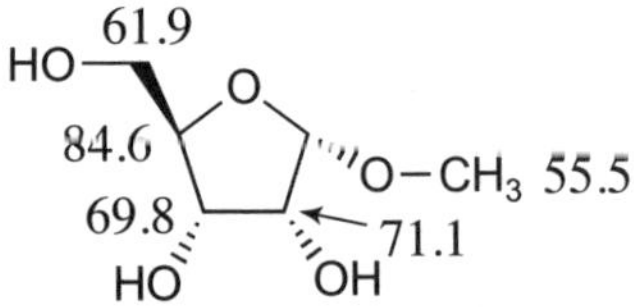

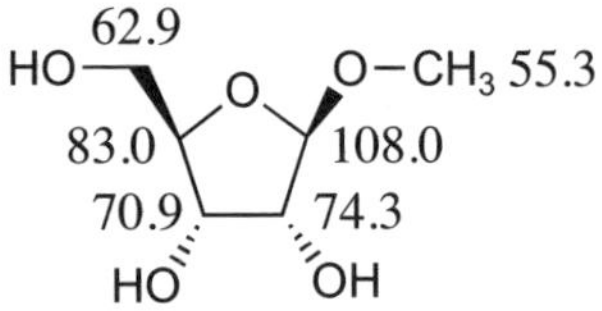

68.1 63.8 O
HO
70.1 94.3
OH 70.8 OH

68.2 63.8 O
HO
69.7 OH
OH 94.7
OH 71.9

62.1
HO O 97.1
83.8 'OH
70.8 71.7
HO OH

63.3
HO O OH
83.3 101.7
71.2 76.0
HO OH

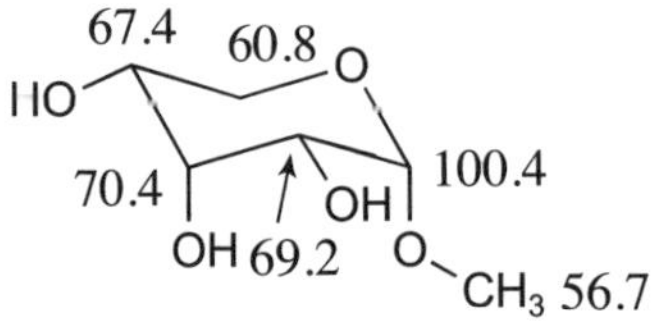

67.4 60.8 O
HO
70.4 100.4
OH 69.2 O–CH$_3$ 56.7

61.9
HO O
84.6 ''O–CH$_3$ 55.5
69.8 71.1
HO OH

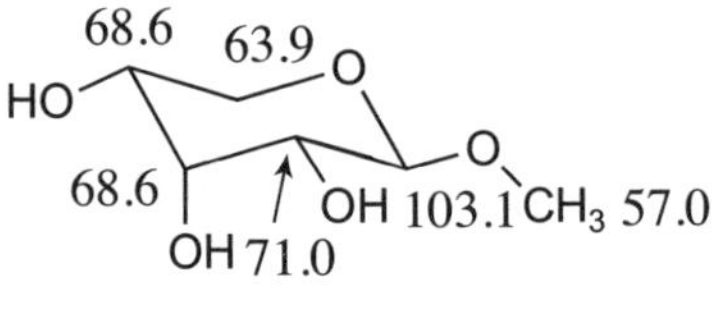

68.6 63.9 O
HO
68.6 OH 103.1 CH$_3$ 57.0
OH 71.0 O

62.9
HO O O–CH$_3$ 55.3
83.0 108.0
70.9 74.3
HO OH

Glucose

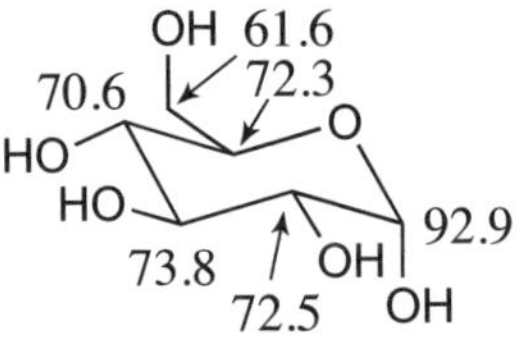

OH 61.6
70.6 72.3
HO O
HO 92.9
73.8 OH
72.5 OH

OH 61.7
70.6 76.8
HO O
HO OH
76.7 OH 96.7
75.1

OH 61.6
70.6 72.5
HO O
HO 100.0
74.1 OH
72.2 O–CH$_3$ 55.9

OH 61.8
70.6 76.8
HO O
HO O
76.8 OH 104.0 CH$_3$ 58.1
74.1

Natural
Products

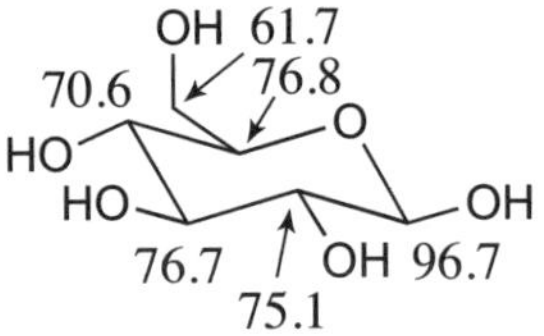

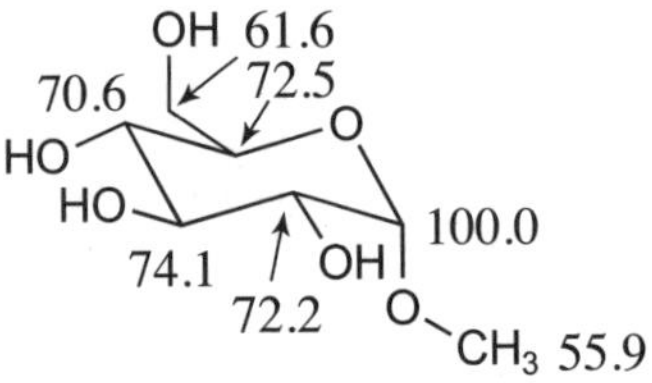

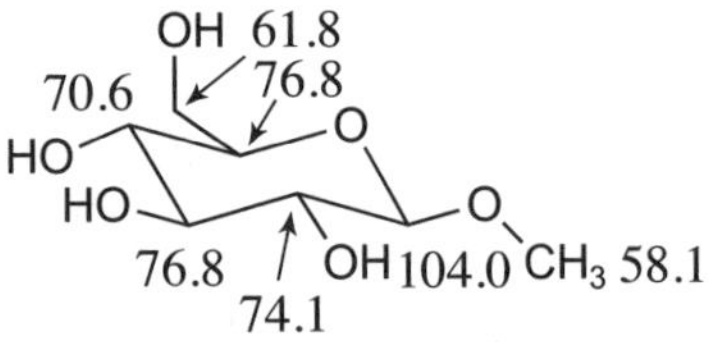

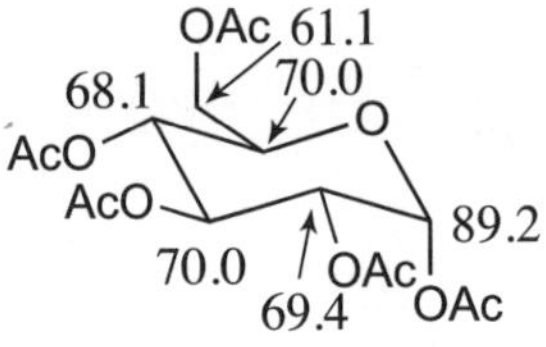

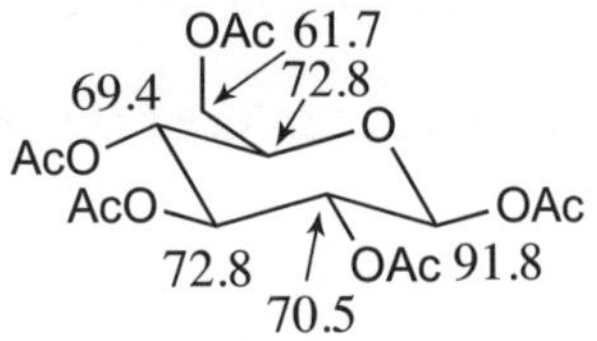

Fructose

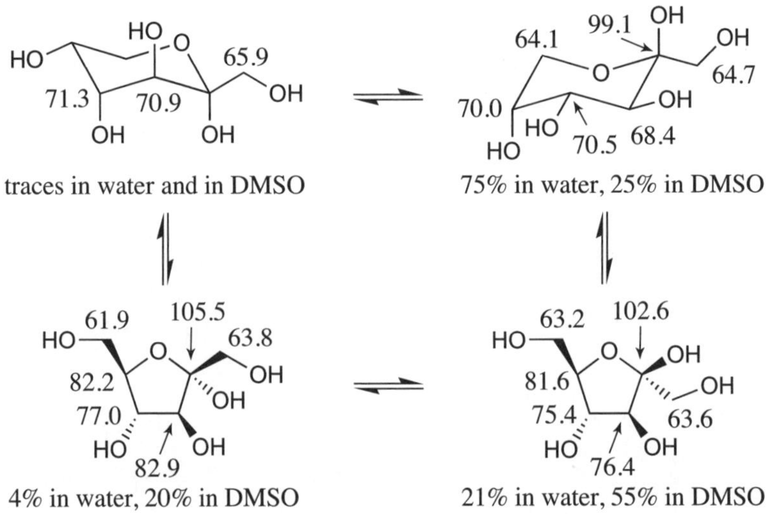

traces in water and in DMSO

75% in water, 25% in DMSO

4% in water, 20% in DMSO

21% in water, 55% in DMSO

^{13}C-^{1}H Coupling Constants through One Bond ($^{1}J_{CH}$ in Hz)

4.13.3 Nucleotides and Nucleosides

^{13}C Chemical Shifts of Nucleotides (δ in ppm)

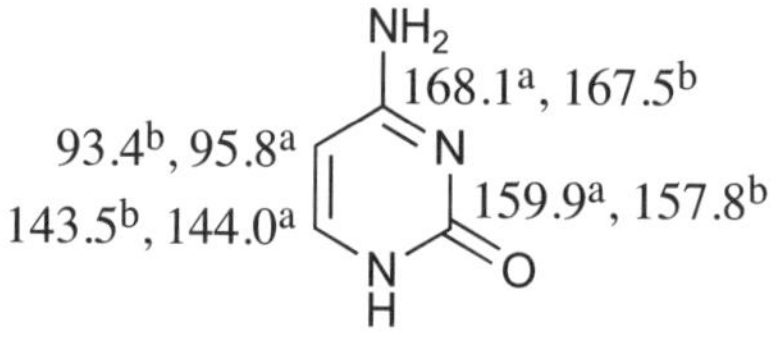

NH2
168.1[a], 167.5[b]
93.4[b], 95.8[a]
143.5[b], 144.0[a]
159.9[a], 157.8[b]
N
N
H
O

([a] in D_2O, [b] in DMSO)

NH2 • HCl
160.8
N
149.7
146.9
N
H
O

(in D_2O)

O
165.1
101.0
NH
142.9
152.3
N
H
O

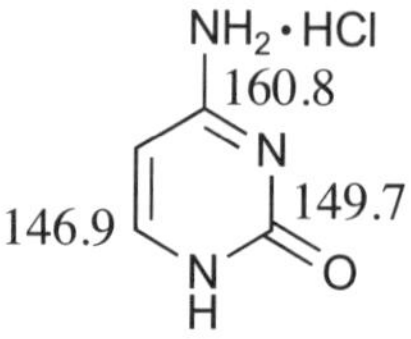

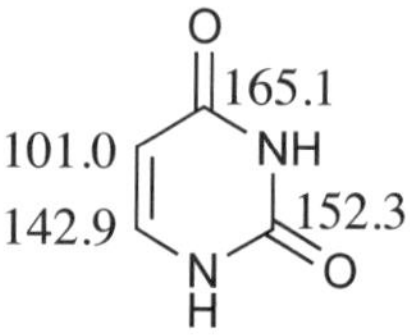

(in DMSO)

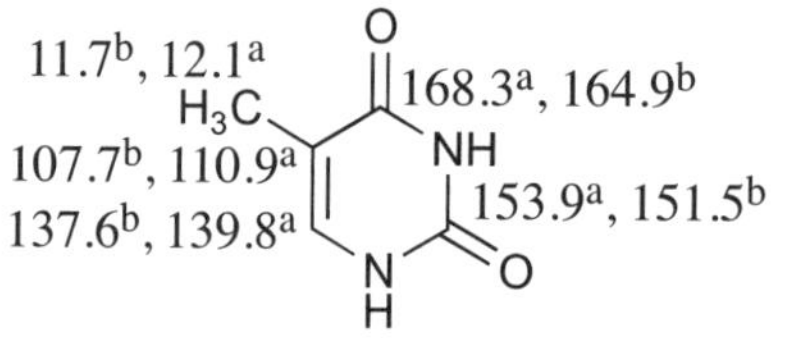

11.7[b], 12.1[a]
O
H3C
168.3[a], 164.9[b]
107.7[b], 110.9[a]
NH
137.6[b], 139.8[a]
153.9[a], 151.5[b]
N
H
O

([a] in D_2O, [b] in DMSO)

119.1 NH2
156.4
N
N
140.3
153.4
N
H
N
151.7

(in DMSO)

115.9 NH2 • HCl
151.4
N
N
144.5
145.8
N
H
N
149.8

(in D_2O)

119.6 O
168.8
N
NH
150.1
N
H
N
NH2
162.2
160.0

(in D_2O)

108.4 O
155.7
N
NH
137.9
N
H
N
NH2 • HCl
150.5
153.8

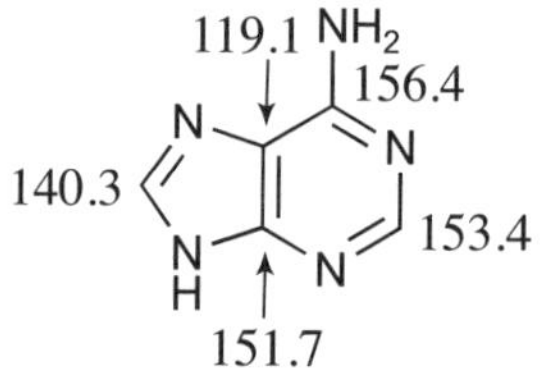

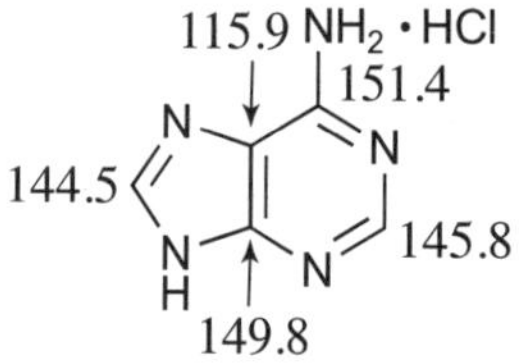

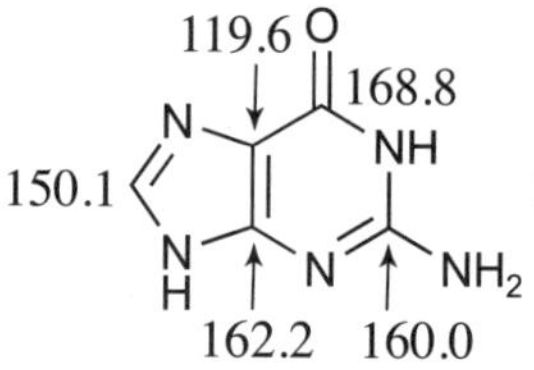

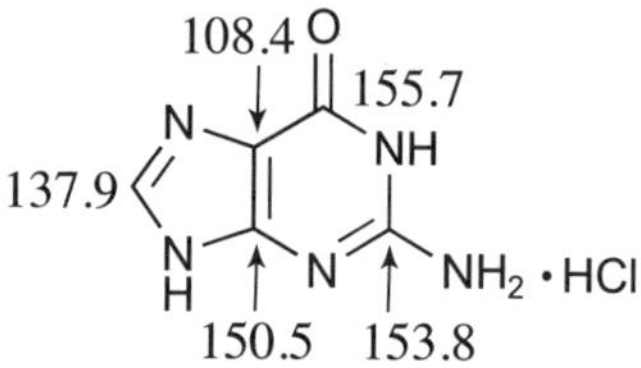

(in DMSO)

Nucleosides (δ in ppm)

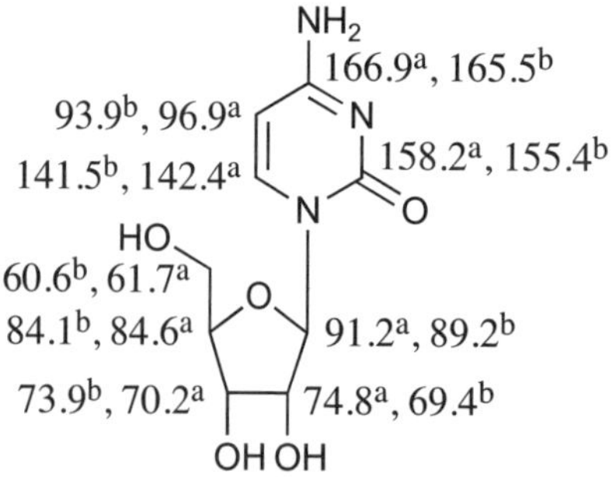

NH2
166.9[a], 165.5[b]
93.9[b], 96.9[a]
N
141.5[b], 142.4[a]
158.2[a], 155.4[b]
N
O
HO
60.6[b], 61.7[a]
O
84.1[b], 84.6[a]
91.2[a], 89.2[b]
73.9[b], 70.2[a]
74.8[a], 69.4[b]
OH OH

([a] in D_2O, [b] in DMSO)

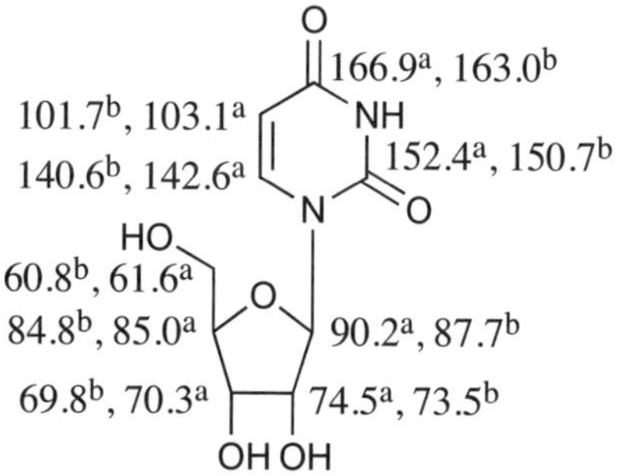

O
166.9[a], 163.0[b]
101.7[b], 103.1[a]
NH
140.6[b], 142.6[a]
152.4[a], 150.7[b]
N
O
HO
60.8[b], 61.6[a]
O
84.8[b], 85.0[a]
90.2[a], 87.7[b]
69.8[b], 70.3[a]
74.5[a], 73.5[b]
OH OH

([a] in D_2O, [b] in DMSO)

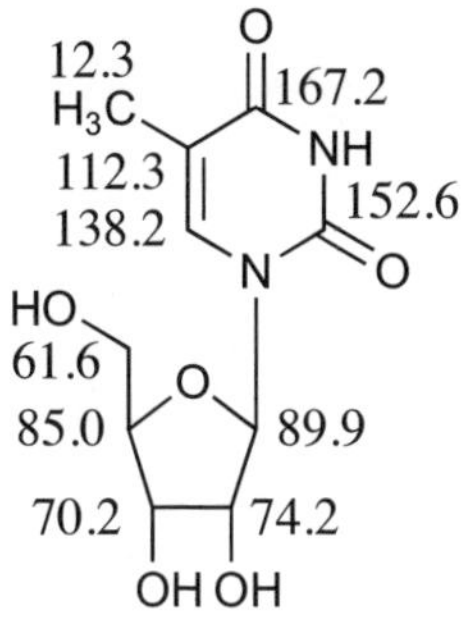

12.3
H₃C — 167.2
NH
112.3
138.2 — 152.6
O
HO
61.6
85.0 — 89.9
70.2 — 74.2
OH OH

(in D₂O)

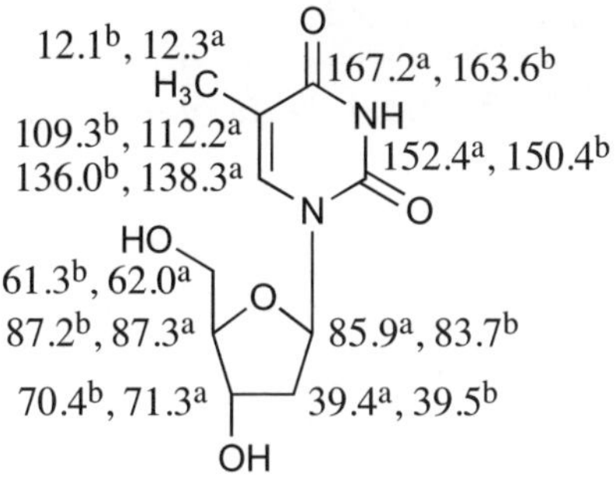

12.1ᵇ, 12.3ᵃ
H₃C — 167.2ᵃ, 163.6ᵇ
NH
109.3ᵇ, 112.2ᵃ
136.0ᵇ, 138.3ᵃ — 152.4ᵃ, 150.4ᵇ
O
HO
61.3ᵇ, 62.0ᵃ
87.2ᵇ, 87.3ᵃ — 85.9ᵃ, 83.7ᵇ
70.4ᵇ, 71.3ᵃ — 39.4ᵃ, 39.5ᵇ
OH

(ᵃ in D₂O, ᵇ in DMSO)

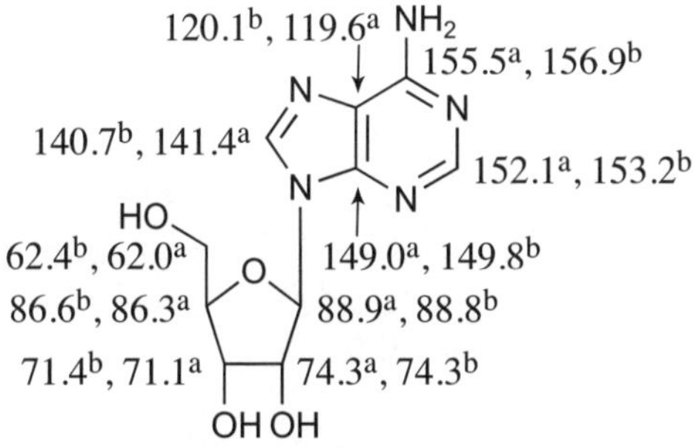

120.1ᵇ, 119.6ᵃ NH₂
155.5ᵃ, 156.9ᵇ
140.7ᵇ, 141.4ᵃ
152.1ᵃ, 153.2ᵇ
HO
62.4ᵇ, 62.0ᵃ — 149.0ᵃ, 149.8ᵇ
86.6ᵇ, 86.3ᵃ — 88.9ᵃ, 88.8ᵇ
71.4ᵇ, 71.1ᵃ — 74.3ᵃ, 74.3ᵇ
OH OH

(ᵃ in D₂O, ᵇ in DMSO)

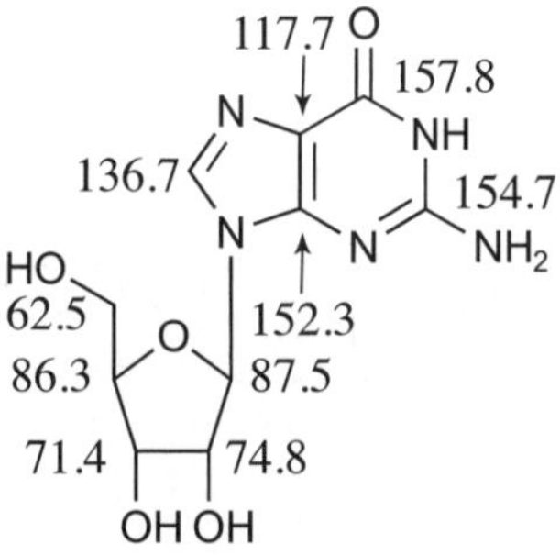

117.7 O
157.8
NH
136.7 — 154.7
NH₂
HO
62.5 — 152.3
86.3 — 87.5
71.4 — 74.8
OH OH

(in DMSO)

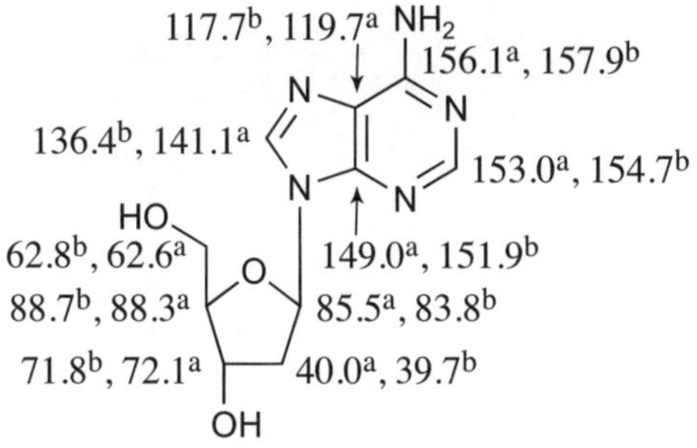

117.7ᵇ, 119.7ᵃ NH₂
156.1ᵃ, 157.9ᵇ
136.4ᵇ, 141.1ᵃ
153.0ᵃ, 154.7ᵇ
HO
62.8ᵇ, 62.6ᵃ — 149.0ᵃ, 151.9ᵇ
88.7ᵇ, 88.3ᵃ — 85.5ᵃ, 83.8ᵇ
71.8ᵇ, 72.1ᵃ — 40.0ᵃ, 39.7ᵇ
OH

(ᵃ in D₂O, ᵇ in DMSO)

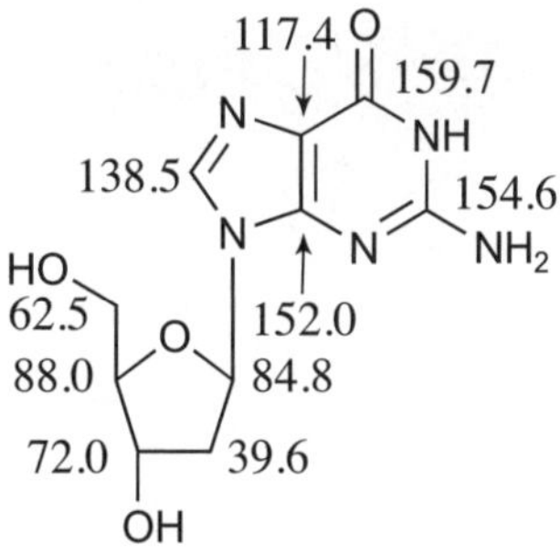

117.4 O
159.7
NH
138.5 — 154.6
NH₂
HO
62.5 — 152.0
88.0 — 84.8
72.0 — 39.6
OH

(in D₂O)

Natural
Products

4.13.4 Steroids

¹³C Chemical Shifts (δ in ppm)

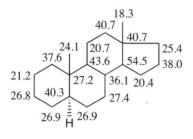

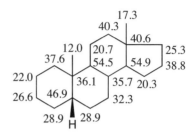

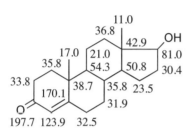

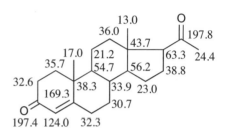

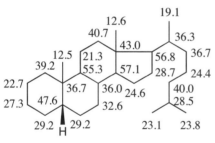

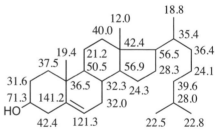

4.14   Spectra of Solvents and Reference Compounds

4.14.1 ^{13}C NMR Spectra of Common Deuterated Solvents (125 MHz, δ in ppm)

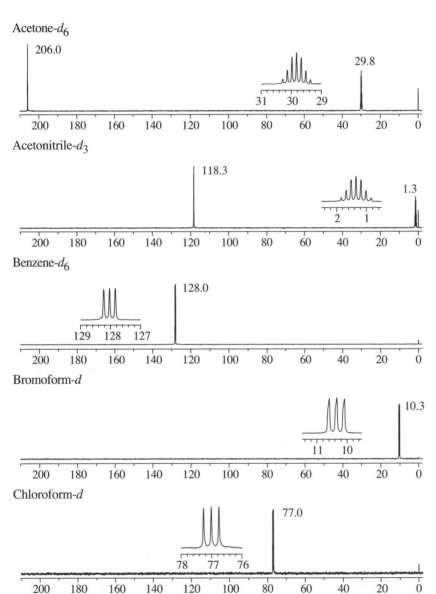

Acetone-d_6

Acetonitrile-d_3

Benzene-d_6

Bromoform-d

Chloroform-d

Solvents

Cyclohexane-d_{12}

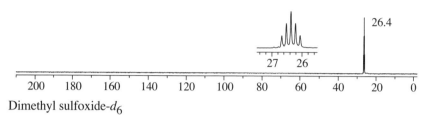

Dimethyl sulfoxide-d_6

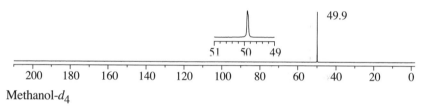

Methanol-d_1

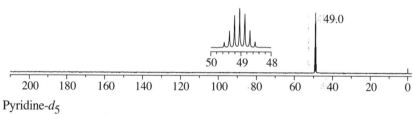

Methanol-d_4

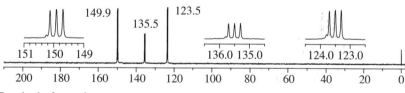

Pyridine-d_5

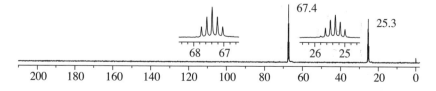

Tetrahydrofuran-d_8

Solvents

4.14.2 ^{13}C NMR Spectra of Secondary Reference Compounds

Chemical shifts in ^{13}C NMR spectra are usually reported relative to the peak position of tetramethylsilane (TMS), which is added as an internal reference. If TMS is not sufficiently soluble in the sample, the use of a capillary with TMS as external reference is recommended. In this case, owing to the difference in volume susceptibilities, the local magnetic fields in the solvent and reference are different. Therefore, the position of the reference must be corrected. For a D_2O solution in a cylindrical sample with TMS in a capillary, the correction amounts to +0.68 and -0.34 ppm for superconducting and electromagnets, respectively. These values must be subtracted from the ^{13}C chemical shifts relative to the external TMS signal if its position is set to 0.00 ppm. Alternatively, secondary references with $(CH_3)_3SiCH_2$ groups may be used. The following spectra of two secondary reference compounds in D_2O were measured at 125 MHz with a superconducting magnet and TMS as external reference. Chemical shifts are reported in ppm relative to TMS upon correction for the difference in the volume susceptibilities of D_2O. As a result, the peak for the external TMS appears at 0.68 ppm.

3-(Trimethylsilyl)-1-propanesulfonic acid sodium salt (sodium 4,4-dimethyl-4-silapentane-1-sulfonate; DSS)

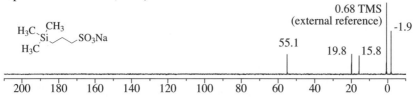

2,2,3,3-D$_4$-3-(Trimethylsilyl)-propionic acid sodium salt

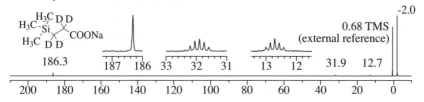

Solvents

4.14.3 ^{13}C NMR Spectrum of a Mixture of Common Nondeuterated Solvents

The broad-band-decoupled ^{13}C NMR spectrum (125 MHz, δ in ppm relative to TMS) of a CDCl$_3$ sample with 20 common solvents (0.05–0.4 vol%) shown below serves as a guide to identify possible solvent impurities. Chemical shifts of signals marked with an asterisk (*) may change up to a few ppm if the sample contains solutes having functional groups that can form hydrogen bonds.

DMF: dimethyl formamide; THF: tetrahydrofuran; EGDME: ethylene glycol dimethyl ether.

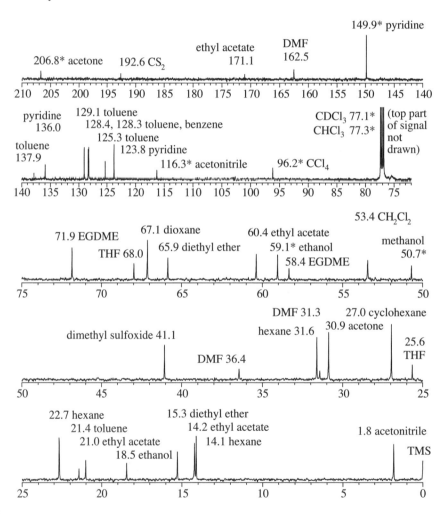

Solvents

5 ¹H NMR Spectroscopy

5.1 Alkanes

5.1.1 Chemical Shifts

¹H Chemical Shifts and Coupling Constants (δ in ppm, J in Hz)

CH_4 0.23
$^2J_{gem}$ -12.4

CH_3 0.86
|
CH_3

CH_3 0.91 $^3J_{vic}$ 7.4
|
CH_2 1.33
|
CH_3

$CH_3 a$ 0.91 $^3J_{ab}$ 7.3
|
$CH_2 b$ 1.31 $^2J_{bb'}$ -12.4
| $^3J_{bc}$ 5.7
CH_2 $^3J_{bc'}$ 8.5
|
CH_3

CH_3 0.89 $^3J_{vic}$
| 6.8
CH 1.74
/   \
H_3C CH_3

CH_3 0.90
H₃C—⧼—CH₃
     |
     CH_3

CH_3 0.88
|
CH_2 1.27
|
CH_2 1.27
|
CH_2
|
CH_3

In long-chain alkanes, the methyl groups at δ ca. 0.8 ppm typically show distorted triplets due to second-order effects:

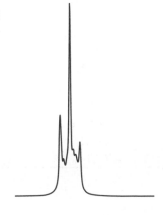

^{1}H Chemical Shifts of Monosubstituted Alkanes (δ in ppm)

Substituent	Methyl $-CH_3$	Ethyl $-CH_2$	$-CH_3$	1-Propyl $-CH_2$	$-CH_2$	$-CH_3$
C −H	0.23	0.86	0.86	0.91	1.33	0.91
−CH=CH$_2$	1.71	2.00	1.00	2.02	1.43	0.91
−C≡CH	1.80	2.16	1.15	2.10	1.50	0.97
−phenyl	2.35	2.63	1.21	2.59	1.65	0.95
X −F	4.27	4.55	1.35	4.30	1.68	0.97
−Cl	3.06	3.47	1.33	3.47	1.81	1.06
−Br	2.69	3.37	1.66	3.35	1.89	1.06
−I	2.16	3.16	1.88	3.16	1.88	1.03
O −OH	3.48	3.71	1.24	3.59	1.59	0.94
−O−alkyl	3.24	3.37	1.15	3.27	1.55	0.93
−OCH=CH$_2$	3.16	3.66	1.21			
−O−phenyl	3.73	3.98	1.38	3.86	1.70	1.05
−OCOCH$_3$	3.67	4.12	1.26	4.02	1.65	0.95
−OCO−phenyl	3.88	4.37	1.38	4.25	1.76	1.07
−OS(O)$_2$−4-tolyl	3.70	4.07	1.30	3.94	1.60	0.95
N −NH$_2$	2.47	2.66	1.11	2.65	1.46	0.91
−NHCH$_3$	2.30					
−N(CH$_3$)$_2$	2.22	2.32	1.06			
−NHCOCH$_3$	2.79	3.26	1.14	3.18	1.55	0.96
−NO$_2$	4.29	4.37	1.58	4.28	2.01	1.03
−C≡N	1.98	2.35	1.31	2.34	1.70	1.08
−NC	2.85	3.39	1.28			
S −SH	2.00	2.44	1.31	2.50	1.63	0.99
−S−alkyl	2.09	2.49	1.25	2.43	1.59	0.98
−SS−alkyl	2.30	2.67	1.35	2.63	1.71	1.03
−S(O)CH$_3$	2.50					
−S(O)$_2$CH$_3$	2.84	2.94	2.80			
O‖C −CHO	2.20	2.46	1.13	2.37	1.64	0.97
−COCH$_3$	2.17	2.44	1.06	2.40	1.60	0.93
−CO−phenyl	2.55	2.92	1.18	2.86	1.72	1.02
−COOH	2.10	2.36	1.16	2.31	1.68	1.00
−COOCH$_3$	2.01	2.32	1.15	2.22	1.65	0.98
−CONH$_2$	2.02	2.23	1.13	2.19	1.68	0.99
−COCl	2.66	2.93	1.24	2.87	1.74	1.00

^{1}H Chemical Shifts of Monosubstituted Alkanes (δ in ppm, contd.)

Substituent	2-Propyl		n-Butyl				tert-Butyl
	–CH	–CH$_3$	–CH$_2$	–CH$_2$	–CH$_2$	–CH$_3$	–CH$_3$
C –H	1.33	0.91	0.91	1.31	1.31	0.91	0.89
–CH=CH$_2$			2.06	≈1.5	≈1.2	0.90	1.02
–C≡C–	2.59	1.15	2.18	1.52	1.41	0.92	1.24
–phenyl	2.89	1.25	2.61	1.60	1.34	0.93	1.32
X –F	4.84	1.34	4.34	1.65		0.95	1.34
–Cl	4.14	1.55	3.42	1.68	1.41	0.92	1.60
–Br	4.21	1.73	3.42	1.84	1.46	0.93	1.76
–I	4.24	1.89	3.20	1.80	1.42	0.93	1.95
O –OH	4.02	1.21	3.64	1.56	1.39	0.94	1.26
–O–alkyl	3.55	1.08	3.40	1.54	1.38	0.92	1.24
–OCH=CH$_2$	4.06	1.23	3.68	1.61	1.39	0.94	
–O–phenyl	4.51	1.31	3.94	1.76	1.47	0.97	
–OCOCH$_3$	4.99	1.23	4.06	1.60	1.39	0.94	1.45
–OCO–phenyl	5.22	1.37					1.58
–OS(O)$_2$–4-tolyl	4.70	1.25	4.03	1.62	1.36	0.88	
N –NH$_2$	3.07	1.03	2.69	1.43	1.35	0.92	1.15
–NHCOCH$_3$	4.01	1.13	3.21	1.49	1.35	0.92	1.28
–NO$_2$	4.44	1.53	4.47	2.07	1.50	1.07	1.59
–C≡N	2.67	1.35	2.34	1.63	1.50	0.96	1.37
–NC	3.87	1.45					1.44
S –SH	3.16	1.34	2.52	1.59	1.43	0.92	1.43
–S–alkyl	2.93	1.25	2.49	1.56	1.42	0.92	1.39
–SS–alkyl			2.69	1.64	1.42	0.93	1.32
–S(O)$_2$CH$_3$	3.13	1.41					1.44
O ‖ C –CHO	2.39	1.13	2.42	1.59	1.35	0.93	1.08
–COCH$_3$	2.58	1.11					1.13
–CO–phenyl	3.58	1.22	2.95	1.72	1.41	0.96	
–COOH	2.56	1.21	2.35	1.62	1.39	0.93	1.23
–COOCH$_3$	2.56	1.17	2.31	1.61	1.33	0.92	1.20
–CONH$_2$	2.44	1.18	2.22	1.60	1.37	0.93	1.22
–COCl	2.97	1.31	2.88	1.67	1.40	0.93	

Estimation of ^{1}H Chemical Shifts of Substituted Alkanes (δ in ppm)

CH$_3$	CH$_2$	CH

$$\delta_{CH_3R^1} = 0.86 + Z_\alpha \qquad \delta_{CH_2} = 1.37 + \sum_i Z_{\alpha_i} + \sum_j Z_{\beta_j} \quad \delta_{CH} = 1.50 + \sum_i Z_{\alpha_i} + \sum_j Z_{\beta_j}$$

$$\delta_{CH_3CR^1R^2R^3} = 0.86 + \sum_i Z_{\beta_i}$$

	Substituent (R^1, R^2, R^3)	CH$_3$ Z_α	CH$_3$ Z_β	CH$_2$ Z_α	CH$_2$ Z_β	CH Z_α	CH Z_β
C	–C≤	0.00	0.05	0.00	-0.06	0.17	-0.01
	–C=C<	0.85	0.20	0.63	0.00	0.68	0.03
	–C≡C–	0.94	0.32	0.70	0.13	1.04	
	–phenyl	1.51	0.38	1.22	0.29	1.28	0.38
X	–F	3.41	0.41	2.76	0.16	1.83	0.27
	–Cl	2.20	0.63	2.05	0.24	1.98	0.31
	–Br	1.83	0.83	1.97	0.46	2.44	0.41
	–I	1.30	1.02	1.80	0.53	2.46	0.15
O	–OH	2.53	0.25	2.20	0.15	1.73	0.08
	–O–C≤	2.38	0.25	2.04	0.13	1.85	0.32
	–OC=C<	2.64	0.36	2.63	0.33	2.00	0.30
	–O–phenyl	2.87	0.47	2.61	0.38	2.20	0.50
	–O–CO–	2.81	0.44	2.83	0.24	2.47	0.59
N	–N<	1.61	0.14	1.32	0.22	1.13	0.23
	–N$^+$≤	2.44	0.39	1.91	0.40	1.78	0.56
	–N–CO–	1.88	0.34	1.63	0.22	2.10	0.62
	–NO$_2$	3.43	0.65	3.08	0.58	2.31	
	–C≡N	1.12	0.45	1.08	0.33	1.00	
	–NCS	2.51	0.54	2.20	0.36	1.94	0.60
S	–S–	1.14	0.45	1.23	0.26	1.06	0.31
	–S–CO–	1.41	0.37	1.54	0.63	1.31	0.19
	–S(O)–	1.64	0.36	1.24	0.30	1.25	
	–S(O)$_2$–	1.98	0.42	2.08	0.52	1.50	0.40
	–SCN	1.75	0.66	1.62		1.64	
O‖C	–CHO	1.34	0.21	1.07	0.29	0.86	0.22
	–CO–	1.23	0.20	1.12	0.24		
	–COOH	1.22	0.23	0.90	0.23	0.87	0.32
	–COO–	1.15	0.28	0.92	0.35	0.83	0.63
	–CO–N<	1.16	0.28	0.85	0.24	0.94	0.30
	–COCl	1.94	0.22	1.51	0.25		

¹H Chemical Shifts of Aromatically Substituted Alkanes (δ in ppm)

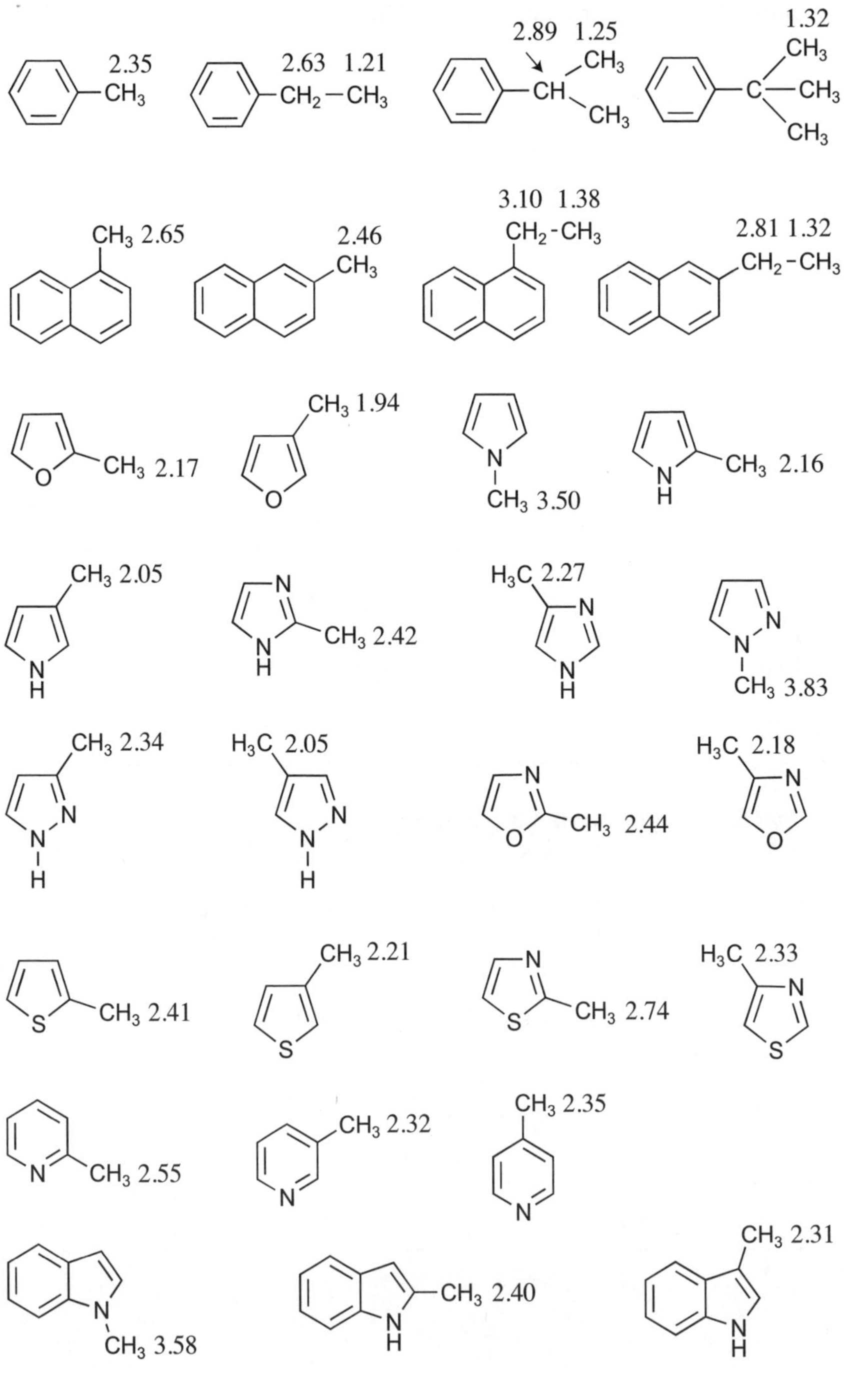

Benzene–CH₃ 2.35

Benzene–CH₂–CH₃ 2.63 1.21

Benzene–CH(CH₃)₂ 2.89 1.25

Benzene–C(CH₃)₃ 1.32

Naphthalene CH₃ 2.65 (1-position)

Naphthalene CH₃ 2.46 (2-position)

Naphthalene CH₂–CH₃ 3.10 1.38 (1-position)

Naphthalene CH₂–CH₃ 2.81 1.32 (2-position)

Furan–CH₃ 2.17

Furan CH₃ 1.94

N-methylpyrrole CH₃ 3.50

Pyrrole–CH₃ 2.16

Pyrrole CH₃ 2.05

Imidazole–CH₃ 2.42

Imidazole H₃C 2.27

Pyrazole CH₃ 3.83

Pyrazole CH₃ 2.34

Pyrazole H₃C 2.05

Oxazole–CH₃ 2.44

Oxazole H₃C 2.18

Thiophene–CH₃ 2.41

Thiophene CH₃ 2.21

Thiazole–CH₃ 2.74

Thiazole H₃C 2.33

Pyridine–CH₃ 2.55 (2-position)

Pyridine CH₃ 2.32 (3-position)

Pyridine CH₃ 2.35 (4-position)

N-methylindole CH₃ 3.58

Indole–CH₃ 2.40 (2-position)

Indole CH₃ 2.31 (3-position)

5.1.2 Coupling Constants

Geminal Coupling Constants (^2J in Hz)

    $^2J_{HCH}$ -8 to -18

Electronegative substituents cause a decrease in $|J|_{gem}$ while a double or triple bond next to the CH$_2$ group causes an increase. The latter effect is strongest if one of the C–H bonds is parallel to the π orbitals:

Influence of Substituents on the Geminal Coupling Constant

Compound	J_{gem}	Compound	J_{gem}
CH$_4$	-12.4	CH$_3$COCH$_3$	-14.9
CH$_3$Cl	-10.8	CH$_3$COOH	-14.5
CH$_2$Cl$_2$	-7.5	CH$_3$CN	-16.9
CH$_3$OH	-10.8	CH$_2$(CN)$_2$	-20.3
⬡—CH$_3$	-14.3	⬡—CH$_2$-CN	-18.5

Vicinal Coupling Constants (^3J in Hz)

conformation not fixed: $^3J_{HCCH} \approx 7$

fixed: $^3J_{HCCH} \approx 0$–18

Influence of Substituents on the Vicinal Coupling Constant

Compound	J_{vic}	Compound	J_{vic}	Compound	J_{vic}
CH$_3$CHF$_2$	4.5	CH$_3$CH$_2$OH	6.9	CH$_3$CH$_2$CN	7.6
CH$_3$CHCl$_2$	6.1	(CH$_3$CH$_2$)$_3$O$^+$ BF$_4^-$	7.2	(CH$_3$CH$_2$)$_2$S	7.4
CH$_3$CH$_2$F	6.9	(CH$_3$CH$_2$)$_3$N	7.1	(CH$_3$CH$_2$)$_4$Si	8.0
CH$_3$CH$_2$Cl	7.2	(CH$_3$CH$_2$)$_4$N$^+$ I$^-$	7.3	CH$_3$CH$_2$Li	8.4

Vicinal coupling constants strongly depend on the dihedral angle, ϕ (Karplus equation):

$$^3J = J^0 \cos^2 \phi - 0.3 \qquad 0° \le \phi \le 90°$$
$$^3J = J^{180} \cos^2 \phi - 0.3 \qquad 90° \le \phi \le 180°$$

The same relationship between torsional angle and vicinal coupling constant holds for substituted alkanes if appropriate values are used for J^0 and J^{180}. These limiting values depend on the electronegativity and orientation of substituents, the hybridization of carbon atoms, bond lengths, and bond angles.

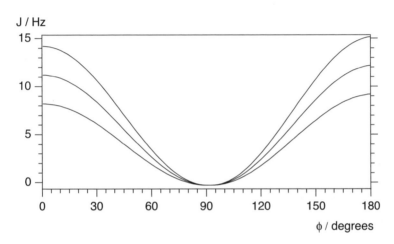

Long-Range Coupling Constants (|J| in Hz)

Coupling constants through more than three bonds (long-range coupling) in alkanes are generally much smaller than 1 Hz and, thus, not visible in routine 1D NMR spectra. They are, however, much larger than 1 Hz for fixed conformations (e.g., in condensed alicyclic systems, see Chapter 5.4) and in unsaturated compounds (see Chapter 5.2). They are also significant when electronegative substituents are present between the coupling partners, as e.g.:

$$\text{RO}_{\prime\prime\prime\prime}\diagup^{CH_3} \qquad ^4J_{HH}\ 0.7$$
$$\text{RO}\diagup\diagdown_{CH_3}$$

5.2 Alkenes

5.2.1 Substituted Ethylenes

^{1}H Chemical Shifts and Coupling Constants (δ in ppm, J in Hz)

C = C

$^2J_{gem}$	2.5	
$^3J_{cis}$	11.6	
$^3J_{trans}$	19.1	

4.88 Hb   Ha 5.73

4.97 Hc   CH₃ 1.72   d

$^3J_{ab}$	10.0
$^3J_{ac}$	16.8
$^3J_{ad}$	6.4
$^2J_{bc}$	2.1
$^4J_{bd}$	-1.3
$^4J_{cd}$	-1.8

b
H₃C   Ha 5.55
d
Hc   CH₃ 1.58

$^4J_{ab}$	-1.7
$^3J_{ac}$	15.1
$^3J_{ad}$	6.5
$^5J_{bd}$	1.6

Hb   Ha 5.37

H₃C   CH₃ 1.54
c   d

$^3J_{ab}$	10.9
$^4J_{ac}$	-1.8
$^3J_{ad}$	6.8
$^5J_{cd}$	1.2

4.87 Hb   Ha 5.78

4.94 Hc   CH₂ - CH₃
d   e
2.00   1.00

$^3J_{ab}$	10.3	$^4J_{bd}$	-1.3
$^3J_{ac}$	17.2	$^4J_{cd}$	-1.7
$^3J_{ad}$	6.2		
$^2J_{bc}$	2.0		

Geminal and Vicinal Couplings of Alkenes (J in Hz)

The values of the coupling constants strongly depend on the electronegativity of the substituents (see Table on pp 166, 167). They decrease with increasing electronegativity and number of electronegative substituents. The same trend holds for the signed values of geminal coupling constants but not for the absolute values because J_{gem} can be positive or negative. Although the total ranges of cis and trans vicinal coupling constants overlap, $J_{trans} > J_{cis}$ always holds for given substituents.

Typical ranges:	J_{gem}	-4 to 4
	J_{cis}	4 to 12
	J_{trans}	14 to 19

Coupling Over More than Three Bonds in Alkenes (Long-Range Coupling, J in Hz)

Allylic Coupling

$$cisoid: \quad {}^{4}J_{ab} \text{ -3.0 to +2.0}$$
$$transoid: \quad {}^{4}J_{ac} \text{ -3.5 to +2.5}$$

In acyclic systems, the coupling constants range from ca. -0.8 to -1.8 Hz and, usually, $|J|_{cisoid}$ is larger than $|J|_{transoid}$. The magnitudes of the coupling constants depend on the conformation. Largest absolute values are observed if the C–H bond of the substituents overlaps with the π electrons of the double bond (ϕ = 0 or 180°):

ϕ	${}^{4}J_{ab}$	${}^{4}J_{ac}$
0°	-3.0	-3.5
90°	+1.8	+2.2
180°	-3.0	-3.5
270°	0.0	0.8

Homoallylic Coupling

$$cisoid: \quad |{}^{5}J_{ab}| \text{ 0–3}$$
$$transoid: \quad |{}^{5}J_{ac}| \text{ 0–3}$$

The values of homoallylic coupling constants between methyl groups and of allylic ones are comparable:

$$^{5}J_{H_3C-C=C-CH_3} \approx {}^{4}J_{H-C=C-CH_3}$$

In acyclic systems, $|J_{cisoid}| < |J_{transoid}|$ usually holds. Large homoallylic coupling constants are occasionally observed in cyclic systems with fixed conformation between the protons:

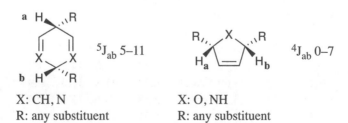

$^{5}J_{ab}$ 5–11

X: CH, N
R: any substituent

$^{4}J_{ab}$ 0–7

X: O, NH
R: any substituent

¹H Chemical Shifts and Coupling Constants of Monosubstituted Ethylenes
(δ in ppm, J in Hz)

	Substituent R	H_a	H_b	H_c	J_{ab}	J_{ac}	J_{bc}	Other
	–H	5.28	5.28	5.28	11.6	19.1	2.5	
C	–CH₃	5.73	4.88	4.97	10.0	16.8	2.1	CH₃ 1.72
	–CH₂CH=CH₂	5.71	4.92	4.95	10.3	16.9	2.2	CH₂ 2.72
	–CH₂–phenyl	5.89	5.00	5.01	10.0	17.0	1.9	CH₂ 3.19
	–cyclopropyl	5.32	4.84	5.04	10.4	17.1	1.8	
	–cyclohexyl	5.79	4.88	4.95	10.5	17.6	1.9	
	–CF₃	5.90	5.56	5.85	11.1	17.5	0.2	
	–CH=C=CH₂	6.31	4.99	5.19	10.1	17.2	1.6	
	–C≡C–CH₃	5.62	5.24	5.39	11.1	17.0	2.3	
	–phenyl	6.72	5.20	5.72	11.1	17.9	1.0	
	–2-naphthyl	6.87	5.32	5.86				
	–2-nitrophenyl	7.19	5.45	5.68	10.7	17.4	1.1	
	3 nitrophenyl	6.74	5.42	5.86	10.9	17.5	0.4	
	–4-nitrophenyl	6.77	5.48	5.90	10.9	17.4	0.8	
	–2-pyridyl	6.84	5.45	6.22	11.3	18.5	1.4	
	–4-pyridyl	6.61	5.42	5.91	10.8	17.6	0.7	
X	–F	6.17	4.03	4.37	4.7	12.8	-3.2	
	–Cl	6.26	5.39	5.48	7.5	14.5	-1.4	
	–Br	6.44	5.97	5.84	7.1	14.9	-1.9	
	–I	6.53	6.23	6.57	7.8	15.9	-1.5	
O	–OH	6.45	3.82	4.18	6.4	14.2	-1.0	
	–OCH₃	6.44	3.88	4.03	7.0	14.1	-2.0	CH₃ 3.16
	–OCH=CH₂	6.49	4.21	4.52	6.4	14.0	-1.8	
	–O–phenyl	6.64	4.40	4.74	6.1	13.7	-1.6	
	–OCHO	7.33	4.66	4.96	6.4	13.9	-1.7	CHO 8.07
	–OCOCH₃	7.28	4.56	4.88	6.3	14.1	-1.6	CH₃ 2.13
	–OCOCH=CH₂	7.39	4.62	4.96	6.4	14.2	-1.6	
	–OCO–phenyl	7.52	4.67	5.04	6.3	13.8	-1.7	
	–OP(O)(O–ethyl)₂	6.58	4.59	4.91	6.0	13.8	-2.1	
N	–NH₂	≈6.05	≈3.99	≈4.04				
	–N⁺(CH₃)₃Br⁻	6.50	5.54	5.76	8.2	15.1	-4.3	
	–NHCOCH₃	≈7.33	≈4.68	≈4.53				
	–NO₂	7.12	5.87	6.55	7.0	14.6	1.4	
	–C≡N	5.69	6.11	6.24	11.8	17.9	0.9	
	–NC	5.90	5.35	5.58	8.6	15.6	-0.5	
	–NCO	6.12	4.77	5.01	7.6	15.2	-0.1	

C = C

	Substituent R	H_a	H_b	H_c	J_{ab}	J_{ac}	J_{bc}	Other
S	–SCH$_3$	6.43	5.18	4.95	10.3	16.4	-0.3	CH$_3$ 2.25
	–S–phenyl	6.53	5.32	5.32	9.6	16.7	-0.2	
	–S(O)CH$_3$	6.77	5.92	6.08	9.8	16.7	-0.6	CH$_3$ 2.61
	–S(O)$_2$CH$_3$	6.76	6.14	6.43	10.0	16.5	-0.5	CH$_3$ 2.96
	–S(O)$_2$CH=CH$_2$	6.67	6.17	6.41	10.0	16.4	-0.6	
	–S(O)$_2$OH	6.73	6.13	6.41	10.2	16.8	-1.2	
	–S(O)$_2$OCH$_3$	6.57	6.22	6.43	10.1	16.9	-0.6	CH$_3$ 3.85
	–S(O)$_2$NH$_2$	6.93	5.98	6.17	10.0	16.3	0	NH$_2$ 6.7
	–S(O)$_2$NH–phenyl	6.56	5.86	6.18	10.1	16.7	-0.3	NH 9.07
	–SF$_5$	6.63	5.64	5.96	9.8	16.6	0.4	
	–SCN	6.19	5.70	5.66				
O ‖ **C**	–CHO	6.37	6.52	6.35	10.0	17.4	1.0	CHO 9.59
	–COCH$_3$	6.30	5.91	6.21	10.7	18.7	1.3	CH$_3$ 2.29
	–COCH=CH$_2$	6.67	5.82	6.28	11.0	17.9	1.4	
	–CO–phenyl	7.20	5.81	6.52	9.9	17.7	2.3	
	–COOH	6.15	5.95	6.53	10.5	17.2	1.8	COOH 12.8
	–COOCH$_3$	6.12	5.83	6.41	10.6	17.4	1.5	CH$_3$ 3.77
	–CONH$_2$	6.48	5.71	6.17	7.9	17.3	5.0	NH$_2$ 7.55
	–CON(CH$_3$)$_2$	6.64	5.55	6.12	9.8	17.0	3.4	
	–COF	6.14	6.25	6.60	10.7	17.3	0.8	
	–COCl	6.35	6.16	6.63	10.6	17.4	0.2	
P	–P(CH$_3$)$_2$	6.23	5.51	5.39	11.8	18.3	2.0	CH$_3$ 0.95
	–P(CH=CH$_2$)$_2$	6.16	5.64	5.59	11.8	18.4	2.0	
	–P(phenyl)$_2$	7.38	6.31	7.07	12.5	18.2	0	
	–PCl$_2$	7.48	6.68	6.64	11.7	18.6	0.4	
	–P(O)(phenyl)$_2$	6.72	6.21	6.25	12.9	18.9	1.8	
	–PSCl$_2$	6.42	5.90	6.13	11.0	17.5	0.3	
	–P(S)(CH$_3$)$_2$	6.60	6.14	6.26	11.8	17.9	1.8	
	–P(S)(phenyl)$_2$	6.82	6.17	6.34	11.7	17.9	1.6	
M	–Li	7.29	6.65	5.91	19.3	23.9	7.1	
	–MgBr	6.66	6.15	5.51	17.7	23.3	7.6	
	–Si(CH$_3$)$_3$	6.11	5.88	5.63	14.6	20.2	3.8	CH$_3$ 0.06
	–Sn(CH=CH$_2$)$_3$	6.39	6.21	5.75	13.4	20.7	3.1	
	–Pb(CH=CH$_2$)$_3$	6.70	6.19	5.46	12.2	19.8	2.1	
	–HgBr	6.45	5.92	5.52	11.9	18.7	3.1	

C = C

Estimation of ^{1}H Chemical Shifts of Substituted Ethylenes (δ in ppm)

$$\delta_{C=CH} = 5.25 + Z_{gem} + Z_{cis} + Z_{trans}$$

Substituent R	Z_{gem}	Z_{cis}	Z_{trans}
–H	0.00	0.00	0.00
C –alkyl	0.45	-0.22	-0.28
–alkyl ring[1]	0.69	-0.25	-0.28
–CH$_2$–aromatic	1.05	-0.29	-0.32
–CH$_2$X, X: F, Cl, Br	0.70	0.11	-0.04
–CHF$_2$	0.66	0.32	0.21
–CF$_3$	0.66	0.61	0.32
–CH$_2$O–	0.64	-0.01	-0.02
–CH$_2$N<	0.58	-0.10	-0.08
–CH$_2$CN	0.69	-0.08	0.06
–CH$_2$S–	0.71	-0.13	-0.22
–CH$_2$CO–	0.69	-0.08	-0.06
–C=C<	1.00	-0.09	-0.23
–C=C< conjugated[2]	1.24	0.02	-0.05
–C≡C–	0.47	0.38	0.12
–aromatic	1.38	0.36	-0.07
–aromatic, fixed[3]	1.60	–	-0.05
–aromatic, o-substituted	1.65	0.19	0.09
X –F	1.54	-0.40	-1.02
–Cl	1.08	0.18	0.13
–Br	1.07	0.45	0.55
–I	1.14	0.81	0.88
O –OC≤ (sp^3)	1.22	-1.07	-1.21
–OC= (sp^2)	1.21	-0.60	-1.00
–OCO–	2.11	-0.35	-0.64
–OP(O)(OCH$_2$CH$_3$)$_2$	1.33	-0.34	-0.66
N –NR$_2$; R: H, C≤ (sp^3)	0.80	-1.26	-1.21
–NR–; R: C= (sp^2)	1.17	-0.53	-0.99
–NCO–R	2.08	-0.57	-0.72
–N=N–phenyl	2.39	1.11	0.67
–NO$_2$	1.87	1.30	0.62
–C≡N	0.27	0.75	0.55

C = C

Substituent R	Z_{gem}	Z_{cis}	Z_{trans}
S –S–	1.11	-0.29	-0.13
–S(O)–	1.27	0.67	0.41
–S(O)$_2$–	1.55	1.16	0.93
–SCO–	1.41	0.06	0.02
–SCN	0.94	0.45	0.41
–SF	1.68	0.61	0.49
O –CHO	1.02	0.95	1.17
‖ –CO–	1.10	1.12	0.87
C –CO– conjugated[2]	1.06	0.91	0.74
–COOH	0.97	1.41	0.71
–COOH conjugated[2]	0.80	0.98	0.32
–COOR	0.80	1.18	0.55
–COOR conjugated[2]	0.78	1.01	0.46
–CON<	1.37	0.98	0.46
–COCl	1.11	1.46	1.01
–P(O)(OCH$_2$CH$_3$)$_2$	0.66	0.88	0.67

$C = C$

1) The increment "alkyl ring" is to be used if the substituent and the double bond are part of a cyclic structure.
2) The increment "conjugated" is to be used if either the double bond or the substituent is conjugated to other substituents.
3) The increment "aromatic, fixed" is to be used if the double bond conjugated to an aromatic ring is part of a fused ring (such as in 1,2-dihydronaphthalene).

Influence of cis- *and* trans-*Substituents on the* ^1H *Chemical Shift of Methyl Groups at the Double Bond in Isobutenes (δ in ppm)*

1.70 H$_3$C   H 4.63
H$_3$C   H

1.68 H$_3$C   H 5.13
1.62 H$_3$C

1.80 H$_3$C   H 5.17
1.88 H$_3$C

1.75 H$_3$C   H 5.78
1.75 H$_3$C   Br

1.65 H$_3$C   H 6.79
1.65 H$_3$C   OCOCH$_3$

1.91 H$_3$C   H 5.63
2.11 H$_3$C   CHO

1.86 H$_3$C   H 5.97
2.06 H$_3$C   COCH$_3$

1.84 H$_3$C   H 5.62
2.12 H$_3$C   COOCH$_3$

1.97 H$_3$C   H 6.01
2.12 H$_3$C   COCl

^{1}H Chemical Shifts and Coupling Constants of Enols (δ in ppm, J in Hz)

C=C

≈16

8.40 H$_a$ ... H$_c$ ≈9.3

H$_b$
5.04

$^3J_{ab}$ 9.7
$^3J_{bc}$ ≈8

≈16

7.90 H$_a$ ... CH$_3$ 2.11

H$_b$
5.60

$^3J_{ab}$ 5.1

5.2.2 Conjugated Dienes

^{1}H Chemical Shifts and Coupling Constants (δ in ppm, J in Hz)

6.27
5.06 H$_a$ H$_f$
H$_b$ H$_e$
H$_c$ H$_d$
5.16

$^3J_{ab}$ 10.2	$^2J_{bc}$ 1.8
$^3J_{ac}$ 17.1	$^5J_{be}$ 1.3
$^3J_{ad}$ 10.4	$^5J_{bf}$ 0.6
$^4J_{ae}$ -0.9	$^5J_{cf}$ 0.7
$^4J_{af}$ -0.8	

6.21 5.61
4.86 H$_a$ H$_f$
H$_b$ CH$_3$
                     e
H$_c$ H$_d$ 1.71
4.98   5.98

$^3J_{ab}$ 10.2	$^2J_{bc}$ 1.9	$^6J_{ce}$ -0.7
$^3J_{ac}$ 16.9	$^4J_{bd}$ -0.8	$^5J_{cf}$ 0.7
$^3J_{ad}$ 10.3	$^6J_{be}$ -0.7	$^4J_{de}$ -1.6
$^5J_{ae}$ 0.4	$^5J_{bf}$ 0.7	$^3J_{df}$ 15.1
$^4J_{af}$ -0.8	$^4J_{cd}$ -0.8	$^3J_{ef}$ 6.6

6.59 1.72
5.03 H$_a$ CH$_3$f
H$_b$ H$_e$
                     5.45
H$_c$ H$_d$
5.11   5.92

$^3J_{ab}$ 10.2	$^2J_{bc}$ 2.1	$^5J_{ce}$ 0.7
$^3J_{ac}$ 16.9	$^4J_{bd}$ -0.8	$^6J_{cf}$ -0.6
$^3J_{ad}$ 10.9	$^5J_{be}$ -0.7	$^3J_{de}$ 10.8
$^4J_{ae}$ -1.1	$^6J_{bf}$ 0.7	$^4J_{df}$ -1.8
$^5J_{af}$ 0.2	$^4J_{cd}$ -0.8	$^3J_{ef}$ 7.0

5.2.3 Allenes

^{1}H Chemical Shifts and Coupling Constants (δ in ppm, J in Hz)

$$^2J_{ab}\ -9 \qquad ^4J_{ac}\ -6$$

$$^5J_{ab}\ 3.2 \qquad ^6J_{ac}\ 0 \qquad ^3J_{ad}\ 6.8 \qquad ^4J_{bd}\ -6.4$$

$C=C$

^{1}H Chemical Shifts and Coupling Constants of Monosubstituted Allenes (δ in ppm, J in Hz)

Substituent R	H_a	H_b	J_{ab}	Other
–H	4.67	4.67	-9.0	
–CH$_3$	4.94	4.50	-6.7	CH$_3$ (**c**) 1.59, $^3J_{ac}$ 7.2, $^5J_{bc}$ 3.4
–CH$_2$CH$_3$	5.03	4.55	-6.8	CH$_2$ (**c**) 1.95, $^3J_{ac}$ 6.2, $^5J_{bc}$ 3.5
–CH$_2$Cl	5.4.3	4.92	-6.6	CH$_2$ (**c**) 4.11, $^3J_{ac}$ 7.7, $^5J_{bc}$ 2.2
–CH=CH$_2$	5.96	4.92	-6.6	CH (**c**) 6.31, $^3J_{ac}$ 10.4, $^5J_{bc}$ 1.1[a]
–Cl	5.76	5.17	-6.1	
–Br	5.85	4.83	-6.1	
–I	5.63	4.48	-6.3	
–phenyl	5.91	4.92		
–OCH$_3$	6.77	5.48	-5.9	
–COCH$_3$	5.77	5.25	-6.4	
–C≡N	4.97	5.04		
–Si(CH$_3$)$_3$	4.92	4.31		
–SiCl$_3$	5.35	4.92	-5.9	
–SnCl$_3$	4.98	4.11	-7.2	

[a] =CH$_2$, H$_{cis}$ (**d**) 5.19, $^4J_{ad}$ -0.8, $^6J_{bd}$ -1.5, $^3J_{cd}$ 17.2; CH$_2$, H$_{trans}$ (**e**) 4.99, $^4J_{ae}$ -0.9, $^6J_{be}$ -1.8, $^3J_{ce}$ 10.1, $^2J_{de}$ 1.6

5.3 Alkynes

^{1}H Chemical Shifts of Substituted Alkynes (δ in ppm)

$$R-C\equiv C-H_a$$

C≡C

	Substituent R	H_a		Substituent R	H_a
C	–H	1.91	**S**	–SCH$_2$CH$_3$	2.79
	–CH$_3$	1.91		–SCH=CH$_2$	3.26
	–CH$_2$CH$_3$	1.97		–S–phenyl	3.28
	–C(CH$_3$)$_3$	2.07		–S(O)$_2$–n-butyl	3.95
	–CF$_3$	2.95	**O**	–COCH$_3$	3.65
	–CH=CH$_2$	3.07	**‖**	–CO–phenyl	3.48
	–C≡CH	2.16	**C**	–COOH	3.17
	–phenyl	3.07		–COOCH$_2$CH$_3$	2.90
	–1-naphthyl	3.43		–CONH$_2$	3.05
X	–F	1.74	**Si**	–Si(CH$_3$)$_3$	2.34
	–Cl	2.05		–Si(phenyl)$_3$	2.47
	Br	2.32	**P**	–P(CH$_2$CH$_3$)$_2$	2.85
	–I	2.34		–P(phenyl)$_2$	3.22
O	–OCH$_2$CH$_3$	1.48		–P(O)(CH$_2$CH$_3$)$_2$	3.33
	–OCH=CH$_2$	2.04		–P(O)(phenyl)$_2$	3.48
	–O–phenyl	2.07			
N	–N(CH$_2$CH$_3$)$_2$	2.30			
	–N(phenyl)$_2$	2.86			
	–C≡N	2.63			

^{1}H,^{1}H Coupling Constants of Substituted Alkynes (J in Hz)

$$\overset{a}{H}=\!\!=\!\!=\overset{b}{CH_3} \qquad {}^4J_{ab}\ 2.9$$

$$\overset{a}{H_3C}=\!\!=\!\!=\overset{b}{CH_3} \qquad |{}^5J_{ab}|\ 2.7$$

$$\overset{a}{H}=\!\!=\!\!=\overset{b}{CH_2}-\overset{c}{CH_3} \qquad \begin{array}{l}{}^4J_{ab}\ 2.6\\ {}^5J_{ac}\ 0\\ {}^3J_{bc}\ 7.4\end{array}$$

$$\overset{a}{H_3C}=\!\!=\!\!=\overset{b}{CH_2}-CH_3 \qquad |{}^5J_{ab}|\ 2.5$$

$$\begin{array}{l}{}^4J_{ab}\ 2.0\\ {}^5J_{ac}\ 1.0\\ {}^6J_{ad}\ 0.6\\ {}^3J_{bc}\ 10.5\\ {}^4J_{bd}\ 1.6\\ {}^3J_{cd}\ 6.5\end{array}$$

$$\begin{array}{l}{}^5J_{ab}\ 0.3\\ {}^6J_{ac}\ -0.1\\ {}^7J_{ad}\ 0.2\end{array}$$

5.4 Alicyclics

¹H Chemical Shifts and Coupling Constants of Saturated Alicyclic Hydrocarbons (δ in ppm, J in Hz)

0.20 $^2J_{gem}$ -4.3 In derivatives:
$^3J_{cis}$ 9.0 $^2J_{gem}$ -3 to -9
$^3J_{trans}$ 5.6 $^3J_{cis}$ 6 to 12
  $^3J_{trans}$ 2 to 9
Throughout:
$J_{cis} > J_{trans}$

1.94 In derivatives:
$^2J_{gem}$ -10 to -17
$^3J_{cis}$ 4 to 12
$^3J_{trans}$ 2 to 10
$^4J_{cis}$ ≈ 0
$^4J_{trans}$ ≈ -1

1.51 In derivatives:
$^2J_{gem}$ -8 to -18
$^3J_{cis}$ 5 to 10
$^3J_{trans}$ 5 to 10

1.44 In derivatives:
$^2J_{gem}$ -11 to -14
$^3J_{ax,ax}$ 8 to 13
At -100 °C $^3J_{eq,ax}$ 2 to 6
H_{ax} 1.12 $^3J_{eq,eq}$ 2 to 5
H_{eq} 1.60 Generally:
$J_{eq,ax}$ ≈ $J_{eq,eq}$ +1

c b 7.01 In derivatives:
a 0.92 $^3J_{ab}$ 1.5 to 2.0
$^3J_{bc}$ 0.5 to 1.5

5.95
c b $^2J_{gem}$ -13.7 $^3J_{ad,cis}$ 1.8
d a $^3J_{ab}$ 1.0 $^3J_{ad,trans}$ 4.6
2.57 $^4J_{ac}$ -0.3 $^3J_{bc}$ 2.8

d c 5.66 $^2J_{gem,a}$ -12.8 $^4J_{bd}$ -2.3
e b 2.27 $^3J_{ab,cis}$ 9.3 $^5J_{be,cis}$ 2.1
a $^3J_{ab,trans}$ 5.7 $^5J_{be,trans}$ 3.0
1.79 $^2J_{gem,b}$ -16.1 $^3J_{cd}$ 5.8
$^3J_{bc}$ 2.3

d c 6.43 $^3J_{ab}$ 1.3 $^4J_{bd}$ 1.1
e b 6.28 $^4J_{ac}$ -1.5 $^5J_{be}$ 2.0
a $^3J_{bc}$ 5.0 $^3J_{cd}$ 1.9
2.80

f a 6.53 $^3J_{ab}$ 5.1 $^4J_{bc}$ -0.2
e b 6.23 $^5J_{ac}$ 0.5 $^4J_{bd}$ -0.4
$^5J_{ad}$ 1.4 $^4J_{be}$ 2.0
d c $^4J_{ae}$ 1.3 $^2J_{cd}$ 0.1
H H 5.89 $^3J_{af}$ 2.0

a
b 5.69 $^3J_{ab}$ ≈10
c 1.99 $^3J_{bc}$ 1.5
d 1.60

a 5.67 $^3J_{ab}$ 9.7
2.08 b 5.79 $^4J_{ac}$ 1.0
c $^5J_{ad}$ 1.1
d $^3J_{bc}$ 5.1

a b 5.79 $^3J_{ab}$ ≈10
c 2.12 $^3J_{bc}$ 3.7
d 1.50
e
1.75

a   b 6.50
g
f
e, e'
2.22
c 6.09
d 5.26

$^3J_{ab}$ 11.2 $^5J_{cg}$ -0.6
$^4J_{ac}$ 0.8 $^3J_{de}$ 6.7
$^3J_{bc}$ 5.5 $^2J_{ee'}$ -13.0
$^3J_{cd}$ 8.9
$^5J_{cf} = {}^4J_{df} = {}^5J_{dg} = 0$

a   b 5.56
c 2.11
d 1.5
e 1.5

$^3J_{ab}$ ≈10
$^3J_{bc}$ 3.7

a'   a 5.79
b'
b 5.59
c 2.14
d 1.47

$^3J_{ab}$ 11.3
$^3J_{aa'}$ 4.1
$^4J_{ab'}$ -0.6
$^5J_{bb'}$ 0.5

1.76
1.87

7.90
7.38
8.31
7.12
7.52

1.18
H   H
H 1.47
H   H 1.16
2.19

1.34 H   H 1.12
6.02 H   H
2.87   H 0.98
H 1.63

2.00
H   H
H   H 6.75
3.58

1.50   1.50

6.25 H
H   H 1.23
2.48
H 1.50

4.60   6.63

In condensed alicyclics, couplings over four bonds are often observed. Such long-range couplings are particularly large if the arrangement of the bonds between the two protons is W-shaped (cf. J_{ac} vs. J_{ad} and J_{bd} below left and J_{ac} vs. J_{bc} below right). Owing to the rigid arrangement, vicinal coupling constants (^3J) may assume unusually small values when the torsional angles are close to 90° (J_{ce} below right).

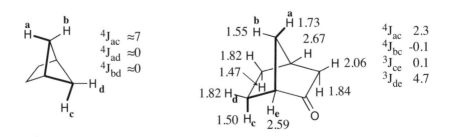

a   b
H   H
H$_d$
H$_c$

$^4J_{ac}$ ≈7
$^4J_{ad}$ ≈0
$^4J_{bd}$ ≈0

b   a
1.55 H   H 1.73
1.82 H   2.67   H
1.47   H 2.06
1.82 H$_d$ H 1.84
1.50 H$_c$ H$_e$ 2.59

$^4J_{ac}$ 2.3
$^4J_{bc}$ -0.1
$^3J_{ce}$ 0.1
$^3J_{de}$ 4.7

¹H Chemical Shifts and Coupling Constants of Monosubstituted Cyclopropanes (δ in ppm, J in Hz)

	Substituent R	H_a	$H_{b;d}$	$H_{c;e}$	$^3J_{ab}$	$^3J_{ac}$	$^2J_{bc}$	$^3J_{bd}$	$^3J_{be}$	$^3J_{ce}$
	–H	0.20	0.20	0.20	9.0	5.6	-4.3	9.0	5.6	9.0
C	–CH₃	1.00	0.35	0.15						
	–CH₂OH	1.14	0.40	0.30						
	–CH=CH₂	1.35	0.64	0.34	8.2	4.9	-4.5	9.3	6.2	9.0
	–phenyl	1.83	0.89	0.65	9.5	6.3	-4.5	9.5	5.2	8.9
X	–F	4.32	0.69	0.27	5.9	2.4	-6.7	10.8	7.7	12.0
	–Cl	2.55	0.87	0.74	7.0	3.6	-6.0	10.3	7.1	10.6
	–Br	2.83	0.96	0.81	7.1	3.8	-6.1	10.2	7.0	10.5
	–I	2.31	1.04	0.76	7.5	4.4	-5.9	9.9	6.6	10.0
O	–OH	3.35	0.40	0.48	6.2	2.9	-5.4	10.3	6.8	10.9
N	–NH₂	2.23	0.32	0.20	6.6	3.6	-4.3	9.7	6.2	9.9
	–NH₃⁺	1.06	0.52	0.34						
	–NO₂	4.21	1.13	1.60	7.0	3.4	-5.5	10.1	8.3	11.3
	–C≡N	1.29	0.96	1.04	8.4	5.1	-4.7	9.2	7.1	9.5
O	–CHO	1.79	0.99	1.03	8.0	4.6	-4.5	8.8	7.0	9.6
‖	–COCH₃	1.83	0.77	0.93	7.9	4.6	-3.5	9.2	7.0	9.5
C	–CO–cyclopropyl	1.70	0.56	1.02	7.9	4.6	-3.5	9.1	7.0	9.5
	–CO–phenyl	2.65	1.01	1.23						
	–COOH	1.59	0.91	1.05	8.0	4.6	-4.0	9.3	7.1	9.7
	–COOCH₃	1.61	0.86	0.98	8.0	4.6	-3.4	8.8	6.9	9.6
	–CONH₂	1.39	0.70	0.95						
	–COF	1.66	1.11	1.20	8.0	4.6	-4.5	10.1	7.5	9.3
	–COCl	2.11	1.18	1.28	7.9	4.4	-4.5	9.2	7.6	10.0
M	–Li	-2.53	0.43	-0.12	10.3	9.1	-1.6	7.7	3.2	6.5
	–MgBr	-2.04	0.25	-0.13	11.0	8.5	-1.7	7.8	3.5	6.6
	–B(cyclopropyl)₂	-0.25	0.61	0.66	8.9	5.8	-3.3	8.2	5.9	8.4
	–Si(cyclopropyl)₃	-0.67	0.49	0.36	9.7	6.9	-3.4	8.4	5.1	8.1
	–P⁺(phenyl)₃	3.28	1.82	0.63						
	–Hg–cyclopropyl	0.00	0.75	0.47	9.6	6.9	-3.7	8.5	4.8	7.9

^{1}H Chemical Shifts and Coupling Constants of Equatorially and Axially Substituted Cyclohexanes (δ in ppm, J in Hz)

Eq. substituent R	H$_{1,ax}$	H$_{2,ax}$	H$_{2,eq}$	H$_{3,ax}$	H$_{3,eq}$	H$_{4,ax}$	H$_{4,eq}$
–D*	1.12	1.12	1.60	1.12	1.60	1.12	1.60
C –C*	1.27	0.81	1.57	1.15	1.60	1.06	1.58
–C≡C*	2.25	1.36	1.98	1.20	1.73	1.17	1.67
–phenyl*	2.46						
X –F*	4.49	1.42	2.15	1.28	1.86	1.12	1.65
–Cl*	3.88	1.58	2.22	1.33	1.84	1.18	1.68
–Br*	4.09	1.75	2.33	1.35	1.80	1.22	1.72
–I*	4.18	1.97	2.45	1.36	1.67	1.30	1.80
O –OH**	3.52	1.22	2.01	1.05	1.78	0.97	
–OCOCH$_3$*	4.74	1.72	1.85	1.35	1.41	1.25	1.55
N –NH$_2$**	2.55	1.03	1.89	1.03	1.76	0.96	
–NHCOCH$_3$**	3.67	1.07	2.01	1.11	1.78	1.01	
–NO$_2$*	4.38	2.23	1.85	1.38	1.85	1.28	1.67
–C≡N**	2.31	1.53	2.16	0.98	1.86	1.03	
S –SH*	2.79	1.34	2.01	1.31	1.75	1.22	1.61
–COOCH$_3$*	2.30	1.44	1.90	1.27	1.75	1.24	1.64

Ax. substituent R	H$_{1,eq}$	H$_{2,ax}$	H$_{2,eq}$	H$_{3,ax}$	H$_{3,eq}$	H$_{4,ax}$	H$_{4,eq}$
–D*	1.60	1.12	1.60	1.12	1.60	1.12	1.60
C –C*	1.93	1.37	1.40	1.39	1.34	1.06	1.58
–C≡C*	2.87	1.48	1.78				
–phenyl*	3.16		2.42				
X –F*	4.94	1.43	2.03	1.63	1.75	1.28	1.58
–Cl*	4.59	1.76	2.00	1.77	1.75	1.26	1.75
–Br*	4.80	1.81	2.08	1.79	1.60	1.24	1.78
–I*	4.96	1.53	2.06	1.72	1.62	1.26	1.73
O –OH**	4.03	1.49	1.83	1.35	1.54	0.99	
–OCOCH$_3$*	5.31	1.49	2.51				
N –NH$_2$**	3.15	1.54	1.65	1.27	1.53	0.96	
–NHCOCH$_3$**	4.11	1.51	1.85	1.03	1.66	1.04	
–NO$_2$**	4.43	1.6	2.6				
–C≡N**	2.96	1.54	2.00	1.50	1.70	1.20	
S –SH**	3.43	1.5	1.9				

* R': –H; ** R': –*tert*-butyl

5.5 Aromatic Hydrocarbons

¹H Chemical Shifts and Coupling Constants (δ in ppm, J in Hz)

7.34 In derivatives:

$^3J_{ortho}$ 6.5–8.5
$^4J_{meta}$ 1.0–3.0
$^5J_{para}$ 0.0–1.0

7.84
7.48

In derivatives:

$^3J_{ab}$ 8–9 $^6J_{af}$ ≈-0.1
$^4J_{ac}$ 1–2 $^5J_{ag}$ ≈0.2
$^5J_{ad}$ ≈1 $^4J_{ah}$ ≈-0.5
$^3J_{bc}$ 5–7 $^7J_{bf}$ ≈0.3
$^5J_{ae}$ ≈0.9 $^6J_{bg}$ ≈0.1

8.01
7.47
8.43

In derivatives:

$^3J_{ab}$ 8.5–9.5
$^4J_{ac}$ 0.8–1.5
$^5J_{ad}$ 0.6–0.9
$^5J_{ae}$ ≈0.8
$^3J_{bc}$ 6.5–8.0
$^4J_{de}$ ≈0.4

8.00
7.39
8.67

7.67
8.70 a c 7.61
d 7.90
e 7.75
f

$^3J_{ab}$ 8.4
$^4J_{ac}$ 1.2
$^5J_{ad}$ 0.7
$^3J_{bc}$ 7.2
$^4J_{bd}$ 1.3
$^3J_{cd}$ 8.1

In derivatives:
$^3J_{ef}$ ≈9

7.68
8.84 7.65
8.13 9.17 7.85
7.56
7.54 7.62
8.05 8.37 7.80

8.08 8.19
8.01

7.66
7.47
8.20

8.90

Weak long-range couplings between aromatic protons and aliphatic substituents are usually not resolved but lead to a characteristic broadening of the corresponding lines.

CH₃ a $^4J_{ab}$ -0.7
$^5J_{ac}$ 0.3
b $^6J_{ad}$ -0.6
c
d

CH₃ a $^5J_{ab}$ ≈0.8
b

Effect of Substituents on ^{1}H Chemical Shifts of Monosubstituted Benzenes (in ppm)

$$\delta_{H_i} = 7.34 + Z_i$$

Substituent R	Z_2	Z_3	Z_4
C –CH$_3$	-0.17	-0.09	-0.17
–CH$_2$CH$_3$	-0.14	-0.05	-0.18
–CH(CH$_3$)$_2$	-0.13	-0.08	-0.18
–C(CH$_3$)$_3$	0.05	-0.04	-0.18
–CF$_3$	0.19	-0.07	0.00
–CCl$_3$	0.55	-0.07	-0.09
–CH$_2$OH	-0.07	-0.07	-0.07
–CH=CH$_2$	0.08	-0.02	-0.09
–CH=CH–phenyl (*trans*)	0.16	0.00	-0.15
–C≡CH	0.16	-0.01	-0.01
–C≡C–phenyl	0.20	-0.04	-0.07
–phenyl	0.22	0.06	-0.04
–2-pyridyl	0.73	0.09	0.02
X –F	-0.31	-0.03	-0.21
–Cl	-0.01	-0.06	-0.12
–Br	0.15	-0.12	-0.06
–I	0.36	-0.24	-0.02
O –OH	-0.51	-0.10	-0.41
–OCH$_3$	-0.44	-0.05	-0.40
–OCH$_2$CH=CH$_2$	-0.45	-0.13	-0.43
–O–phenyl	-0.33	-0.02	-0.25
–OCOCH$_3$	-0.26	0.03	-0.12
–OCO–phenyl	-0.12	0.10	-0.06
–OS(O)$_2$CH$_3$	-0.05	0.07	-0.01
N –NH$_2$	-0.67	-0.20	-0.59
–NHCH$_3$	-0.73	-0.16	-0.64
–N(CH$_3$)$_2$	-0.60	-0.10	-0.62
–N(phenyl)$_2$	-0.26	-0.10	-0.34
–N$^+$(CH$_3$)$_3$ I$^-$	0.72	0.40	0.34
–NHCHO (*trans* to O)	-0.25	0.03	-0.13
–NHCHO (*cis* to O)	-0.20	0.21	-0.01
–N(CH$_3$)CHO	-0.16	0.07	-0.05
–NHCOCH$_3$	0.15	-0.02	-0.23
–NHCSNH$_2$	0.14	0.07	-0.14

	Substituent R	Z_2	Z_3	Z_4
	$-NHNH_2$	-0.60	-0.08	-0.55
	$-N=N-phenyl$	0.67	0.20	0.20
	$-NO$	0.55	0.29	0.35
	$-NO_2$	0.93	0.26	0.39
	$-C\equiv N$	0.32	0.14	0.28
	$-NCS$	-0.11	0.04	-0.02
S	$-SH$	-0.08	-0.16	-0.22
	$-SCH_3$	-0.08	-0.10	-0.24
	$-S-phenyl$	-0.06	-0.20	-0.26
	$-S-S-phenyl$	0.13	-0.05	-0.10
	$-S(O)-CH=CH_2$	0.28	0.15	0.15
	$-S(O)-phenyl$	0.29	0.09	0.13
	$-S(O)_2CH_3$	0.70	0.37	0.41
	$-S(O)_2OCH_3$	0.60	0.26	0.28
	$-S(O)_2Cl$	0.68	0.27	0.37
	$-S(O)_2NH_2$	0.51	0.28	0.24
O	$-CHO$	0.54	0.19	0.29
‖	$-COCH_3$	0.62	0.12	0.22
C	$-COCH_2CH_3$	0.61	0.11	0.21
	$-CO-phenyl$	0.56	0.12	0.23
	$-CO-(2-pyridyl)$	0.86	0.11	0.20
	$-COOH$	0.79	0.14	0.28
	$-COOCH_3$	0.70	0.09	0.21
	$-COOCH(CH_3)_2$	0.73	0.11	0.20
	$-COO-phenyl$	0.87	0.18	0.30
	$-CONH_2$	0.48	0.11	0.19
	$-COF$	0.71	0.21	0.38
	$-COCl$	0.77	0.15	0.35
	$-COBr$	0.70	0.15	0.32
	$-CH=N-phenyl$	0.64	0.24	0.24
M	$-Li$	0.77	0.26	-0.29
	$-MgBr$	0.40	-0.19	-0.26
	$-Mg-phenyl$	-0.49	0.18	0.25
	$-Si(CH_3)_3$	0.19	0.00	0.00
	$-Si(phenyl)_2Cl$	0.32	0.07	0.12
	$-SiCl_3$	0.52	0.20	0.20
	$-P(phenyl)_2$	0.0	0.0	0.0
	$-P(O)(OCH_3)_2$	0.46	0.14	0.22
	$-Pb^+(phenyl)\ Cl^-$	0.30	0.49	0.61
	$-Zn-phenyl$	-0.36	0.02	0.05
	$-Hg-phenyl$	0.06	0.10	-0.10

¹H-¹H Coupling Constants in Selected Monosubstituted Benzenes (|J| in Hz)

	Substituent R	J_{23}	J_{24}	J_{25}	J_{26}	J_{34}	J_{35}
C	–CH₃	7.7	1.3	0.6	2.0	7.5	1.5
	–CH=CH₂	7.8	1.1	0.6	1.9	7.4	1.5
	–C≡CH	7.8	1.3	0.6	1.7	7.6	1.3
	–phenyl	7.8	1.2	0.6	2.0	7.5	1.4
X	–F	8.4	1.1	0.4	2.7	7.5	1.8
	–Cl	8.1	1.1	0.5	2.3	7.5	1.7
	–Br	8.0	1.1	0.5	2.2	7.4	1.8
	–I	7.9	1.1	0.5	1.9	7.5	1.8
O	–OH	8.2	1.1	0.5	2.7	7.4	1.7
	–OCH₃	8.3	1.0	0.4	2.7	7.4	1.8
	–O–phenyl	8.3	1.1	0.5	2.6	7.4	1.7
	–OCOCH₃	8.2	1.1	0.5	2.5	7.5	1.7
N	–NH₂	8.0	1.1	0.5	2.5	7.4	1.6
	–NHCOCH₃	8.2	1.2	0.5	2.4	7.4	1.5
	–NO₂	8.4	1.2	0.5	2.4	7.5	1.5
	–C≡N	7.8	1.3	0.7	1.8	7.7	1.3
S	–SH	7.9	1.2	0.6	2.1	7.5	1.5
	–S(O)₂OCH₃	8.0	1.2	0.6	2.0	7.6	1.4
O ‖ C	–CHO	7.7	1.3	0.6	1.8	7.5	1.3
	–COCH₃	8.0	1.3	0.6	1.8	7.5	1.3
	–COOH	7.9	1.3	0.6	1.9	7.4	1.4
	-COOCH₃	7.9	1.4	0.6	1.8	7.5	1.3
	–CONH₂	7.9	1.2	0.6	2.0	7.5	1.3
	–COCl	8.0	1.2	0.6	2.0	7.5	1.4
M	–Li	6.7	1.5	0.8	0.7	7.4	1.3
	–MgBr	6.9	1.5	0.7	0.7	7.4	1.4
	–P(phenyl)₂	7.6	1.2	0.6	1.7	7.4	1.4
	–PO(OCH₃)₂	7.7	1.4	0.6	1.6	7.6	1.4
	–Zn–phenyl	6.6	2.1	0.7	0.8	7.4	1.5
	–Hg–phenyl	7.5	1.4	0.6	1.1	7.5	1.5

¹H Chemical Shifts and Coupling Constants of Condensed Aromatic-Alicyclic Hydrocarbons (δ in ppm, J in Hz)

7.08   2.91
6.99
2.04

7.40
d
7.26
7.19
7.47

c 3.39
b
6.55
a 6.88

$^3J_{ab}$ 5.5
$^4J_{ac}$ 2.0
$^5J_{ad}$ 0.7
$^3J_{bc}$ 1.9

7.01   2.85
6.93
1.60

7.84
7.38
7.28
3.87   7.55

3.91   7.31
7.19

2.87
7.23
7.22
7.75 7.29

a   3.34
b 7.11
c 7.31
d   7.46

$^4J_{ab}$ 1.5
$^5J_{ac}$ 0
$^6J_{ad}$ 0.5
$^3J_{bc}$ 6.7
$^4J_{bd}$ 1.2
$^3J_{cd}$ 8.1

a 7.15
b 7.90
c 7.58
d 7.79

$^4J_{ab}$ 0
$^5J_{ac}$ 0
$^6J_{ad}$ 0
$^3J_{bc}$ 7
$^4J_{bd}$ 0.6
$^3J_{cd}$ 8

7.04   0.73
1.54
0.51   1.08   2.62

-4.03
0.72
H
1.45
2.23; 2.91
6.86

Effect of Substituents in Position 1 on the ^{1}H Chemical Shifts of Monosubstituted Naphthalenes (in ppm)

for R: H $\delta_{H_1}, \delta_{H_4}, \delta_{H_5}, \delta_{H_8} = 7.84$

$\delta_{H_2}, \delta_{H_3}, \delta_{H_6}, \delta_{H_7} = 7.48$

	Substituent R	H-2	H-3	H-4	H-5	H-6	H-7	H-8
C	–CH$_3$	-0.20	-0.14	-0.17	-0.03	-0.03	0.00	0.12
	–CH$_2$CH$_3$	-0.15	-0.08	-0.15	0.00	-0.02	0.01	0.21
	–CH$_2$C≡CH	0.09	-0.23	-0.23	-0.17	-0.13	-0.03	0.52
	–CH$_2$Cl	-0.10	-0.18	-0.11	-0.07	-0.05	0.02	0.22
	–CF$_3$	0.51	-0.01	0.01	0.06	0.07	0.13	0.35
	–CH$_2$OH	-0.07	-0.10	-0.09	-0.01	-0.02	0.01	0.18
	–CH$_2$NH$_2$	-0.14	-0.13	-0.14	-0.05	-0.07	-0.03	0.10
	–C≡CH	0.22	-0.14	-0.08	-0.08	-0.04	0.05	0.51
	–phenyl	-0.11	-0.04	-0.06	0.05	-0.13	-0.07	-0.02
X	–F	-0.35	-0.10	-0.23	0.00	0.05	0.03	0.27
	–Cl	0.06	-0.14	-0.12	-0.02	0.02	0.09	0.42
	–Br	0.29	-0.25	-0.04	-0.01	0.04	0.11	0.39
	–I	0.20	-0.46	0.13	-0.22	-0.09	-0.02	0.18
O	–OH	-0.75	-0.22	-0.42	-0.05	-0.03	-0.01	0.32
	–OCH$_3$	-0.84	-0.25	-0.55	-0.18	-0.12	-0.13	0.33
	–O–phenyl	-0.53	-0.10	-0.22	0.03	0.04	0.00	0.37
	–OCOCH$_3$	-0.31	-0.05	-0.27	-0.14	-0.23	-0.09	-0.01
N	–NH$_2$	-0.76	-0.22	-0.55	-0.07	-0.05	-0.08	-0.09
	–N(CH$_3$)$_2$	-0.46	-0.13	-0.36	-0.06	-0.03	-0.06	0.38
	–NHCOCH$_3$	0.24	0.01	-0.12	0.08	0.04	0.08	0.27
	–NO$_2$	0.71	0.02	0.24	0.08	0.04	0.16	0.69
	–C≡N	0.42	0.04	0.23	0.07	0.14	0.21	0.38
	–NCO	-0.25	-0.13	-0.18	-0.03	0.05	0.01	0.21
	–NCS	-0.13	-0.13	-0.14	-0.04	0.03	0.08	0.21
O‖C	–CHO	0.46	0.07	0.20	0.05	0.06	0.17	1.39
	–COCH$_3$	0.40	-0.02	0.09	0.01	0.02	0.09	0.90
	–COOH	0.72	0.14	0.34	0.20	0.13	0.20	1.09
	–COOCH$_3$	0.65	-0.08	0.08	-0.05	-0.02	0.09	1.09
	–COCl	1.03	0.06	0.22	0.02	0.07	0.17	0.88

Effect of Substituents in Position 2 on the ^{1}H Chemical Shifts of Monosubstituted Naphthalenes (in ppm)

for R: H $\delta_{H_1}, \delta_{H_4}, \delta_{H_5}, \delta_{H_8} = 7.84$

$\delta_{H_2}, \delta_{H_3}, \delta_{H_6}, \delta_{H_7} = 7.48$

	Substituent R	H-1	H-3	H-4	H-5	H-6	H-7	H-8
C	$-CH_3$	-0.24	-0.18	-0.11	-0.06	-0.09	-0.05	-0.10
	$-CH_2CH_3$	-0.22	-0.14	-0.08	-0.05	-0.08	-0.06	-0.08
	$-CH(CH_3)_2$	-0.24	-0.15	-0.12	-0.10	-0.12	-0.10	-0.10
	$-CF_3$	0.28	0.14	0.06	-0.10	0.09	0.06	-0.10
	$-CH_2OH$	-0.13	-0.08	-0.07	-0.05	-0.04	-0.03	-0.08
	$-CH=CH_2$	-0.11	0.14	-0.06	-0.06	-0.06	-0.04	-0.06
	$-C \equiv CH$	0.19	0.04	-0.05	-0.03	0.02	0.02	-0.03
	$-$phenyl	0.20	0.25	0.06	0.01	0.02	-0.02	0.05
X	$-Cl$	-0.04	-0.08	-0.10	-0.05[a]	-0.03[b]	-0.01[b]	-0.12[a]
	$-Br$	0.14	0.05	-0.16	-0.12	-0.02	0.00	-0.06
O	$-OH$	-0.72	-0.39	-0.10	-0.09	-0.16	-0.06	-0.18
	$-OCH_3$	-0.76	-0.33	-0.14	-0.10	-0.14	-0.06	-0.14
	$-O$–phenyl	-0.53	-0.22	-0.01	-0.02	-0.08	-0.04	-0.15
	$-OCOCH_3$	-0.30	-0.27	-0.04	-0.04	-0.04	-0.02	-0.08
N	$-NH_2$	-0.93	-0.62	-0.23	-0.19	-0.27	-0.15	-0.27
	$-N(CH_3)_2$	-1.07	-0.49	-0.30	-0.29	-0.39	-0.24	-0.33
	$-NHCOCH_3$	0.33	-0.02	-0.10	-0.11	-0.09	-0.06	-0.09
	$-NO_2$	0.90	0.70	0.05	0.05	0.19	0.15	0.14
	$-C \equiv N$	0.40	0.13	0.08	0.06	0.19	0.13	0.06
S	$-SH$	-0.14	-0.19	-0.17	-0.11	-0.09	-0.06	-0.19
O **‖** **C**	$-CHO$	0.44	0.45	0.05	0.03	0.14	0.08	0.12
	$-COCH_3$	0.58	0.51	0.01	0.01	0.08	0.03	0.01
	$-CO$–phenyl	0.42	0.46	0.09	0.06	0.13	0.06	0.06
	$-COOH$	0.83	0.57	0.20	0.19	0.20	0.16	0.31
	$-COOCH_3$	0.66	0.50	-0.08	-0.07	-0.01	-0.05	0.00
	$-COCl$	0.85	0.58	0.22	0.32	0.17	0.21	0.20

[a] interchangeable; [b] interchangeable

5.6 Heteroaromatic Compounds

5.6.1 Non-Condensed Heteroaromatic Rings

^{1}H Chemical Shifts and Coupling Constants (δ in ppm, $|J|$ in Hz)

Furan:
c—b 6.38 $^3J_{ab}$ 1.8
d—a 7.42 $^4J_{ac}$ 0.9
O $^4J_{ad}$ 1.5
$^3J_{bc}$ 3.4

Pyrrole:
c—b 6.23 $^3J_{ab}$ 2.6
d—a 6.71 $^4J_{ac}$ 1.3
N $^4J_{ad}$ 2.1
H e 8.1, $^3J_{ae}$ 2.6
broad $^3J_{bc}$ 3.5
$^4J_{be}$ 2.6

Thiophene:
c—b 7.09 $^3J_{ab}$ 4.8
d—a 7.31 $^4J_{ac}$ 1.0
S $^4J_{ad}$ 2.8
$^3J_{bc}$ 3.5

Selenophene:
c—b 7.23 $^3J_{ab}$ 5.4
d—a 7.88 $^4J_{ac}$ 1.5
Se $^4J_{ad}$ 2.3
$^3J_{bc}$ 3.7

Isoxazole:
N—b 7.15 $^3J_{ab}$ 0.8
c a 7.68 $^4J_{ac}$ 0.5
7.90 O $^4J_{bc}$ 0.0

Imidazole:
N—b 7.13 $^3J_{ab}$ 1–2
c a 7.13 $^4J_{ac}$ 1.0
7.74 N $^4J_{bc}$ 1.0
H 13.4

Thiazole:
N—b 7.98 $^3J_{ab}$ 3.2
c a $^4J_{ac}$ 1.9
8.88 S 7.41 $^4J_{bc}$ 0.0

Oxazole:
8.31 c—b 6.38 $^3J_{ab}$ 1.7
N a 8.49 $^4J_{ac}$ 0.3
O $^3J_{bc}$ 1.8

Pyrazole:
7.74 c—b 6.10 $^3J_{ab}$ 2.1
N a 7.74 $^4J_{ac}$ 0.0
N $^3J_{bc}$ 2.1
H 13.7

Isothiazole:
8.54 c—b 7.26 $^3J_{ab}$ 4.7
N a 8.72 $^4J_{ac}$ 0.4
S $^3J_{bc}$ 1.7

Oxadiazole:
8.19
N O N

Triazole:
7.75
N N
N
H ≈12

Oxadiazole:
N—N
O 8.33

Triazole:
N—N
8.27
N
H 13.5

Thiadiazole:
8.58
N S N

Imidazolium:
HN—b 7.58 $^4J_{ac}$ 1.4
9.18 c a 7.58 $^3J_{ad} + {}^4J_{bd} = 4.4$
N $^3J_{cd}$ 2.4
d H 13.5
Cl⁻

Solvent: $^3J_{ab}$ 6.0

	CDCl$_3$	DMSO	$^4J_{ac}$ 1.9
a	8.59	8.58	$^5J_{ad}$ 0.9
b	7.25	7.38	$^4J_{ae}$ 0.4
c	7.62	7.75	$^3J_{bc}$ 7.6
			$^4J_{bd}$ 1.6

Solvent: $^3J_{ab}$ 6.0

	CDCl$_3$*	DMSO**	$^4J_{ac}$ 1.6
a	9.00	8.98	$^5J_{ad}$ 0.8
b	7.97	8.14	$^4J_{ae}$ 1.0
c	8.43	8.67	$^3J_{bc}$ 7.9
			$^4J_{bd}$ 1.4

* p-toluylsulfonate

** HSO$_3^-$

7.32 **c**

d **b** 7.40

e **a** 8.19

$^3J_{ab}$ 6.5
$^4J_{ac}$ 1.1
$^5J_{ad}$ 0.6
$^4J_{ae}$ 1.9
$^3J_{bc}$ 7.7
$^4J_{bd}$ 2.1

(in acetone)

9.88 **c** **b** 9.48 $^3J_{ab}$ 2.7

a 8.84 $^4J_{ac}$ 0.0

$^5J_{bc}$ 2.2

9.23

c

d **b** 7.56

a 9.22

$^3J_{ab}$ 4.9
$^4J_{ac}$ 2.0
$^5J_{ad}$ 3.5
$^3J_{bc}$ 8.4

7.22
8.54 **c**
d **b** 7.83
a 8.26

$^3J_{ab}$ 5.3
$^4J_{ac}$ 1.0
$^5J_{ad}$ 1.0
$^3J_{bc}$ 8.0
$^4J_{bd}$ 2.5
$^3J_{cd}$ 6.5

c

b 7.38

d

9.27 **a** 8.78

$^3J_{ab}$ 5.0
$^4J_{ac}$ 2.5
$^4J_{ad}$ 0
$^5J_{bd}$ 1.5

8.24
c
b 7.34
d
8.98 **a** 8.43

$^3J_{ab}$ 6.8
$^4J_{ac}$ 1.6
$^4J_{ad}$ 2.0
$^3J_{bc}$ 4.9
$^5J_{bd}$ 1.0
$^4J_{cd}$ 0

8.63

c **b** 8.11
d **a** 8.44

$^3J_{ab}$ 4.1
$^5J_{ac}$ 0.8
$^4J_{ad}$ 0.6
$^4J_{bc}$ 0.4

Effect of Substituents on the ^{1}H Chemical Shifts of Monosubstituted Furans (in ppm)

$$\delta_{H-2} = 7.42 + Z_{i2}$$
$$\delta_{H-3} = 6.38 + Z_{i3}$$
$$\delta_{H-4} = 6.38 + Z_{i4}$$
$$\delta_{H-5} = 7.42 + Z_{i5}$$

Substituent	H_3	H_4	H_5	H_2	H_4	H_5
	in position 2 or 5			in position 3 or 4		
	Z_{23} Z_{54}	Z_{24} Z_{53}	Z_{25} Z_{52}	Z_{32} Z_{45}	Z_{34} Z_{43}	Z_{35} Z_{42}
C –CH$_3$	-0.45	-0.15	-0.17	-0.25	-0.17	-0.12
–CH$_2$CH$_3$	-0.42	-0.12	-0.14			
–CH$_2$OH	-0.12	-0.07	-0.05	-0.07	0.00	-0.06
–CH$_2$SH	-0.22	-0.09	-0.09			
–CH$_2$SCH$_3$	-0.21	-0.09	-0.08			
–CH=CHCOCH$_3$ (*trans*)	0.29	0.11	0.08			
X –Br	-0.23	-0.17	-0.17			
–I	0.04	-0.21	-0.05	-0.17	-0.04	-0.26
O –OCH$_3$	-1.26	-0.14	-0.57	-0.50	-0.36	-0.41
N –NO$_2$	1.13	0.47	0.47			
–C≡N	0.48	-0.02	-0.04	0.41	0.14	-0.06
S –SCH$_3$	0.05	0.01	0.13	-0.22	-0.13	-0.19
–SCN	0.32	-0.02	0.06	0.15	0.11	-0.01
O –CHO	0.92	0.25	0.31	0.92	0.47	0.19
‖ –COCH$_3$	0.81	0.16	0.18	0.42	0.28	-0.16
C –COCO–2-furyl	1.26	0.27	0.37			
–COOH	0.97	0.19	0.24	0.70	0.40	0.03
–COOCH$_3$	0.81	0.14	0.18	0.60	0.37	0.01
–COCl	1.14	0.32	0.46			
M –P(–x-furyl)$_2$	0.25[a]	-0.12[a]	0.03[a]	-0.16[b]	-0.10[b]	-0.09[b]
–P(O)(–x-furyl)$_2$	0.76[a]	0.15[a]	0.30[a]	0.14[b]	0.19[b]	0.31[b]
–P(S)(–x-furyl)$_2$	0.77[a]	0.12[a]	0.27[a]	0.10[b]	0.18[b]	0.30[b]
–P$^+$(CH$_3$)(2-furyl)$_2$ I$^-$	1.53	0.49	0.77			
–HgCl				-0.09	0.02	0.25
–Hg–x-furyl	0.18[a]	0.24[a]	0.47[a]	-0.10[b]	0.10[b]	-0.10[b]

[a] x = 2, [b] x = 3

Effect of Substituents on the ^{1}H Chemical Shifts of Monosubstituted Pyrroles (in ppm)

$\delta_{H\text{-}1} \approx 8$, broad, solvent-dependent

$\delta_{H\text{-}2} = 6.71 + Z_{i2}$

$\delta_{H\text{-}3} = 6.23 + Z_{i3}$

$\delta_{H\text{-}4} = 6.23 + Z_{i4}$

$\delta_{H\text{-}5} = 6.71 + Z_{i5}$

Substituent in position 1	H_3 Z_{12} Z_{15}	H_4 Z_{13} Z_{14}
–CH$_3$	-0.13	-0.11
–CH$_2$CH$_3$	-0.16	-0.12
–CH$_2$CH$_2$CN	-0.05	-0.07
–CH$_2$–phenyl	-0.12	-0.04
–phenyl	0.36	0.11
–N(CH$_3$)$_2$	0.11	-0.19
–COCH$_3$	0.56	0.12
–CO–phenyl	0.57	0.18
–Si(CH(CH$_3$)$_2$)$_3$	0.08	0.08

	Substituent	H_3 Z_{23} Z_{54}	H_4 Z_{24} Z_{53}	H_5 Z_{25} Z_{52}	H_2 Z_{32} Z_{45}	H_4 Z_{34} Z_{43}	H_5 Z_{35} Z_{42}
		in position 2 or 5			in position 3 or 4		
C	–CH$_3$	5.72	5.89	6.36	-0.33	-0.16	-0.26
N	–NO$_2$	7.11	6.29	7.05	1.06	0.24	0.43
	–C≡N	6.88	6.28	7.13	0.83	0.23	0.51
S	–SCH$_3$	6.23	6.10	6.72	0.18	0.05	0.10
	–SCN	6.53	6.15	6.90	0.48	0.10	0.28
O	–CHO	7.01	6.34	7.18	0.78	0.11	0.47
‖	–COCH$_3$	6.93	6.26	7.06	0.70	0.03	0.35
C	–COOCH$_3$	6.84	6.18	6.91	0.79	0.13	0.29

Effect of Substituents on the ¹H Chemical Shifts of Monosubstituted Thiophenes (in ppm)

$$\delta_{H\text{-}2} = 7.31 + Z_{i2}$$
$$\delta_{H\text{-}3} = 7.09 + Z_{i3}$$
$$\delta_{H\text{-}4} = 7.09 + Z_{i4}$$
$$\delta_{H\text{-}5} = 7.31 + Z_{i5}$$

	Substituent	H_3	H_4	H_5	H_2	H_4	H_5
		in position 2 or 5			in position 3 or 4		
		Z_{23} Z_{54}	Z_{24} Z_{53}	Z_{25} Z_{52}	Z_{32} Z_{45}	Z_{34} Z_{43}	Z_{35} Z_{42}
C	–CH₃	-0.34	-0.20	-0.24	-0.45	-0.22	-0.15
	–C≡C	0.02	-0.29	-0.23			
	–phenyl				0.11	0.28	0.05
	–2-thienyl	0.08	-0.09	-0.11			
	–2-pyridyl	0.48	0.01	0.06			
X	–F	-0.78	-0.54	-0.86	-0.80	-0.40	-0.31
	–Cl	-0.30	-0.35	-0.39	-0.25	-0.17	-0.09
	–Br	-0.05	-0.23	-0.10	-0.23	-0.21	-0.21
	–I	0.11	-0.34	-0.01	-0.05	-0.13	-0.30
O	–OH*	-0.85	0.44	-3.21			
	–OCH₃	-0.93	-0.41	-0.82	-1.10	-0.36	-0.17
N	–NH₂	-1.08	-0.58	-0.96	-1.36	-0.66	-0.36
	–NO₂	0.69	-0.16	0.19	0.84	0.47	-0.08
	–C≡N	0.34	-0.13	0.17	0.52	0.07	0.04
S	–SH	-0.13	-0.33	-0.18	-0.33	-0.33	-0.21
	–SCH₃	-0.16	-0.31	-0.16	-0.44	-0.23	-0.14
	–S(O)₂CH₃	0.90	0.07	0.68	0.85	0.35	0.35
	–S(O)₂Cl	0.60	-0.07	0.34			
	–SCN	0.17	-0.18	0.17	0.14	-0.08	-0.06
O‖C	–CHO	0.69	0.13	0.47	0.81	0.44	0.06
	–COCH₃	0.60	0.03	0.32	0.74	0.45	0.01
	–CO–phenyl	0.55	0.06	0.40			
	–COOH	0.67	-0.05	0.29	0.93	0.48	0.03
	–COOCH₃	0.70	0.00	0.22	0.67	0.34	-0.16
	–CONHNH₂	0.63	0.04	0.41	-7.31	-7.09	-7.31
	–COCl	0.75	-0.07	0.33	0.94	0.37	-0.08

* Keto form

Effect of Substituents on the ^{1}H Chemical Shifts of Monosubstituted Pyridines (in ppm)

$$\delta_{H\text{-}2} = 8.59 + Z_{i2}$$
$$\delta_{H\text{-}3} = 7.25 + Z_{i3}$$
$$\delta_{H\text{-}4} = 7.62 + Z_{i4}$$
$$\delta_{H\text{-}5} = 7.25 + Z_{i5}$$
$$\delta_{H\text{-}6} = 8.59 + Z_{i\,6}$$

	Substituent in position 2 or 6	H_3 Z_{23} Z_{65}	H_4 Z_{24} Z_{63}	H_5 Z_{25} Z_{63}	H_6 Z_{26} Z_{62}
C	–CH$_3$	-0.11	-0.08	-0.15	-0.11
	–CH$_2$CH$_3$	-0.09	0.01	-0.15	0.03
	–CH$_2$–phenyl	0.03	-0.06	0.04	-0.04
	–CH$_2$OH	0.14	0.03	-0.08	-0.14
	–CH=CH$_2$	-0.07	-0.14	-0.23	-0.12
	–phenyl	0.42	0.02	-0.09	0.07
	–2-pyridyl	1.27	0.04	-0.11	0.00
X	–F	-0.30	0.16	-0.05	-0.36
	–Cl	0.09	0.02	0.00	-0.10
	–Br	0.26	-0.06	0.03	-0.23
	–I	0.49	-0.29	0.04	-0.23
O	–OH*	-0.63	-0.13	-0.93	-1.17
	–OCH$_3$	-0.51	-0.10	-0.41	-0.43
N	–NH$_2$	-0.76	-0.24	-0.63	-0.54
	–NHCH$_2$CH$_3$	-0.87	-0.22	-0.69	-0.52
	–N(CH$_3$)$_2$	-0.77	-0.23	-0.73	-0.44
	–NHNH$_2$	-0.55	-0.17	-0.58	-0.48
	–NHCOCH$_3$	1.00	0.09	-0.19	-0.32
	–NHN=CH–2-pyridyl	0.21	-0.01	-0.42	-0.36
	–NO$_2$	0.93	0.44	0.45	0.00
	–C≡N	0.52	0.26	0.35	0.15
S	–SH	0.34	-0.20	-0.42	-0.91
O ‖ C	–CHO	0.73	0.26	0.31	0.21
	–COCH$_3$	0.80	0.22	0.24	0.10
	–CO–phenyl	0.81	0.27	0.25	0.13
	–COOH	0.87	0.41	0.44	0.17
	–COOCH$_3$	0.91	0.24	0.27	0.17
	–CONH$_2$	0.98	0.24	0.22	-0.01
	–CH=N–NH–2-pyridyl	0.76	0.05	-0.06	-0.03
	–Si(CH$_3$)$_3$	0.15	-0.22	-0.24	0.09

* Keto form (2-pyridone)

Substituent	H_2	H_4	H_5	H_6	H_2	H_3
	in position 3 or 5				in position 4	
	Z_{32} Z_{56}	Z_{34} Z_{54}	Z_{35} Z_{53}	Z_{36} Z_{52}	Z_{42} Z_{46}	Z_{43} Z_{45}
C –CH$_3$	-0.15	-0.16	-0.07	-0.17	-0.13	-0.13
–CH$_2$CH$_3$	-0.13	-0.14	-0.06	-0.17	-0.12	-0.14
–CH$_2$–phenyl	-0.08	-0.18	-0.04	-0.14	0.00	-0.15
–phenyl	0.25	0.20	0.08	-0.03		
–CH=CH$_2$					-0.12	-0.08
X –F	-0.05	-0.21	0.04	-0.13	-0.07	-0.03
–Cl	0.09	0.00	0.05	-0.05	0.00	0.05
–Br	0.09	0.18	-0.04	-0.07	0.09	0.35
O –OH	-0.31	-0.29	0.06	-0.50		
–OCH$_3$	-0.27	-0.37	-0.04	-0.40	-0.16	-0.42
–OCOCH$_3$	-0.15	-0.15	0.08	-0.13		
N –NH$_2$	-0.51	-0.65	-0.20	-0.60	-0.15	-0.74
–N(CH$_3$)$_2$					-0.38	-0.77
–NHCOCH$_3$	0.37	0.50	0.06	0.16	-0.19	0.16
–C≡N	0.32	0.38	0.25	0.26	0.24	0.32
S –S–phenyl					0.05	-0.16
–S(O)$_2$OH	0.70	1.14	0.81	0.70		
O‖C –CHO	0.52	0.58	0.30	0.28	0.31	0.49
–COCH$_3$	0.58	0.61	0.20	0.20	0.21	0.50
–CO–phenyl					0.23	0.35
–COOH	0.54	0.57	0.20	0.24	0.20	0.45
–COOCH$_3$	0.64	0.67	0.16	0.19	0.19	0.61
–COO–phenyl					0.24	0.75
–CONH$_2$	0.49	0.50	0.15	0.15		
–CSNH$_2$	0.68	0.67	0.24	0.26	0.35	0.68
–CH=NOH	0.39	0.43	0.19	0.15	0.06	0.32
–Si(CH$_3$)	0.08	0.00	-0.21	-0.11	-0.08	0.01

5.6.2 Condensed Heteroaromatic Rings

^1H Chemical Shifts and Coupling Constants (δ in ppm, $|J|$ in Hz)

Benzofuran (first structure)

7.55 c
b 6.69
7.20 d
a 7.54
7.25 e
f
7.47

$^3J_{ab}$ 2.5
$^5J_{bf}$ 0.9
$^3J_{cd}$ 7.9
$^4J_{ce}$ 1.2
$^5J_{cf}$ 0.8
$^3J_{de}$ 7.3
$^4J_{df}$ 0.9
$^3J_{ef}$ 8.4

(all other coupling constants negligible)

Indole (second structure)

7.64 c
b 6.52
7.12 d
a 7.05
7.18 e
f
N
7.27
H g 7.81

(chemical shifts in CDCl$_3$, coupling constants in acetone)

$^3J_{ab}$ 3.1
$^3J_{ag}$ 2.5
$^5J_{bf}$ 0.7
$^4J_{bg}$ 2.0
$^3J_{cd}$ 7.8
$^4J_{ce}$ 1.2
$^5J_{cf}$ 0.9
$^5J_{cg}$ 0.8
$^3J_{de}$ 7.1
$^4J_{df}$ 1.3
$^3J_{ef}$ 8.1

(all other coupling constants negligible)

Benzothiophene (third structure)

7.82 c
b 7.33
7.36 d
a 7.42
7.33 e
f
7.88

$^3J_{ab}$ 5.5
$^5J_{bf}$ 0.8
$^3J_{cd}$ 8.0
$^4J_{ce}$ 1.1
$^5J_{cf}$ 0.9
$^3J_{de}$ 7.2
$^4J_{df}$ 1.0
$^3J_{ef}$ 8.0

(all other coupling constants negligible)

Benzoxazole (fourth structure)

7.79 b
7.41 c
N
a 8.10
7.34 d
O
e
7.58

(chemical shifts in CDCl$_3$, coupling constants in acetone)

$^5J_{ab}$ 0.2
$^6J_{ac}$ -0.1
$^6J_{ad}$ 0.4
$^5J_{ae}$ 0.0
$^3J_{bc}$ 8.2
$^4J_{bd}$ 1.0
$^5J_{be}$ 0.7
$^3J_{cd}$ 7.4
$^4J_{ce}$ 1.2
$^3J_{de}$ 8.3

Benzimidazole (fifth structure)

7.70 b
7.26 c
N
a 8.08
7.26 d
N
e
H
7.70 12.5*

* in DMSO

$^3J_{bc}$ 8.2
$^4J_{bd}$ 1.4
$^5J_{be}$ 0.7
$^3J_{cd}$ 7.1

(all other coupling constants negligible)

Benzothiazole (sixth structure)

7.94 b
7.51 c
N
a 8.97
7.46 d
S
e
8.14

(chemical shifts in CDCl$_3$, coupling constants in acetone)

$^5J_{ab}$ 0.1
$^6J_{ac}$ -0.2
$^6J_{ad}$ 0.4
$^5J_{ae}$ 0.1
$^3J_{bc}$ 8.2
$^4J_{bd}$ 1.1
$^5J_{be}$ 0.6
$^3J_{cd}$ 7.2
$^4J_{ce}$ 1.1
$^3J_{de}$ 8.2

7.78 8.10
b   **a**
7.13 **c**
7.36 **d**
e   H**f** 13.1
7.58

(in DMSO)

$^5J_{ae}$ 0.8
$^3J_{bc}$ 7.8
$^4J_{bd}$ 1.2
$^5J_{be}$ 1.0
$^3J_{cd}$ 7.0
$^4J_{ce}$ 1.2
$^3J_{de}$ 7.9
(all other coupling constants negligible)

7.83
a
7.42 **b**
c
d

$^3J_{ab}$ 9.1
$^4J_{ac}$ 1.1
$^5J_{ad}$ 0.9
$^3J_{bc}$ 6.4

7.95
a
7.39 **b**
c
d   H ≈16

$^3J_{ab}$ 8.3
$^4J_{ac}$ 1.0
$^5J_{ad}$ 0.9
$^3J_{bc}$ 7.0

(chemical shifts in CDCl$_3$, coupling constants in acetone)

7.97
a
7.53 **b**
c
d

$^3J_{ab}$ 8.9
$^4J_{ac}$ 1.2
$^5J_{ad}$ 0.9
$^3J_{bc}$ 6.7

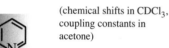

8.61
a
7.61 **b**
7.66 **c**
d
8.08

$^3J_{ab}$ 8.4
$^4J_{ac}$ 1.0
$^5J_{ad}$ 0.8
$^3J_{bc}$ 7.0
$^4J_{bd}$ 1.0
$^3J_{cd}$ 7.9

9.21
b
8.99 **c**
a 8.70
H**d** 13.5

(in DMSO)

7.25
d   **c** 6.28
6.50 **e**
6.31 **f**
g   **a** 7.14
7.76   **b** 6.64

$^3J_{ab}$ 2.7
$^4J_{ac}$ 1.2
$^5J_{ad}$ 0.5
$^3J_{bc}$ 3.9
$^6J_{bf}$ 0.5
$^5J_{cg}$ 1.0
$^3J_{de}$ 9.0
$^4J_{df}$ 1.0
$^5J_{dg}$ 1.2
$^3J_{ef}$ 6.4
$^4J_{eg}$ 1.0
$^3J_{fg}$ 6.8
(all other coupling constants negligible)

7.68
c
6.97 **d**
6.55 **e**
f   **a** 7.60   **b** 7.67
8.05

$^3J_{ab}$ 1.2
$^4J_{af}$ 0.7
$^3J_{cd}$ 9.0
$^4J_{ce}$ 1.3
$^5J_{cf}$ 1.3
$^3J_{de}$ 6.6
$^4J_{df}$ 1.3
$^3J_{ef}$ 6.8
(all other coupling constants negligible)

Imidazo[1,5-a]pyridine (top left)

Shifts: c 7.34, b 7.27, d 6.58, e 7.41, f 7.88, a 7.97

$^5J_{ac}$ 1.0
$^5J_{bf}$ 0.5
$^3J_{cd}$ 9.2
$^4J_{ce}$ 1.1
$^5J_{cf}$ 1.1
$^3J_{de}$ 6.4
$^4J_{df}$ 0.9
$^3J_{ef}$ 7.1
(all other coupling constants negligible)

Pyrazolo[1,5-a]pyridine (top right)

Shifts: c 7.44, b 6.58, d 6.97, e 6.62, f 8.39, a 7.80

$^3J_{ab}$ 2.2
$^6J_{ad}$ 0.5
$^5J_{bf}$ 0.9
$^3J_{cd}$ 8.9
$^4J_{ce}$ 1.2
$^5J_{cf}$ 1.0
$^3J_{de}$ 6.8
$^4J_{df}$ 1.0
$^3J_{ef}$ 6.9
(all other coupling constants negligible)

1,4-Benzodioxine (second row left)

Shifts: a 6.52, b 6.71, c, d, 5.77

$^3J_{ab}$ 7.9
$^4J_{ac}$ 1.5
$^5J_{ad}$ 0.4
$^3J_{bc}$ 7.9

1,4-Benzodithiine (second row right)

Shifts: a 7.19, b 7.12, c, d, 6.42

$^3J_{ab}$ 7.8
$^4J_{ac}$ 1.3
$^5J_{ad}$ 1.1
$^3J_{bc}$ 7.1

Coumarin (2H-chromen-2-one) (third row left)

Shifts: c 7.46, b 7.70, d 7.23, e 7.47, f 7.22, a 6.36

$^3J_{ab}$ 9.6
$^3J_{cd}$ 7.7
$^4J_{ce}$ 1.6
$^3J_{de}$ 7.4
$^4J_{df}$ 1.1
$^3J_{ef}$ 8.4
(all other coupling constants negligible)

Chromone (4H-chromen-4-one) (third row right)

Shifts: c 8.21, d 7.42, e 7.68, f 7.47, b 6.34, a 7.88

$^3J_{ab}$ 6.0
$^3J_{cd}$ 8.0
$^4J_{ce}$ 1.8
$^5J_{cf}$ 0.5
$^3J_{de}$ 7.0
$^4J_{df}$ 1.1
$^3J_{ef}$ 8.4

Quinoline (bottom left)

Shifts: d 7.82, c 8.12, e 7.55, b 7.39, f 7.72, a 8.92, g 8.15

$^3J_{ab}$ 4.2
$^4J_{ac}$ 1.8
$^3J_{bc}$ 8.2
$^5J_{cg}$ 0.8
$^3J_{de}$ 8.2
$^4J_{df}$ 1.4
$^5J_{dg}$ 0.7
$^3J_{ef}$ 6.9
$^4J_{eg}$ 1.2
$^3J_{fg}$ 8.5
(all other coupling constants negligible)

Isoquinoline (bottom right)

Shifts: d 7.72, c 7.55, e 7.60, b 8.50, f 7.51, g 7.86, a 9.22

$^5J_{ac}$ 1.0
$^5J_{ad}$ 0.9
$^3J_{bc}$ 5.8
$^5J_{cg}$ 0.9
$^3J_{de}$ 8.3
$^4J_{df}$ 1.2
$^5J_{dg}$ 0.8
$^3J_{ef}$ 6.9
$^4J_{eg}$ 1.2
$^3J_{fg}$ 8.3
(all other coupling constants negligible)

7.90   7.77
d   c
7.68 e   b 7.32
7.79 f
        g
      8.77 O (N→O)

$^3J_{ab}$ 6.1
$^4J_{ac}$ 1.0
$^3J_{bc}$ 8.5
$^5J_{cg}$ 0.9
$^3J_{de}$ 8.2
$^4J_{df}$ 1.2
$^5J_{dg}$ 0.3
$^3J_{ef}$ 7.2
$^4J_{eg}$ 1.4
$^3J_{fg}$ 8.6

(all other coupling constants negligible)

a 8.55

7.72   7.68
d   c
7.62* e   b 8.14
7.60* f
        g   a
      7.79   8.78   (N→O)

$^4J_{ab}$ 1.7
$^3J_{bc}$ 7.0

* assignment uncertain

8.01   8.18
c   b
7.86 d   a 9.29
7.95 e
      f
     8.44
N–N

$^3J_{ab}$ 5.9
$^5J_{bf}$ 0.8
$^3J_{cd}$ 7.8
$^4J_{ce}$ 1.5
$^5J_{cf}$ 0.8
$^3J_{de}$ 6.9
$^4J_{df}$ 1.3
$^3J_{ef}$ 8.6

7.93   9.41
c   b
7.93* d   a 9.35
7.67* e
      f
     8.06
N, N

$^4J_{ab}$ 0.0
$^5J_{bf}$ 0.5
$^3J_{cd}$ 8.0
$^4J_{ce}$ 1.3
$^5J_{cf}$ 0.9
$^3J_{de}$ 7.0
$^4J_{df}$ 1.3
$^3J_{ef}$ 8.6

* assignment uncertain

8.13
c
7.79 d   N   b
7.95 e
      f   N
                a 8.85

$^3J_{cd}$ 8.4
$^4J_{ce}$ 1.4
$^5J_{cf}$ 0.7
$^3J_{de}$ 6.9

7.93   9.44
c   b
7.85 d   N
     e   N
      f   a

$^5J_{ac}$ 0.4
$^3J_{cd}$ 8.2
$^4J_{ce}$ 1.2
$^5J_{cf}$ 0.6
$^3J_{de}$ 6.8

d
     N
        c 8.97
     N
   a   b 7.58
  8.40

$^3J_{ab}$ 8.0
$^4J_{ac}$ 1.8
$^5J_{ad}$ 0.6
$^3J_{bc}$ 4.1

7.93
d
8.76 e   N   c 9.10
     N
   f   a   b 7.52
  9.28 8.28

$^3J_{ab}$ 8.2
$^4J_{ac}$ 1.9
$^5J_{ad}$ 0.9
$^3J_{bc}$ 4.1
$^3J_{de}$ 6.0
$^5J_{df}$ 0.9

(all other coupling constants negligible)

9.66
d
N   N
8.73 e   c 9.14
   f   a   b 7.67
  7.72 8.26

$^3J_{ab}$ 8.4
$^4J_{ac}$ 1.6
$^5J_{ad}$ 0.9
$^3J_{bc}$ 4.2
$^5J_{df}$ 0.9
$^3J_{ef}$ 5.6

(all other coupling constants negligible)

N   N
        c 9.13
     a   b 7.50
    8.21

$^3J_{ab}$ 8.2
$^4J_{ac}$ 2.0
$^3J_{bc}$ 4.3

7.96
d
c 7.35
b 7.46
a
7.57

$^3J_{ab}$ 8.5
$^4J_{ac}$ 0.9
$^5J_{ad}$ 0.6
$^3J_{bc}$ 7.3
$^4J_{bd}$ 1.3
$^3J_{cd}$ 7.6

8.08
e
d 7.24
c 7.42
b
a 7.42
8.03

$^5J_{ae}$ 0.7
$^3J_{bc}$ 8.1
$^4J_{bd}$ 1.0
$^5J_{be}$ 0.7
$^3J_{cd}$ 7.2
$^4J_{ce}$ 1.2
$^3J_{de}$ 7.9

8.16
d
c 7.45
b 7.45
a
7.85

$^3J_{ab}$ 8.0
$^4J_{ac}$ 1.1
$^5J_{ad}$ 0.7
$^3J_{bc}$ 7.2
$^4J_{bd}$ 1.2
$^3J_{cd}$ 8.1

7.79
d
c 7.58
b 7.48
a
O 7.97

$^3J_{ab}$ 7.7
$^4J_{ac}$ 1.1
$^5J_{ad}$ 0.7
$^3J_{bc}$ 7.5
$^4J_{bd}$ 1.1
$^3J_{cd}$ 7.6

7.77
d
c 7.61
b 7.51
a
7.80

$^3J_{ab}$ 7.7
$^4J_{ac}$ 1.1
$^5J_{ad}$ 0.6
$^3J_{bc}$ 7.5
$^4J_{bd}$ 1.0
$^3J_{cd}$ 7.8

8.59 7.86
e **d**
c 7.43
b 7.71
a
8.22

$^3J_{ab}$ 8.9
$^4J_{ac}$ 1.1
$^5J_{ad}$ 0.8
$^5J_{ae}$ 0.9
$^3J_{bc}$ 6.7
$^4J_{bd}$ 1.5
$^3J_{cd}$ 8.4
$^4J_{de}$ -0.5

O 8.32
d
c 7.36
b 7.70
a
7.46

$^3J_{ab}$ 8.4
$^4J_{ac}$ 1.1
$^5J_{ad}$ 0.5
$^3J_{bc}$ 7.1
$^4J_{bd}$ 1.8
$^3J_{cd}$ 8.0

O 8.27
e
d 7.27
c 7.74
b 7.57
a 11.70

$^5J_{ae}$ 0.4
$^3J_{bc}$ 8.6
$^4J_{bd}$ 1.0
$^5J_{be}$ 0.4
$^3J_{cd}$ 7.0
$^4J_{ce}$ 1.4
$^3J_{de}$ 8.2

5.7 Halogen Compounds

5.7.1 Fluoro Compounds

^{19}F (natural abundance 100%) has a spin quantum number I of 1/2. The signals of ^{1}H atoms are split by coupling to ^{19}F up to a distance of about four bonds.

^{1}H Chemical Shifts and Coupling Constants (δ in ppm, |J| in Hz)

4.27
CH$_3$F ^{2}J$_{HF}$ 46.4

5.45
CH$_2$F$_2$ ^{2}J$_{HF}$ 50.2

6.25
CHF$_3$ ^{2}J$_{HF}$ 79.2

1.24
b

F ^{2}J$_{aF}$ 46.4
        ^{3}J$_{bF}$ 25.2
a
4.36 ^{3}J$_{ab}$ 6.9

1.56
b

F ^{2}J$_{aF}$ 57.3
        ^{3}J$_{bF}$ 20.9
a
F 5.94 ^{3}J$_{ab}$ 4.5

1.68

0.97   4.30
F

Hal

1.34

F

4.37
H b   F ^{2}J$_{aF}$ 87.4 ^{3}J$_{ab}$ 12.8
              ^{3}J$_{bF}$ 20.1 ^{3}J$_{ac}$ 4.7
H$_c$ H$_a$ ^{3}J$_{cF}$ 52.4 ^{2}J$_{bc}$ -3.2
4.03   6.17

1.45
H——————F
        ^{3}J$_{HF}$ 15.3

0.27
c
H
e H$_b$ F
H
H$_d$ H$_a$
0.69   4.32

^{2}J$_{aF}$ 64.9
^{3}J$_{bF}$ 9.9
^{3}J$_{cF}$ 21.0
^{3}J$_{ab}$ 5.9
^{3}J$_{ac}$ 2.4
^{2}J$_{bc}$ -6.7
^{3}J$_{bd}$ 10.8
^{3}J$_{be}$ 7.7
^{3}J$_{ce}$ 12.0

(in benzene/CFCl$_3$)

1.28   4.49
H   H
1.65   F
H   H 2.15
1.12 H 1.86 H 1.42

1.63   F
H
1.58   H 4.94
H   H
              2.03
1.28 H 1.75 H 1.43

F
e   a 7.03
d   b 7.31
c
7.13

^{3}J$_{aF}$ 8.9 ^{3}J$_{ab}$ 8.4
^{4}J$_{bF}$ 5.7 ^{4}J$_{ac}$ 1.1
^{5}J$_{cF}$ 0.2 ^{5}J$_{ad}$ 0.4
                   ^{4}J$_{ae}$ 2.7
                   ^{3}J$_{bc}$ 7.5
                   ^{4}J$_{bd}$ 1.8

CF$_3$
e   a 7.53
d   b 7.27
c
7.34

^{4}J$_{aF}$ -0.8 ^{3}J$_{ab}$ 7.9
^{5}J$_{bF}$ 0.8 ^{4}J$_{ac}$ 1.2
^{6}J$_{cF}$ -0.7 ^{5}J$_{ad}$ 0.6
                    ^{4}J$_{ae}$ 2.0
                    ^{3}J$_{bc}$ 7.6
                    ^{4}J$_{bd}$ 1.3

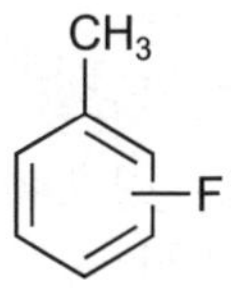

CH₃

Couplings
with CH₃:
^4J_o 2.5
^5J_m 0.0
^6J_p 1.5

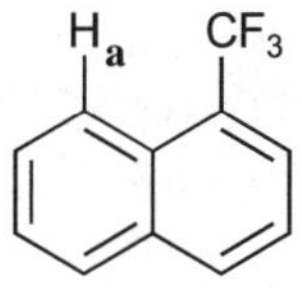

—F

H$_a$ CF₃

$^5J_{aF}$ 2.2

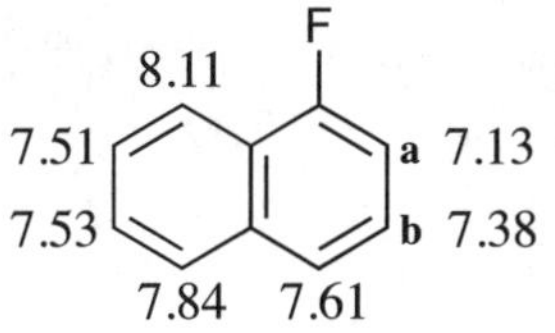

F
8.11
7.51
7.53
7.84 7.61
a 7.13
b 7.38
$^3J_{aF}$ 10.7
$^4J_{bF}$ 5.4

5.7.2 Chloro Compounds

¹H Chemical Shifts and Coupling Constants (δ in ppm, J in Hz)

3.06
CH₃Cl

5.33
CH₂Cl₂

7.26
CHCl₃

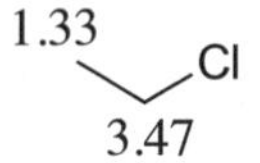

1.33

Cl ^3J 7.2
3.47

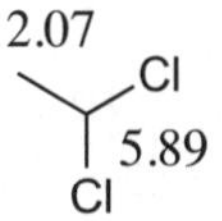

2.07
Cl
5.89
Cl ^3J 6.1

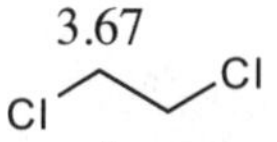

3.67
Cl
Cl ^3J 6.8

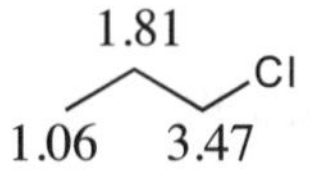

1.81
Cl
1.06 3.47

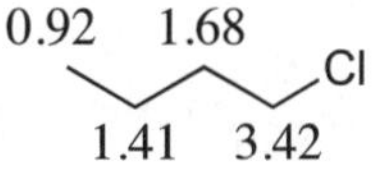

0.92 1.68
Cl
1.41 3.42

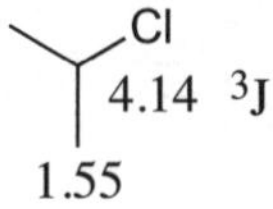

Hal

Cl
4.14 ^3J 6.4
1.55

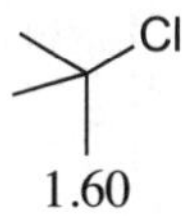

Cl
1.60

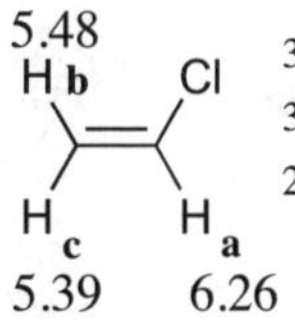

5.48
H b
Cl
H$_c$ H$_a$
5.39   6.26
$^3J_{ab}$ 14.5
$^3J_{ac}$ 7.5
$^2J_{bc}$ -1.4

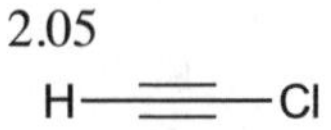

2.05
H——Cl

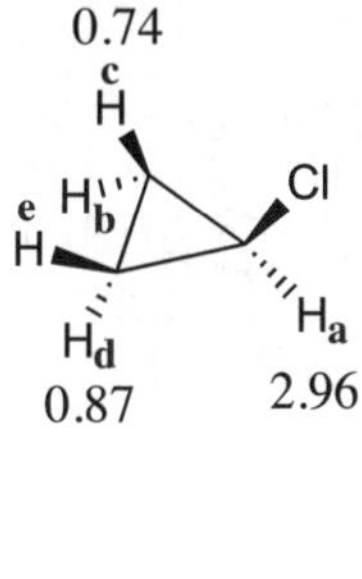

0.74
c
H
e H$_b$ Cl
H
H$_d$ H$_a$
0.87   2.96

$^3J_{ab}$ 7.0
$^3J_{ac}$ 3.6
$^2J_{bc}$ -6.0
$^3J_{bd}$ 10.3
$^3J_{be}$ 7.1
$^3J_{ce}$ 10.6

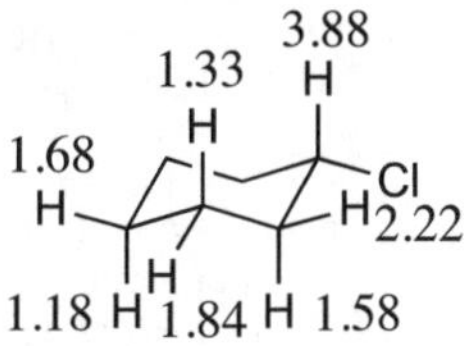

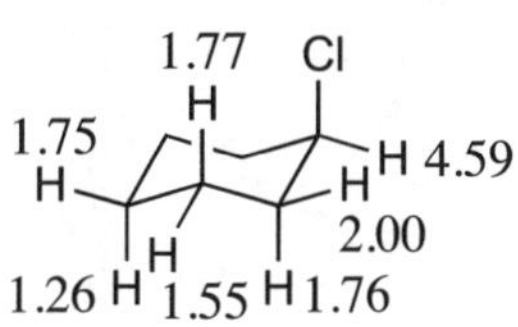

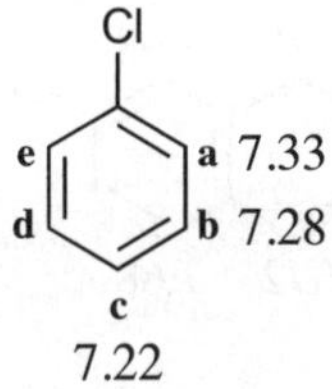

3.88
1.33 H
1.68 H
H——   Cl
H   H 2.22
1.18 H 1.84 H 1.58

1.77 Cl
1.75 H
H——   H 4.59
2.00
1.26 H 1.55 H 1.76

Cl
e b   a 7.33
d   b 7.28
c
7.22

$^3J_{ab}$ 8.1
$^4J_{ac}$ 1.1
$^5J_{ad}$ 0.5
$^4J_{ae}$ 2.3
$^3J_{bc}$ 7.5
$^4J_{bd}$ 1.7

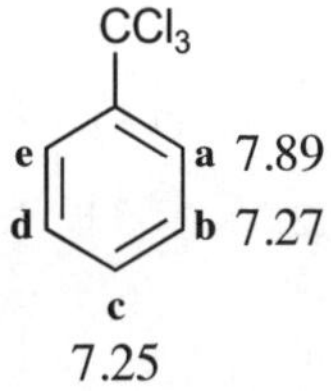

CCl₃
e   a 7.89
d   b 7.27
c
7.25

$^3J_{ab}$ 8.1
$^4J_{ac}$ 1.1
$^5J_{ad}$ 0.5
$^4J_{ae}$ 2.4
$^3J_{bc}$ 7.5
$^4J_{bd}$ 1.4

8.21   Cl
g
7.48 f
7.43 e   a 7.49
               b 7.28
d   c
7.74   7.65

$^3J_{ab}$ 7.7 $^4J_{df}$ 1.0
$^4J_{ac}$ 1.1 $^5J_{dg}$ 0.2
$^3J_{bc}$ 7.8 $^3J_{ef}$ 6.8
$^5J_{cg}$ 0.7 $^4J_{eg}$ 0.6
$^3J_{de}$ 8.1 $^3J_{fg}$ 7.8

7.72* 7.65
               Cl
7.47**
7.45**   7.40
7.79* 7.74

*; **: assignments
interchangeable

5.7.3 Bromo Compounds

^{1}H Chemical Shifts and Coupling Constants (δ in ppm, J in Hz)

2.68   4.94   6.82   1.65
CH_3Br CH_2Br_2 $CHBr_3$ Br
                                    3.36

Hal

2.47
   Br ^3J 6.4
   5.86
   Br

3.63
      Br
Br

1.89
      Br
1.04 3.36

0.93 1.84
         Br
1.46 3.42

   Br
   4.20
1.73

      Br
1.76

5.84
H b   Br $^3J_{ab}$ 14.9
             $^3J_{ac}$ 7.1
H   H $^2J_{bc}$ -1.9
c   a
5.97   6.44

2.32
H━━━━Br

0.81
c
H
e H   Br
  b
H
Hd   Ha
0.96   2.83

$^3J_{ab}$ 7.1
$^3J_{ac}$ 3.8
$^2J_{bc}$ -6.1
$^3J_{bd}$ 10.2
$^3J_{be}$ 7.0
$^3J_{ce}$ 10.5

         4.09
1.35 H
   H
1.72   Br
H   H 2.33
1.22 H 1.80 H 1.75

1.79 Br
   H
1.78   H 4.81
H   H
              2.08
1.24 H 1.60 H 1.81

   Br
e   a 7.49
       b 7.22
d
   c
7.28

$^3J_{ab}$ 8.0
$^4J_{ac}$ 1.1
$^5J_{ad}$ 0.5
$^4J_{ae}$ 2.2
$^3J_{bc}$ 7.4
$^4J_{bd}$ 1.8

            Br
8.19
7.51   7.71
7.44   7.21
7.73 7.70

7.78 7.98
            Br
7.48
7.46   7.53
7.72 7.68

5.7.4 Iodo Compounds

^{1}H Chemical Shifts and Coupling Constants (δ in ppm, J in Hz)

2.16
CH$_3$I

3.90
CH$_2$I$_2$

4.91
CHI$_3$

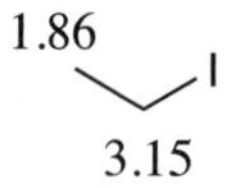

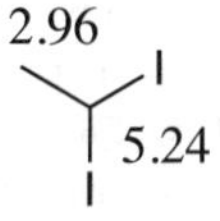

1.86

3.15

2.96

5.24

^{3}J 7.0

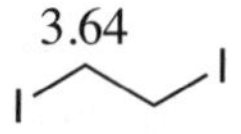

3.64

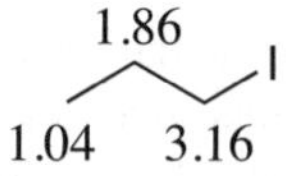

1.86

1.04   3.16

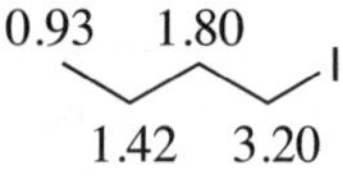

0.93   1.80

1.42   3.20

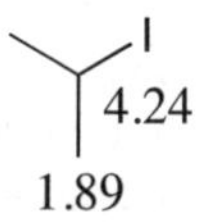

4.24

1.89

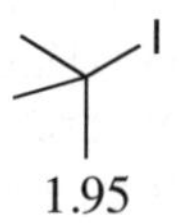

1.95

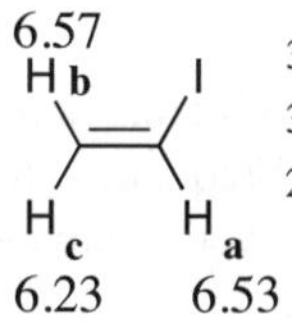

6.57
H b

H$_c$ H$_a$
6.23   6.53

^{3}J$_{ab}$ 15.9
^{3}J$_{ac}$ 7.8
^{2}J$_{bc}$ -1.5

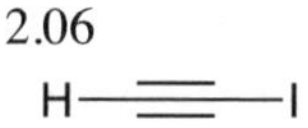

2.06

H━━━━I

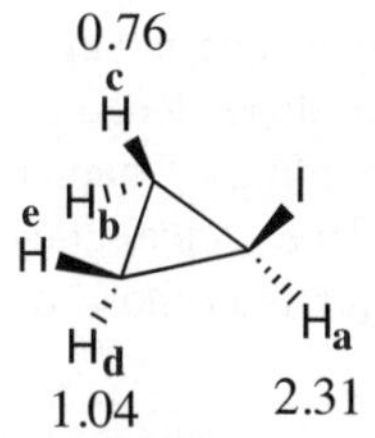

Hal

0.76
H$_c$

e H$_b$
H

H$_d$ H$_a$
1.04   2.31

^{3}J$_{ab}$ 7.5
^{3}J$_{ac}$ 4.4
^{2}J$_{bc}$ -5.9
^{3}J$_{bd}$ 9.9
^{3}J$_{be}$ 6.6
^{3}J$_{ce}$ 10.0

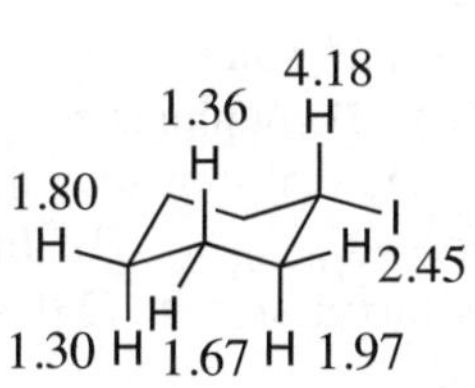

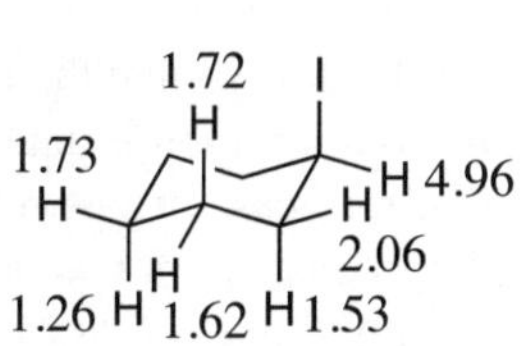

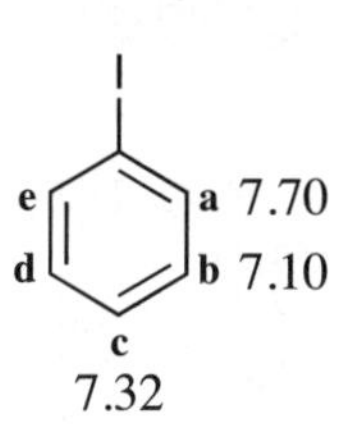

4.18
1.36   H
H
1.80
H
2.45
1.30 H 1.67 H 1.97

1.72
1.73   H
H
H 4.96
1.26 H 1.62 H 1.53
2.06

e

d   b 7.10
c
7.32

a 7.70

^{3}J$_{ab}$ 8.0
^{4}J$_{ac}$ 1.1
^{5}J$_{ad}$ 0.5
^{4}J$_{ae}$ 2.2
^{3}J$_{bc}$ 7.4
^{4}J$_{bd}$ 1.8

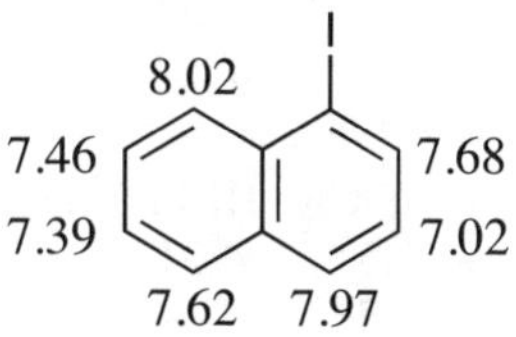

8.02
7.46
7.39

7.68
7.02
7.62   7.97

5.8 Alcohols, Ethers, and Related Compounds

5.8.1 Alcohols

^{1}H Chemical Shifts and Coupling Constants (δ in ppm, J in Hz)

Aliphatic and alicyclic alcohols: δ_{OH} = 0.5–3.0 (in DMSO: 4–6) ppm
Phenols: δ_{OH} = 4.0–8.0 (in DMSO: 8–12) ppm

Hydrogen bonds strongly deshield hydroxyl protons. The position of the signal may depend heavily on the experimental conditions including the concentration of the sample. If a compound contains several kinds of hydroxyl protons (–OH, –COOH, H_2O), in general only one signal at an average position is seen because of rapid exchange. In dimethyl sulfoxide (DMSO) as solvent, this exchange in most cases is so slow that isolated signals are obtained. In this case, the chemical shifts of hydroxyl protons are characteristic. However, if the sample contains strong acids or amine bases, the exchange rate increases and, also in DMSO, a single signal at an average position is observed. Frequently, intermediate exchange rates lead to very broad signals extending over several ppm and, therefore, sometimes not discernible in routine spectra.

As a consequence of fast intermolecular exchange of the hydroxyl protons, their coupling with the protons on the adjacent carbon atoms is usually not observed. However, in very pure (acid-free) solutions or in DMSO, the exchange is sufficiently slow so that the H–O–C–H couplings become visible. Their dependence on the conformation is analogous to that shown by the H–C–C–H couplings (Chapter 5.1.2). In case of fast rotation: $^3J_{HOCH}$ ≈ 5 Hz. In cyclohexanols, the vicinal coupling constants for axial hydroxyl protons (3.0–4.2 Hz) are lower than those of equatorial ones (4.2–5.7 Hz).

b a	c a	c a
CH_3OH	(OH) b	(OH) d b

CDCl$_3$	DMSO	D$_2$O		CDCl$_3$	DMSO	D$_2$O		CDCl$_3$	DMSO	D$_2$O
a 1.13	4.05		a	1.51	4.31		a	1.51	4.31	
b 3.49	3.17	3.34	b	3.71	3.44	3.65	b	3.59	3.34	3.61
$^3J_{ab}$ 5.2			c	1.24	1.06	1.17	c	1.59	1.42	1.57
			$^3J_{ab}$ 4.8				d	0.94	0.84	0.89
			$^3J_{bc}$ 6.9							

	CDCl$_3$	DMSO	D$_2$O
a	1.36	4.30	
b	4.04	3.78	4.02
c	1.22	1.04	1.17
	$^3J_{ab}$ 6.2		

	CDCl$_3$	DMSO	D$_2$O
a	1.50	4.30	
b	3.64	3.38	3.61
c	1.56	1.40	1.51
d	1.39	1.30	1.35
e	0.94	0.87	0.91

	CDCl$_3$	DMSO	D$_2$O
a	1.37	4.19	
b	1.28	1.11	1.24

5.2

⟨benzene⟩—CH$_2$—OH

(in DMSO)

5.6

(⟨benzene⟩)$_2$—CH—OH

(in DMSO)

6.4

(⟨benzene⟩)$_3$—C—OH

(in DMSO)

0.34

$^3J_{ab}$ 6.2
$^3J_{ac}$ 2.9
$^2J_{bc}$ -5.4
$^3J_{bd}$ 10.3
$^3J_{be}$ 6.8
$^3J_{ce}$ 10.9

0.59   3.35

h 1.88   2.46

$^3J_{bc}$ 7.0 $^3J_{cf}$ 2.3
$^3J_{bd}$ 8.1 $^4J_{cg}$ 5.2
$^4J_{be}$ 0 $^3J_{de}$ 10.4
$^4J_{bf}$ -1.1 $^3J_{df}$ 9.7
$^2J_{cd}$ -11.0 $^4J_{dg}$ -0.9
$^3J_{ce}$ 7.9 $^4J_{dh}$ 0
$^3J_{ef}$ -11.0

O

	CDCl$_3$	DMSO
a	1.28	4.33
b	4.32	4.09
c	1.56	1.44
d	1.76	1.61
e	1.76	1.61
f	1.56	1.44

1.05   H 3.52
H
1.26
H OH
H 2.01
0.85   H 1.78 H
0.97   1.22

Derivatives
in DMSO:
δ_{OH} 4.0–4.5
$J_{CH,OH}$ 4.2–5.7

1.35   OH 1.25
H
H 4.03
H
1.83
0.86   H 1.54 H
0.99   1.49

Derivatives
in DMSO:
δ_{OH} 3.8–4.2
$J_{CH,OH}$ 3.0–4.2

OH a

	CDCl$_3$	DMSO
a	4.69	9.29
b	6.83	6.75
c	7.24	7.15
d	6.93	6.76

$^3J_{bc}$ 8.0
$^4J_{bd}$ 1.1
$^5J_{be}$ 0.5
$^4J_{bf}$ 2.2
$^3J_{cd}$ 7.4
$^4J_{ce}$ 1.8

OH 11.1

6.96

8.14

NO$_2$

in DMSO

OH 10.57

NO$_2$

7.16

7.58   8.10

7.00

in CDCl$_3$

¹H Chemical Shifts and Coupling Constants of Enols (δ in ppm, J in Hz)

≈16
H
O
O
$^3J_{ab}$ 9.7
$^3J_{bc}$ ≈8
H_a
H_c
8.40
H_b
≈9.3
5.04

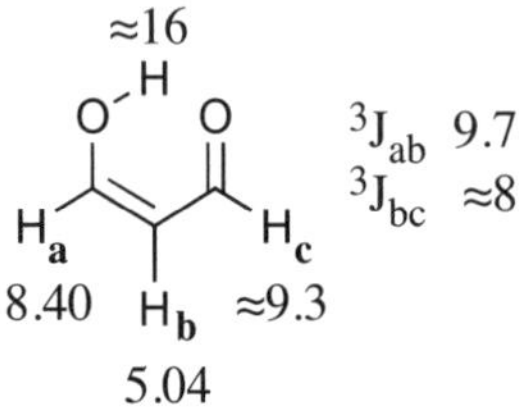

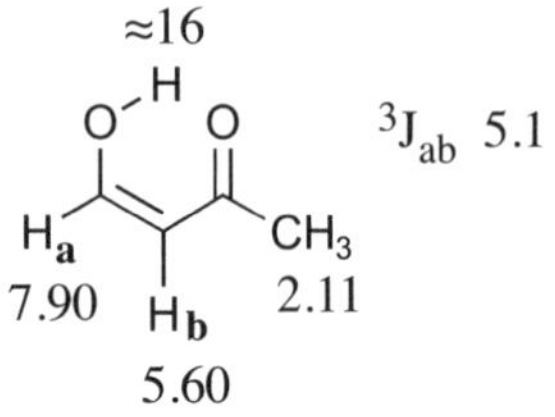

≈16
H
O
O
$^3J_{ab}$ 5.1
H_a
CH_3
7.90
H_b
2.11
5.60

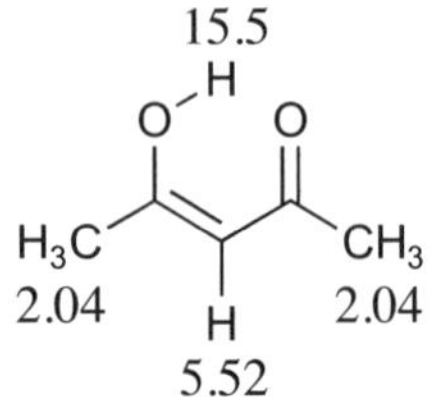

15.5
H
O
O
H_3C
CH_3
2.04
H
2.04
5.52

(in $CDCl_3$, partly enolized; for the keto form, see Chapter 5.11.2)

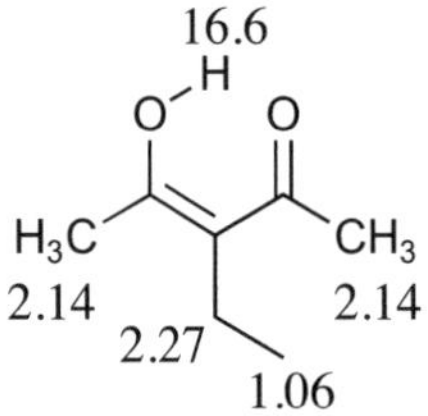

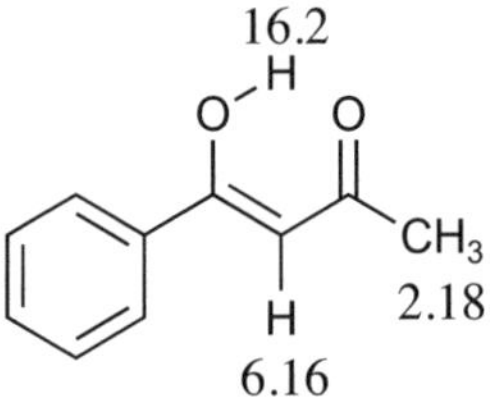

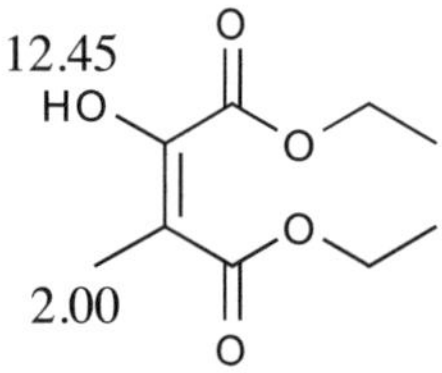

16.6
H
O
O
H_3C
CH_3
2.14
2.14
2.27
1.06

16.2
H
O
O
CH_3
H
2.18
6.16

12.45
HO
O
O
O
O
2.00

(in $CDCl_3$, partly enolized)

5.8.2 Ethers

¹H Chemical Shifts and Coupling Constants (δ in ppm, J in Hz)

3.24
O
$^2J_{gem}$ -10.6

3.48
O
$^3J_{vic}$ 7.0
1.20

3.34
3.34
0.93
O
1.59

3.65
1.13
O
$^3J_{vic}$ 6.1

3.40
1.38
alk
O
1.54
0.92

3.21
1.19
O

b
a
O
O

	$CDCl_3$	DMSO	D_2O
a	3.40	3.24	3.37
b	3.55	3.43	3.60

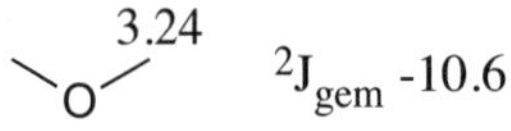

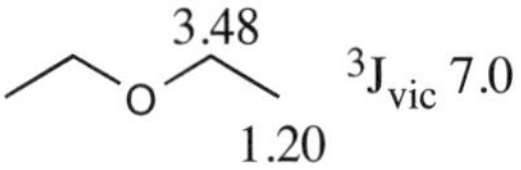

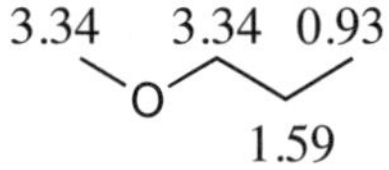

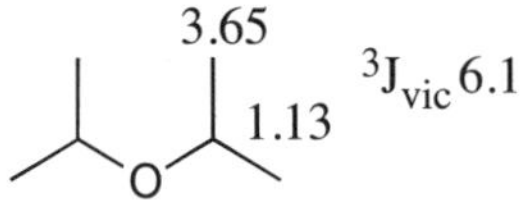

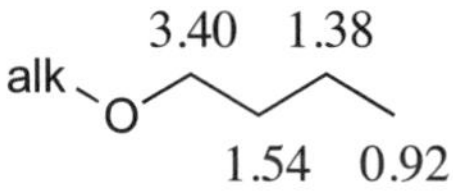

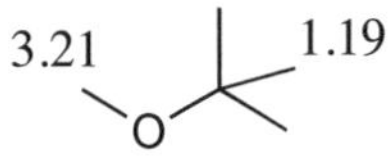

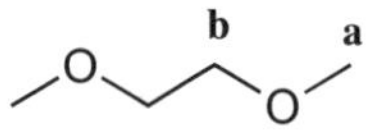

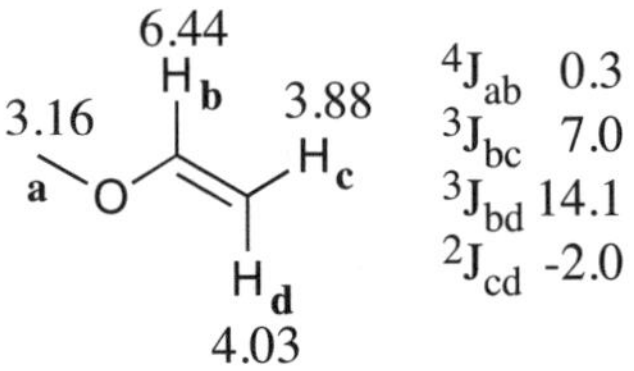

6.44
H_b
3.16
3.88
H_c
a
O
H_d
4.03
$^4J_{ab}$ 0.3
$^3J_{bc}$ 7.0
$^3J_{bd}$ 14.1
$^2J_{cd}$ -2.0

^1H Chemical Shifts and Coupling Constants of Cyclic Ethers (δ in ppm, J in Hz)

(oxirane) 2.54

In derivatives:
$^2J_{gem}$ 5–6
$^3J_{cis}$ 4.5
$^3J_{trans}$ 3.1
Throughout:
$J_{cis} > J_{trans}$

2.78 H$_d$ O CH$_3$ 1.47 (a)
3.07 H$_c$ H$_b$ 3.34

$^3J_{ab}$ 5.2
$^3J_{bc}$ 4.2
$^3J_{bd}$ 3.3
$^2J_{cd}$ 4.2

c   a 4.73
b 2.72
$^2J_{a,gem}$ -5.8
$^2J_{b,gem}$ -11.0
$^3J_{cis}$ 8.7
$^3J_{trans}$ 6.6
$|^4J_{ac}|$ <0.3

	CDCl$_3$	DMSO
a	3.74	3.60
b	1.85	1.76

	CDCl$_3$	DMSO
a	3.65	3.53
b	1.57	1.47
c	1.64	1.58

CDCl$_3$	DMSO	D$_2$O
3.71	3.57	3.75

3.67
2.87
N
H 1.92

3.88
2.57

6.31 d   a 4.31
4.95 c   b 2.58

$^3J_{ab,cis}$ 8.3
$^3J_{ab,trans}$ 10.7
$^3J_{bc}$ 2.5
$^4J_{bd}$ 2.6
$^3J_{cd}$ 2.6

d   a 4.63
c   b 5.89

$^3J_{ab}$ 1.6
$^4J_{ac}$ -2.5
$^4J_{ad,cis}$ 7.1
$^4J_{ad,trans}$ 4.6
$^3J_{bc}$ 6.3

3.96 e   a 6.34
1.85 d   b 4.64
c
1.98

$^3J_{ab}$ 6.2
$^4J_{ac}$ 2.0
$^3J_{bc}$ 3.8
$^4J_{bd}$ 0.6

e   a 6.17
d   b 4.63
c
2.66

$^3J_{ab}$ 7.0
$^4J_{ac}$ 1.7
$^4J_{ae}$ 1.5
$^3J_{bc}$ 3.4

a 7.77
6.38 d   b 6.43
c
7.56

$^3J_{ab}$ 5.0
$^4J_{ac}$ 2.4
$^5J_{ad}$ 1.2
$^3J_{bc}$ 6.3
$^4J_{bd}$ 1.5
$^3J_{cd}$ 9.4

d   a 7.89
c   b 6.34
O

$^3J_{ab}$ 6.0
$^5J_{ac}$ 0.3
$^4J_{ad}$ 2.7
$^4J_{bc}$ 1.1

O

^{1}H Chemical Shifts and Coupling Constants of Aromatic Ethers (δ in ppm, J in Hz)

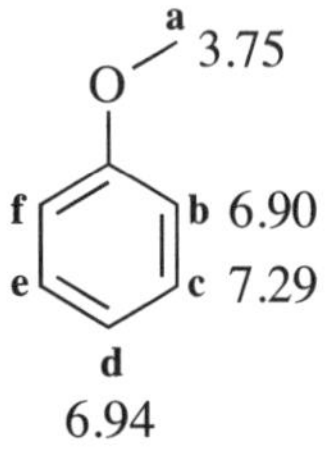

a 3.75
$^5J_{ab} \approx 0.8$
$^3J_{bc}$ 8.3
$^4J_{bd}$ 1.0
f
b 6.90
$^5J_{be}$ 0.4
e
c 7.29
$^4J_{bf}$ 2.7
d
$^3J_{cd}$ 7.4
6.94
$^4J_{ce}$ 1.8

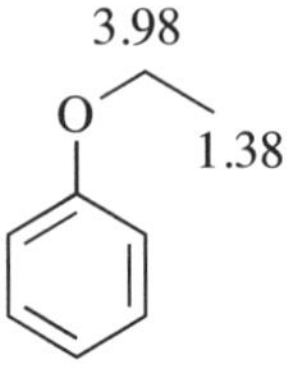

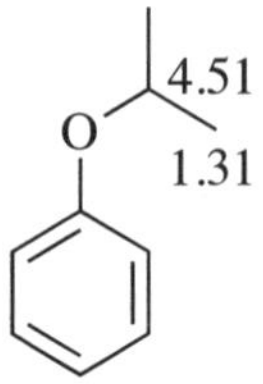

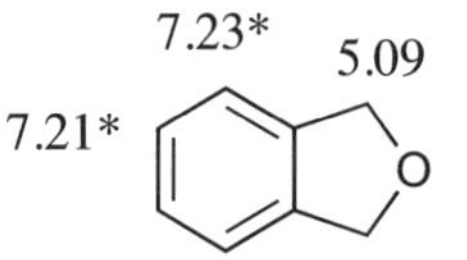

3.98
1.38

4.51
1.31

7.23*   5.09
7.21*

* assignment uncertain

e a 7.01
d b 7.32
c
7.09

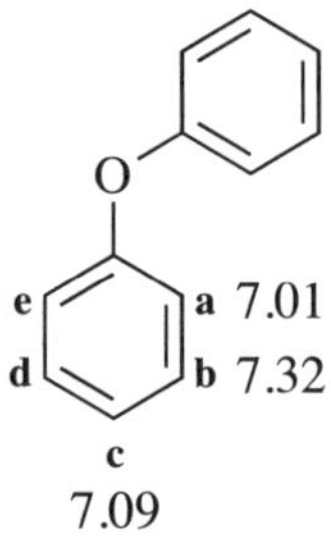

$^3J_{ab}$ 8.3
$^4J_{ac}$ 1.1
$^5J_{ad}$ 0.5
$^4J_{ae}$ 2.6
$^3J_{bc}$ 7.4
$^4J_{bd}$ 1.7

^{1}H Chemical Shifts and Coupling Constants of Acetals, Ketals, and Ortho Esters (δ in ppm, J in Hz)

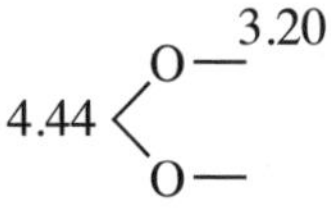

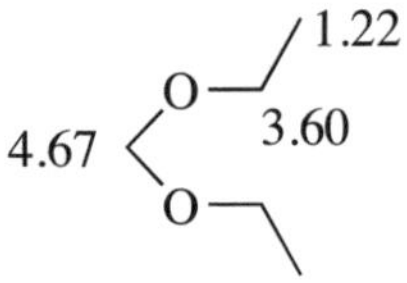

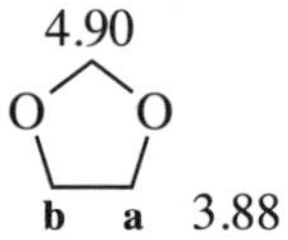

O

4.44   3.20
O—

4.67   1.22
3.60

4.90
b a 3.88

$^2J_{a,gem}$ -7.5
$^3J_{ab,cis}$ 7.3
$^3J_{ab,trans}$ 6.0

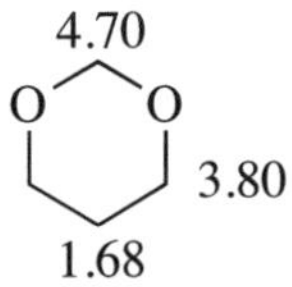

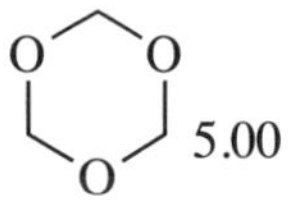

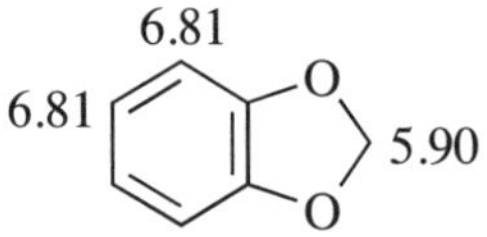

4.70
3.80
1.68

5.00

6.81
6.81
5.90
$^2J_{gem}$ 1.5

1.28   O— 3.31
4.57
O—

1.62   3.32
O—
4.29
O—
0.91

O—
O— 4.96
O— 3.33

O— 5.16
1.23 3.61

—O   O— 3.29
—O   O—

1.22
O—
3.58
O—

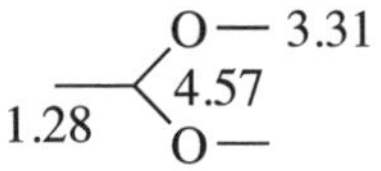

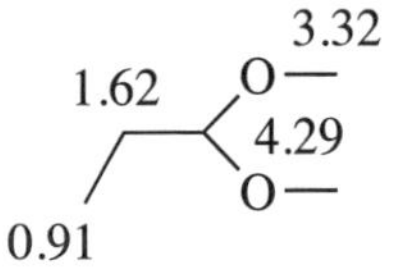

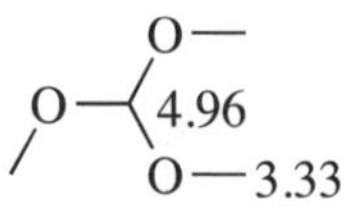

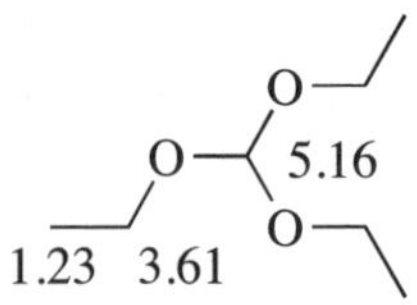

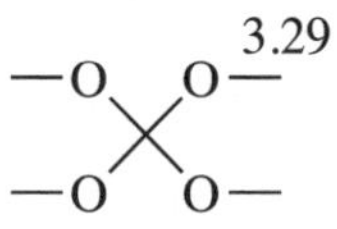

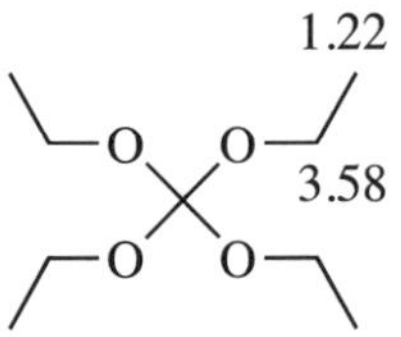

5.9 Nitrogen Compounds

5.9.1 Amines

Amine and Ammonium Protons (δ in ppm, |J| in Hz)

Chemical shifts of amine protons lie around 0.5–6 ppm depending on solvent, concentration, and hydrogen bonding. Those of ammonium protons are found between ca. 7 and 12 ppm. Neighboring H bond acceptors lead to deshielding in all cases.

			in $CDCl_3$	in DMSO
Amines:	$\delta_{NH_2}, \delta_{NH}$	aliphatic	<1–2	2–4
		aromatic	3–4	4–7
Ammonium:	$\delta_{NH_3^+}, \delta_{NH_2^+}, \delta_{NH^+}$	aliphatic	7–11	7–11
		aromatic	8–12	8–12

Coupling of amine protons with vicinal H atoms is usually not seen in aliphatic amines because of their rapid intermolecular exchange. However, for =C–NH–CH moieties (enamines, aromatic amines, amides, etc.), the exchange rate is slower and splitting (or line broadening at intermediate rates) is often observed. The H–C–N–H coupling depends on the conformation in a similar way as the H–C–C–H coupling (see Chapter 5.1.2). For N–CH$_3$ and N–CH$_2$ groups: $^3J_{HCNH} \approx 5$–6 Hz.

In acidic media (e.g., in trifluoroacetic acid as solvent), the exchange of the ammonium protons is slowed down to such an extent that the vicinal coupling H–N$^+$–C–H generally becomes observable. In other media, signals are usually broad owing to intermediate exchange rates.

The signals of amine and especially of ammonium protons are often broadened additionally because the ^{14}N–^1H coupling is only partly eliminated by the quadrupole relaxation of ^{14}N (spin quantum number, I = 1; natural abundance, 99.6%; $^1J_{NH} \approx 60$ Hz). This line broadening has no effect on the vicinal H–C–N–H coupling so that sharp multiplets can be observed for neighboring H atoms even when the NH proton exhibits a broad signal. In ammonium compounds of high symmetry, the quadrupole relaxation is slow and the coupling with ^{14}N leads to triplets of equal intensity for all three lines.

N

NH_4^+ $^1J_{NH}$ 52.8

$^2J_{aN}$ 0.5
$^3J_{bN}$ 1.6
$^4J_{cN}$ 0.0

^{1}H Chemical Shifts and Coupling Constants of Amines and Ammonium Salts (δ in ppm, J in Hz)

2.47
CH₃NH₂

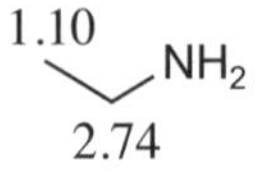

1.10
NH₂
2.74

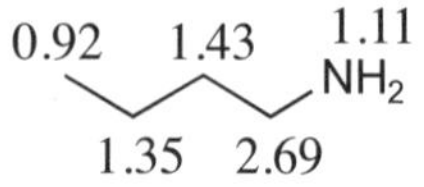

0.92   1.43   1.11
NH₂
1.35   2.69

≈2.3
(CH₃)₂NH

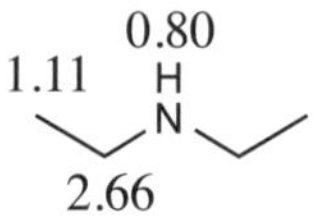

0.80
1.11   H
N
2.66

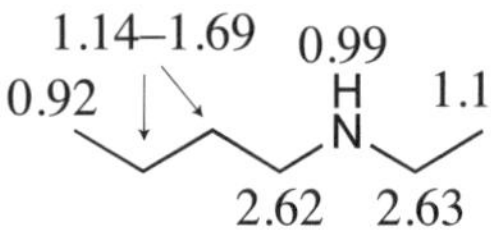

1.14–1.69   0.99
0.92   H   1.10
N
2.62   2.63

2.22
(CH₃)₃N
$^2J_{gem}$ -11.7

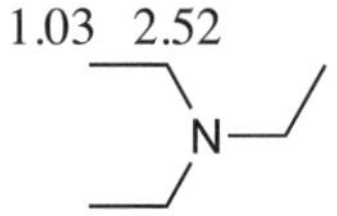

1.03   2.52
N

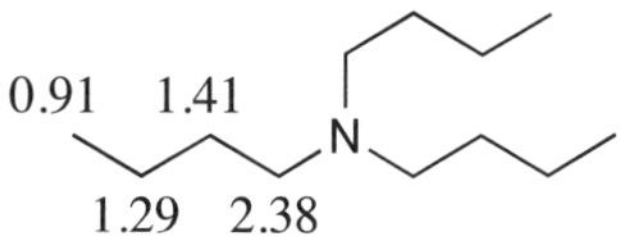

0.91   1.41
N
1.29   2.38

3.21 (in D₂O)
(CH₃)₄N⁺ I⁻

3.46 (in CDCl₃)
(CH₃)₄N⁺ AcO⁻

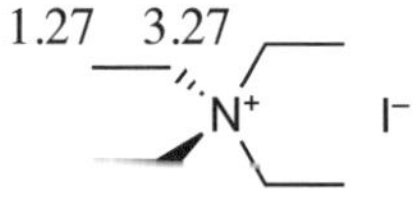

1.27   3.27
N⁺   I⁻
(in D₂O)

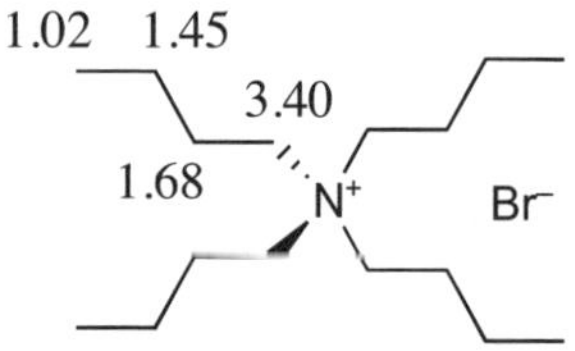

1.02   1.45
3.40
1.68   N⁺   Br⁻

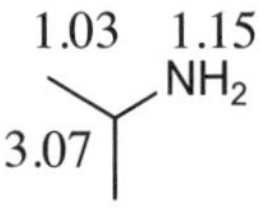

1.03   1.15
NH₂
3.07

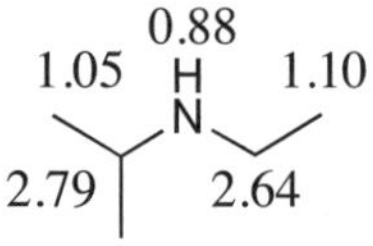

1.05   0.88
H   1.10
N
2.79   2.64

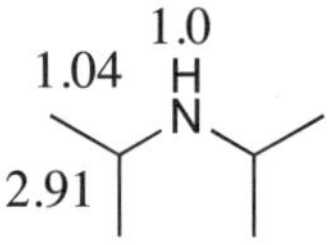

1.04   1.0
H
N
2.91

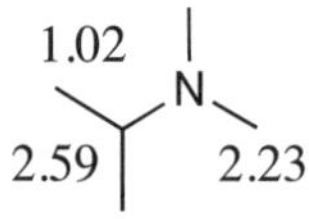

1.02
N
2.59   2.23

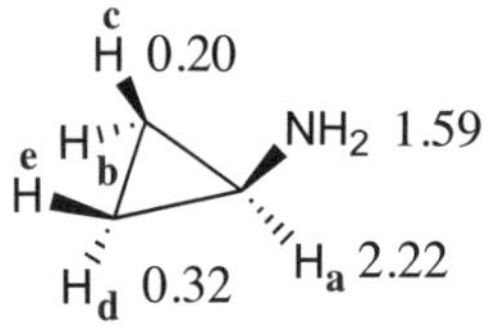

c
H 0.20
e   H
H   b   NH₂ 1.59
Hd 0.32   Ha 2.22
(neat)

$^3J_{ab}$ 6.6
$^3J_{ac}$ 3.6
$^2J_{bc}$ -4.3
$^3J_{bd}$ 9.7
$^3J_{be}$ 6.2
$^3J_{ce}$ 9.9

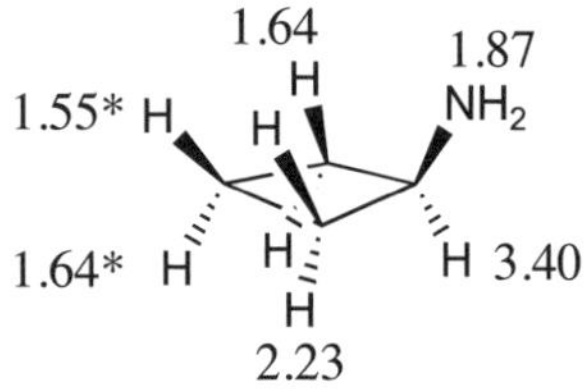

1.64
H   1.87
1.55* H   H   NH₂
1.64* H   H   H 3.40
H
2.23
* assignment uncertain

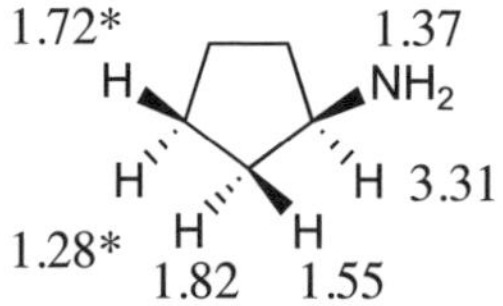

1.72*   1.37
H   NH₂
H   H 3.31
1.28*   H   H
1.82   1.55
* assignment uncertain

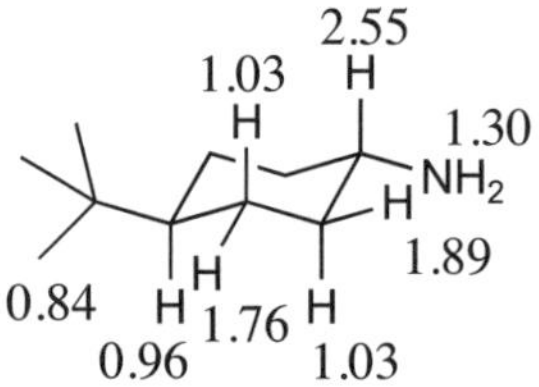

2.55
1.03   H
H   1.30
NH₂
H
1.89
0.84   H 1.76 H
0.96   1.03

1.30
1.27   NH₂
H   3.15
H
0.86   H 1.53 H   1.65
0.96   1.54

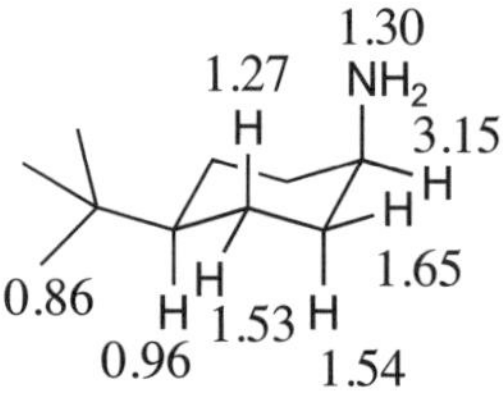

NH₂ 3.61 $^3J_{ab}$ 8.0
$^4J_{ac}$ 1.1

e   a 6.67 $^5J_{ad}$ 0.5
d   b 7.14 $^4J_{ae}$ 2.5
   c $^3J_{bc}$ 7.4
  6.75 $^4J_{bd}$ 1.6

δ_{NH_2} in DMSO: 4.94

3.66 ⌐ 2.83
HN
$^3J_{ab}$ 8.2
$^4J_{ac}$ 1.1
e   a 6.61 $^5J_{ad}$ 0.4
d   b 7.18 $^4J_{ae}$ 2.5
   c $^3J_{bc}$ 7.3
  6.70 $^4J_{bd}$ 1.7

δ_{NH} in DMSO: 5.52

⌐N⌐ 2.94
$^3J_{ab}$ 8.4
$^4J_{ac}$ 1.0
e   a 6.74 $^5J_{ad}$ 0.4
d   b 7.24 $^4J_{ae}$ 2.8
   c $^3J_{bc}$ 7.3
  6.72 $^4J_{bd}$ 1.8

3.70
$N^+(CH_3)_3$ Cl⁻

8.08
7.64
7.59
(in DMSO)

4.62 HN ⌐ 2.94
6.53
8.09
NO_2
δ_{NH} in DMSO: 7.32

⌐N⌐ 3.10
6.59
8.09
NO_2

^{1}H Chemical Shifts and Coupling Constants of Cyclic Amines (δ in ppm, J in Hz)

H   0.9 $^2J_{gem}$ ≈1
N $^3J_{ab,cis}$ 6.3
b △ a 1.61 $^3J_{ab,trans}$ 3.8
(neat)

H   2.08
N
3.63
2.33

H   2.01
N
2.75
1.59

H   2.18
N
2.79
1.53
1.53

| 2.23
N
2.33
1.59
1.41

H 2.59
N
2.86
3.67
O

|
N
2.32
3.62
O

| 2.26
N
2.37
2.88
N
H 1.72

5.9.2 Nitro and Nitroso Compounds

^{1}H Chemical Shifts and Coupling Constants (δ in ppm, J in Hz)

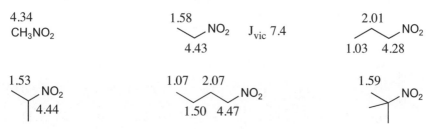

4.34
CH_3NO_2

1.58
NO_2 J_{vic} 7.4
4.43

2.01
NO_2
1.03   4.28

1.53
NO_2
4.44

1.07   2.07
NO_2
1.50   4.47

1.59
NO_2

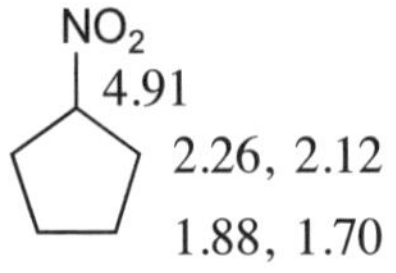

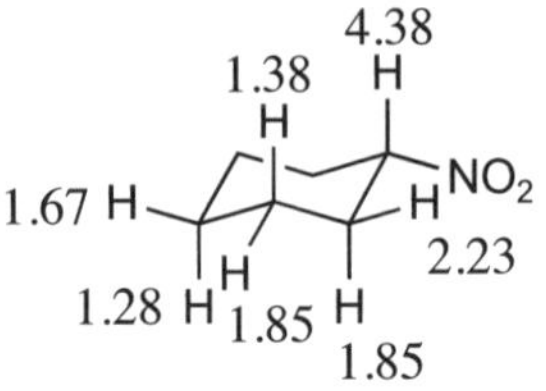

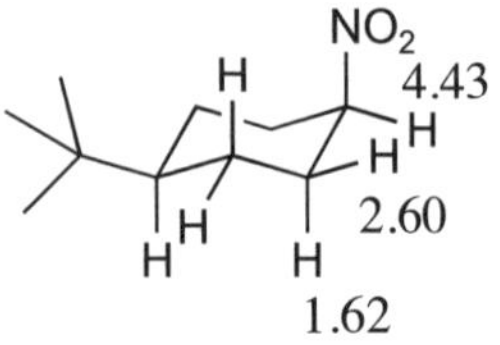

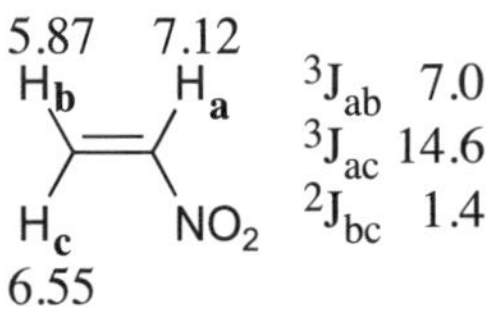

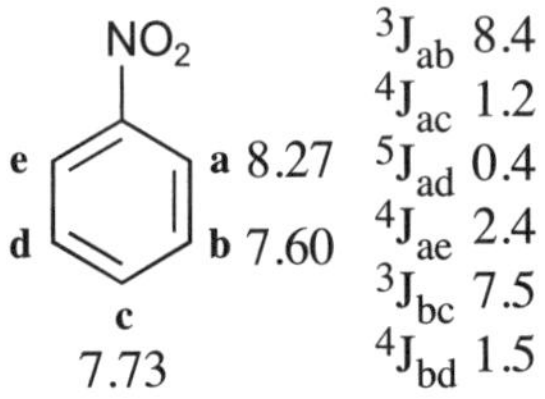

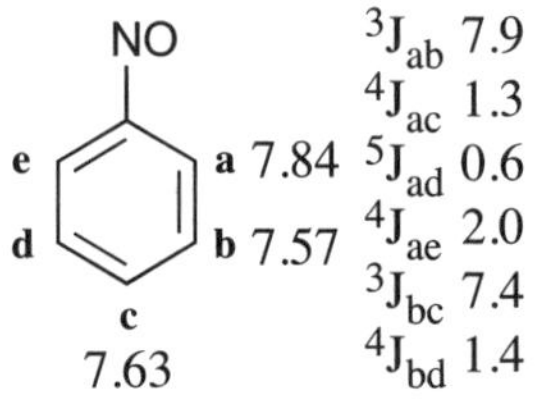

NO₂
4.91
2.26, 2.12
1.88, 1.70

4.38
1.38 H
H
1.67 H
1.28 H H
1.85 H
1.85
NO₂
2.23

NO₂
4.43
H
H
H H H
1.62
2.60

5.87 7.12
H$_b$ H$_a$ $^3J_{ab}$ 7.0
               $^3J_{ac}$ 14.6
H$_c$ NO₂ $^2J_{bc}$ 1.4
6.55

NO₂
e   a 8.27
d   b 7.60
c
7.73

$^3J_{ab}$ 8.4
$^4J_{ac}$ 1.2
$^5J_{ad}$ 0.4
$^4J_{ae}$ 2.4
$^3J_{bc}$ 7.5
$^4J_{bd}$ 1.5

NO
e   a 7.84
d   b 7.57
c
7.63

$^3J_{ab}$ 7.9
$^4J_{ac}$ 1.3
$^5J_{ad}$ 0.6
$^4J_{ae}$ 2.0
$^3J_{bc}$ 7.4
$^4J_{bd}$ 1.4

5.9.3 Nitrites and Nitrates

¹H Chemical Shifts (δ in ppm)

4.78
O—NO
1.39

1.41 4.71
O—NO
0.96 1.72

1.59
O—NO

4.52
O—NO₂
1.37

1.06 4.39
O—NO₂
1.78

5.19
O—NO₂
1.37

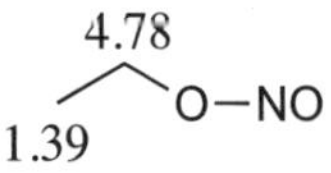

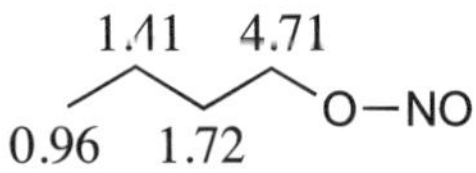

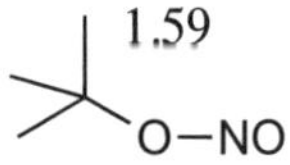

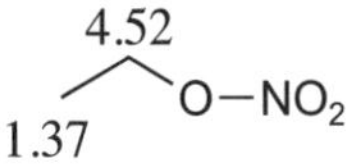

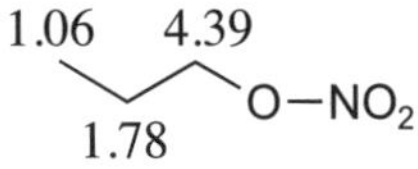

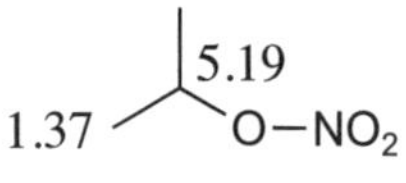

5.9.4 Nitrosamines, Azo and Azoxy Compounds

¹H Chemical Shifts (δ in ppm)

Owing to hindered rotation around the N–NO bond, corresponding protons in *cis* and *trans* positions have different chemical shifts in the neighborhood of the N=O group.

In general: $\delta_{cis} < \delta_{trans}$ for α-CH₃, α-CH₂, and β-CH₃
               $\delta_{cis} > \delta_{trans}$ for α-CH

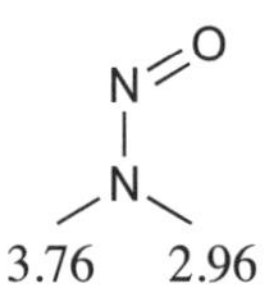

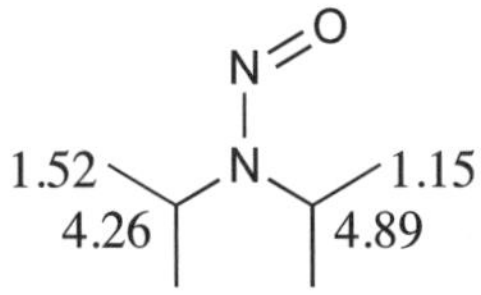

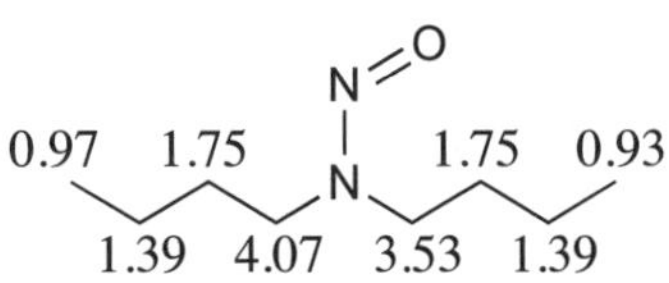

N=O
N
3.76   2.96

N=O
1.52   N   1.15
4.26   4.89

N=O
0.97 1.75   N   1.75 0.93
1.39 4.07 3.53 1.39

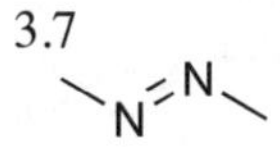

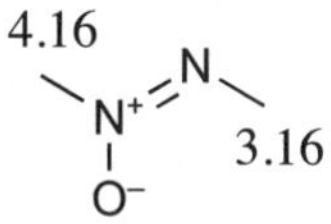

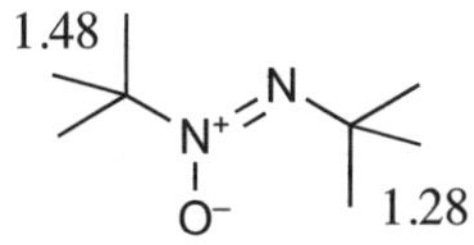

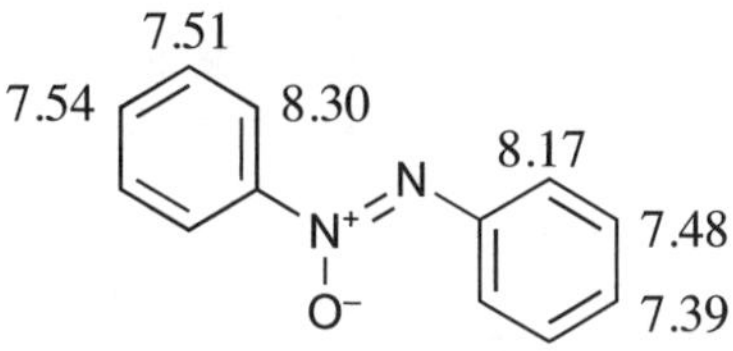

3.7
$\overset{\diagdown}{N}{=}\overset{N}{\diagup}$

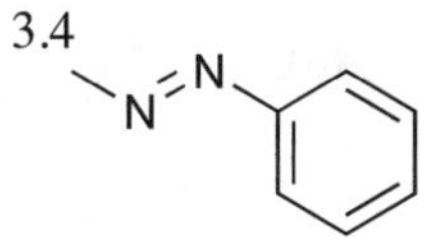

3.4
$\overset{\diagdown}{N}{=}\overset{N}{\diagup}$ C₆H₅

7.92
7.50
7.45
(azobenzene)

4.16
$\overset{\diagdown}{\underset{O^-}{N^+}}{=}\overset{N}{\diagup}$ 3.16

1.48
$\overset{(CH_3)_3C\diagdown}{\underset{O^-}{N^+}}{=}\overset{N}{\diagup}$ C(CH₃)₃ 1.28

7.54 7.51 8.30
8.17
$\overset{N^+}{\underset{O^-}{}}{=}N$ 7.48
7.39

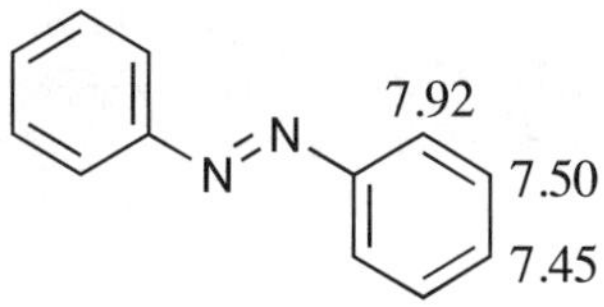

5.9.5 Imines, Oximes, Hydrazones, and Azines

¹H Chemical Shifts (δ in ppm)

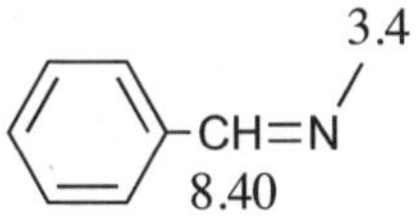

3.4
⬡–CH=N⟍
8.40

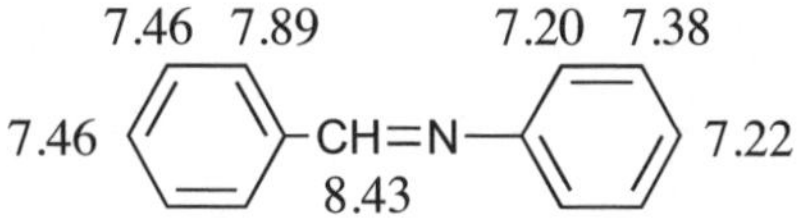

7.46 7.89   7.20 7.38
7.46 ⬡–CH=N–⬡ 7.22
8.43

In aldoximes and ketoximes, the chemical shift difference between *syn* and *anti* protons at the α-CH group, $\Delta\delta = \delta_{syn} - \delta_{anti}$, depends on the dihedral angle, $\phi_{H-C-C=N}$:

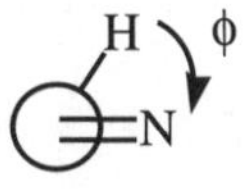

ϕ	$\Delta\delta = \delta_{syn} - \delta_{anti}$
0°	1
60°	0
115°	-0.3

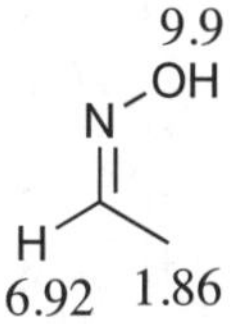

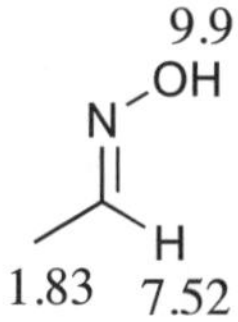

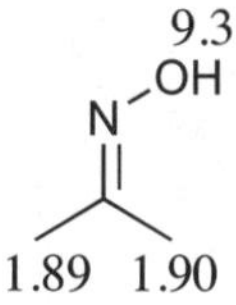

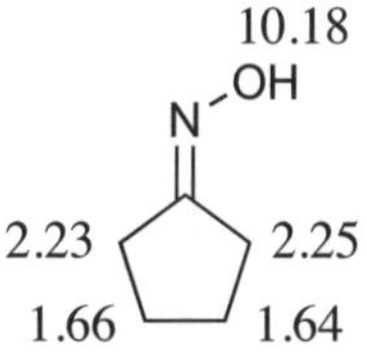

9.9
N–OH
H
6.92 1.86

9.9
N–OH
1.83 H 7.52

9.3
N–OH
1.89 1.90

10.18
N–OH
2.23 ⬠ 2.25
1.66   1.64
(in DMSO)

8.9
N–OH
2.22 ⬡ 2.50
} 1.74–1.55

6.9
N–OH
2.60

7.0
N–OH
2.30

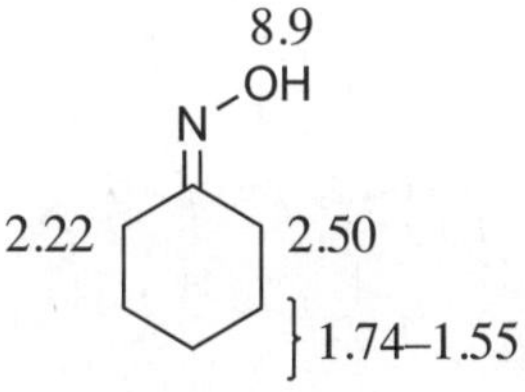

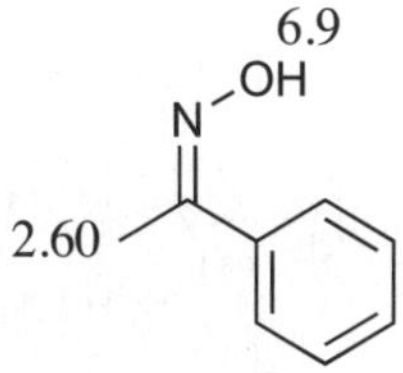

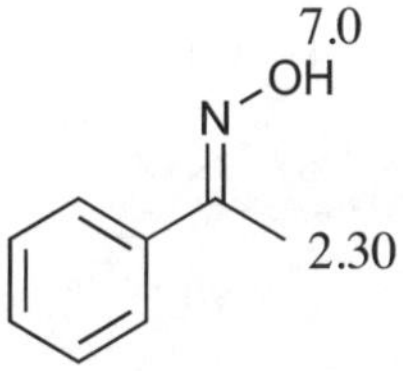

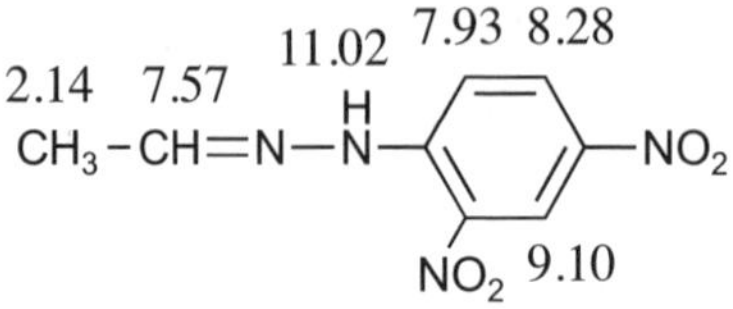

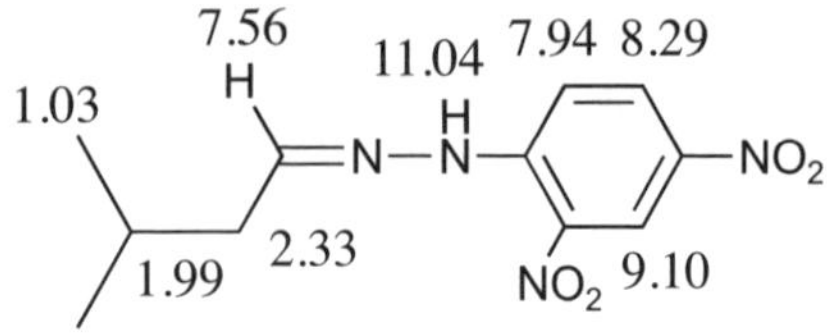

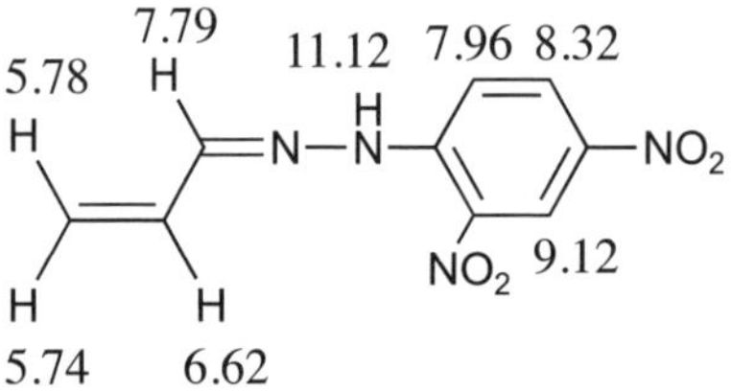

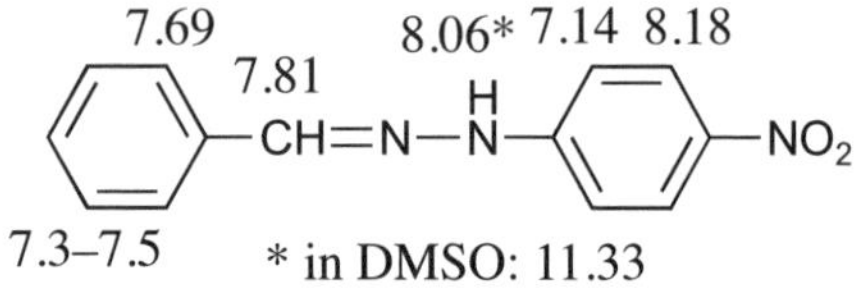

2.14 7.57 11.02 7.93 8.28
CH$_3$–CH=N–N
H (11.02)
–NO$_2$
NO$_2$ 9.10

7.56
1.03
H
11.04 7.94 8.29
=N–N
H
–NO$_2$
1.99 2.33
NO$_2$ 9.10

7.79
5.78 H 11.12 7.96 8.32
H
=N–N
H
–NO$_2$
NO$_2$ 9.12
H   H
5.74   6.62

7.69
7.81
8.06* 7.14 8.18
CH=N–N
H
–NO$_2$
7.3–7.5   * in DMSO: 11.33

2.03
=N
H   N
7.89

2.00
=N
N
1.83

7.44 7.83
8.66
7.44   CH=N
N=CH

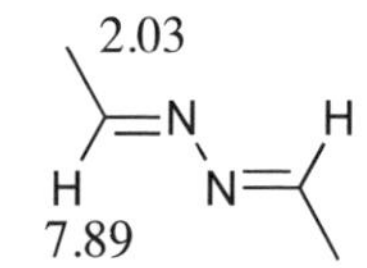

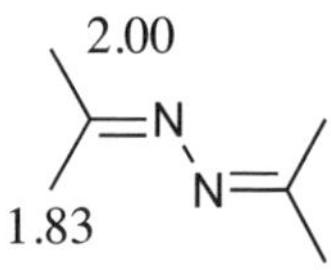

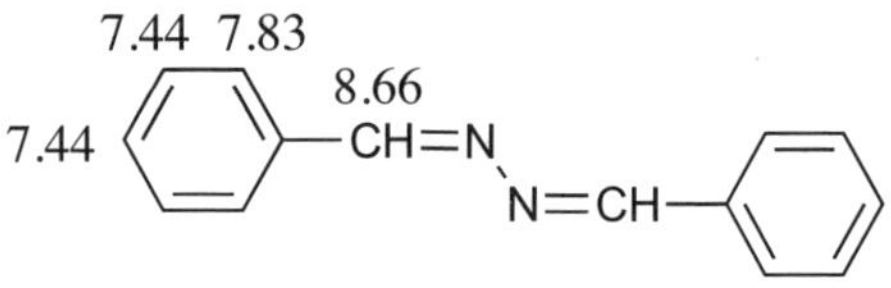

5.9.6 Nitriles and Isonitriles

^{1}H Chemical Shifts and Coupling Constants of Nitriles (δ in ppm, J in Hz)

N

1.98
CH$_3$CN

1.31
CN
2.35
$^3J_{vic}$ 7.6

1.70
CN
1.08 2.34

1.35
CN
2.67

0.96 1.63
CN
1.50 2.34

1.37
CN

6.11 5.69
H$_b$ H$_a$
H$_c$ CN
6.24
$^3J_{ab}$ 11.8
$^3J_{ac}$ 17.9
$^2J_{bc}$ 0.9

CN
e   a 7.66
d   b 7.48
c
7.62
$^3J_{ab}$ 7.8
$^4J_{ac}$ 1.3
$^5J_{ad}$ 0.7
$^4J_{ae}$ 1.8
$^3J_{bc}$ 7.7
$^4J_{bd}$ 1.3

c
H 1.04
e H$_b$
H
H$_d$ 0.96   H$_a$
1.29
CN
$^3J_{ab}$ 8.4
$^3J_{ac}$ 5.1
$^2J_{bc}$ -4.7
$^3J_{bd}$ 9.2
$^3J_{be}$ 7.1
$^3J_{ce}$ 9.5

1.22 H 2.39
H
H   CN
1.70   H 2.08
H
H 1.76 H 1.52
1.20

1.50 CN
H   2.96
H
H   H
1.70   2.00
H
H 1.70 H 1.54
1.20

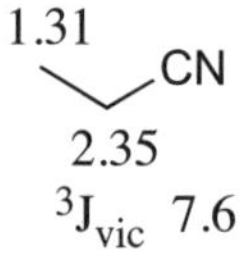

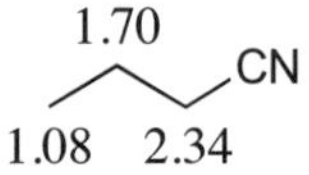

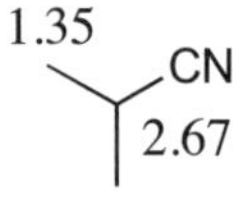

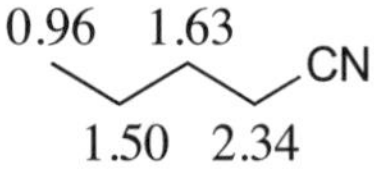

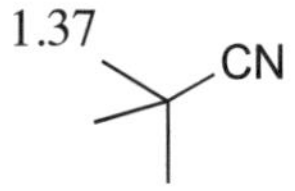

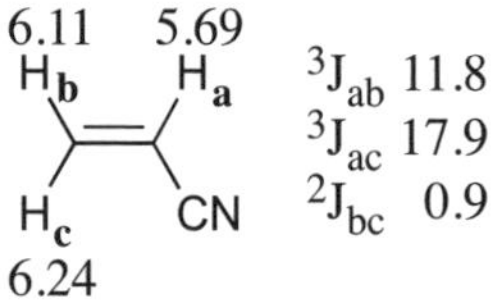

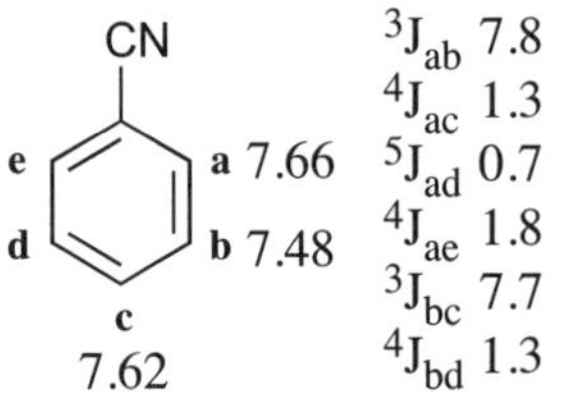

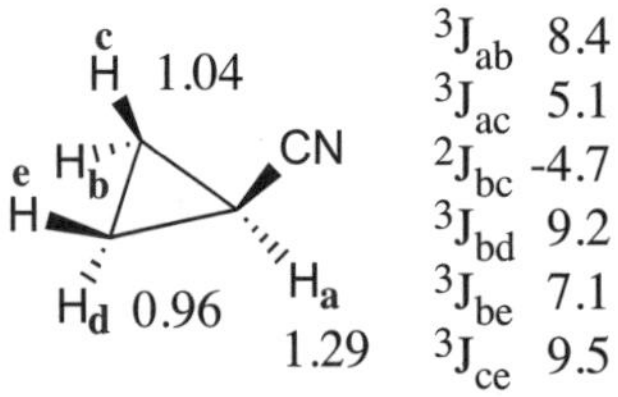

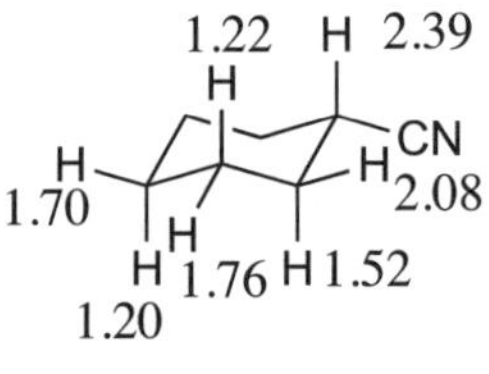

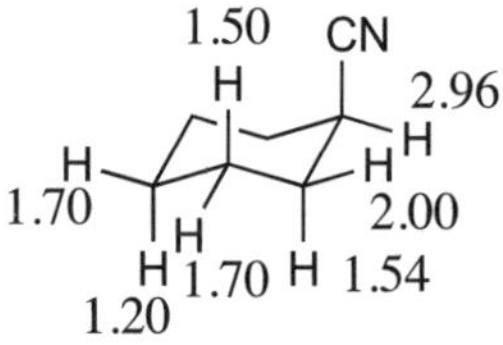

¹H Chemical Shifts and Coupling Constants of Isonitriles (δ in ppm, |J| in Hz)

Because of the symmetrical electron distribution around the N atom, the quadrupole relaxation of the nitrogen nucleus is so slow that the ^{14}N-^{1}H coupling becomes observable and leads to triplets with relative intensities of 1:1:1 (spin quantum number of ^{14}N: I = 1; natural abundance, 99.6%):

$$\overset{b}{-CH_2}-\overset{a}{CH_2}-^{14}NC \qquad \begin{array}{ll} |^2J_{aN}| & 1.8\text{–}2.8 \\ |^3J_{bN}| & 1.5\text{–}3.5 \end{array}$$

2.85
a
$CH_3NC \quad |^2J_{aN}| \ 2.3$

1.28
b ＼＼＼ NC
3.89 (a)
$|^2J_{aN}| \ 2.0$
$|^3J_{bN}| \ 2.4$
$|^3J_{ab}| \ 7.3$

1.45
b ＼＼ NC
a | 3.87
$|^2J_{aN}| \ 1.8$
$|^3J_{bN}| \ 2.6$
$|^3J_{ab}| \ 7.0$

1.45
a ＞＜ NC
$|^3J_{aN}| \ 2.1$

5.35 H$_b$ H$_a$ 5.90
＞＝＜
5.58 H$_c$ NC

$|^2J_{aN}| \ 2.3 \quad ^3J_{ab} \ 8.6$
$|^3J_{bN}| \ 6.1 \quad ^3J_{ac} 15.6$
$|^3J_{cN}| \ 3.1 \quad ^2J_{bc} -0.5$

5.9.7 Cyanates, Isocyanates, Thiocyanates, and Isothiocyanates

¹H Chemical Shifts and Coupling Constants (δ in ppm, J in Hz)

1.45
＼＼OCN
4.54

3.02
CH_3NCO

1.20
＼＼NCO
3.37

1.63
＼＼＼NCO
0.99 3.26

1.29
＼NCO
3.72

0.94   1.58
＼／＼／NCO
1.42   3.29

4.77 6.12
H$_b$ H$_a$
＞＝＜
H$_c$ NCO
5.01

$^3J_{ab} \ 7.6$
$^3J_{ac} 15.2$
$^2J_{bc} -0.1$

2.60
CH_3SCN

1.53
＼＼SCN
2.98

＼＼SCN
3.48

3.37
CH_3NCS

1.40
＼＼NCS
3.64

＼＼NCS
3.98

NCS
7.15
7.30
7.24

5.10 Sulfur Compounds

5.10.1 Thiols

¹H Chemical Shifts and Coupling Constants (δ in ppm, J in Hz)

Typical ranges of SH chemical shifts:

alk–SH   1–2   〈benzene ring〉—SH   2–4

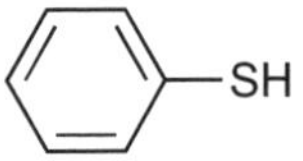

The exchange with other SH, OH, NH, or COOH protons is generally so slow that the chemical shift is characteristic and the vicinal coupling with SH protons becomes visible (5–9 Hz in aliphatic systems with fast rotation).

2.00 1.26
CH₃SH J_{vic} 7.4
(in benzene)

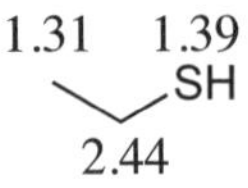

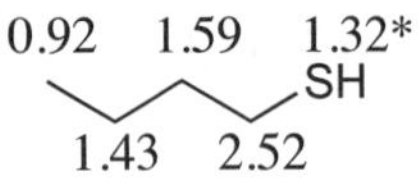

1.31   1.39
∖∕ SH
2.44

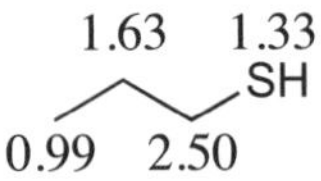

1.63   1.33
∕∖∕ SH
0.99   2.50

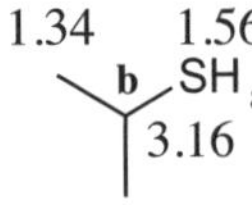

1.34   1.56
∖ᵇ SHᵃ $^3J_{ab}$ 5.7
3.16

0.92 1.59 1.32*
∖∕∖ SH
1.43   2.52
* in DMSO: 2.17

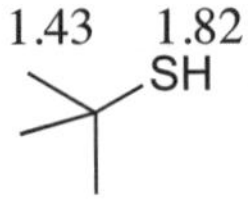

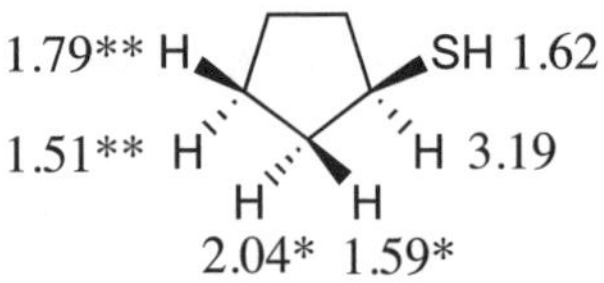

1.43   1.82
∖∕ SH
(t-butyl)

1.79** H╱╲ SH 1.62
1.51** H╱ ╲H 3.19
       H   H
      2.04* 1.59*
*; **: assignments uncertain

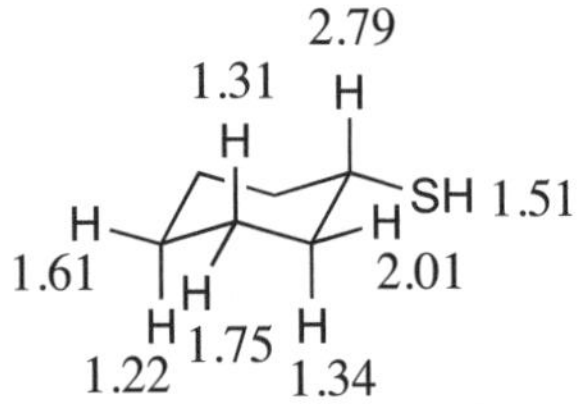

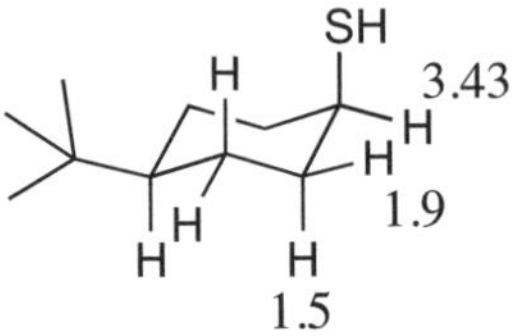

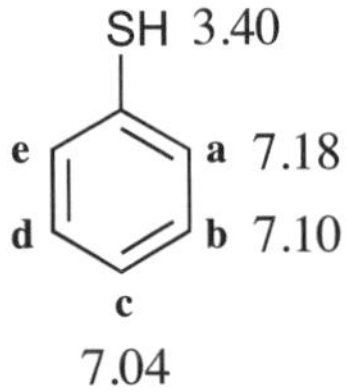

2.79
1.31 H
H
H— ╱╲ ╱H SH 1.51
1.61 ╲H╱ ╲H 2.01
    H 1.75 H
   1.22   1.34

        SH
    H ╱╲ 3.43
      ╲ ╱H
   ╲H╱ ╲H 1.9
    H   H
         1.5

SH 3.40
e ⟨benzene⟩ a 7.18
d   b 7.10
      c
    7.04

$^3J_{ab}$ 7.9
$^4J_{ac}$ 1.2
$^5J_{ad}$ 0.6
$^4J_{ae}$ 2.1
$^3J_{bc}$ 7.5
$^4J_{bd}$ 1.5

S

5.10.2   Sulfides

¹H Chemical Shifts and Coupling Constants (δ in ppm, J in Hz)

2.12

2.10 2.51
1.26

2.48 0.99
1.60

2.55 2.93
1.25 1.26

2.09 2.49 1.42
1.56 0.92

1.42

2.25

H_a 6.43
H_b 5.18
H_c 4.95

$^3J_{ab}$ 10.3
$^3J_{ac}$ 16.4
$^2J_{bc}$ -0.3

2.47

7.18
7.16

7.02

e a 7.20
d b 7.06
c
7.20

$^3J_{ab}$ 7.8
$^4J_{ac}$ 1.2
$^5J_{ad}$ 0.6
$^4J_{ae}$ 2.0
$^3J_{bc}$ 7.4
$^4J_{bd}$ 1.5

¹H Chemical Shifts and Coupling Constants of Cyclic Sulfides (δ in ppm, J in Hz)

2.36

$^2J_{gem}$ 0
$^3J_{cis}$ 7.2
$^3J_{trans}$ 5.7

c a 3.21
b
2.94

$^2J_{a,gem}$ -8.7
$^2J_{b,gem}$ -11.7
$^3J_{ab,cis}$ 8.9
$^3J_{ab,trans}$ 6.3
$^4J_{ac,cis}$ 1.2
$^4J_{ac,trans}$ -0.2

2.82
1.94

3.81

3.11

S

3.67
5.81

6.06 d a 3.08
5.48 c b 2.62

$^3J_{ab}$ 9.2
$^3J_{bc}$ 2.5
$^4J_{bd}$ 2.2
$^3J_{cd}$ 6.1

2.57
≈1.7

≈1.7

2.57
3.88

2.85

2.08

3.79
2.83

5.97
5.55

2.84

4.28

(in DMSO)

5.10.3   Disulfides and Sulfonium Salts

^{1}H Chemical Shifts and Coupling Constants (δ in ppm, J in Hz)

1.32
2.30

1.70
2.69
1.00   2.66

1.31

$^3J_{ab}$ 7.5
$^4J_{ac}$ 1.4
$^5J_{ad}$ 0.5
$^4J_{ae}$ 2.0
$^3J_{bc}$ 7.2
$^4J_{bd}$ 1.6

e
d   a 7.47
        b 7.29
c
7.24

2.7
1.9

2.94
$(CH_3)_3S^+$ I$^-$

(in DMSO)

1.01   1.57
1.84   3.75
S$^+$ I$^-$

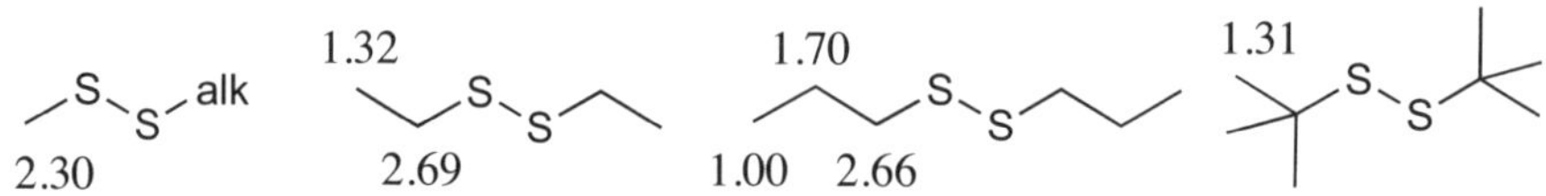

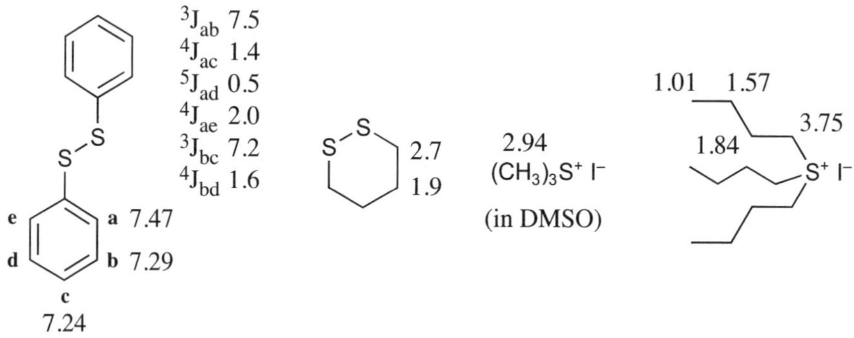

5.10.4   Sulfoxides and Sulfones

^{1}H Chemical Shifts and Coupling Constants (δ in ppm, J in Hz)

2.62
(in DMSO)
neat: 2.50

2.61
H 6.08
H 5.92
H 6.77

2.42
1.92

2.88
2.43
2.05

2.64

2.84

1.47
2.80 2.94

2.85
3.13
1.41

1.44

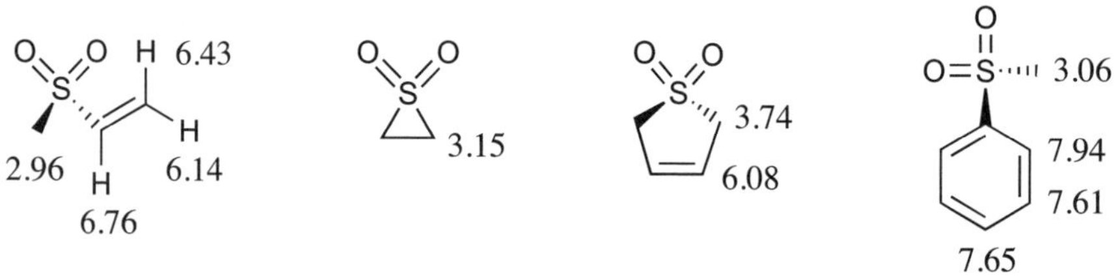

O H 6.43
2.96 H   6.14
H 6.76

3.15

3.74
6.08

O=S 3.06
7.94
7.61
7.65

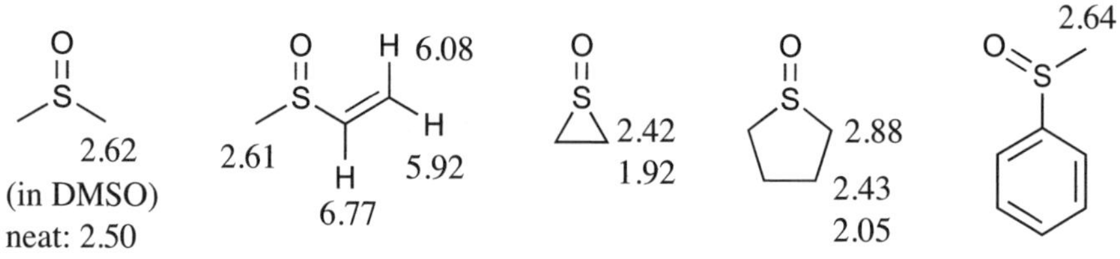

5.10.5 Sulfonic, Sulfurous, and Sulfuric Acids and Derivatives

^{1}H Chemical Shifts and Coupling Constants (δ in ppm, J in Hz)

1.42 3.17   11.1
$CH_3-CH_2-SO_2-OH$

3.02   3.91
$CH_3-SO_2-O-CH_3$

2.78
CH_3-SO_2-N

2.86
$\diagdown CH_3$
$\diagup CH_3$

Apparently lower δ_{OH}
values in DMSO due to
fast exchange with H_2O

OH 11–12
|
SO_2

3.70
OCH_3
|
SO_2

Cl
|
SO_2
7.94
7.60
7.62

8.02
7.61
7.71

NH_2 7.37
|
SO_2
7.85
7.58
7.58
(in DMSO)

5.07   2.63
$NHCH_3$
|
SO_2
7.86
7.4–7.7

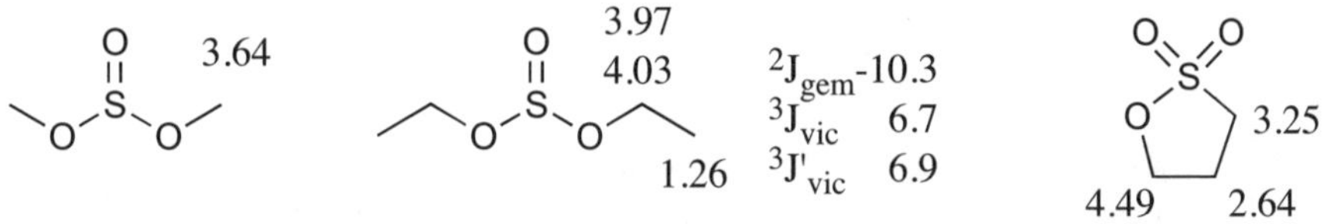

3.64

$^2J_{gem}$ -10.3
$^3J_{vic}$ 6.7
$^3J'_{vic}$ 6.9

3.97
4.03
1.26

3.25
4.49   2.64

3.94

4.34
1.43

4.68

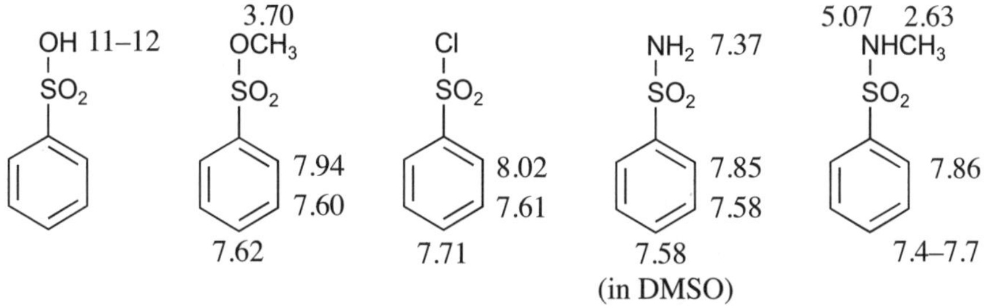

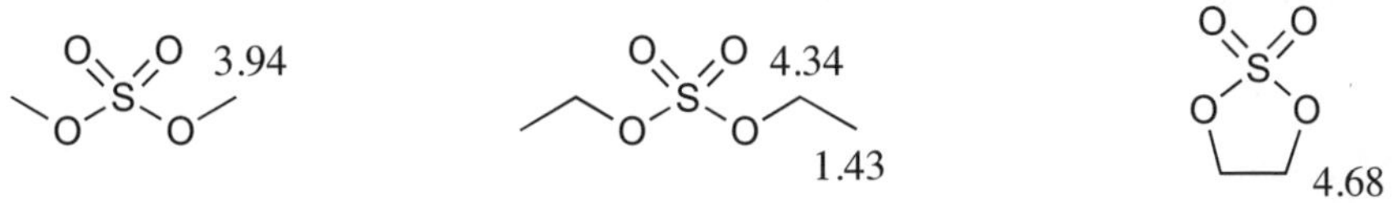

5.10.6 Thiocarboxylate Derivatives

^{1}H Chemical Shifts and Coupling Constants (δ in ppm, J in Hz)

2.41 H_3C SH 5.09

$^3J_{ab}$ 5.9
$^4J_{ac}$ 2.0
$^3J_{bc}$ 2.8
a 6.47
b 7.84
4.29 c

7.7
7.1
6.9
6.4

2.30 H_3C SCH_3 2.27

9.56 HN
3.38
1.79   1.65

2.98
1.72

S

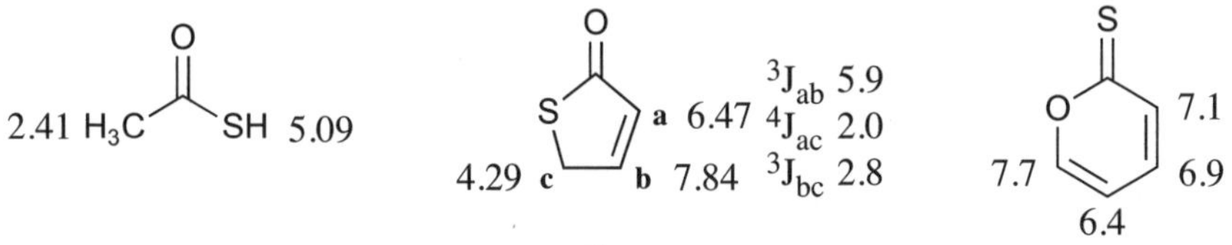

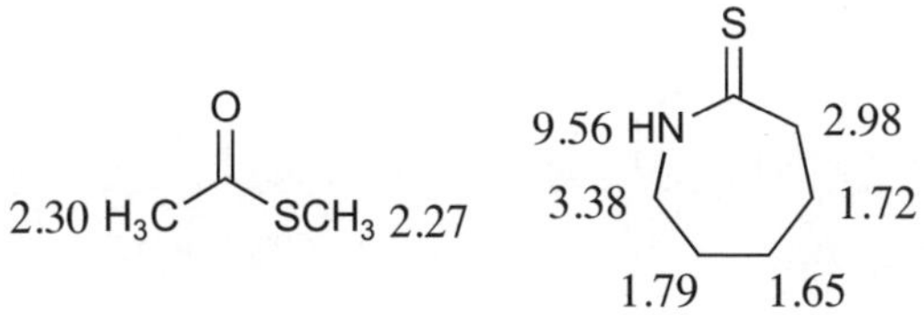

5.11 Carbonyl Compounds

5.11.1 Aldehydes

^{1}H Chemical Shifts and Coupling Constants (δ in ppm, J in Hz)

alk−CHO   9–10 $^3J_{vic}$ 0–3

alken−CHO   9–10 $^3J_{vic}$ ≈8

ortho-substituted: 10–10.5

meta-, *para*-substituted: 9.5–10.2

9.60
$CH_2=O$
(in TMS)
$|^2J_{gem}|$ 42.4

2.20   9.79
H_3C-CHO
  a   b
$^3J_{ab}$ 3.0

1.13   9.79
a CHO
                  b
2.46
$^3J_{ab}$ 1.4

1.67   9.74
a CHO
                  b
0.97   2.42
$^3J_{ab}$ 2.0

1.13   9.57
a CHO
                  b
2.39
$^3J_{ab}$ 1.1

1.08   9.48
CHO

0.93   1.59   9.76
a CHO
                      b
1.35   2.42
$^3J_{ab}$ 1.9

C = X

6.52   6.37
H_c H_b
H_d =O
6.35 H_a 9.59

$^3J_{ab}$ 4.7 $^3J_{bc}$ 10.0
$^4J_{ac}$ <1 $^4J_{bd}$ 17.4
$^4J_{ad}$ <1 $^2J_{cd}$ 1.0

7.43   9.67
H_c CHO
              a
7.4   7.53 H_b 6.69
    7.4

$^3J_{ab}$ 7.8
$^3J_{bc}$ 16.0

O   H 10.03
7.88
7.53
7.63

O   H 10.24
7.77   2.64
7.33   7.23
7.45

O   H 9.95
7.76
7.32
2.42

O   H 11.54
        8.99
7.69
7.56
8.71   8.07

5.11.2 Ketones

¹H Chemical Shifts and Coupling Constants (δ in ppm, J in Hz)

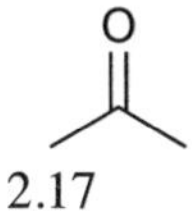

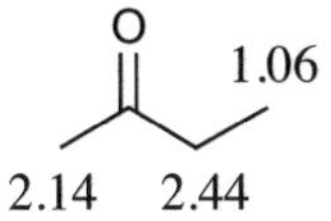

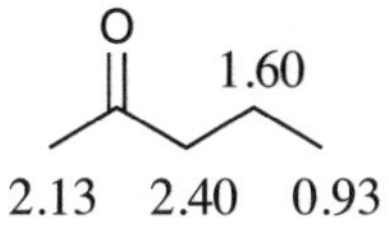

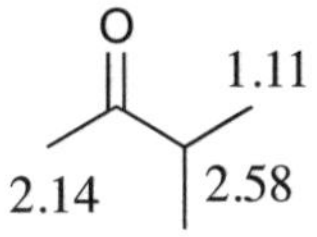

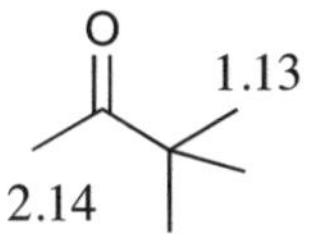

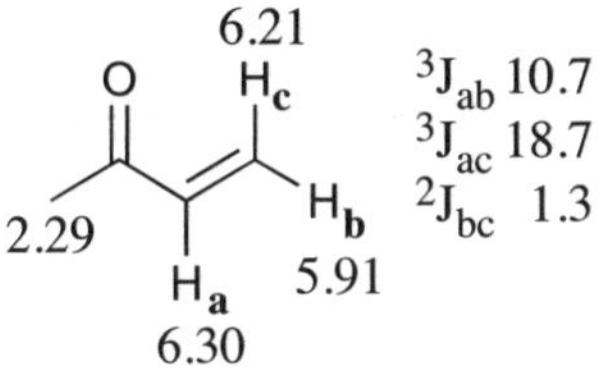

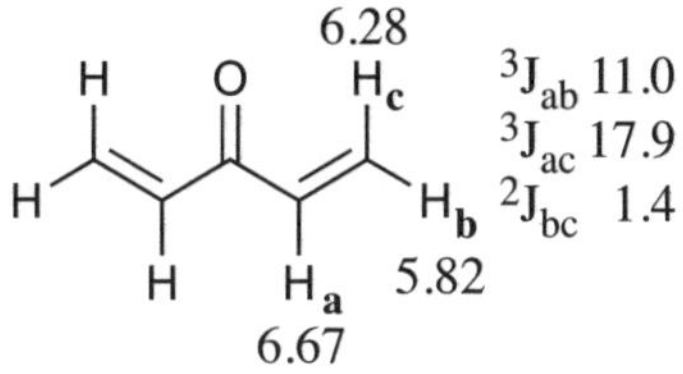

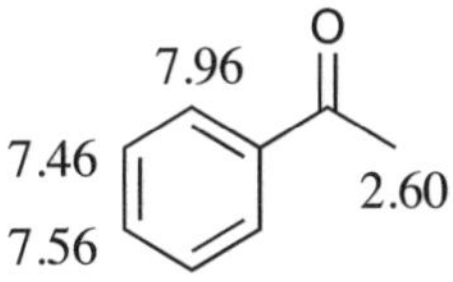

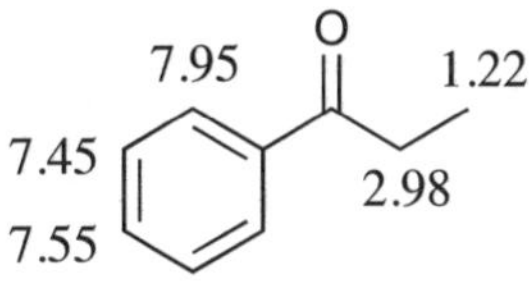

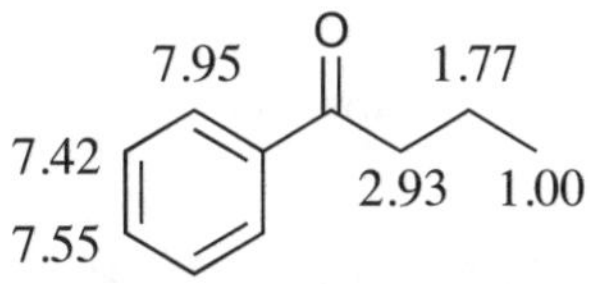

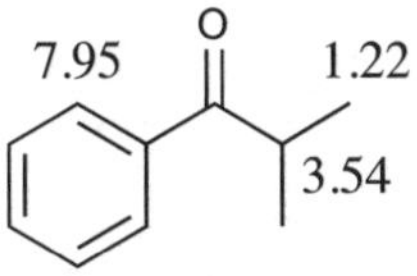

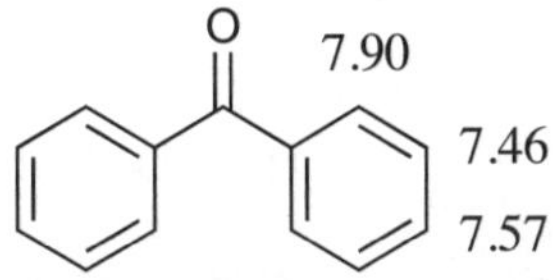

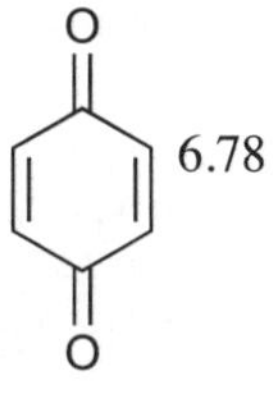

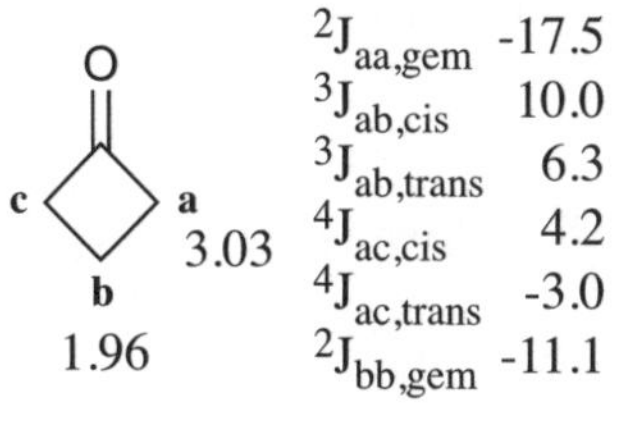

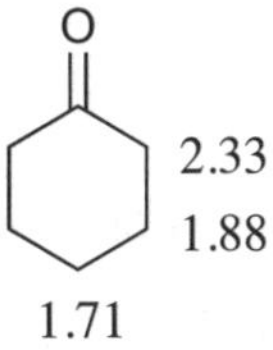

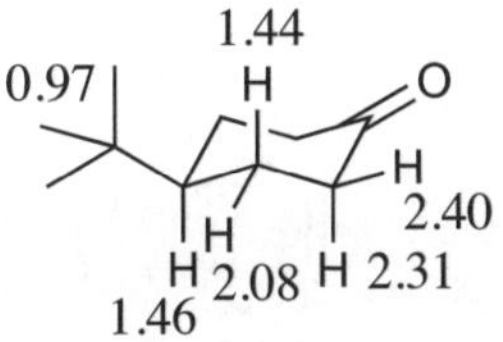

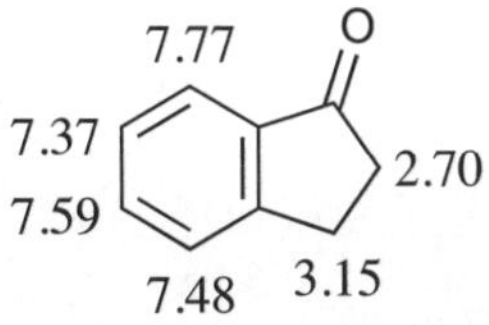

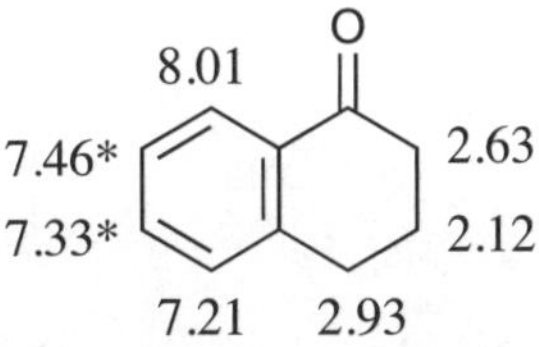

2.17

2.14   2.44   1.06

2.13   2.40   0.93   1.60

2.14   2.58   1.11

2.14   1.13

2.29

6.21
H$_c$
H$_b$ $^3J_{ab}$ 10.7 $^3J_{ac}$ 18.7 $^2J_{bc}$ 1.3
H$_a$ 5.91
6.30

6.28
H$_c$
H$_b$ $^3J_{ab}$ 11.0 $^3J_{ac}$ 17.9 $^2J_{bc}$ 1.4
H$_a$ 5.82
6.67

7.96   7.46   7.56   2.60

7.95   7.45   7.55   1.22   2.98

7.95   7.42   7.55   1.77   2.93   1.00

7.95   1.22   3.54

7.90   7.46   7.57

6.78

1.65

c   a   3.03
b
1.96

$^2J_{aa,gem}$ -17.5
$^3J_{ab,cis}$ 10.0
$^3J_{ab,trans}$ 6.3
$^4J_{ac,cis}$ 4.2
$^4J_{ac,trans}$ -3.0
$^2J_{bb,gem}$ -11.1

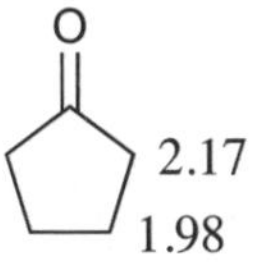

2.17   1.98

2.33   1.88   1.71

C = X

1.44
0.97   H
H
H   2.08   H 2.31
1.46
2.40

7.77   7.37   7.59   7.48   3.15   2.70

8.01   7.46*   7.33*   7.21   2.93   2.63   2.12

* assignment uncertain

¹H Chemical Shifts of Diketones (δ in ppm)

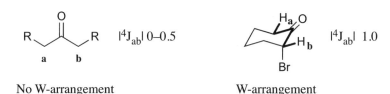

For the enol form,
see Chapter 5.8.1

Long-Range Coupling in Ketones (|J| in Hz)

For fixed conformations, the coupling over the C=O group is often detectable for
W-arrangement of the coupling path.

R ⟍⟋⟍ R $|^4J_{ab}|$ 0–0.5

No W-arrangement

H_a O $|^4J_{ab}|$ 1.0
H_b
Br

W-arrangement

5.11.3 Carboxylic Acids and Carboxylates

¹H Chemical Shifts (δ in ppm)

	a b H−COOH		
	CDCl₃	DMSO	D₂O
a	8.05	8.13	8.26
b	10.85	12.50	

	a b H₃C−COOH	
	CDCl₃	DMSO
a	2.10	1.91
b	11.51	11.91

a ⟍⟋ COOH c
b

	CDCl₃	DMSO
a	1.16	1.00
b	2.39	2.21
c	10.35	11.90

C = X

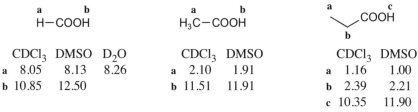

1.68   11.51
∕∖∕ COOH
1.00   2.31

1.21   11.88
∕ COOH
2.56

0.93 1.62   11.96
∕∖∕∖ COOH
1.39   2.35

1.23   11.49
✕ COOH

1.06
∕ COONa
2.18

(in D₂O)

12.2
2.43 ∕ COOH

⟍ COOH

(in DMSO)

6.53
H   O
H   OH
5.95   H   12.8
6.15

COOH 12.09
8.12
7.45
7.62

5.11.4 Esters and Lactones

^{1}H Chemical Shifts and Coupling Constants of Carboxylic Acid Esters (δ in ppm, J in Hz)

3.76
O—CH$_3$
H$_a$ b
8.07 $|^4J_{ab}|$ 0.8

7.33
H$_b$
O   H$_c$
H$_a$ O   4.66
8.07   H$_d$
4.96

$^4J_{ab}$ -0.7 $^3J_{bc}$ 6.4
$^5J_{ac}$ 1.6 $^3J_{bd}$ 13.9
$^5J_{ad}$ 0.8 $^2J_{cd}$ -1.7

3.67
2.01

4.12
2.04   1.26

4.02 0.95
2.05   1.65

4.99
2.02   1.23

4.06 1.39
2.04   1.60 0.94

1.45

7.28
H$_a$ 4.56
H$_b$
2.13
H$_c$
4.88

$^3J_{ab}$ 6.3
$^3J_{ac}$ 14.1
$^2J_{bc}$ -1.6

d
e   c 7.22
b 7.37
2.29   a
7.08

$^3J_{ab}$ 8.2
$^4J_{ac}$ 1.1
$^5J_{ad}$ 0.5
$^4J_{ae}$ 2.5
$^3J_{bc}$ 7.5
$^4J_{bd}$ 1.7

1.15   3.67
2.32

1.65
0.98 2.22

1.17   3.67
2.56

C = X

0.92 1.61   3.66
1.33   2.31

1.20   3.66

F
F   4.41
F   1.40

6.41 H O 3.77
H
5.83 H
6.12

6.37 H O 4.21
H 1.30
5.81 H
6.13

O O 3.92
8.04
7.43
7.55

1.38
O O
4.36
8.05
7.41
7.52

1.37
O O
5.22
7.99
7.37
7.46

7.22
O O 7.44
7.28
8.21
7.52
7.64

O
O 3.56
4.29

O H_a
O 2.49
H_b
H_e H_c
4.32 H_f H_d 2.26

$^2J_{ab}$ -17.5 $^2J_{cd}$ -12.7
$^3J_{ac}$ 9.5 $^3J_{ce}$ 7.9
$^3J_{ad}$ 6.9 $^3J_{cf}$ 6.3
$^4J_{ae}$ 0.3 $^2J_{ef}$ -8.8
$^4J_{af}$ -0.5

O
O 6.18
4.91 7.59

O
O 3.31
4.09 ≈1.6
≈1.6

O
O 6.05
4.44 7.10
2.50

O
O a 6.31
7.48 d b 7.33
c
6.25

$^3J_{ab}$ 9.4
$^4J_{ac}$ 1.5
$^5J_{ad}$ 1.3
$^3J_{bc}$ 6.3
$^4J_{bd}$ 2.4
$^3J_{cd}$ 5.0

C = X

5.11.5 Amides and Lactams

Amide Protons (δ in ppm, J in Hz)

O
R NH_2
5–7
R: alk or ar

O
R N alk
H
6–8.5
R: alk or ar

O
R N
H
7.5–10
R: alk or ar

Higher values in DMSO or with H bond acceptors in the neighborhood.

The signals of the NH protons are often broad because the $^{14}N-^{1}H$ coupling is only partly eliminated by the quadrupole relaxation of ^{14}N (spin quantum number, $I = 1$; $^{1}J_{NH} \approx 60$). In primary amides, the hindered rotation around the CO–N bond is another reason for line broadening. At slow rotation, the chemical shifts of the two primary amide protons differ by about 0.4–1 ppm. Therefore, at intermediate rotation rates, line widths of up to 1 ppm may be observed.

Due to the slow intermolecular exchange of amide protons, their coupling to neighboring hydrogen atoms is usually detectable. The splitting of the C–H signal is clearly observed even in those cases where the signal of the NH proton is broad and featureless. The H–N–C–H coupling depends on the conformation in a similar way as the H–C–C–H coupling (see Chapter 5.1.2). For N–CH$_3$ and N–CH$_2$ groups: $^{3}J_{HNCH} \approx 7$ Hz.

Tertiary Alkylamides

The rotation around the CO–N bond is usually so slow that, for identical substituents, two separate signals are observed for *cis* and *trans* positions. With different N-substituents, two separate pairs of signals are observed for the two conformers. In general, the following relationships hold:

for NCH$_3$, NCH$_2$CH$_3$, and NCH(CH$_3$)$_2$ $\delta_{cis\ to\ O} \le \delta_{trans\ to\ O}$
for NCH(CH$_3$)$_2$ and NC(CH$_3$)$_3$ $\delta_{trans\ to\ O} \le \delta_{cis\ to\ O}$
for NCH$_2$ $\delta_{cis\ to\ O} \approx \delta_{trans\ to\ O}$

Formamides (δ in ppm, J in Hz)

In the more stable conformer of monosubstituted formamides, the substituent occupies the *cis* position relative to the carbonyl oxygen. In the more stable conformer of asymmetrically disubstituted formamides, the larger substituent occupies the *trans* position relative to the carbonyl oxygen.

C = X

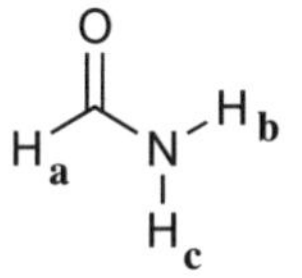

	CDCl$_3$	DMSO
a	8.23	7.98
b	5.80	7.14
c	5.48	7.41

$\approx 90\%$ in CDCl$_3$

	CDCl$_3$	DMSO
a	8.19	8.01
b	2.86	2.59
c	5.55	7.90

$\approx 10\%$ in CDCl$_3$

	CDCl$_3$	DMSO
a	8.06	7.81
b	2.94	2.72
c	5.86	7.90

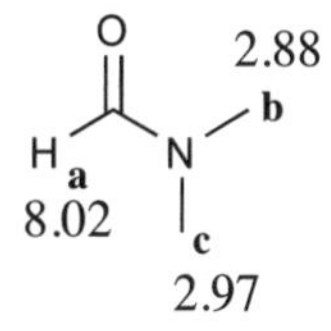

2.88
8.02
2.97

$^{4}J_{ab} \approx 0.3$
$^{4}J_{ac} \approx 0.7$

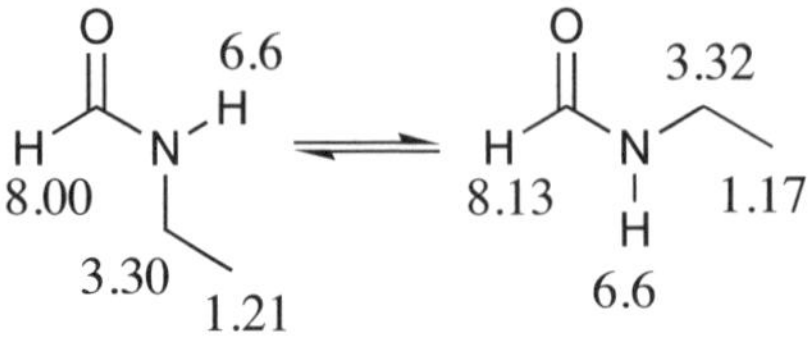

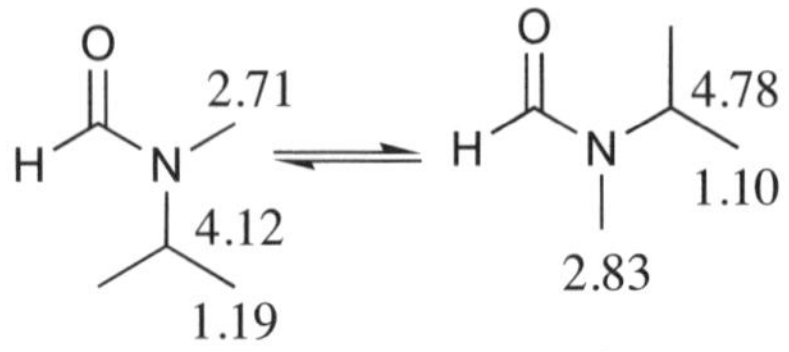

H–CO–N(H)(Et): 6.6 (NH), 8.00 (CHO), 3.30, 1.21 ⇌ 8.13 (CHO), 3.32, 6.6 (NH), 1.17

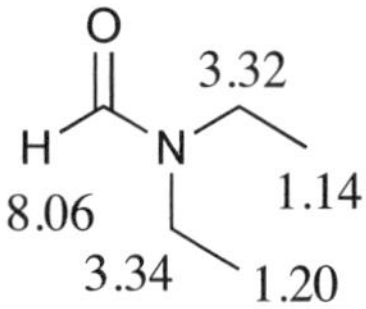

H–CO–N(H)(iPr): 2.71, 4.12, 1.19, ≈ 70% ⇌ 4.78, 1.10, 2.83, ≈ 30%

H–CO–NEt$_2$: 8.06 (CHO), 3.32, 1.14, 3.34, 1.20

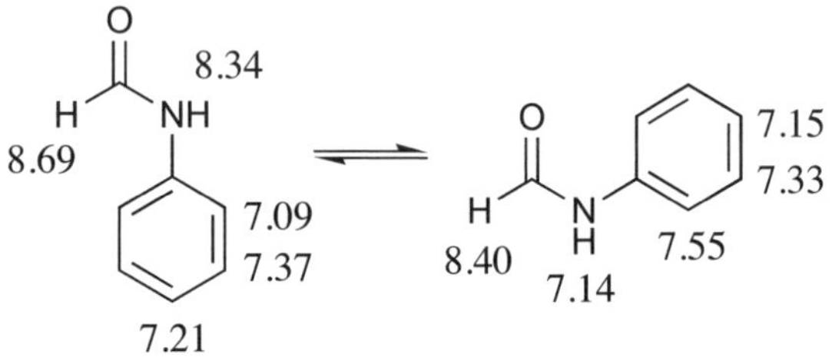

Formanilide: 8.69 (CHO), 8.34 (NH), 7.09, 7.37, 7.21, ≈ 35%, $^3J_{HCNH}$ 10.6 ⇌ 8.40 (CHO), 7.14, 7.55, 7.15, 7.33, ≈ 65%, $^3J_{HCNH} \approx 0$

Amides of Aliphatic Carboxylic Acids (δ in ppm, J in Hz)

In *monosubstituted* acetamides, the substituent of the only observable conformer is *cis* to the carbonyl oxygen. In *disubstituted* acetamides, the more stable conformer has the larger substituent *cis* to the carbonyl oxygen.

C = X

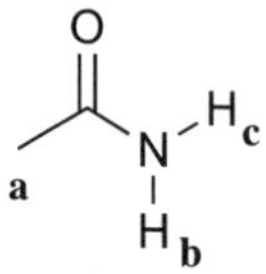

	CDCl$_3$	DMSO
a	2.03	1.76
b	5.42	7.30
c	5.42	6.70

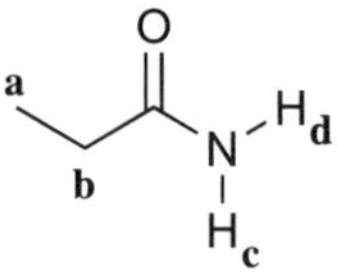

	CDCl$_3$	DMSO
a	1.17	0.97
b	2.26	2.04
c	5.38	7.16
d	6.14	6.62

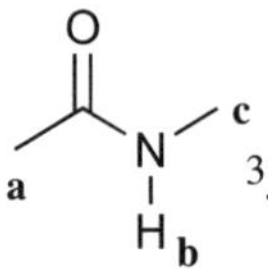

$^3J_{bc}$ 4.8

	CDCl$_3$	DMSO
a	1.98	1.78
b	5.53	7.70
c	2.80	2.50

3.26 (a), 1.98, 1.14, 6.7 (Hb), $^3J_{ab}$ 5.9

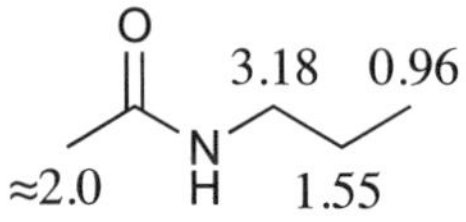

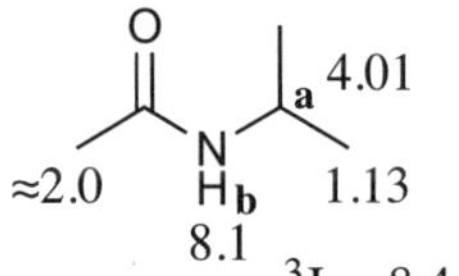

3.18 0.96, ≈2.0, 1.55

4.01 (a), ≈2.0, 1.13 (Hb), 8.1, $^3J_{ab}$ 8.4

3.21 1.35, 1.98, 1.49 0.92, 7.05

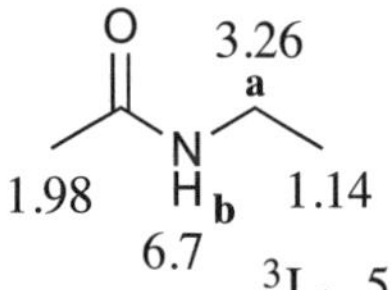

≈2.0, 1.28, 7.3

2.08, 3.01, 2.94

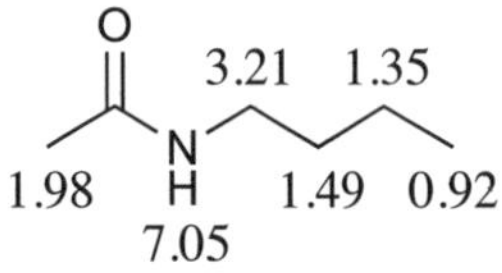

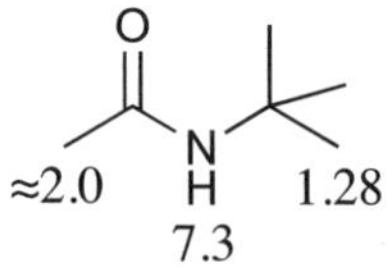

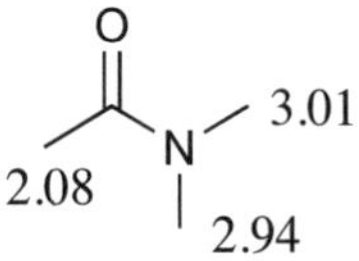

O
‖
2.70

N
3.92

1.15

≈ 40%

⇌

O
‖
1.03
4.52

N
2.83

≈ 60%

O
‖
3.46

N
2.05
3.42 1.96
1.86

O
‖
1.99

N
H
5.56 1.85

4.11
H 1.03
H
0.87

H
H 1.66 H
1.51   1.04

O
‖
5.42

NH 1.11
H

3.67 H
H
2.01

H 1.78 H
1.07   1.01

0.84

O
‖
3.40

2.08

N
3.54

1.57*

1.54*

1.64*

* assignment uncertain

O
‖
2.18

N
H
a
7.49

e
b 7.32
c 7.11
d

$^3J_{ab}$ 8.2
$^4J_{ac}$ 1.2
$^5J_{ad}$ 0.5
$^4J_{ae}$ 2.4
$^3J_{bc}$ 7.4
$^4J_{bd}$ 1.5

Lactams (δ in ppm, J in Hz)

O
‖
2.75

—N
3.20
2.82

O
‖
6.06
HN
3.40 2.14
2.30

$^3J_{HNCH_2}$ 0.8

O
‖
2.85

N
3.39 2.03
2.37

O
‖
6.33
HN
3.31
1.79
2.36
1.81

C = X

O
‖
2.94

N
3.29
1.81
2.37
1.82

O
‖
HN

NH
8.00
3.35
O

$^3J_{HNCH_2}$ 2.2

O
‖
6.35
HN
3.21
1.65 1.75
2.46
1.69

O
‖
2.98
N
3.36
1.65 1.70
2.52
1.66

5.11.6 Miscellaneous Carbonyl Derivatives

Carboxylic Acid Halides (δ in ppm, J in Hz)

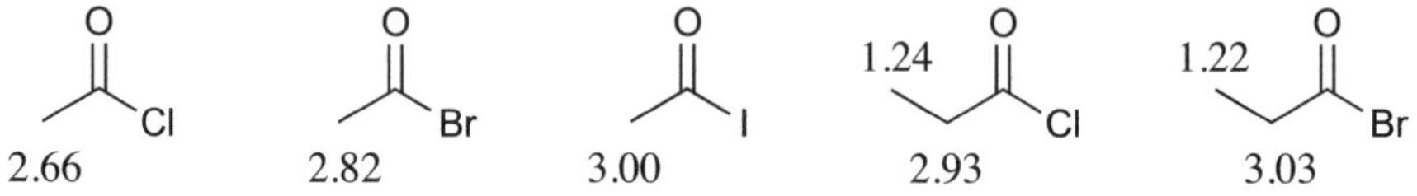

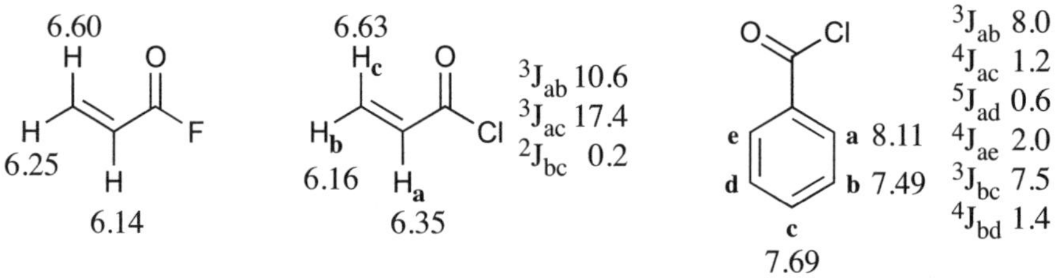

Acetyl chloride: 2.66
Acetyl bromide: 2.82
Acetyl iodide: 3.00
Propionyl chloride: 1.24, 2.93
Propionyl bromide: 1.22, 3.03

Acryloyl fluoride: 6.60 H, 6.25 H, 6.14 H, F

Acryloyl chloride: 6.63 H_c, H_b 6.16, H_a 6.35
$^3J_{ab}$ 10.6
$^3J_{ac}$ 17.4
$^2J_{bc}$ 0.2

Benzoyl chloride:
a 8.11
b 7.49
c 7.69
$^3J_{ab}$ 8.0
$^4J_{ac}$ 1.2
$^5J_{ad}$ 0.6
$^4J_{ae}$ 2.0
$^3J_{bc}$ 7.5
$^4J_{bd}$ 1.4

Carboxylic Acid Anhydrides (δ in ppm)

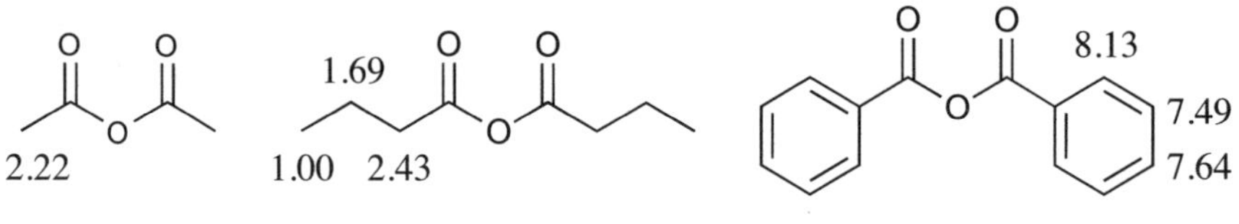

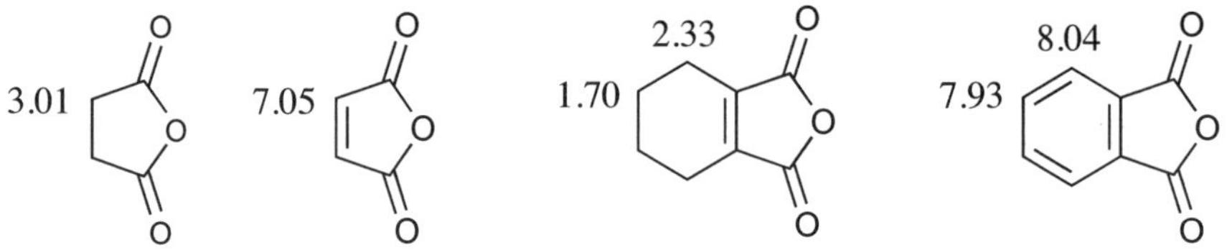

Acetic anhydride: 2.22
Butyric anhydride: 1.69, 1.00, 2.43
Benzoic anhydride: 8.13, 7.49, 7.64

Succinic anhydride: 3.01
Maleic anhydride: 7.05
Tetrahydrophthalic anhydride: 2.33, 1.70 (in DMSO)
Phthalic anhydride: 8.04, 7.93

Carboxylic Acid Imides (δ in ppm, J in Hz)

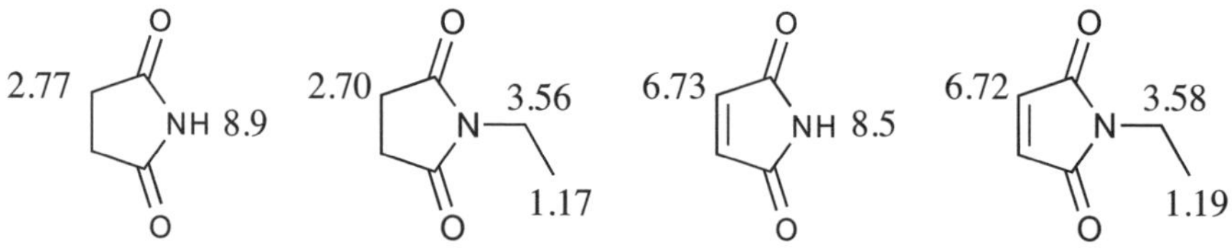

Succinimide: 2.77, NH 8.9
N-ethylsuccinimide: 2.70, 3.56, 1.17
Maleimide: 6.73, NH 8.5
N-ethylmaleimide: 6.72, 3.58, 1.19

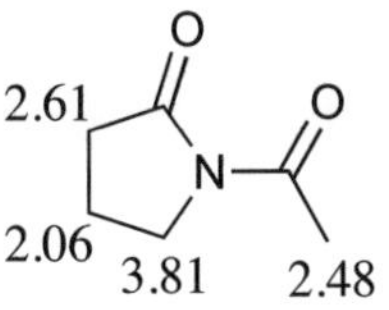

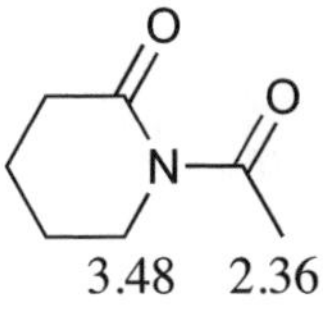

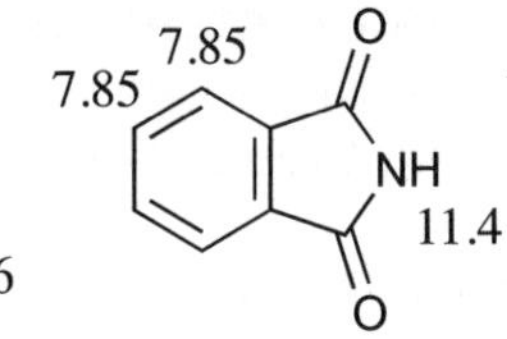

2.61 2.06 3.81 2.48

O O

3.48 2.36

O O

7.85 7.85 O NH 11.4 O

(in DMSO)

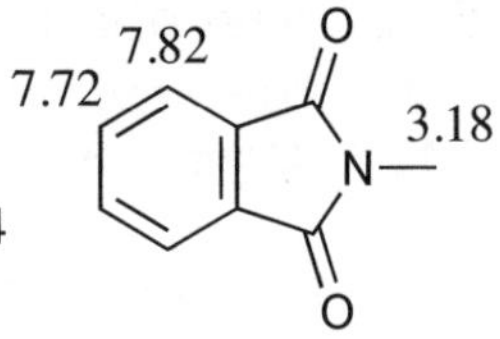

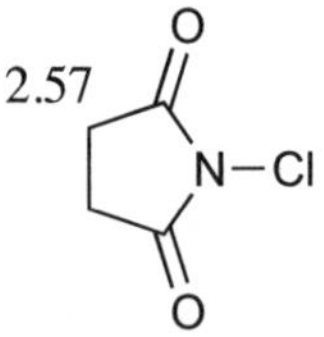

7.72 7.82 O N— 3.18 O

2.57 O N—Cl O

(in DMSO)

2.97 O N—Br O

3.56 1.22 N=C=N

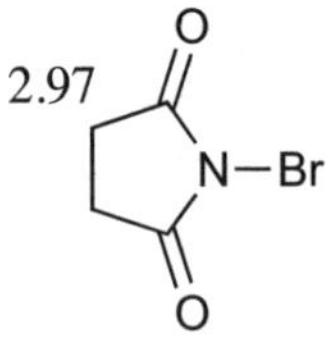

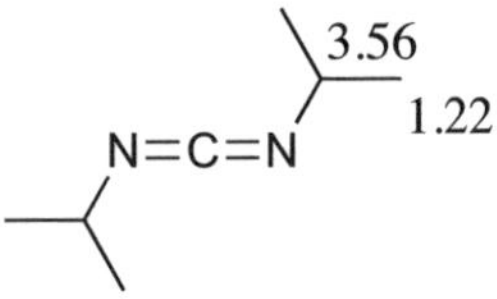

Carbonic Acid Derivatives (δ in ppm, J in Hz)

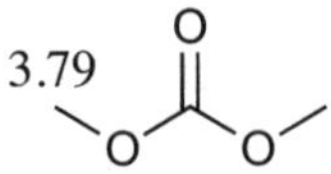

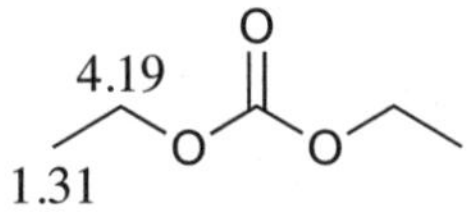

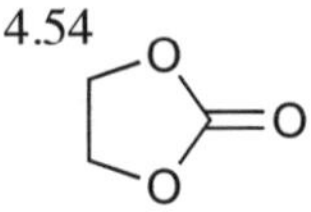

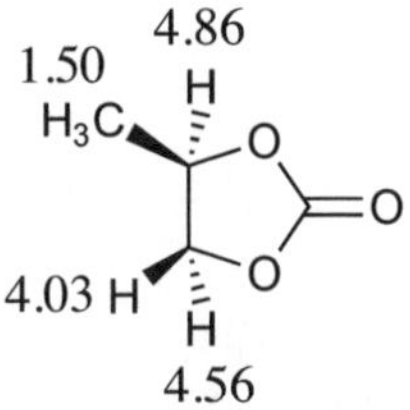

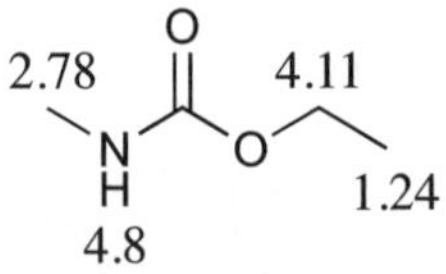

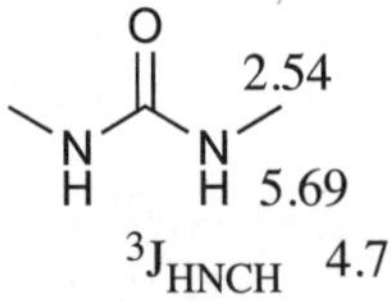

3.79 O O

4.19 O O 1.31

4.54 O O O

1.50 H$_3$C 4.86 H O O 4.03 H H 4.56

2.78 O 4.11 N H 1.24 4.8

O 2.54 N H N H 5.69 $^3J_{HNCH}$ 4.7 (in DMSO)

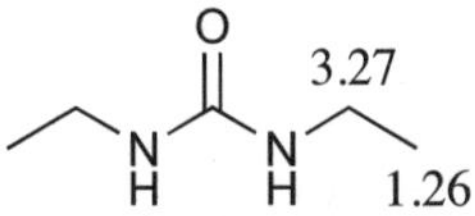

O 3.27 N H N H 1.26

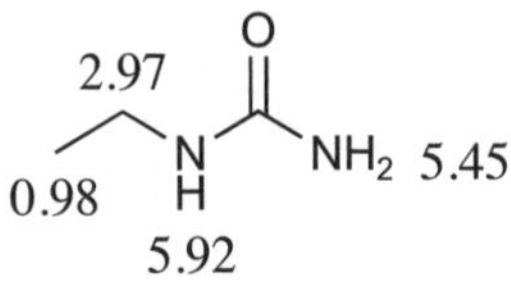

2.97 O N H NH$_2$ 5.45 0.98 5.92 (in DMSO)

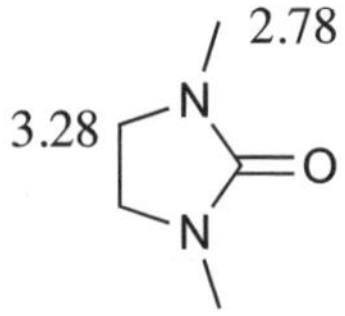

2.78 3.28 N O N

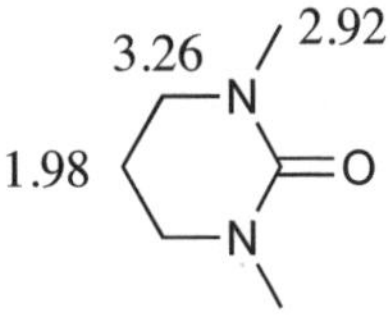

3.26 2.92 C=X 1.98 N O N

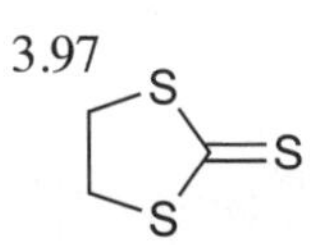

3.97 S S S

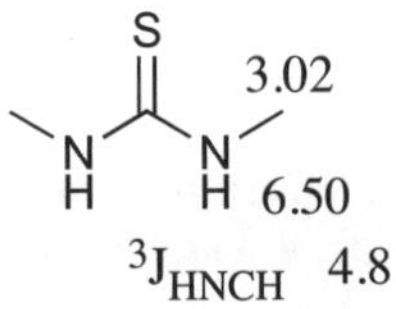

S 3.02 N H N H 6.50 $^3J_{HNCH}$ 4.8

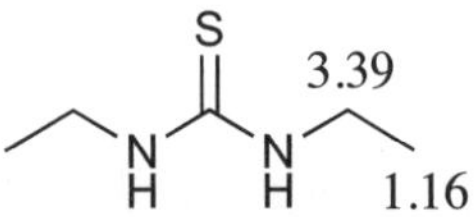

S 3.39 N H N H 1.16 (in D$_2$O)

5.12 Miscellaneous Compounds

5.12.1 Compounds with Group IV Elements

Silicon Compounds (δ in ppm, J in Hz)

Coupling with silicon: The isotope ^{29}Si (natural abundance, 4.7 %) has a spin quantum number I of 1/2. Doublets with the corresponding intensity ("Si satellites") are usually observed. Typical coupling constants: $^1J_{HSi}$ -150 to -380 Hz

$^2J_{HCSi}$ 5 to 10 Hz

H
|
H–Si–H 3.20
|
H

$^1J_{HSi}$ -202.5

0.19   H
|
H$_3$C–Si–H 3.58
|
H

$^1J_{HSi}$ -202.5
$^3J_{HSiCH}$ 4.7

0.14   CH$_3$
|
H$_3$C–Si–H 3.83
|
H

$^1J_{HSi}$ -202.5
$^2J_{HCSi}$ 7.7
$^3J_{HSiCH}$ 4.3

0.08   CH$_3$
|
H$_3$C–Si–H 4.00
|
CH$_3$

$^1J_{HSi}$ -190
$^2J_{HCSi}$ 7.5
$^3J_{HSiCH}$ 3.8

CH$_3$
|
H$_3$C–Si–CH$_3$ 0.00
|
CH$_3$

$^2J_{HCSi}$ 5.5

7.3 7.53

7.3 ⬡—Si–H 4.45
          |
   CH$_3$ 0.33
   CH$_3$

$^1J_{HSi}$ -189
$^2J_{HCSi}$ 7.6
$^3J_{HSiCH}$ 3.6

H   H
| |
H–Si–N–Si–H 4.35
| | |
H | H
      CH$_3$
      2.63

$^1J_{HSi}$ -209.4

Cl   0.42
|
H$_3$C–Si–CH$_3$
|
CH$_3$

Cl   0.79
|
Cl–Si–CH$_3$
|
CH$_3$

Cl   1.14
|
Cl–Si–CH$_3$
|
Cl

7.37

7.31

7.56

–Si–H 5.47

$^1J_{HSi}$ -199.1

5.88   6.11
H$_b$ H$_a$

H$_c$ Si–CH$_3$
              |
              CH$_3$
5.63   CH$_3$ 0.06

$^3J_{ab}$ 14.6
$^3J_{ac}$ 20.2
$^2J_{bc}$ 3.8

1.25   3.85

         O
          |
–O–Si–H 4.29
          |
         O

CH$_3$ OH
| |
H$_3$C–Si–CF$_2$–P=O
| |
CH$_3$ OH

pH-dependent shift of the methyl H atoms relative to DSS. Between pH 4.3 and 8.2 [1]:

$$pH = 6.246 - \log\left[\frac{0.224 - \delta}{\delta - 0.193}\right]$$

P Si

The silanol hydrogen is exchangeable with D_2O. Slow intermolecular exchange is observed in DMSO as solvent so that the vicinal coupling in H–Si–O–H is detectable ($^3J_{HSiOH} \approx 2\text{–}7$ Hz).

$$\begin{array}{l} CH_3 \\ | \\ H\text{-}Si\text{-}OH \quad 4.42 \\ | \\ CH_3 \qquad ^3J_{HSiOH} \quad 1.8 \\ \qquad\qquad (\text{in DMSO}) \end{array}$$

5.45 H-Si-OH 5.78 $^3J_{HSiOH}$ 2.0   (in DMSO)

Germanium, Tin, and Lead Compounds (δ in ppm, J in Hz)

$$\begin{array}{c} CH_3 \\ | \\ H_3C\text{-}Ge\text{-}CH_3 \quad 0.13 \\ | \\ CH_3 \end{array}$$

$$\begin{array}{c} CH_3 \\ | \\ H_3C\text{-}Sn\text{-}CH_3 \quad 0.07 \\ | \\ CH_3 \end{array}$$

$$\begin{array}{c} CH_3 \\ | \\ H_3C\text{-}Pb\text{-}CH_3 \quad 0.71 \\ | \\ CH_3 \end{array}$$

$^2J_{HC^{117}Sn}$ 51.9 (7.6 %)
$^2J_{HC^{119}Sn}$ 54.3 (8.6 %)

5.12.2 Phosphorus Compounds

^{31}P (natural abundance, 100%) has a spin quantum number I of 1/2. Couplings to protons through up to 5 bonds are usually observed.

Phosphines and Phosphonium Compounds (δ in ppm, J in Hz)

PH$_3$ 1.79

$^1J_{HP}$ 184.9

0.98 2.63
H$_3$C–PH$_2$

$^1J_{HP}$ 186.4
$^2J_{HCP}$ 4.1
$^3J_{HCPH}$ 8.2

1.06
$\begin{array}{c} H_3C \\ \diagdown \\ \qquad P\text{-}H \quad 3.13 \\ \diagup \\ H_3C \end{array}$

$^1J_{HP}$ 191.6
$^2J_{HCP}$ 3.6
$^3J_{HCPH}$ 7.7

$\begin{array}{c} H_3C \quad\quad 0.97 \\ \diagdown \\ \qquad P\text{-}CH_3 \\ \diagup \\ H_3C \end{array}$

$^2J_{HCP}$ 2.1

a 1.20 $^2J_{aP}$ 13.7
$^3J_{bP}$ 0.5
b 0.96 $^3J_{ab}$ 7.6

2.52 1.28
a   b

$^2J_{aP}$ 12.8
$^3J_{bP}$ 18.0
$^3J_{ab}$ 7.6

P Si

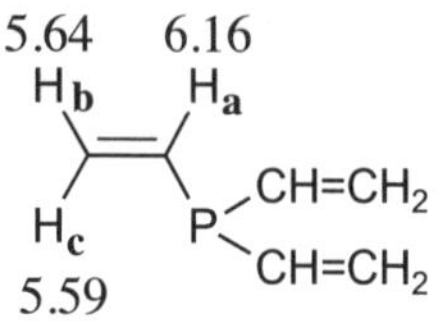

5.64   6.16 $^2J_{aP}$ 11.7
H**b**   H**a** $^3J_{bP}$ 30.2
            $^3J_{cP}$ 13.6
H**c**   P—CH=CH$_2$ $^3J_{ab}$ 11.8
5.59   \—CH=CH$_2$ $^3J_{ac}$ 18.4
            $^2J_{bc}$ 2.0

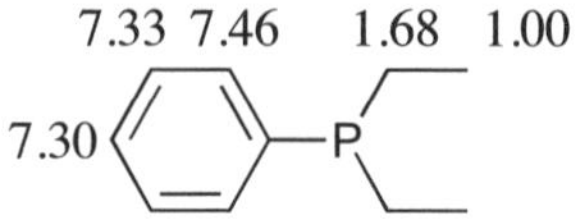

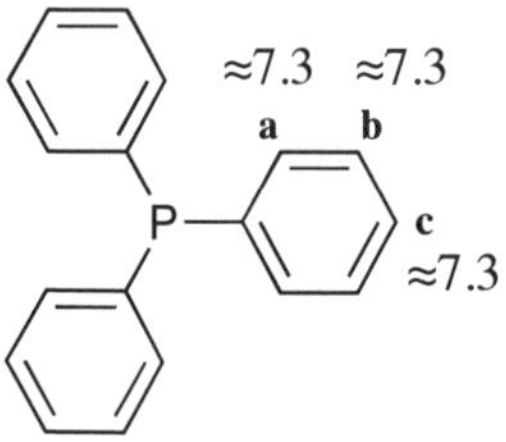

7.33 7.46   1.68   1.00
7.30 [ring]—P

≈7.3 ≈7.3
a   **b** $^3J_{aP}$ 7.5
            $^4J_{bP}$ 1.4
c $^5J_{cP}$ 0.7 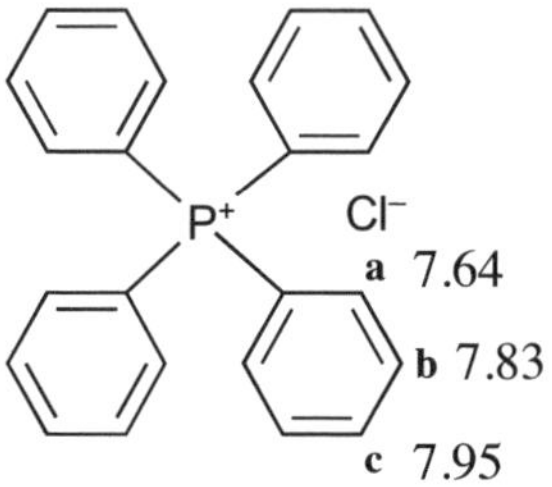
≈7.3

P$^+$ Cl$^-$ $^3J_{aP}$ 13.1
a 7.64 $^4J_{bP}$ 3.7
            $^5J_{cP}$ 2.1
b 7.83
c 7.95

Phosphine Oxides and Sulfides (δ in ppm, J in Hz)

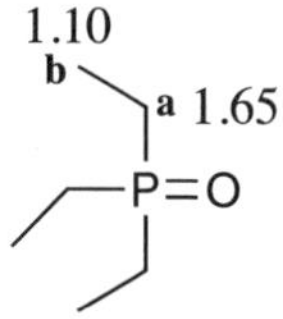

1.10
b   a 1.65
—P=O

$^3J_{aP}$ 11.9
$^4J_{bP}$ 16.3
$^5J_{ab}$ 7.8

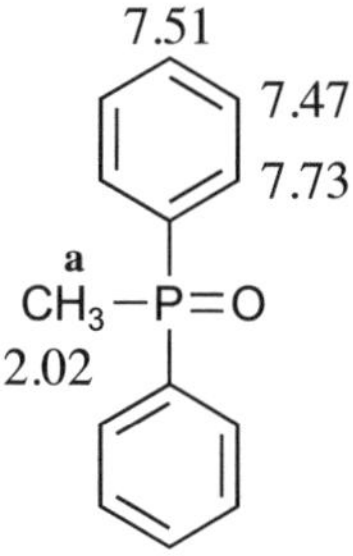

7.51
7.47
7.73
a
CH$_3$—P=O
2.02

$^3J_{aP}$ 13.2

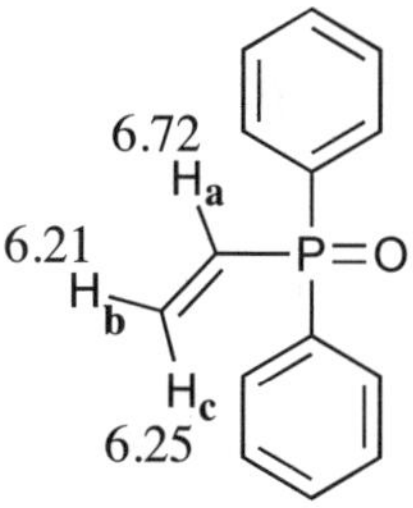

6.72
H**a**
6.21   P=O
H**b**
H**c**
6.25

$^2J_{aP}$ 25.9 $^3J_{ab}$ 12.9
$^3J_{bP}$ 41.8 $^3J_{ac}$ 18.9
$^3J_{cP}$ 22.3 $^2J_{bc}$ 1.8

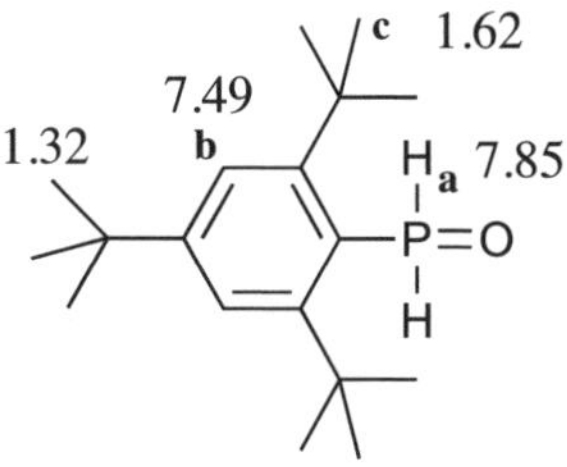

P Si

c 1.62
7.49
b
1.32   H**a** 7.85 $^1J_{aP}$ 481.7
P=O $^4J_{bP}$ 0
H $^5J_{cP}$ 0

7.3–7.6
7.3–7.6
7.71
P=O

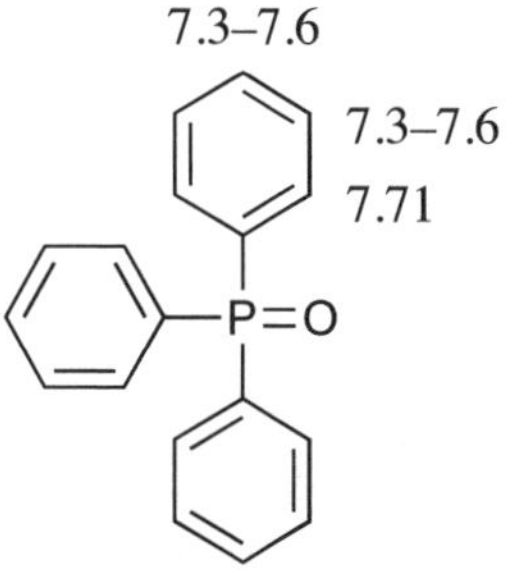

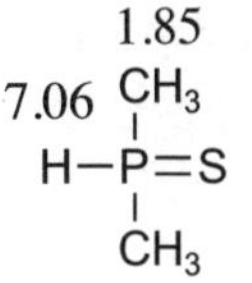

1.85
7.06 CH₃
H−P=S
CH₃

¹J_HP 14.3
²J_HCP 4.4

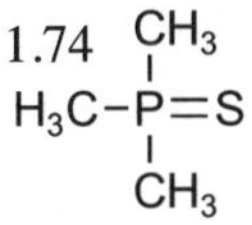

1.74 CH₃
H₃C−P=S
CH₃

²J_HCP 13.0

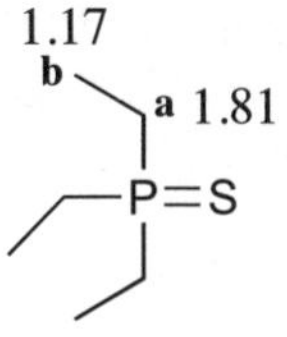

1.17
b   a 1.81
P=S

²J_aP 11.3
³J_bP 18.1
³J_ab 7.5

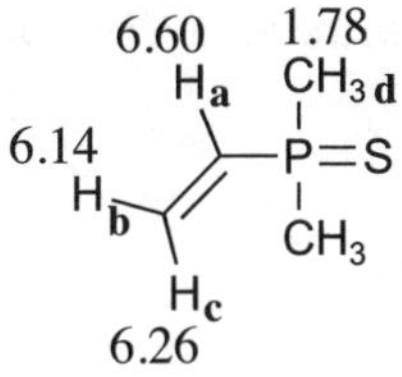

6.60   1.78   ²J_aP 25.9
H_a   CH₃ d ³J_bP 45.3
6.14   P=S   ³J_cP 25.4
H_b   CH₃   ²J_dP 13.5
       H_c   ³J_ab 11.8
6.26   ³J_ac 17.9
              ²J_bc 1.8

6.82
H_a
6.17   P=S
H_b
H_c
6.34

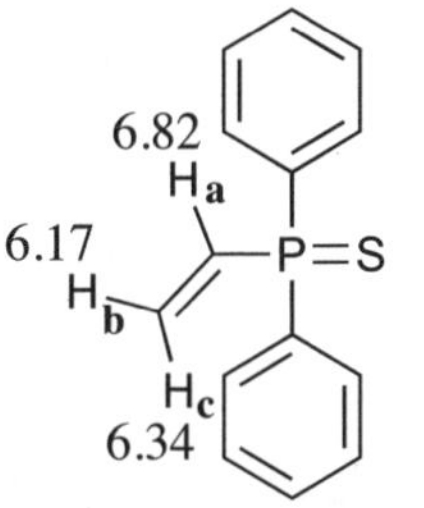

²J_aP 24.9   ³J_ab 11.7
³J_bP 47.0   ³J_ac 17.9
³J_cP 25.5   ²J_bc 1.6

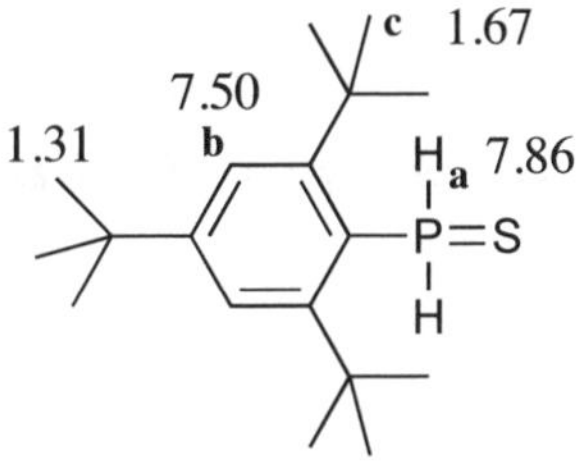

c 1.67
7.50
1.31 b   H_a 7.86
                   P=S
                   H

¹J_aP 466.2
⁴J_bP   4.8
⁵J_cP   1.1

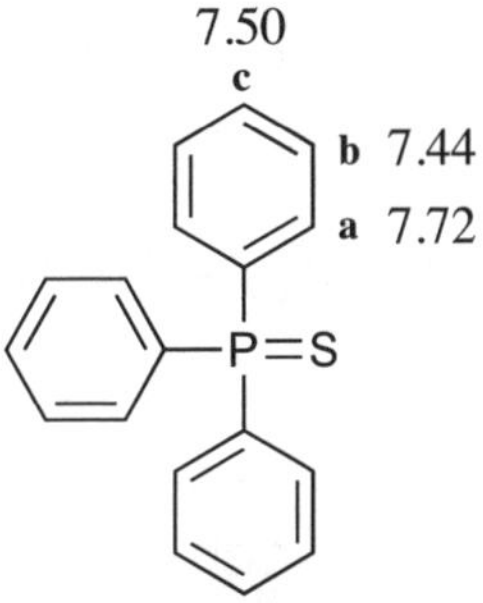

7.50
c
                b 7.44
                a 7.72
        P=S

³J_aP 13.3
⁴J_bP 3.1
⁵J_cP 2.1

Phosphinic and Phosphonic Acid Derivatives (δ in ppm, J in Hz)

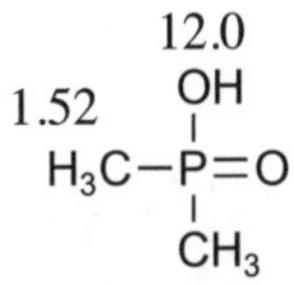

12.0
1.52 OH
H₃C−P=O
CH₃

²J_HP 14.4

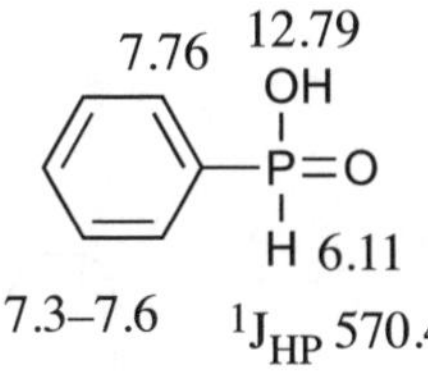

7.76 12.79
      OH
      P=O
      H 6.11

7.3–7.6   ¹J_HP 570.4

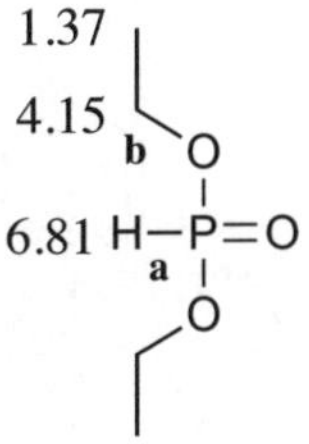

1.37
4.15 b   O
6.81 H−P=O
     a
         O

¹J_aP 691.7
³J_bP   7.1

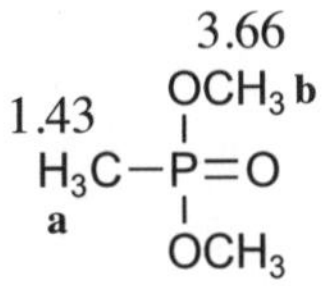

3.66
1.43 OCH₃ b
H₃C−P=O
a   OCH₃

²J_aP 17.3
³J_bP 11.0

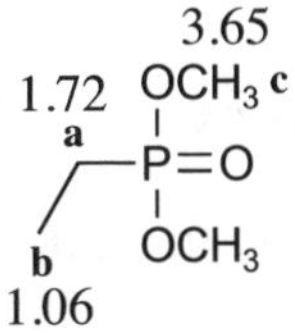

3.65
1.72 OCH₃ c
a
      P=O
b   OCH₃
1.06

²J_aP -18.0
³J_bP 19.5
³J_cP 10.0
³J_ab 7.5

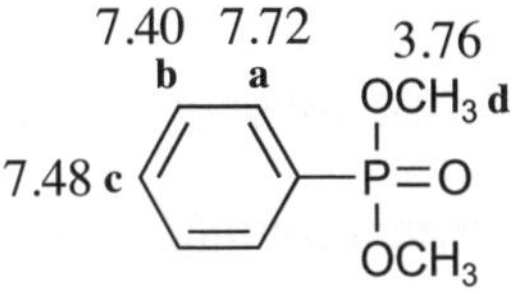

7.40 7.72   3.76
b   a   OCH₃ d
7.48 c   P=O
             OCH₃

³J_aP 13.3
⁴J_bP 4.1
⁵J_cP 1.2
³J_dP 11.1

P Si

Phosphonous and Phosphorous Acid Derivatives (δ in ppm, J in Hz)

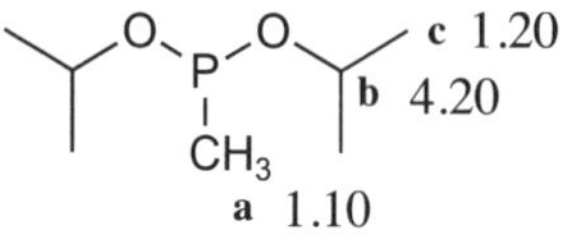

c 1.20 $^2J_{aP}$ 9.7
b 4.20 $^3J_{bP}$ 9.5
CH$_3$ $^4J_{cP}$ 6.0
a 1.10

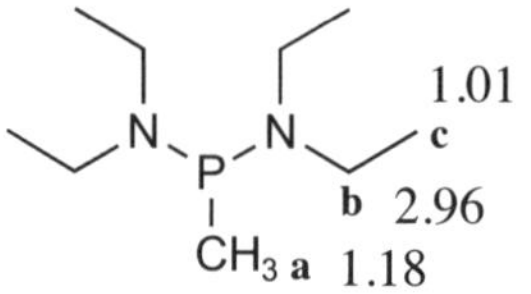

$^2J_{aP}$ 8.7
1.01 $^3J_{bP}$ 8.0
c $^4J_{cP}$ ≈1.0
b 2.96 $^3J_{bc}$ 7.5 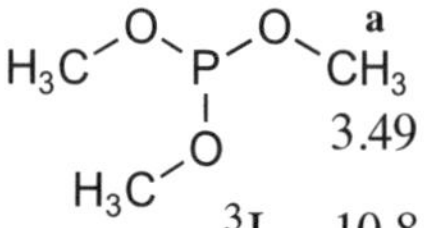
CH$_3$ a 1.18

H$_3$C—O—P—O—CH$_3$ a
3.49
H$_3$C—O
$^3J_{aP}$ 10.8

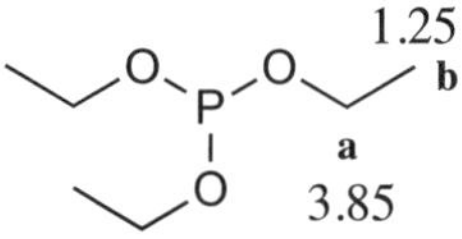

1.25
b $^3J_{aP}$ 8.0
a $^4J_{bP}$ 0.6
3.85 $^3J_{ab}$ 7.1

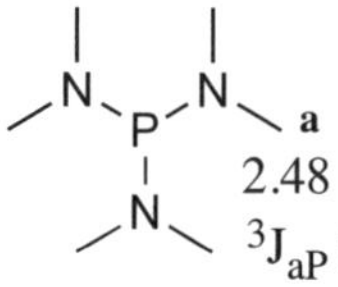

N—P—N a
2.48
$^3J_{aP}$ 9.1

Phosphoric Acid Derivatives (δ in ppm, J in Hz)

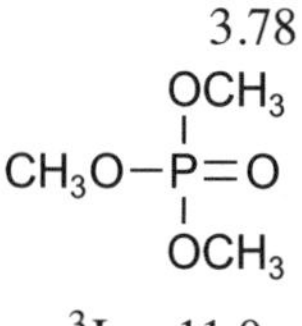

3.78
OCH$_3$
CH$_3$O—P=O
OCH$_3$
$^3J_{HP}$ 11.0

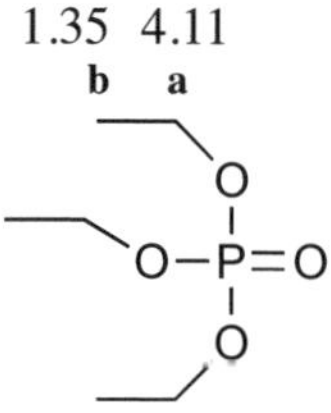

1.35 4.11
b   a
$^3J_{aP}$ 8.0
O—P=O $^4J_{bP}$ 0.9
$^3J_{ab}$ 7.1

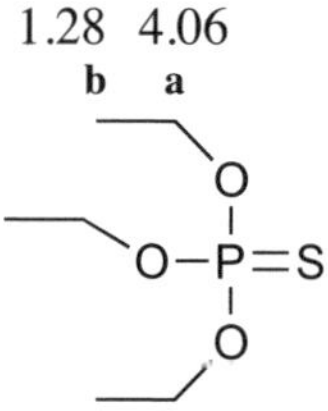

1.28 4.06
b   a
$^3J_{aP}$ 10.0
O—P=S $^4J_{bP}$ 0.7
$^3J_{ab}$ 7

Phosphorus Ylids (δ in ppm, J in Hz)

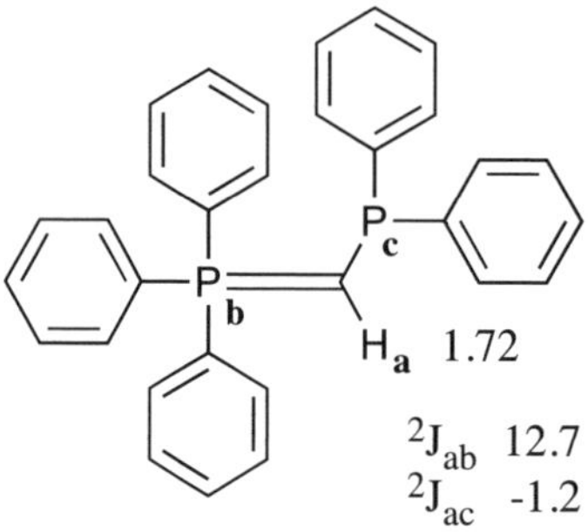

H$_a$ 1.72
$^2J_{ab}$ 12.7
$^2J_{ac}$ -1.2

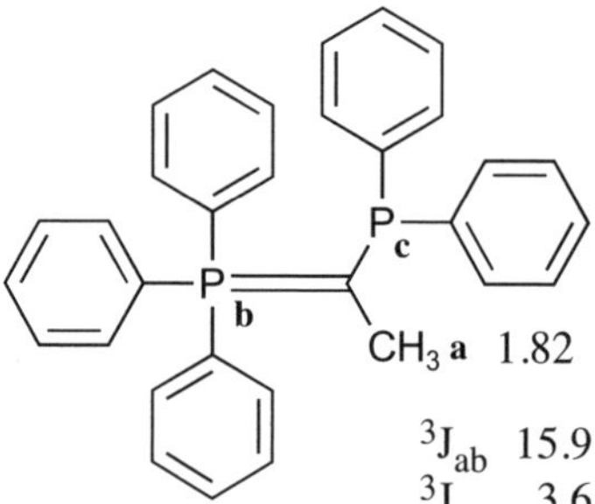

CH$_3$ a 1.82
$^3J_{ab}$ 15.9
$^3J_{ac}$ 3.6

P Si

5.12.3 Miscellaneous Compounds

Organometallic Compounds (δ in ppm, J in Hz)

Li—CH$_3$ -1.32 (in benzene)
-1.74 (in ether)

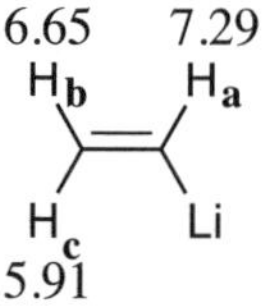

6.65   7.29
H$_b$ H$_a$ $^3J_{ab}$ 19.3
$^3J_{ac}$ 23.9
H$_c$ Li $^2J_{bc}$ 7.1
5.91

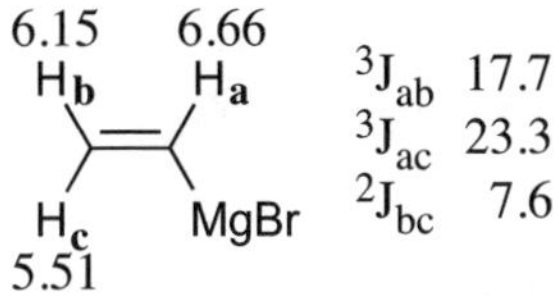

6.15   6.66
H**b**   H**a** $^3J_{ab}$ 17.7
                     $^3J_{ac}$ 23.3
H**c**   MgBr $^2J_{bc}$ 7.6
5.51

(in THF)

7.15 7.74

7.08 ⟨benzene⟩—MgBr

5.92   6.45
H**b**   H**a** $^3J_{ab}$ 11.9
                     $^3J_{ac}$ 18.7
H**c**   HgBr $^2J_{bc}$ 3.1
5.52

1.06 1.35
⟨chain⟩Hg⟨chain⟩
       1.82 0.91

7.40 7.44
⟨phenyl⟩–Hg–⟨phenyl⟩ 7.24

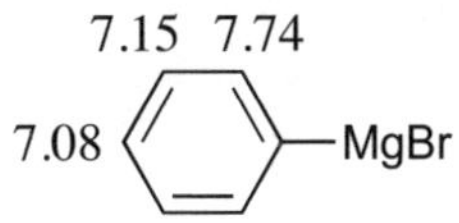

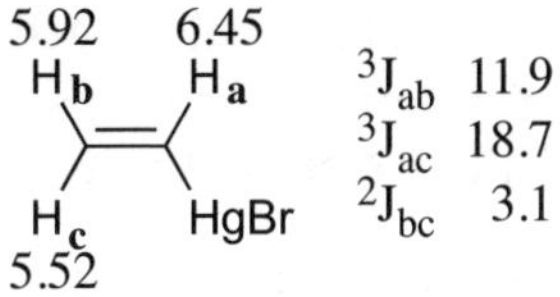

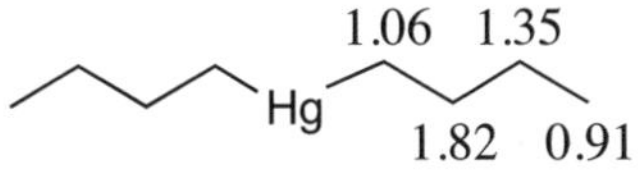

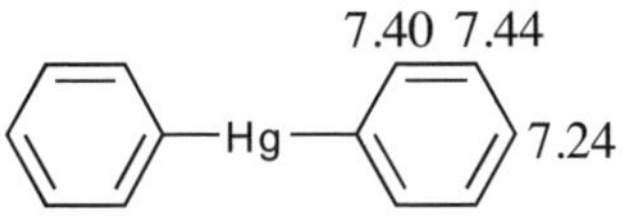

Boron Compounds (δ in ppm, J in Hz)

1.50   2.26
$BH_3 \cdot CH_3SCH_3$

$^1J_{H^{11}B}$ 104
(^{11}B: 80.4%, I = 3/2)

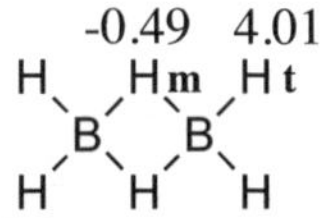

-0.49 4.01
H   H**m** H**t**
 \B / \B/
H   H   H
$^2J_{mt}$ 7.5
$^1J_{m^{11}B}$ 46.2
$^1J_{t^{11}B}$ 133.0

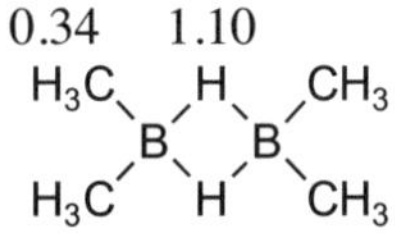

0.34   1.10
H_3C H CH_3
    \B / \B/
H_3C H CH_3

$^3J_{HH}$ 3.0

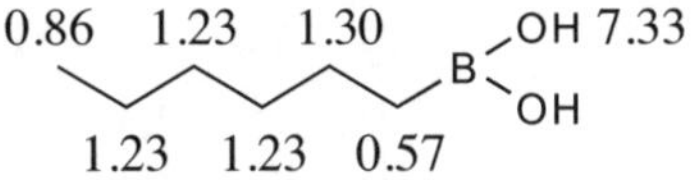

0.86   1.23   1.30   OH 7.33
⟨chain⟩B
                     OH
1.23   1.23   0.57

(in DMSO)

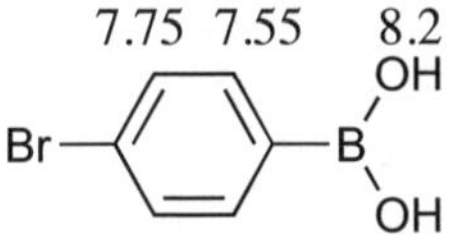

7.75 7.55   8.2
                 OH
Br—⟨phenyl⟩—B
                 OH

(in DMSO)

1.55 0.91
     O—⟨propyl⟩
⟨propyl⟩O–B 3.73
     O—⟨propyl⟩

                         Na⁺

⟨tetraphenylborate⟩B⁻

7.20
6.94
6.80

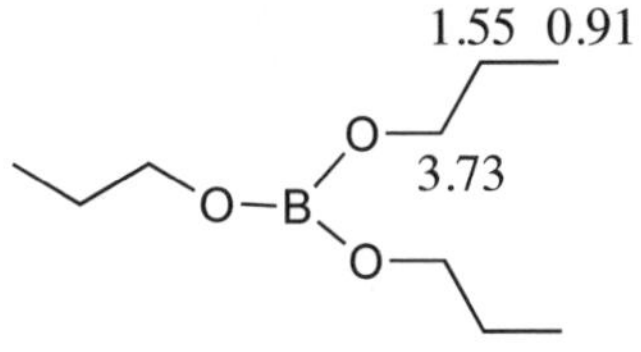

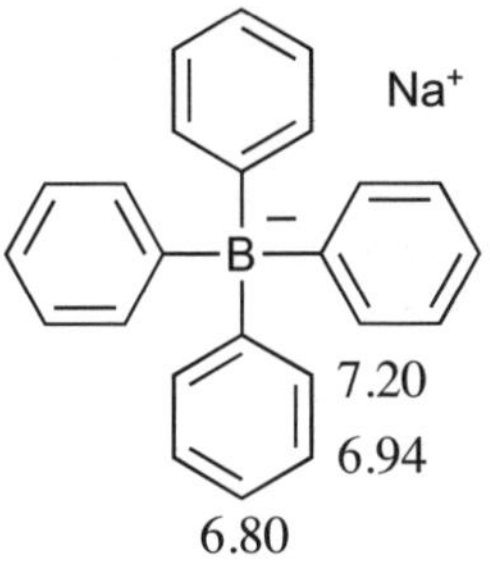

(in DMSO)

P Si

5.12.4 References

[1] M.D. Reily, L.C. Robosky, M.L. Manning, A. Butler, J.D. Baker, R.T. Winters, DFTMP, an NMR reagent for assessing the near-neutral pH of biological samples, *J. Am. Chem. Soc.* **2006**, *128*, 12360.

5.13 Natural Products

5.13.1 Amino Acids

Chemical Shifts and Coupling Constants (δ in ppm, J in Hz)

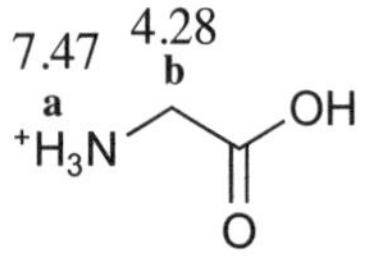

7.47 a, 4.28 b, ^+H_3N, OH, O

$^3J_{ab}$ 5.7
(in TFA)

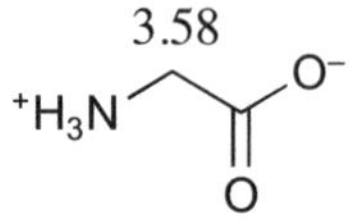

3.58, ^+H_3N, O$^-$, O

(in D$_2$O)

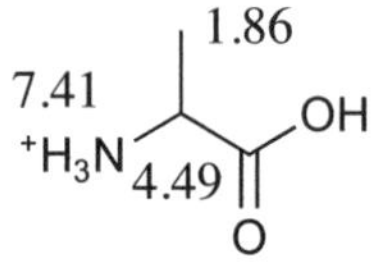

1.86, 7.41, ^+H_3N 4.49, OH, O

(in TFA)

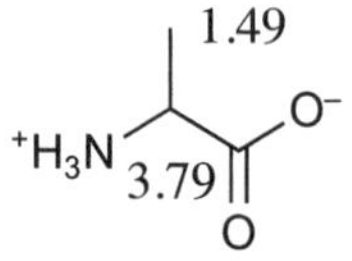

1.49, ^+H_3N 3.79, O$^-$, O

(in D$_2$O)

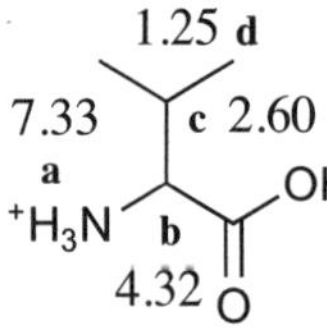

7.33 a, 1.25 d, c 2.60, ^+H_3N b, 4.32 O, OH

$^3J_{ab}$ 5.7
$^3J_{bc}$ 4.2
$^3J_{cd}$ 6.9

(in TFA)

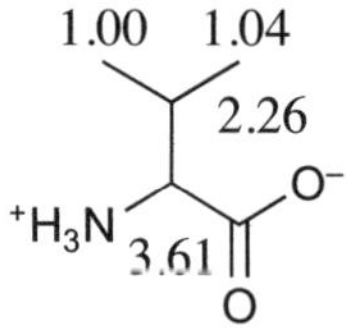

1.00 1.04, 2.26, ^+H_3N 3.61, O$^-$, O

(in D$_2$O)

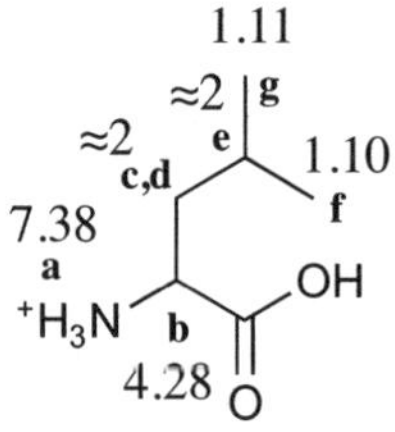

1.11, ≈2 g, ≈2 c,d e, 1.10 f, 7.38 a, ^+H_3N b, 4.28 O, OH

$^3J_{ab}$ 5.5
$^3J_{bc}$ ≈6.7
$^3J_{bd}$ ≈6.7
$^3J_{eg}$ 6.1
$^3J_{ef}$ 5.7

(in TFA)

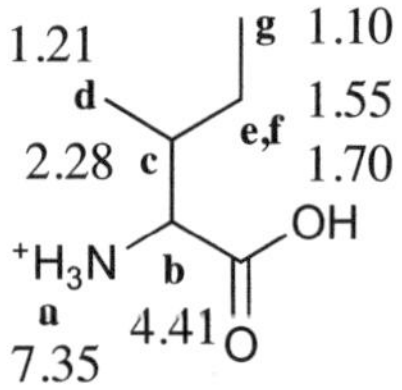

1.21 d, g 1.10, 1.55 e,f, 1.70, 2.28 c, ^+H_3N b, a 7.35, 4.41 O, OH

$^3J_{ab}$ 5.5 $^3J_{cf}$ 8.4
$^3J_{bc}$ 3.6 $^2J_{ef}$ -13.6
$^3J_{cd}$ 7.0 $^3J_{eg}$ 7.0
$^3J_{ce}$ 6.1 $^3J_{fg}$ 7.0

(in TFA)

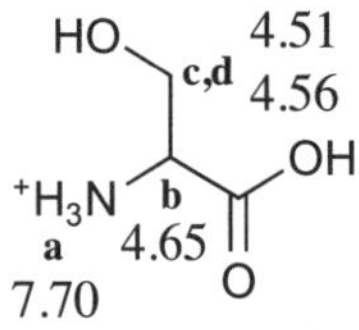

HO, c,d 4.51 4.56, ^+H_3N b, a 4.65, 7.70, OH

$^3J_{ab}$ ≈6
$^3J_{bc}$ 4.0*
$^3J_{bd}$ 4.0*
$^2J_{cd}$ -13.5

(in TFA)
* average value

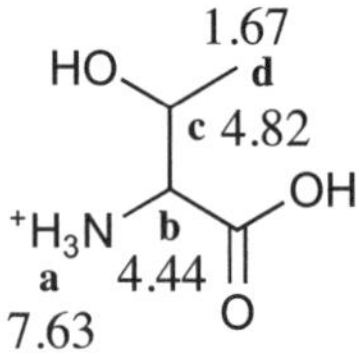

HO, 1.67 d, c 4.82, ^+H_3N b, a 4.44, 7.63, OH

$^3J_{ab}$ 5.5
$^3J_{bc}$ 4.5
$^3J_{cd}$ 6.5

(in TFA)

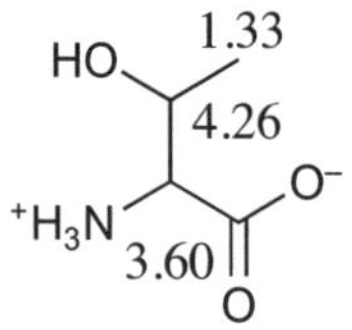

HO, 1.33, 4.26, ^+H_3N 3.60, O$^-$, O

(in D$_2$O)

1.84
e
HS
3.36
c,d 3.41
OH
$^+$H$_3$N b
a 4.68 O
7.58

$^3J_{ab}$ 5.3 $^2J_{cd}$ -15.5
$^3J_{bc}$ 5.0* $^3J_{ce}$ 9.1*
$^3J_{bd}$ 5.0* $^3J_{de}$ 9.1*

(in TFA)
* average value

HS 3.12
3.18
Cl$^-$
OH
$^+$H$_3$N 4.35
O

(in D$_2$O)

2.27
S f
2.50
2.65 c,d
e 2.96
OH
$^+$H$_3$N b
a 4.67 O
7.73

$^3J_{ab}$ 5.5 $^2J_{cd}$ -15.7
$^3J_{bc}$ 7.7 $^3J_{ce}$ 6.5*
$^3J_{bd}$ 4.4 $^3J_{de}$ 6.5*

(in TFA)
* average value

2.14
S
2.09– 2.64
2.23
$^+$H$_3$N 3.86
O$^-$
O

(in D$_2$O)

≈7.45
7.3   ≈7.45
3.37 c,d
3.64
OH
$^+$H$_3$N b
a 4.68 O
7.33

$^3J_{bc}$ 8.5
$^3J_{bd}$ 4.5
$^2J_{cd}$ -14.5

(in TFA)

7.43
7.33   7.38
3.13
3.29
$^+$H$_3$N 3.99
O$^-$
O

(in D$_2$O)

7.03 OH
7.27
3.34 c,d
3.60
OH
$^+$H$_3$N b
a 4.64 O
7.4

$^3J_{bc}$ 8.5
$^3J_{bd}$ 4.5
$^2J_{cd}$ -15.0

(in TFA)

O
OH
3.55 c
OH
$^+$H$_3$N b
a 4.76 O
7.73

$^3J_{ab}$ 5.1
$^3J_{bc}$ 4.7

(in TFA)

O OH
2.55 c,d
2.63
e 3.01
OH
$^+$H$_3$N b
a 4.60 O
7.71

$^3J_{ab}$ 5.5 $^2J_{cd}$ -15.5
$^3J_{bc}$ 5.6* $^3J_{ce}$ 6.2*
$^3J_{bd}$ 5.6* $^3J_{de}$ 6.2*

(in TFA)
* average value

6.97
h
NH$_3$$^+$
3.38 g
f 2.00
2.26 c,d
2.35
e ≈1.8
OH
$^+$H$_3$N b
a 4.52 O
7.60

$^3J_{ab}$ 5.8 $^2J_{cd}$ -15.0
$^3J_{bc}$ 5.6* $^3J_{ce}$ 6.0*
$^3J_{bd}$ 5.6* $^3J_{de}$ 6.0*

(in TFA)
* average value

NH$_3$$^+$
2.98 1.73
1.68 1.43
H$_2$N 3.44
OH
O

(in D$_2$O)

Natural
Products

6.19 $^+$H$_2$N ⟋⟍ NH$_2$ 6.19

6.50 d HN⟍⟍ g 3.43

2.25 c
2.34 d
e ≈2.00
f ≈2.08

$^+$H$_3$N b
a
7.60
OH
4.46 O

$^3J_{ab}$	5.5
$^3J_{bc}$	5.3*
$^3J_{bd}$	5.3*
$^2J_{cd}$	≈ -15.0
$^3J_{eg}$	6.5*
$^3J_{fg}$	6.5*
$^3J_{gh}$	5.3

(in TFA)
* average value

$^+$H$_2$N ⟋⟍ NH$_2$

HN⟍
3.21
1.63

1.63

H$_2$N 3.27
OH
O

(in D$_2$O)

2.06 d   b 2.42
2.04 e ⟍⟋ c 2.14
3.46 f   OH
3.39 g N a
H$_2$$^+$↑ O
4.33

$^3J_{ab}$	8.5	$^3J_{cd}$	5.5
$^3J_{ac}$	6.5	$^3J_{ce}$	7.5
$^2J_{bc}$	-13.5	$^2J_{de}$	-13.0
$^3J_{bd}$	7.5	$^3J_{df}$	5.5
$^3J_{be}$	5.5	$^3J_{dg}$	7.5
$^4J_{bf}$	-0.4	$^3J_{ef}$	7.5
$^4J_{bg}$	0.0	$^3J_{eg}$	5.5
		$^2J_{fg}$	-11.0

(in D$_2$O, pH 2.0)

1.63 d   b 1.96
1.60 e ⟍⟋ c 1.68
3.04 f   O$^-$
2.95 g N a
H$_2$$^+$↑ O
3.74

$^3J_{ab}$	8.4	$^3J_{cd}$	5.6
$^3J_{ac}$	6.2	$^3J_{ce}$	7.8
$^2J_{bc}$	-13.5	$^2J_{de}$	-13.0
$^3J_{bd}$	7.6	$^3J_{df}$	5.7
$^3J_{be}$	5.4	$^3J_{dg}$	7.9
$^4J_{bf}$	-0.4	$^3J_{ef}$	7.9
$^4J_{bg}$	0.0	$^3J_{eg}$	5.7
		$^2J_{fg}$	-11.0

(in D$_2$O, pH 7.0)

1.05 d   b 1.04
1.07 e ⟍⟋ c 1.45
2.08 f   O$^-$
2.36 g N a
H↑ O
2.81

$^3J_{ab}$	8.6	$^3J_{cd}$	6.7
$^3J_{ac}$	6.6	$^3J_{ce}$	8.5
$^2J_{bc}$	-12.0	$^2J_{de}$	-11.0
$^3J_{bd}$	8.1	$^3J_{df}$	5.5
$^3J_{be}$	5.9	$^3J_{dg}$	8.1
$^4J_{bf}$	-0.6	$^3J_{ef}$	7.7
$^4J_{bg}$	0.0	$^3J_{eg}$	5.7
		$^2J_{fg}$	-10.5

(in D$_2$O, pH 13.0)

≈5
HO,,, d   b 2.95
c 2.56
3.9* e   OH
3.9* f N a
8.00 g H$_2$$^+$↑ O
8.60 h   ≈5

$^3J_{ab}$	8.2
$^3J_{ac}$	10.4
$^2J_{bc}$	-15.0
$^3J_{bd}$	<2
$^3J_{cd}$	4.2

(in TFA)
* average value

HO,,, 4.35   2.17
2.44
3.37
3.49 N
H$_2$$^+$↑
O$^-$
O
4.67

(in D$_2$O)

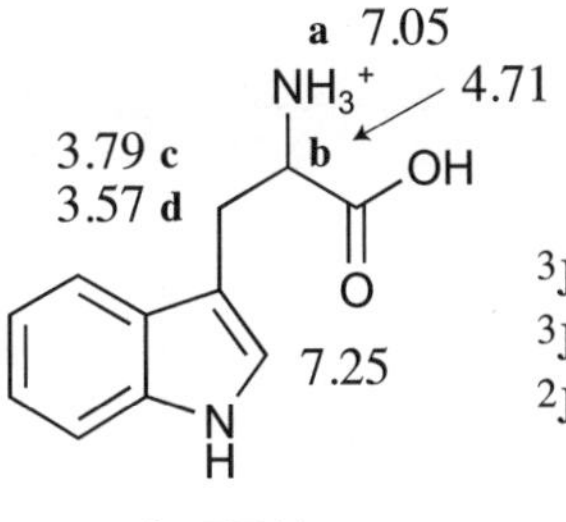

a   7.05
NH₃⁺   4.71
3.79 c
3.57 d   b   OH
                O

7.25

$^3J_{bc}$ 4.0
$^3J_{bd}$ 8.0
$^2J_{cd}$ -15.5

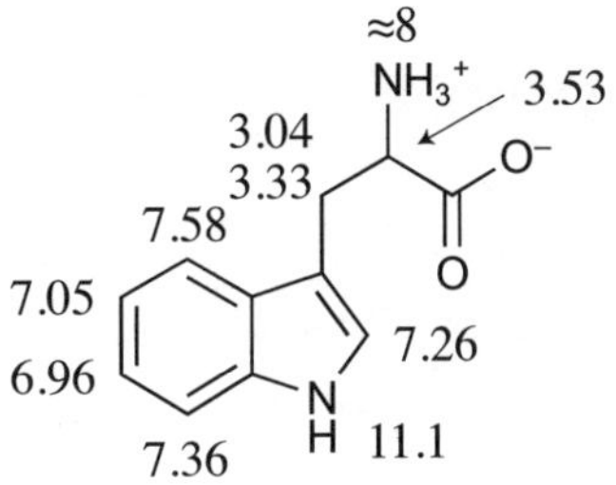

(in TFA)

≈8
NH₃⁺   3.53
3.04   O⁻
3.33
7.58
                O
7.05
6.96   7.26
7.36   H 11.1

(in DMSO)

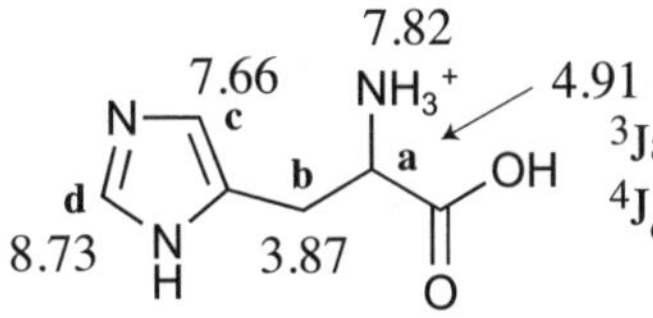

7.82
7.66   NH₃⁺   4.91
        c
     d   b   a   OH
8.73   N   OH
        H   3.87   O

^3Jab 6.4
$^4J_{cd}$ 1.4

(in TFA)

7.42   NH₃⁺   4.06
N   O⁻
8.69
        N   3.36   O
        H   3.38

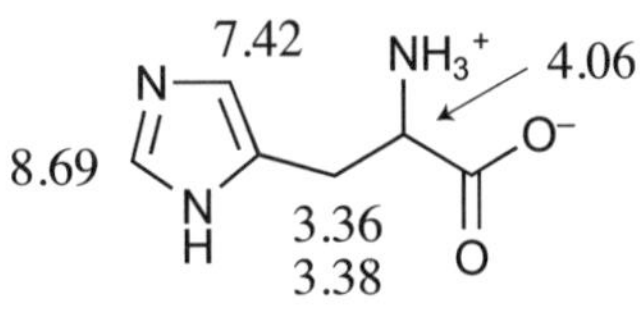

(in D₂O)

5.13.2 Carbohydrates

Chemical Shifts and Coupling Constants (δ in ppm, J in Hz)

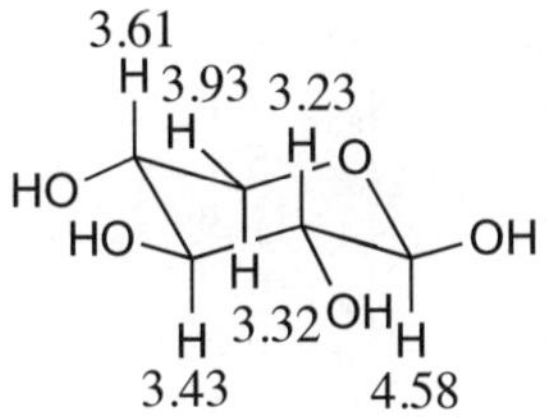

3.61
H 3.93 3.23
   H   H O
HO
HO
            OH
   H   3.32 OH
   H   H
3.43   4.58

(in D₂O)

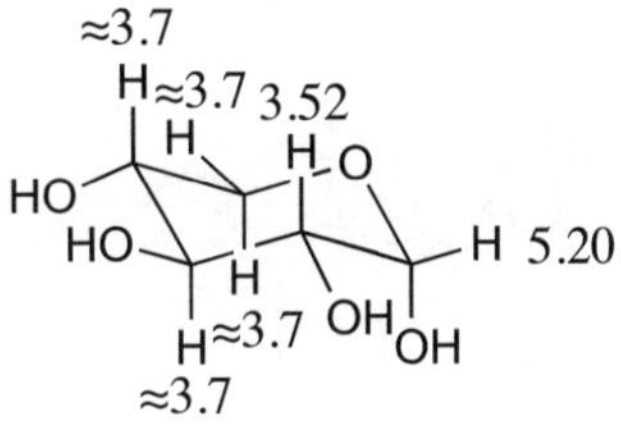

≈3.7
H ≈3.7 3.52
   H   H O
HO
HO   H 5.20
   H   ≈3.7 OH
   H   OH
≈3.7

(in D₂O)

g
H
         HO   c   H
h
HO   OH
HO
   f   H   H_b
           OH
       H_e   d OH_a

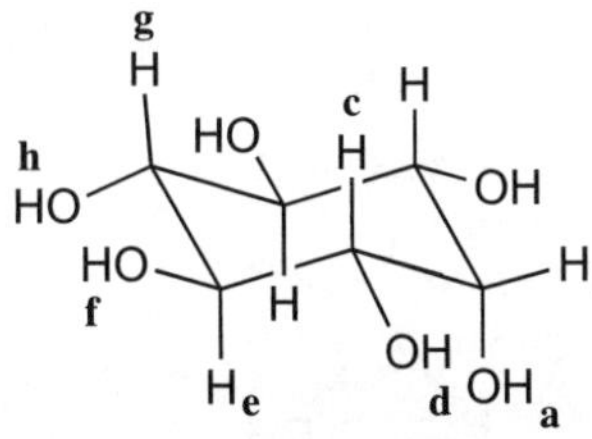

	D₂O	DMSO
a		4.55
b	4.07	3.72
c	3.54	3.14
d		4.51
e	3.64	3.37
f		4.46
g	3.27	2.93
h		4.31

(structure, in DMSO ca. 80%)

Labels: g, OH$_l$, H, j,k, c, h, HO, H, O, HO, f, OH$_a$, H$_i$, H$_e$, OH, d, H$_b$

	D$_2$O*	DMSO		D$_2$O		DMSO
a		6.58	$^3J_{bc}$	7.8	$^3J_{ab}$	6.5
b	4.51	4.27	$^3J_{ce}$	9.5	$^3J_{cd}$	4.5–6
c	3.13	2.89	$^3J_{eg}$	9.5	$^3J_{ef}$	4.5–6
d		4.84	$^3J_{gi}$	9.5	$^3J_{gh}$	4.5–6
e	3.37	3.10	$^3J_{ij}$	2.8	$^3J_{jl}$	5.5
f		4.84	$^3J_{ik}$	5.7	$^3J_{kl}$	6.0
g	3.30	3.10	$^2J_{jk}$	-12.8		
h		4.84				
i	3.35	3.04				
j	3.60	3.42				
k	3.75	3.66	(* relative to internal			
l		4.48	acetone at $\delta = 2.12$)			

(structure, in DMSO ca. 20%)

Labels: g, OH$_l$, H, j,k, c, h, HO, H, O, HO, f, H$_b$, H$_i$, H$_e$, OH, OH$_a$, d

	D$_2$O*	DMSO		D$_2$O		DMSO
a		6.20	$^3J_{bc}$	3.6	$^3J_{ab}$	4.5
b	5.09	4.91	$^3J_{ce}$	9.5	$^3J_{cd}$	6.8
c	3.41	3.10	$^3J_{eg}$	9.5	$^3J_{ef}$	4.8
d		4.84	$^3J_{gi}$	9.5	$^3J_{gh}$	5.5
e	3.61	3.42	$^3J_{ij}$	2.8	$^3J_{jl}$	5.7
f		4.64	$^3J_{ik}$	5.7	$^3J_{kl}$	6.2
g	3.29	3.04	$^2J_{jk}$	-12.8		
h		4.77				
i	3.72	3.57				
j	3.72	3.57	(* relative to internal			
k	3.63	3.42	acetone at $\delta = 2.12$)			
l		4.37				

(structure)

Labels: k, l, H, OH, a, j, f, OH, h, H, H, O, b,c, H, OH, HO, e, HO g, H$_d$, i

	D$_2$O^a	DMSO^b	DMSO	
a		4.48	$^3J_{ab}$	7.4^c
b	3.68	3.39	$^3J_{ac}$	5.4^c
c	3.53	3.25	$^2J_{bc}$	-11.3^d
d	3.76	3.55	$^3J_{de}$	6.8^c
e		4.23	$^3J_{df}$	10.1^d
f	3.86	3.58	$^3J_{fg}$	5.8^c
g		4.38	$^3J_{fh}$	4.0^d
h	3.96	3.62	$^3J_{hi}$	3.8^c
i		4.32	$^3J_{hj}$	1.9^d
j	4.00	3.77	$^3J_{hk}$	1.6^d ^{c} at 25 °C
k	3.68	3.41	$^2J_{jk}$	-12.1^d ^{d} at 70 °C
l		5.14		

^{a} 25% β-D
^{b} 75% β-D

(structure, left) 3.52, 3.40 HO H 3.69 H 3.72 OH H 3.77 O OH OH (in DMSO, 20% α-D)

(structure, right) f,g 3.48, 3.37 HO O OH 3.53 e H OH a,b 3.40, 3.23 H OH H OH 3.79 d c 3.80 (in DMSO, 55% β-D)

$^2J_{ab}$	-11.0
$^3J_{cd}$	7.1
$^3J_{de}$	5.9
$^3J_{ef}$	2.3
$^3J_{eg}$	3.6
$^2J_{fg}$	-11.3

5.13.3 Nucleotides and Nucleosides

Chemical Shifts and Coupling Constants (δ in ppm, J in Hz)

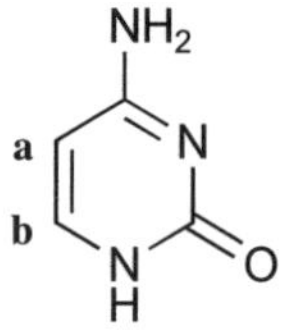

Cytosine:

	D_2O	DMSO
a	5.97	5.57
b	7.50	7.31
J_{ab}	7	

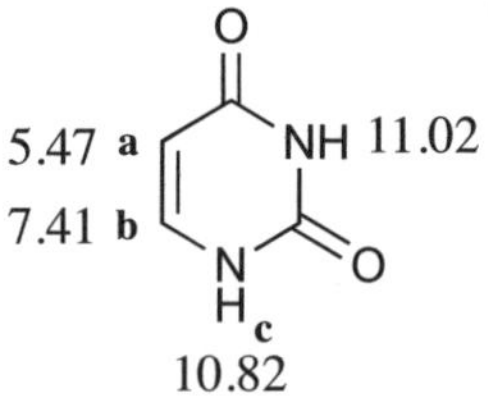

Uracil (in DMSO):

5.47 a, 7.41 b, NH 11.02, H_c 10.82

$^3J_{ab}$ 7.5
$^3J_{bc}$ 5.7

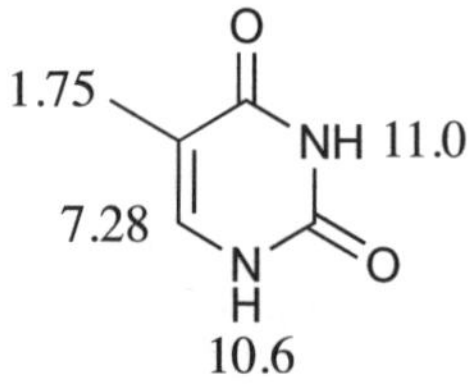

Thymine (in DMSO):

1.75, 7.28, NH 11.0, NH 10.6

Guanine (in TFA): 8.98

Adenine:

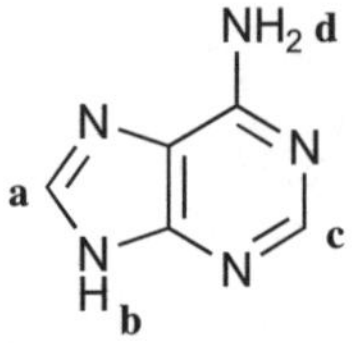

	D_2O	DMSO	$CDCl_3$	TFA
a	8.62	8.12	8.11	8.88
b		12.8		
c	8.57	8.10	8.14	9.31
d		7.09		

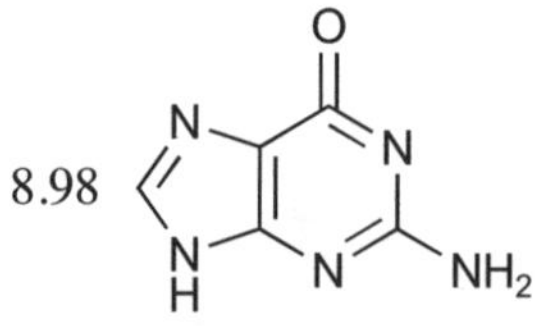

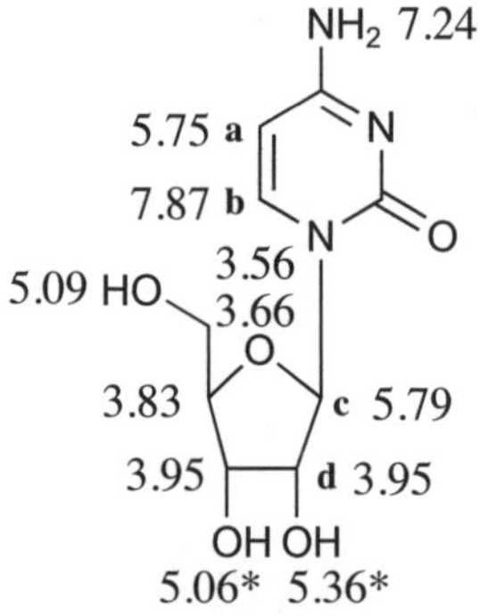

Cytidine (in DMSO):

NH₂ 7.24
5.75 a, 7.87 b
5.09 HO, 3.56, 3.66
3.83, c 5.79
3.95, d 3.95
OH OH 5.06* 5.36*

$^2J_{ab}$ 7.4
$^3J_{cd}$ 3.8

* interchangeable

Thymidine (in DMSO):

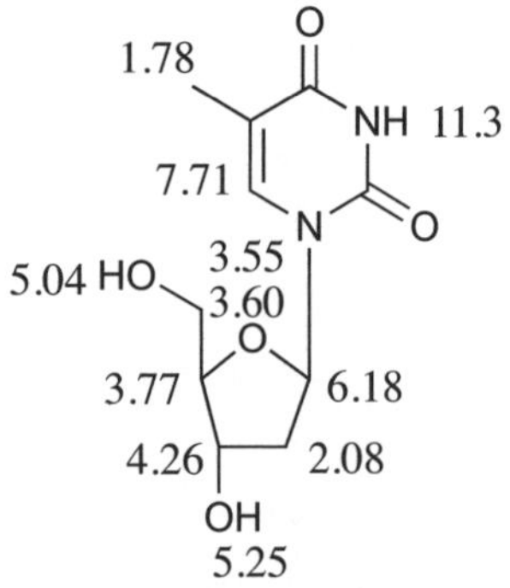

1.78, NH 11.3
7.71
5.04 HO, 3.55, 3.60
3.77, 6.18
4.26, 2.08
OH 5.25

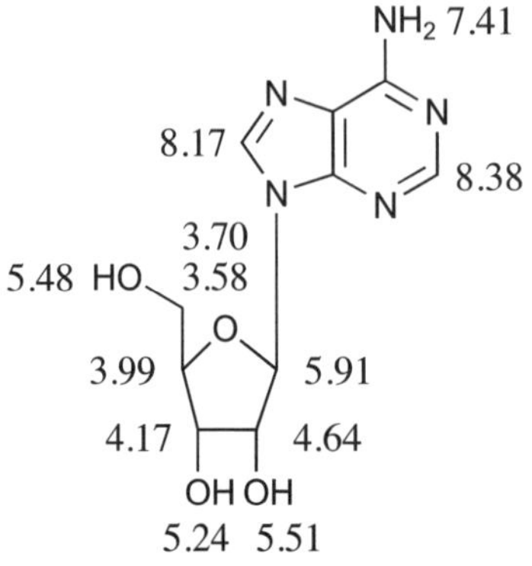

NH$_2$ 7.41

8.17

8.38

3.70
5.48 HO 3.58

O

3.99   5.91

4.17   4.64

OH OH
5.24 5.51

(in DMSO)

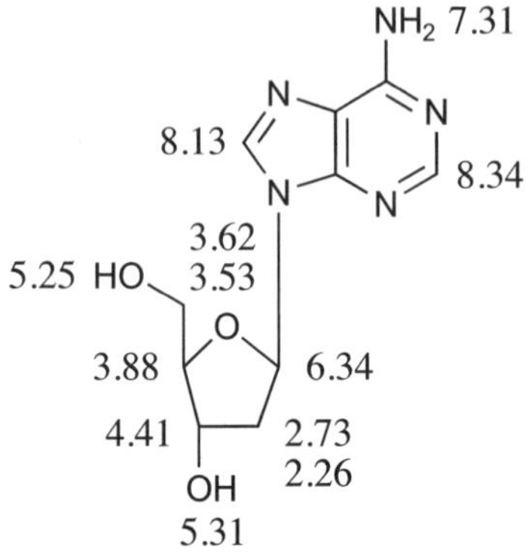

NH$_2$ 7.31

8.13

8.34

3.62
5.25 HO 3.53

O

3.88   6.34

4.41   2.73
              2.26

OH
5.31

(in DMSO)

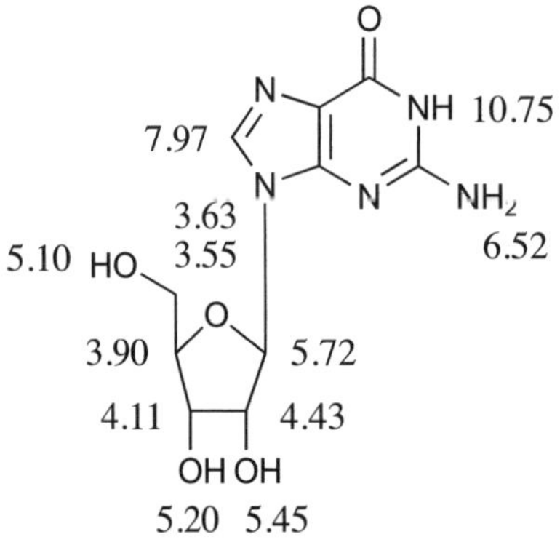

O

NH 10.75
7.97

NH$_2$

3.63   6.52
5.10 HO 3.55

O

3.90   5.72

4.11   4.43

OH OH
5.20 5.45

(in DMSO)

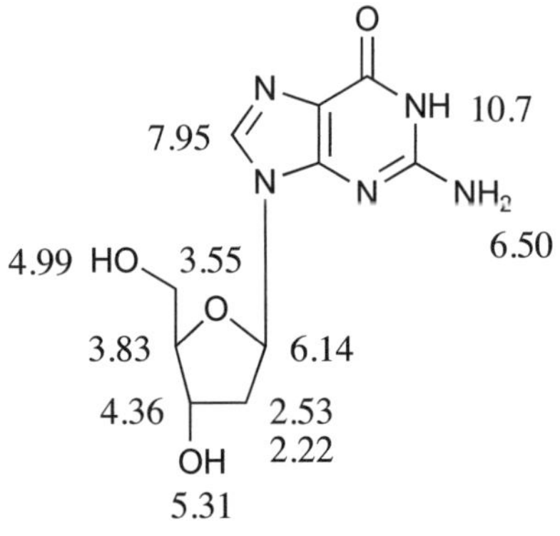

O

NH 10.7
7.95

NH$_2$

6.50
4.99 HO 3.55

O

3.83   6.14

4.36   2.53
              2.22

OH
5.31

(in DMSO)

5.14 Spectra of Solvents and Reference Compounds

5.14.1 ^{1}H NMR Spectra of Common Deuterated Solvents

500 MHz; ≈1 000 data points per 1 ppm; δ in ppm relative to TMS

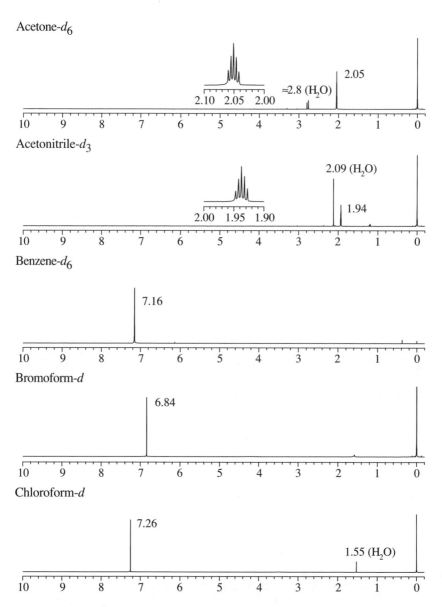

Acetone-d_6

Acetonitrile-d_3

Benzene-d_6

Bromoform-d

Chloroform-d

Solvents

Cyclohexane-d_{12}

Dimethyl sulfoxide-d_6

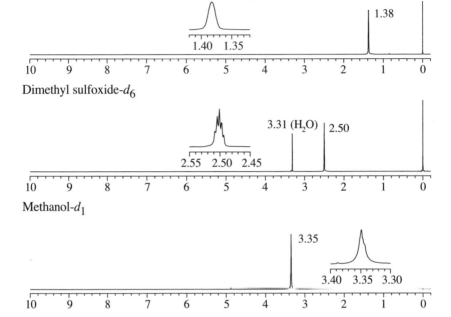

Methanol-d_1

Methanol-d_4

Pyridine-d_5

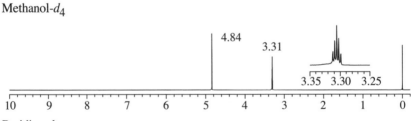

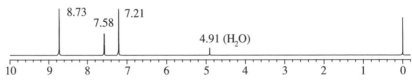

Tetrahydrofuran-d_8

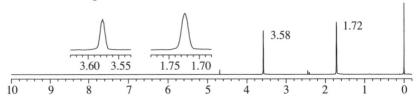

Solvents

Water-d_2

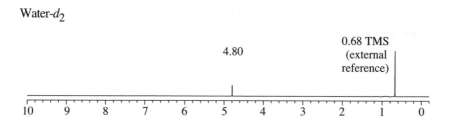

0.68 TMS
(external reference)

4.80

5.14.2 ¹H NMR Spectra of Secondary Reference Compounds

Chemical shifts in ¹H NMR spectra are usually reported relative to the peak position of tetramethylsilane (TMS) added to the sample as an internal reference. If TMS is not sufficiently soluble, a capillary with TMS may be used as external reference. In this case, owing to the different volume susceptibilities, the local magnetic fields in the sample and reference differ, and the peak position of the reference must be corrected. For a D_2O solution in a cylindrical sample and neat TMS in a capillary, the correction amounts to +0.68 and -0.34 ppm for superconducting and electromagnets, respectively. These values must be subtracted from the chemical shifts relative to the external TMS signal if its position is set to 0.00 ppm. Alternatively, secondary references with $(CH_3)_3SiCH_2$ groups may be used. The following spectra of two such secondary reference compounds in D_2O were measured at 500 MHz with TMS as external reference. Chemical shifts are reported in ppm relative to TMS upon correction for the difference in the volume susceptibilities of D_2O and TMS. As a result, the peak for the external TMS appears at 0.68 ppm.

3-(Trimethylsilyl)-1-propanesulfonic acid sodium salt (sodium 4,4-dimethyl-4-silapentane-1-sulfonate; DSS)

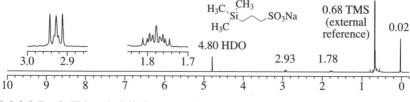

2,2,3,3-D₄-3-(Trimethylsilyl)propionic acid sodium salt

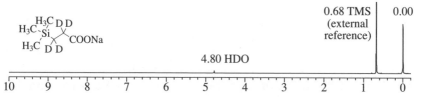

Solvents

5.14.3 ¹H NMR Spectrum of a Mixture of Common Nondeuterated Solvents

The following ¹H NMR spectrum (500 MHz, δ in ppm relative to TMS) of CDCl₃ containing 18 common solvents (0.05–0.4 vol%) is shown as a guide for the identification of possible impurities. Where the signals of several solvents overlap, insets show signals for the individual compounds from separate spectra. Peaks in these insets are labeled with the corresponding chemical shifts from their main spectrum but their values may differ by up to 0.03 ppm. Signals that are particularly prone to vary in their position are marked with *. THF: tetrahydrofuran; EGDME: ethylene glycol dimethyl ether.

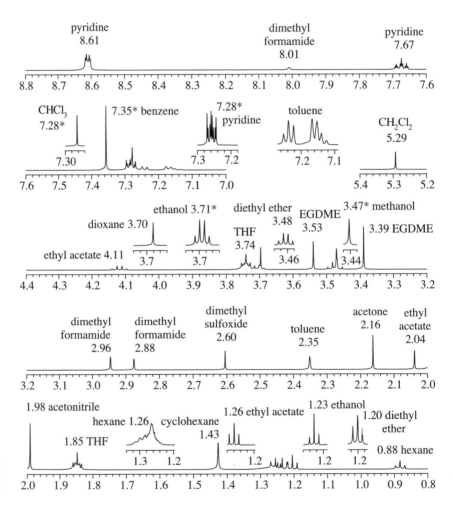

Solvents

6 Heteronuclear NMR Spectroscopy

6.1 ^{19}F NMR Spectroscopy

6.1.1 ^{19}F Chemical Shifts of Perfluoroalkanes (δ in ppm relative to CFCl$_3$)

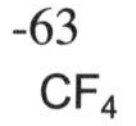

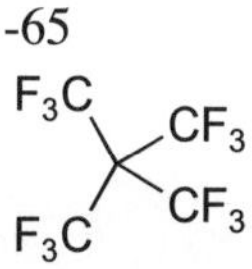

-63
CF$_4$

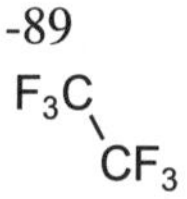

-89
F$_3$C
  CF$_3$

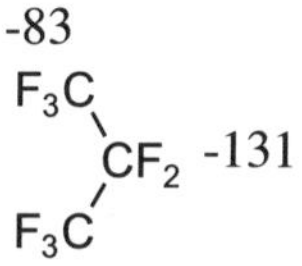

-83
F$_3$C
  CF$_2$ -131
F$_3$C

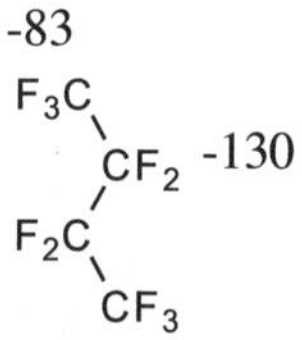

-83
F$_3$C
  CF$_2$ -130
F$_2$C
  CF$_3$

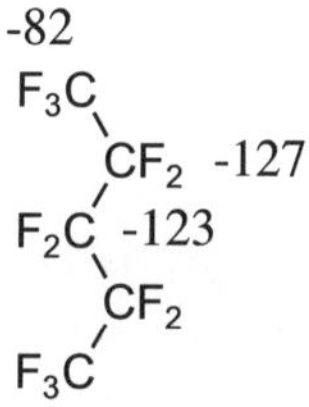

-82
F$_3$C
  CF$_2$ -127
F$_2$C -123
  CF$_2$
F$_3$C

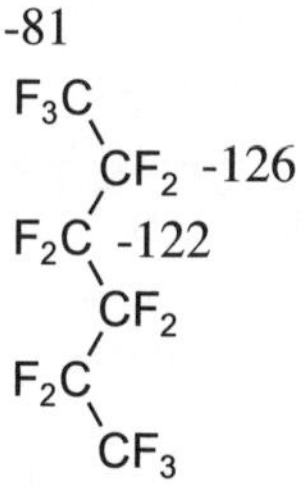

-81
F$_3$C
  CF$_2$ -126
F$_2$C -122
  CF$_2$
F$_2$C
  CF$_3$

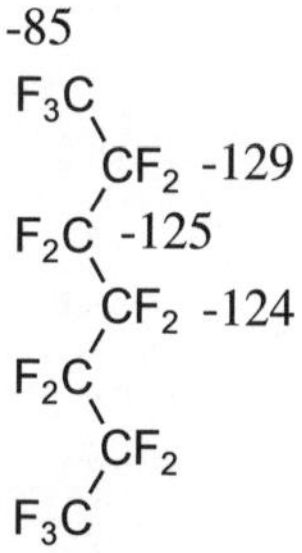

-85
F$_3$C
  CF$_2$ -129
F$_2$C -125
  CF$_2$ -124
F$_2$C
  CF$_2$
F$_3$C

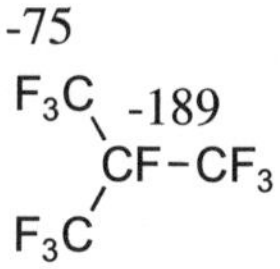

-75
F$_3$C -189
  CF-CF$_3$
F$_3$C

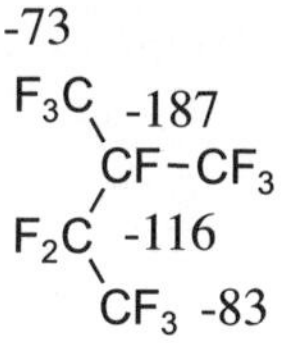

-73
F$_3$C -187
  CF-CF$_3$
F$_2$C -116
  CF$_3$ -83

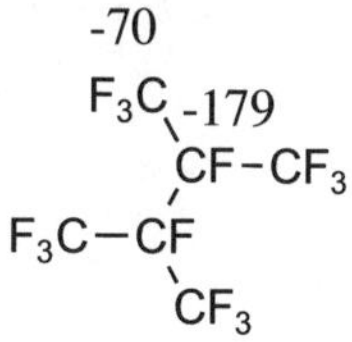

-70
F$_3$C -179
  CF-CF$_3$
F$_3$C-CF
  CF$_3$

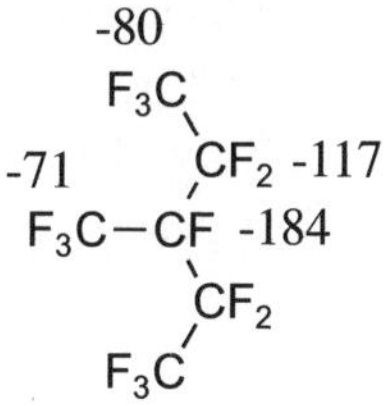

-80
F$_3$C
  CF$_2$ -117
-71
F$_3$C-CF -184
  CF$_2$
F$_3$C

-65
F$_3$C
      CF$_3$
      CF$_3$
F$_3$C

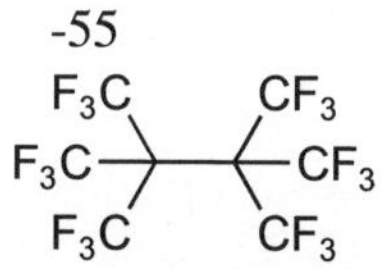

-55
F$_3$C   CF$_3$
F$_3$C-   -CF$_3$
F$_3$C   CF$_3$

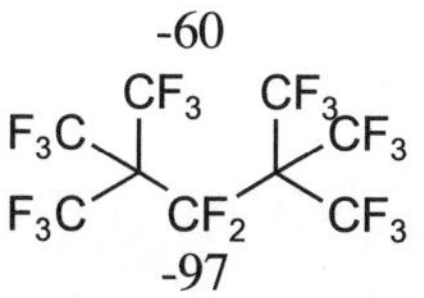

-60
    CF$_3$ CF$_3$
F$_3$C   CF$_3$
F$_3$C   CF$_2$ CF$_3$
       -97

^{19}F Chemical Shifts of CF$_3$ Groups (δ in ppm)

	Substituent	δ		Substituent	δ
	–H	-78	**X**	–F	-63
C	–CH$_3$	-62		–Cl	-29
	–CH$_2$CH$_3$	-70		–Br	-18
	–n-C$_7$H$_{15}$	-67		–I	-5
	–CH$_2$OH	-78	**O**	–OH	-55
	–CH$_2$NH$_2$	-72		–O–cyclohexyl	-58
	–CH$_2$COOH	-64		–O–CF$_3$	-58
	–CH$_2$CH$_2$–1-pyridinium	-75		–O–phenyl	-58
	–C(CF$_3$)$_3$	-65		–O–CO–CO–O–CF$_3$	-31
	–CF$_3$	-89	**N**	–NH$_2$	-49
	–CF$_2$CF$_3$	-83		–C≡N	-53
	–perfluorocyclohexyl	-70		–NC	-51
	–CCl$_3$	-82	**S**	–SH	-32
	–CH=CH$_2$	-67		–S–CF$_3$	-39
	–C≡CH	-56		SS CF$_3$	-47
	–phenyl	-64		–SO$_3$H	-79
	–C$_6$F$_5$	-55		–S(O)$_2$–phenyl	-79
	–4-nitrophenyl	-64	**O**	–COCF$_3$	-85
	–4-aminophenyl	-62	**‖**	–CO–phenyl	-58
	–C$_6$(CF$_3$)$_5$	-53		–COOH	-77
	–1-naphthyl	-75	**C**	–COO$^-$	-74
	–2-naphthyl	-73		–COOCH$_2$CH$_3$	-74
	–2-pyridyl	-68		–COF	-76
	–3-pyridyl	-62	**P**	–P(O)(OCH$_2$CH$_3$)$_2$	-73
	–4-pyridyl	-65		–P(CF$_3$)$_2$	-51
				–P$^+$(phenyl)$_3$	-58

^{19}F Chemical Shifts of CHF$_2$ Groups (δ in ppm)

Substituent	δ	Substituent	δ
–H	-144	–CCl$_3$	-122
–CH$_3$	-110	–phenyl	-111
–CH$_2$CH$_3$	-120	–O–CH$_3$	-88
–CH$_2$CH$_2$CH$_3$	-117	–O–CF$_3$	-86
–CH$_2$–phenyl	-115	–C≡N	-120
–CF$_3$	-141	–S–phenyl	-121
–CF$_2$CF$_3$	-138	–COOH	-127
–cyclohexyl	-126	–P(CF$_3$)$_2$	-126

^{19}F Chemical Shifts of CH$_2$F Groups (δ in ppm)

Substituent	δ	Substituent	δ
–H	-268	–CCl$_3$	-198
–CH$_3$	-212	–CH=CH$_2$	-216
–CH$_2$CH$_3$	-212	–C≡CH	-218
–CH$_2$CH$_2$CH$_3$	-219	–phenyl	-206
–CH$_2$CH$_2$CH$_2$CH$_3$	-219	–C≡N	-251
–CH$_2$OH	-226	–CO–phenyl	-226
–CH$_2$–phenyl	-216	–COOH	-229
–CF$_3$	-241	–COO$^-$	-218
–CF$_2$CF$_3$	-243		

^{19}F Chemical Shifts of CF$_2$R$_2$, CHFR$_2$, and CFR$_3$ Groups (δ in ppm)

Substituent	CF$_2$R$_2$	CHFR$_2$	CFR$_3$
–CH$_3$	-85	-165	-131
–CH$_2$CH$_3$	-92	-183	-156
–CF$_3$	-132	-77	-189
–phenyl	-89	-167	-127
–Cl	-7	-81	0

^{19}F Chemical Shifts of Monosubstituted Perfluoroalkanes (δ in ppm) [1]

CF$_3$ -81.7
F$_2$C -125.9
CF$_2$ -123.0
F$_2$C -121.7
CF$_2$ -121.7
F$_2$C -121.7
CF$_2$ -122.6
F$_2$C -116.8
COOH

CF$_3$ -81.8
F$_2$C -127.0
CF$_2$
F$_2$C
CF$_2$
F$_2$C
CF$_2$ -120.9
F$_2$C -68.7
Cl

CF$_3$ -81.8
F$_2$C -126.9
CF$_2$
F$_2$C
CF$_2$
F$_2$C
CF$_2$ -118.0
F$_2$C -64.0
Br

CF$_3$ -81.8
F$_2$C -126.8
CF$_2$
F$_2$C
CF$_2$
F$_2$C
CF$_2$ -113.7
F$_2$C -59.0
I

CF$_3$ -80.5
F$_2$C -126.3
CF$_2$ -122.8
F$_2$C -121.8
CF$_2$ -121.6
F$_2$C -121.6
CF$_2$ -121.7
F$_2$C -121.6
CF$_2$ -123.6
F$_2$C -115.2

CF$_3$ -86.0
F$_2$C -131.0
CF$_2$ -127.4
F$_2$C -126.6
CF$_2$ -126.4
F$_2$C -126.3
CF$_2$ -125.1
F$_2$C -117.8
SO$_2$R

F$_3$C -85.2
-75.6 CF$_2$ -121.1
F$_3$C—CF -189.8
CF$_2$ -117.3
F$_2$C -124.2
CF$_2$ -124.8
F$_2$C -117.8
SO$_2$R

F$_3$C -85.2
-190.7 CF—CF$_3$ -76.9
F$_2$C -119.5
CF$_2$ -125.3
F$_2$C -125.9
CF$_2$ -125.0
F$_2$C -117.9
SO$_2$R

F$_3$C CF$_3$ -66.5
F$_3$C—C
CF$_2$ -108.6
F$_2$C -121.5
CF$_2$ -124.2
F$_2$C -117.6
SO$_2$R

-110.3 CF$_3$ -83.0
F$_2$C CF$_3$ -64.1
-105.5 C
F$_2$C CF$_3$ -64.1
CF$_2$ -120.0
F$_2$C -117.1
SO$_2$R

-74.1 CF$_3$ -74.1
F$_3$C—CF -184.9
-179.8 CF—CF$_3$ -75.0
F$_2$C -113.3
CF$_2$ -122.7
F$_2$C -116.2
SO$_2$R

R: NHCH$_2$C$_6$H$_5$

Halogen Bonding (δ in ppm) [2]

X—$\overset{a\ F_2}{C}$—$\overset{b\ F_2}{C}$—$\overset{c\ F_2}{C}$—$\overset{d\ F_2}{C}$—$\overset{F_2}{C}$—$\overset{F_2}{C}$—$\overset{F_2}{C}$—X

X	in cyclohexane				in pyridine			
	a	b	c	d	a	b	c	d
–F	-81.1	-121.7	-122.5	-126.1	-80.9	-122.1	-122.8	-126.2
–Br	-65.1	-118.1	-122.6	-123.2	-67.7	-117.9	-122.2	-122.8
–I	-60.0	-114.6	-122.5	-123.3	-71.6	-115.2	-122.0	-122.8

6.1.2 Estimation of ^{19}F Chemical Shifts of Substituted Fluoroethylenes (δ in ppm relative to CFCl$_3$) [3]

$$\delta_{C=CF} = -133.9 + Z_{cis} + Z_{trans} + Z_{gem} + S_{cis/trans} + S_{cis/gem} + S_{trans/gem}$$

Substituent R	Z_{cis}	Z_{trans}	Z_{gem}
–H	-7.4	-31.3	49.9
–CH$_3$	-6.0	-43.0	9.5
–CF$_3$	-25.3	-40.7	54.3
–CH=CH$_2$	–	–	47.7
–CF=CF$_2$	-23.8	-38.9	44.7
–phenyl	-15.7	-35.1	38.7
–F	0.0	0.0	0.0
–Cl	-16.5	-29.4	–
–Br	-17.7	-40.0	–
–I	-21.3	-46.3	17.4
–OC$_2$H$_5$	-77.5	–	84.2
–COF	-46.5	-56.8	54.1
–SCH$_3$	-25.1	-43.7	16.6

Substituent	Substituent	$S_{cis/trans}$	$S_{cis/gem}$	$S_{trans/gem}$
–H	–H	-26.6	–	2.8
–H	–CF$_3$	-21.3	–	–
–H	–CH$_3$	–	11.4	–
–H	–OCH$_2$CH$_3$	-47.0	–	–
–H	–phenyl	-4.8	–	5.2
–CF$_3$	–H	-7.5	-10.6	12.5
–CF$_3$	–CF$_3$	-5.9	-5.3	-4.7
–CF$_3$	–CH$_3$	17.0	–	–
–CF$_3$	–phenyl	-15.6	–	-23.4
–CH$_3$	–H	–	-12.2	–
–CH$_3$	–CF$_3$	–	-13.8	-8.9
–CH$_3$	–phenyl	–	-19.5	-19.5
–OCH$_2$CH$_3$	–H	-5.1	–	–
–phenyl	–H	–	–	20.1
–phenyl	–CF$_3$	-23.2	–	–

6.1.3   Coupling Constants in Fluorinated Alkanes and Alkenes (J_{FF} in Hz)

$|^4J_{FF}|$ 8.9

F_3C
CF_2 $|^3J_{FF}|$ <1
F_2C $|^3J_{FF}|$ 1.8
Cl

$|^4J_{FF}|$ 8.5

F_3C $|^3J_{FF}|$ 1.1
CF_2
F_2C $|^3J_{FF}|$ 2.2, 1.1
$COOH$

F_3C CF_3
F_3C-C
CF_2 $|^4J_{FF}|$ 13
$|^5J_{FF}|$ 13 F_2C
CF_2 R: $NHCH_2C_6H_5$
F_2C
SO_2R

$|^5J_{FF}|$ 6.8 F_3C
CF_2 $|^4J_{FF}|$ 13.6
F_3C-CF $|^5J_{FF}|$ 6.8
CF_2
F_2C
CF_2
F_2C
SO_2R

F_3C $|^3J_{FF}|$ 0.3
CF_2
F_2C $|^3J_{FF}|$ 6.0
$|^3J_{FF}|$ 10.3 CF_2
FC $|^3J_{FF}|$ 140
$|^3J_{FF}|$ 8.7 CF
F_3C

$|^4J_{FF}|$ 9.8
F_3C
CF_2
F_2C $|^4J_{FF}|$ 12.2
$|^4J_{FF}|$ 5.8
CF_2
FC $|^4J_{FF}|$ 26.1
$|^4J_{FF}|$ 21.0
CF
F_3C

F_3C
$|^5J_{FF}|$ 2.4 CF_2
F_2C
CF_2 $|^5J_{FF}|$ 3.0
$|^5J_{FF}|$ 7.2 FC
CF $|^5J_{FF}|$ 1.5
F_3C

$|^3J_{FF}|$ 0.8 CF_3
F_2C $|^3J_{FF}|$ 1.6
$|^3J_{FF}|$ 13.3 CF_2
FC $|^3J_{FF}|$ 2.3
$|^3J_{FF}|$ 7.6 $FC-CF_2$
CF_3
$|^3J_{FF}|$ 12.5

CF_3
F_2C $|^4J_{FF}|$ 9.5
$|^4J_{FF}|$ 8.2 CF_2
FC $|^4J_{FF}|$ 3.4
$FC-CF_2$
$|^4J_{FF}|$ 8.2 CF_3
$|^4J_{FF}|$ 7.6

$|^5J_{FF}|$ 2.5
CF_3
F_2C
CF_2 $|^6J_{FF}|$ 4.6
FC
$FC-CF_2$
CF_3

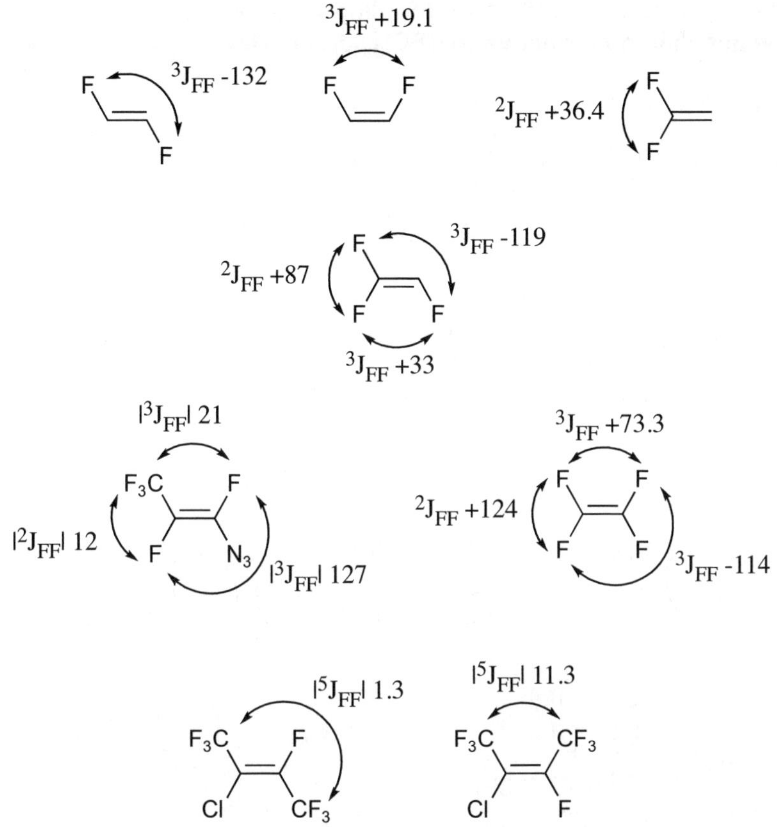

$^3J_{FF}$ +19.1

$^3J_{FF}$ -132

$^2J_{FF}$ +36.4

$^2J_{FF}$ +87

$^3J_{FF}$ -119

$^3J_{FF}$ +33

$|^3J_{FF}|$ 21

$|^2J_{FF}|$ 12

$|^3J_{FF}|$ 127

$^3J_{FF}$ +73.3

$^2J_{FF}$ +124

$^3J_{FF}$ -114

$|^5J_{FF}|$ 1.3

$|^5J_{FF}|$ 11.3

6.1.4 ^{19}F Chemical Shifts of Allenes and Alkynes (δ in ppm relative to CFCl$_3$, $|J_{FF}|$ in Hz)

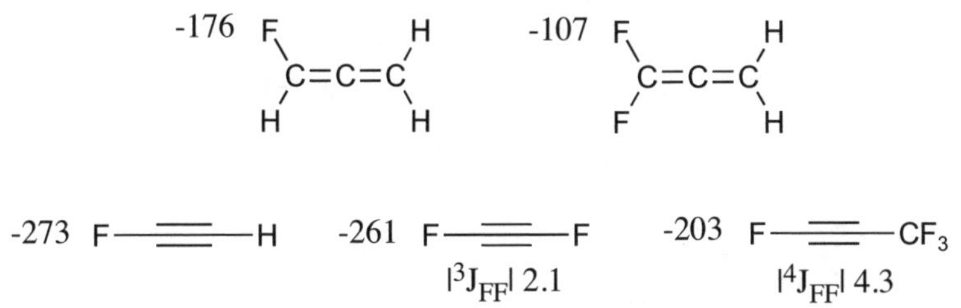

-176

-107

-273 F≡≡—H

-261 F≡≡—F

$|^3J_{FF}|$ 2.1

-203 F≡≡—CF$_3$

$|^4J_{FF}|$ 4.3

6.1.5 ^{19}F Chemical Shifts and Coupling Constants of Fluorinated Alicyclics (δ in ppm relative to CFCl$_3$, |J$_{FF}$| in Hz)

-218
F

-160
F

-171
F

-174
F

-165
F

-160
F

-151
CF$_2$
F$_2$C—CF$_2$
|^{2}J$_{FF}$| 160

-135
CF$_2$
F$_2$C   CF$_2$
CF$_2$
|^{2}J$_{FF}$| 220

-133
CF$_2$
F$_2$C   CF$_2$
F$_2$C—CF$_2$

-133
CF$_2$
F$_2$C   CF$_2$
F$_2$C   CF$_2$
CF$_2$
|^{2}J$_{FF}$| 284

-120
CF$_2$ CF$_2$
F$_2$C   CF$_2$
F$_2$C   CF$_2$
CF$_2$ CF$_2$

F -186

F -166

F -130

-122 -189
F   F   F
-140 F
-142 F   CF$_3$ -70
F   F -132
F   F
-124   -120

F
F   F   -121
F   F
F   F
F   F
F   F -224
F   F
F   F

|^{2}J$_{FF}$| 286
F   F
F   F
F   CF$_3$
F
F   F
F   F
|^{2}J$_{FF}$| 284
F   F |^{2}J$_{FF}$| 291
F   F
|^{3}J$_{FF}$| 13

|^{4}J$_{FF}$| 25
F   F
|^{3}J$_{FF}$| 13.5
F   F
F   CF$_3$
F
F   F
F   F
|^{3}J$_{FF}$| 13.5
F   F |^{4}J$_{FF}$| 3
|^{4}J$_{FF}$| 20

|^{4}J$_{FF}$| 2 F   F
F   F
F   CF$_3$
F
|^{4}J$_{FF}$| 6   F   F
F   F
|^{3}J$_{FF}$| 0

6.1.6 ^{19}F Chemical Shifts and Coupling Constants of Aromatics and Heteroaromatics (δ in ppm relative to CFCl$_3$)

Estimation of ^{19}F Chemical Shifts of Substituted Fluorobenzenes [4]

$$\delta_F = -113.9 + \Sigma\, Z_i$$

Substituent		$Z_{2,6}$	$Z_{3,5}$	Z_4
C	–CH$_3$	-3.9	-0.4	-3.6
	–CF$_3$	0.4	3.1	5.8
	–CH=CH$_2$	-4.4	0.7	-0.6
	–C≡CH	–	–	3.3
X	–F	-23.2	2.0	-6.6
	–Cl	-0.3	3.5	-0.7
	–Br	7.6	3.5	0.1
	–I	19.9	3.6	1.4
O	–OH	-23.5	0.0	-13.3
	–OCH$_3$	-18.9	-0.8	-9.0
	–OCOCH$_3$	–	–	-3.7
N	–NH$_2$	-22.9	-1.3	-17.4
	–NHCOOCH$_3$	–	0.1	-7.1
	–NHCONH$_2$	–	0.9	-8.1
	–N$_3$	-11.4	2.8	-0.3
	–NO$_2$	-5.6	3.8	9.6
	– C≡N	6.9	4.1	10.1
	–NCO	-9.2	2.3	-2.2
S	–SH	10.0	0.9	-3.5
	–SCH$_3$	6.5	1.2	-4.5
	–S(O)$_2$F	7.5	5.8	13.8
	–S(O)$_2$–CF$_3$	9.5	5.5	-14.3
	–S(O)$_2$OCH$_2$CH$_3$	–	3.7	9.1
O‖C	–CHO	-7.4	2.1	10.3
	–COCH$_3$	2.5	1.8	7.6
	–COOH	2.3	1.1	6.5
	-COOCH$_3$	3.3	3.8	7.1
	–CONH$_2$	0.5	-0.8	3.4
	–COF	-14.8	3.0	6.2
	–COCl	3.4	3.5	12.9
	–B(OH)$_2$	6.8	0.8	2.1
	–Si(CH$_3$)$_3$	13.8	0.3	1.6

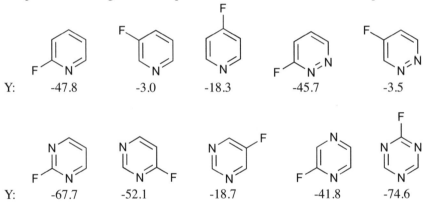

Estimation of [19]F Chemical Shifts of Substituted Pyridines, Pyrimidines, Pyra-zines, and Triazines (δ in ppm) [5]

$$\delta_F = Y + \Sigma\ Z_i$$

To estimate the ^{19}F chemical shifts of substituted 6-ring heteroaromatics, the same increments, Z_i, can be used as for substituted fluorobenzenes (see preceding page). However, different base values, Y (as given below), apply depending on the number and position of nitrogens and the position of the fluorine substituent in question:

Coupling Constants in Aromatics and Heteroaromatics (J_{FF} in Hz)

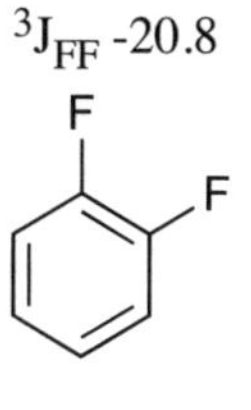

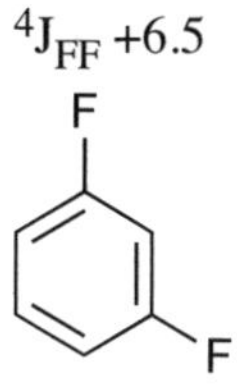

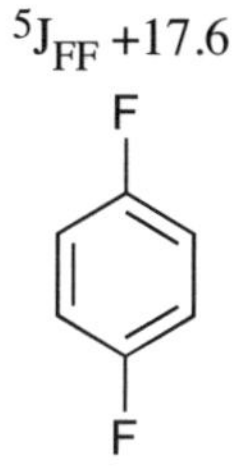

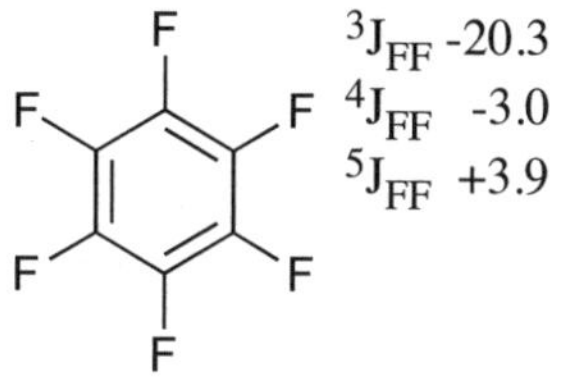

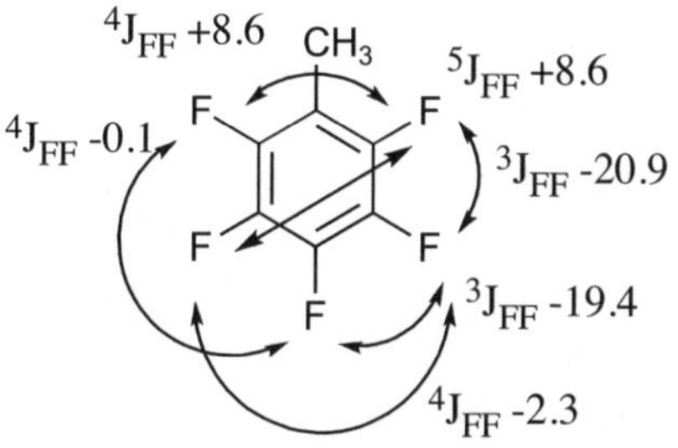

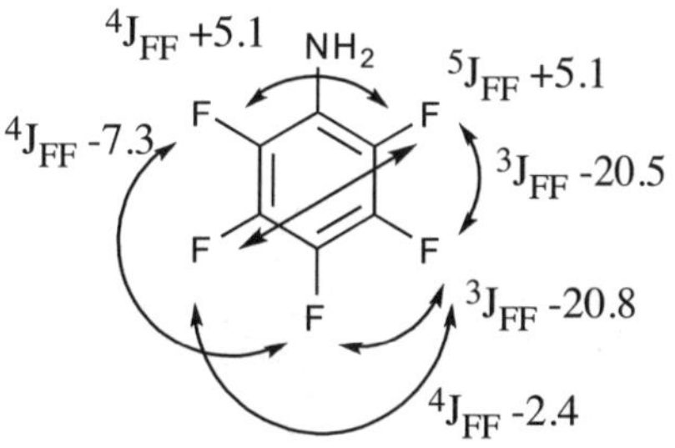

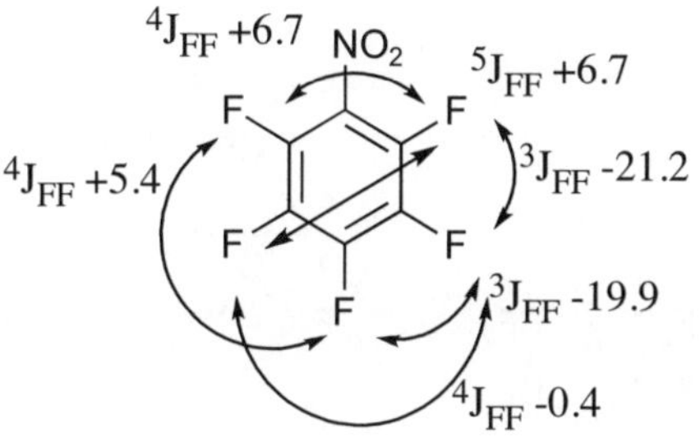

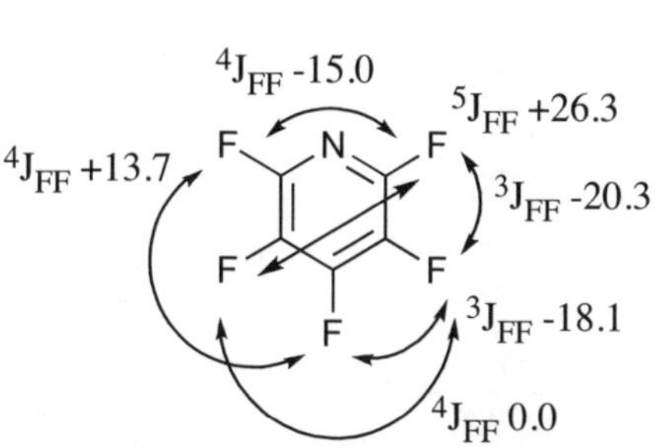

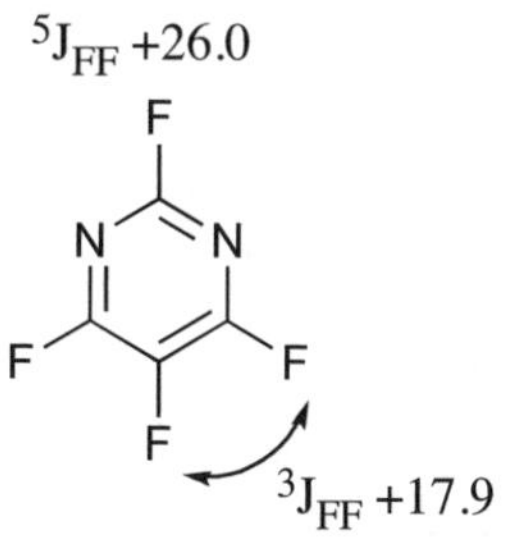

$^3J_{FF}$ -20.8 $^4J_{FF}$ +6.5 $^5J_{FF}$ +17.6

$^3J_{FF}$ -20.3
$^4J_{FF}$ -3.0
$^5J_{FF}$ +3.9

$^4J_{FF}$ +8.6   CH₃ $^5J_{FF}$ +8.6
$^4J_{FF}$ -0.1 $^3J_{FF}$ -20.9
$^3J_{FF}$ -19.4
$^4J_{FF}$ -2.3

$^4J_{FF}$ +5.1   NH₂ $^5J_{FF}$ +5.1
$^4J_{FF}$ -7.3 $^3J_{FF}$ -20.5
$^3J_{FF}$ -20.8
$^4J_{FF}$ -2.4

$^4J_{FF}$ +6.7   NO₂ $^5J_{FF}$ +6.7
$^4J_{FF}$ +5.4 $^3J_{FF}$ -21.2
$^3J_{FF}$ -19.9
$^4J_{FF}$ -0.4

$^4J_{FF}$ -15.0 $^5J_{FF}$ +26.3
$^4J_{FF}$ +13.7 $^3J_{FF}$ -20.3
$^3J_{FF}$ -18.1
$^4J_{FF}$ 0.0

$^5J_{FF}$ +26.0

$^3J_{FF}$ +17.9

6.1.7 ^{19}F Chemical Shifts of Alcohols and Ethers (δ in ppm relative to CFCl$_3$)

OH
|
CF$_3$
-55

F$_3$C$\diagup$^{OH}
-78

F$_3$C$\diagup$$\diagdown$CF$_3$ (OH)
-88

HO$\diagdown$$\diagup$OH
F$_3$C$\diagdown$$\diagup$CF$_3$
-93

F$_3$C -85
CF$_2$ -133
F$_2$C -88
O
F$_2$C -92
CF$_3$ -91

F$_3$C -84
CF$_2$ -129
F$_2$C -129
CF$_2$ -87
O
CF$_2$ -56
O
CF$_2$ -57
O
CF$_2$
O
CF$_2$
F$_2$C
CF$_2$
F$_3$C

F$_3$C   -57
-83   O
F$_3$C—CF -149
CF$_2$ -86
O
CF$_2$ -92
F$_3$C -91

F$_3$C -85
CF$_2$ -133
F$_2$C -86
O
-83
F$_3$C—CF -148
CF$_2$ -84
O
CF$_2$ -92
F$_3$C -91

F$_3$C -84
CF$_2$ -129
F$_2$C -129
CF$_2$ -87
O
CF$_2$ -54
O
CF$_2$
F$_2$C
CF$_2$
F$_3$C

F$_2$C$\diagup^{O}\diagdown$CF$_2$
-91
F$_2$C$\diagdown$$_O$$\diagup$CF$_2$

F$_3$C -82
CF$_2$ -130
F$_2$C -84
O
F$_2$C -83
CF$_2$ -129
F$_2$C -83
O
F$_2$C -88
CF$_3$ -87

F$_3$C -84
CF$_2$ -131
F$_2$C -84
-80   O
F$_3$C—CF -144
CF$_2$ -82
O
CF$_2$ -91
F$_3$C -89

F$_2$C—CF$_2$
F$_2$C$\diagdown$$_O$ CF$_2$
F$_2$C   O$\diagdown$CF$_2$
F$_2$C—O   CF$_2$
F$_2$C   -89   O$\diagup$CF$_2$
O   CF$_2$
F$_2$C   O$\diagup$CF$_2$
F$_2$C—O   O$\diagdown$CF$_2$
F$_2$C$\diagdown$$_O$ O$\diagup$CF$_2$
F$_2$C—CF$_2$

F$_3$C
CF$_2$
F$_2$C
O
F$_3$C—CF
CF$_2$
O
CF—CF$_3$
O=
F -130
+26

F$_3$C
CF$_2$
F$_2$C
O
F$_3$C—CF
CF$_2$
O
CF—CF$_3$
O=
OH -132

6.1.8 ^{19}F Chemical Shifts of Fluorinated Amine, Imine, and Hydroxylamine Derivatives (δ in ppm relative to CFCl$_3$)

F +147
F–N–F
-84
F$_3$C–N–F
F +19

CF$_3$ -57
F$_3$C–N–H

CF$_3$ -58
F$_3$C–$\overset{+}{N}$–H
H

-92
F$_3$C–$\overset{+}{N}$–H
F   H
-75

F$_3$C -82
CF$_2$ -128
F$_2$C -117
N–F
F   +17
F$_3$C

F$_3$C -82
CF$_2$ -121
F$_2$C -106
N–F
F$_2$C–CF$_2$ -91
F$_3$C

F$_3$C -82
CF$_2$ -121
-84 F$_2$C
N–CF$_2$
F$_2$C–CF$_2$ F$_2$C–CF$_3$
F$_3$C

CF$_3$ -56
F$_3$C–N–CF$_3$

CHF$_2$ -96
F$_2$HC–N–CHF$_2$

CF$_3$ -84
-129 F$_2$C
CF$_2$ -120
-86 F$_2$C
N–CF$_2$
F$_2$C–CF$_2$ F$_2$C–CF$_2$
F$_3$C–CF$_2$ CF$_3$

-53 CF$_3$ -93
F$_2$
F$_3$C–N–C–C–N–CF$_3$
F$_2$
CF$_3$

-63
F$_2$ -57
C
F$_3$C–N   N–CF$_3$
F$_2$C–CF$_2$ -93

CF$_3$ -53
F$_2$C–N–CF$_2$-95
F$_2$C–C–CF$_2$-134
F$_2$ -136

CF$_3$ -55
F$_2$C–N–CF$_2$-95
F$_2$C–O–CF$_2$-88

CF$_3$ -54
F$_2$C–N–CF$_2$
F$_2$C   -94   CF$_2$
N
CF$_3$

CF$_3$ -84
F$_2$C   -90
F$_2$C–N–CF$_2$-86
F$_2$C   CF$_2$-123
F$_2$C–CF$_2$ -129

F$_3$C   F -16
-59 N=
H

F$_3$C   H
-62 N=
F -46

H   -71 CF$_3$
-98 N   -99 N
F$_2$C   O   F$_2$C   O
C   C
-81 F$_2$ -83 F$_2$

OH   F$_2$ -104
F$_3$C–N–C–C–N–CF$_3$
F$_2$ -66
OH

O$^-$
F$_3$C–N–CF$_3$ -68

F$_3$C -82
CF$_2$ -125
F$_2$C -96
N–O
F$_2$C–CF$_2$ F$_2$C–CF$_2$
F$_3$C   -89   CF$_3$
-82
-129

O$^{\bullet}$ F$_2$ -105
F$_3$C–N–C–C–N–CF$_3$
F$_2$ -69
O$_{\bullet}$

Ph$_3$P   PPh$_3$
Pt
O   O   -62
F$_3$C–N   -109   N–CF$_3$
F$_2$C–CF$_2$

6.1.9 ^{19}F Chemical Shifts of Sulfur Compounds (δ in ppm relative to CFCl$_3$)

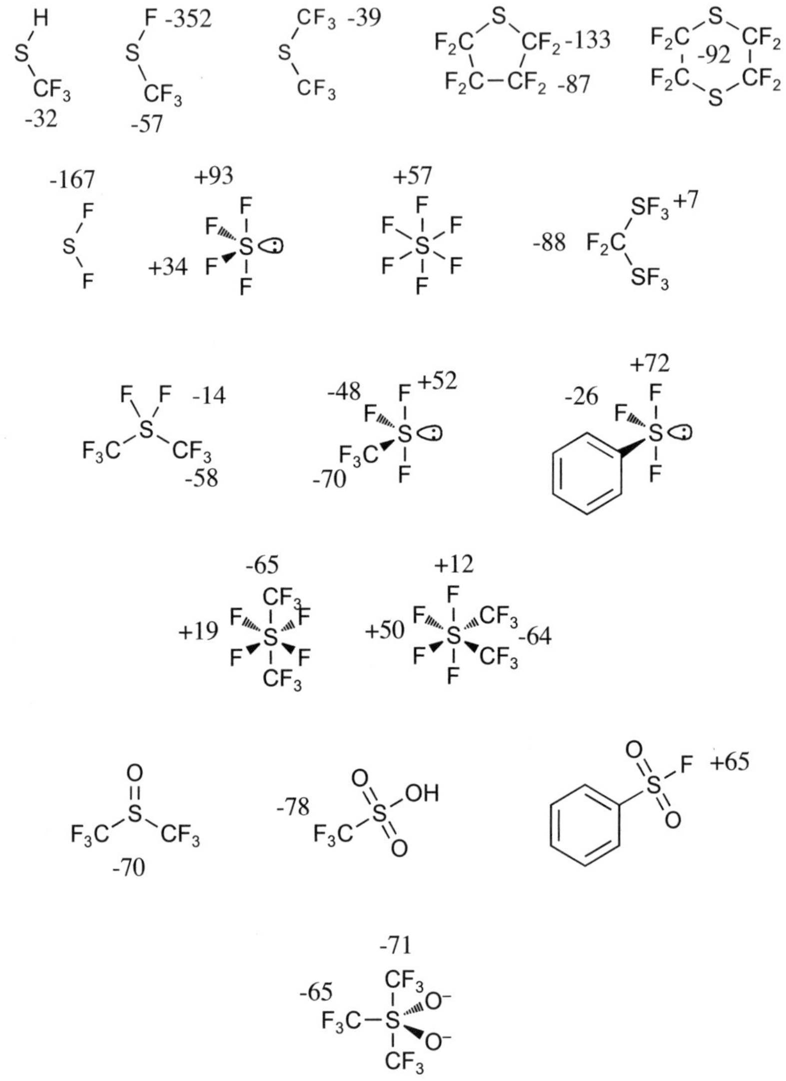

H
S
CF$_3$
-32

F -352
S
CF$_3$
-57

CF$_3$ -39
S
CF$_3$

F$_2$C—S—CF$_2$ -133
F$_2$C—CF$_2$ -87

F$_2$C—S—CF$_2$ -92
F$_2$C—S—CF$_2$

-167
F
S
F

+93
+34 F$_{...}$S$^{\ominus}$
F—F

+57
F—S—F
F—S—F
F

SF$_3$ +7
-88 F$_2$C
SF$_3$

F F -14
F$_3$C—S—CF$_3$
-58

-48 F +52
F$_{...}$S$^{\ominus}$
F$_3$C—F
-70

+72
-26 F
F$_{...}$S$^{\ominus}$
F

-65
CF$_3$
+19 F$_{...}$S$^{...}$F
F—F
CF$_3$

+12
F
+50 F$_{...}$S$^{...}$CF$_3$ -64
F—CF$_3$
F

O
‖
F$_3$C—S—CF$_3$
-70

-78
O
‖
F$_3$C—S—OH
‖
O

O
‖
S—F +65
‖
O

-71
-65 CF$_3$
F$_3$C—S$_{...}$O$^-$
O$^-$
CF$_3$

6.1.10 ^{19}F Chemical Shifts of Carbonyl and Thiocarbonyl Compounds (δ in ppm relative to CFCl$_3$)

Substituent R	δ	Substituent R	δ
–H	+41	–phenyl	+17
–CH$_3$	+49	–F	-23
–C(CH$_3$)$_3$	+22	–NH–CH$_2$CH$_2$CH$_3$	-16
–CH$_2$F	+26	–O–cyclohexyl	-8
–CF$_3$	+15	–O–phenyl	-17
–CF(CF$_3$)$_3$	+31	–S–phenyl	+47
–CH=CH$_2$	+24		

6.1.11 ^{19}F Chemical Shifts of Fluorinated Boron, Phosphorus, and Silicon Compounds (δ in ppm relative to $CFCl_3$, J_{FP} in Hz)

6.1.12 ¹⁹F Chemical Shifts of Natural Product Analogues (δ in ppm relative to CFCl₃, J_FF in Hz)

6.1.13 References

[1] G. Arsenault, B. Chittim, J. Gu, A. McAlees, R. McCrindle, V. Robertson, Separation and fluorine nuclear magnetic resonance spectroscopic (^{19}F NMR) analysis of individual branched isomers present in technical perfluorooctane-sulfonic acid (PFOS), *Chemosphere* **2008**, *73*, S53.

[2] P. Metrangolo, W. Panzeri, F. Recupero, G. Resnati, Perfluorocarbon–hydro-carbon self-assembly, Part 16. ^{19}F NMR study of the halogen bonding between halo-perfluorocarbons and heteroatom containing hydrocarbons, *J. Fluorine Chem.* **2002**, *114*, 27.

[3] R.E. Jetton, J.R. Nanney, C.A.L. Mahaffy, The prediction of the ^{19}F NMR signal positions of fluoroalkenes using statistical methods, *J. Fluorine Chem.* **1995**, *72*, 121.

[4] C.A.L. Mahaffy, J.R. Nanney, The prediction of the ^{19}F NMR spectra of fluoro-arenes using statistical substituent chemical shift values, *J. Fluorine Chem.* **1994**, *67*, 67.

[5] J.R. Nanney, C.A.L. Mahaffy, The use of the ^{19}F NMR spectra of fluoropyri-dines and related compounds to verify the 'statistical' substituent chemical shift values of fluoroarenes, *J. Fluorine Chem.* **1994**, *68*, 181.

6.2 ^{31}P NMR Spectroscopy

6.2.1 ^{31}P Chemical Shifts of Tricoordinated Phosphorus, $PR^1R^2R^3$ (δ in ppm relative to H_3PO_4)

Substituent R^1	R^2	R^3	δ
H₂ –H	–H	–H	-235
–CH₃	–H	–H	-164
–CH₂CH₃	–H	–H	-127
–phenyl	–H	–H	-124
H –CH₃	–CH₃	–H	-99
–CH₂CH₃	–CH₂CH₃	–H	-55
–phenyl	–phenyl	–H	-41
–OCH₃	–OCH₃	–H	171
C –CH₃	–CH₃	–CH₃	-63
–CH₂CH₃	–CH₂CH₃	–CH₂CH₃	-20
–CH₂CH₂CH₃	–CH₂CH₂CH₃	–CH₂CH₂CH₃	-33
–CH(CH₃)₂	–CH(CH₃)₂	–CH(CH₃)₂	20
–C(CH₃)₃	–C(CH₃)₃	–C(CH₃)₃	62
–phenyl	–CH₃	–CH₃	-48
–phenyl	–phenyl	–CH₃	-28
–phenyl	–phenyl	–phenyl	-6
X –CH₃	–CH₃	–F	185
–CH₃	–CH₃	–Cl	92
–CH₃	–CH₃	–Br	88
–CH₃	–F	–F	244
–CH₃	–Cl	–Cl	192
–CH₃	–Br	–Br	184
–CH₃	–I	–I	131
–F	–F	–F	97
–Cl	–Cl	–Cl	220
–Br	–Br	–Br	227
–I	–I	–I	178
O –OCH₃	–CH₃	–CH₃	91
–OCH₃	–OCH₃	–CH₃	183
–OCH₃	–OCH₃	–OCH₃	140
–OCH₂CH₃	–OCH₂CH₃	–OCH₂CH₃	138
N –N(CH₃)₂	–CH₃	–CH₃	39
–N(CH₃)₂	–phenyl	–phenyl	65
–N(CH₃)₂	–N(CH₃)₂	–CH₃	86
–N(CH₃)₂	–N(CH₃)₂	–N(CH₃)₂	123
S –SCH₃	–SCH₃	–SCH₃	125

6.2.2 ^{31}P Chemical Shifts of Tetracoordinated Phosphonium Compounds (δ in ppm relative to H_3PO_4)

^{31}P Chemical Shifts of Symmetrically Substituted Phosphonium Compounds, PR_4^+

Substituent R	δ	Substituent R	δ
$-CH_3$	25	$-n$-butyl	34
$-CH_2CH_3$	41	$-$phenyl	23
$-n$-propyl	31	$-OCH_3$	5

^{31}P Chemical Shifts of Triphenylphosphonium Compounds, $P(phenyl)_3R^+$

Substituent R	δ	Substituent R	δ
$-CH_3$	23	$-CH=CH_2$	19
$-CH_2CH_3$	26	$-CH=C=CH_2$	19
$-CH_2Cl$	24	$-C\equiv C-$phenyl	5
$-CH_2OH$	18	NH_2	36
$-CH_2COCH_3$	26	$-N(CH_3)_2$	48
$-CH_2COOCH_2CH_3$	21	$-OCH_2CH_3$	62

6.2.3 ^{31}P Chemical Shifts of Compounds with a P=C or P=N Bond (δ in ppm relative to H_3PO_4)

6.2.4 [31]P Chemical Shifts of Tetracoordinated P(=O) and P(=S) Compounds (δ in ppm relative to H_3PO_4)

[31]P Chemical Shifts of Tetracoordinated P(=O) Compounds

Substituent R^1		R^2	R^3	δ
C	$-CH_3$	$-CH_3$	$-H$	63
	$-CH_3$	$-CH_3$	$-CH_3$	41
	$-CH_2CH_3$	$-CH_2CH_3$	$-CH_2CH_3$	48
	$-phenyl$	$-phenyl$	$-phenyl$	27
X	$-CH_3$	$-CH_3$	$-F$	66
	$-CH_3$	$-CH_3$	$-Cl$	65
	$-CH_3$	$-CH_3$	$-Br$	51
	$-CH_2CH_3$	$-CH_2CH_3$	$-Cl$	77
	$-phenyl$	$-phenyl$	$-Cl$	43
	$-CH_3$	$-F$	$-F$	27
	$-CH_3$	$-Cl$	$-Cl$	44
	$-CH_3$	$-Br$	$-Br$	9
	$-CH_2CH_3$	$-Cl$	$-Cl$	55
	$-F$	$-F$	$-F$	-36
	$-Cl$	$-Cl$	$-Cl$	2
	$-Br$	$-Br$	$-Br$	-103
N	$-N(CH_3)_2$	$-N(CH_3)_2$	$-N(CH_3)_2$	24
O	$-H$	$-H$	$-OCH_3$	19
	$-CH_3$	$-H$	$-OH$	35
	$-CH_3$	$-CH_3$	$-OH$	31
	$-CH_3$	$-CH_3$	$-OCH_3$	52
	$-phenyl$	$-phenyl$	$-OH$	29
	$-phenyl$	$-phenyl$	$-OCH_3$	32
	$-CH_3$	$-Cl$	$-OCH_2CH_3$	40
	$-Cl$	$-Cl$	$-OCH_3$	6
	$-F$	$-F$	$-OCH_2CH_3$	-21
2 O	$-H$	$-OCH_3$	$-OCH_3$	11
	$-CH_3$	$-OH$	$-OH$	31
	$-CH_3$	$-OCH_3$	$-OCH_3$	32
	$-CCl_3$	$-OCH_2CH_3$	$-OCH_2CH_3$	7
	$-phenyl$	$-OH$	$-OH$	18
	$-phenyl$	$-OCH_3$	$-OCH_3$	21
	$-Cl$	$-OCH_2CH_3$	$-OCH_2CH_3$	3

Substituent R^1	R^2	R^3	δ
3 O –OH	–OH	–OH	0
–OCH$_3$	–OCH$_3$	–OCH$_3$	0
–OCH$_2$CH$_3$	–OCH$_2$CH$_3$	–OCH$_2$CH$_3$	-1
–OCH(CH$_3$)$_2$	–OCH(CH$_3$)$_2$	–OCH(CH$_3$)$_2$	-13
–O–phenyl	–OH	–OH	-4
–O–phenyl	–O–phenyl	–OH	-11
–O–phenyl	–O–phenyl	–O–phenyl	-18
S –S–n-butyl	–S–n-butyl	–OH	37
–S–n-butyl	–S–n-butyl	– S–n-butyl	62

^{31}P Chemical Shifts of Tetracoordinated P(=S) Compounds

$$R^1\!-\!\underset{\underset{R^2}{|}}{\overset{\overset{S}{\|}}{P}}\!-\!R^3$$

Substituent R^1	R^2	R^3	δ
C –CH$_3$	–CH$_3$	–CH$_3$	59
–CH$_2$CH$_3$	–CH$_2$CH$_3$	–CH$_2$CH$_3$	53
–phenyl	–phenyl	–phenyl	43
X –CH$_3$	–CH$_3$	–Cl	87
–phenyl	–phenyl	–Cl	80
–CH$_3$	–CH$_3$	–Br	63
–CH$_2$CH$_3$	–F	–F	111
–CH$_3$	–Cl	–Cl	81
–CH$_2$CH$_3$	–Cl	–Cl	95
–F	–F	–F	32
–Cl	–Cl	–Cl	29
–Br	–Br	–Br	-112
–Br	–I	–I	-315
N –N(CH$_2$CH$_3$)$_2$	–N(CH$_2$CH$_3$)$_2$	–N(CH$_2$CH$_3$)$_2$	78
O –CH$_3$	–OCH$_3$	–OCH$_3$	100
–OCH$_2$CH$_3$	–OCH$_2$CH$_3$	–OCH$_2$CH$_3$	68
S –CH$_3$	–S–n-propyl	–S–n-propyl	78
–S–n-butyl	–S–n-butyl	–OCH$_3$	111
–S–n-propyl	–S–n-propyl	–S–n-propyl	93

6.2.5 ^{31}P Chemical Shifts of Penta- and Hexacoordinated Phosphorus Compounds (δ in ppm relative to H_3PO_4)

6.2.6 ^{31}P Chemical Shifts of Natural Phosphorus Compounds (δ in ppm relative to H$_3$PO$_4$)

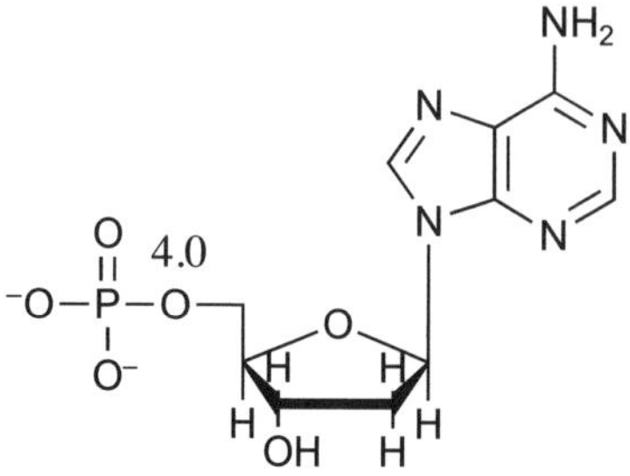
4.0

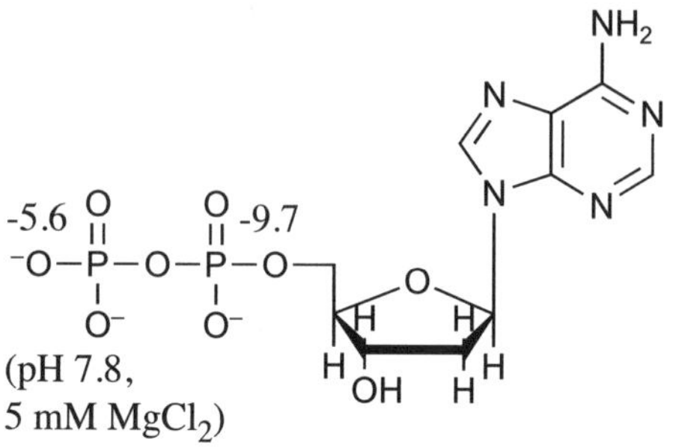
-5.6
-9.7
(pH 7.8,
5 mM MgCl2)

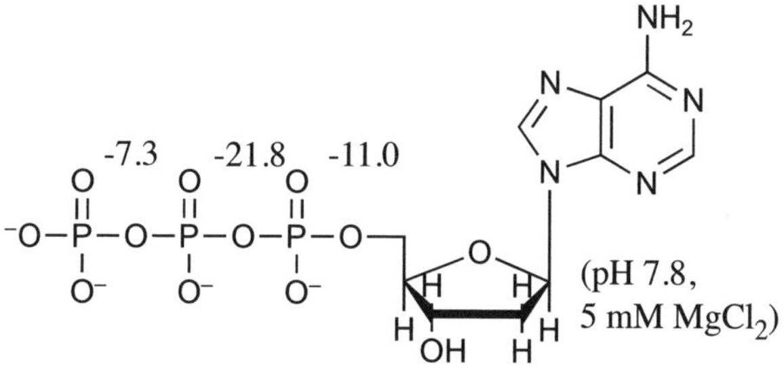
-7.3
-21.8
-11.0
(pH 7.8,
5 mM MgCl2)

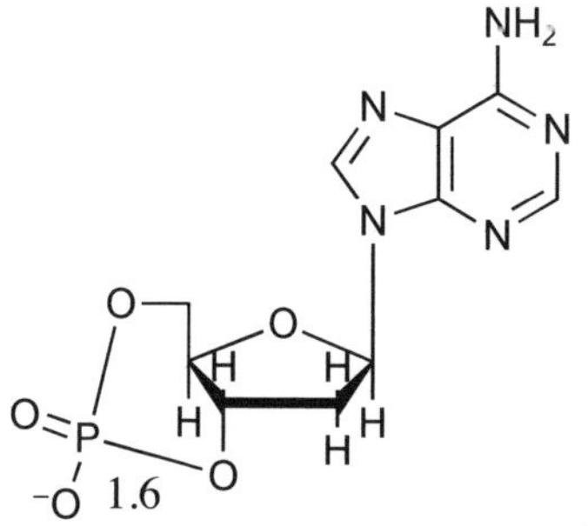
1.6

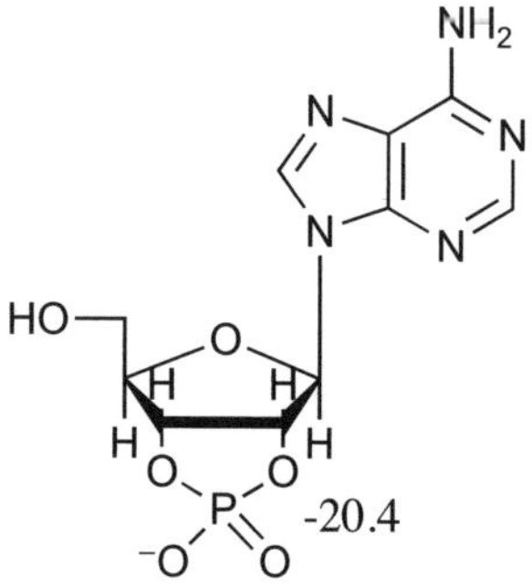
-20.4

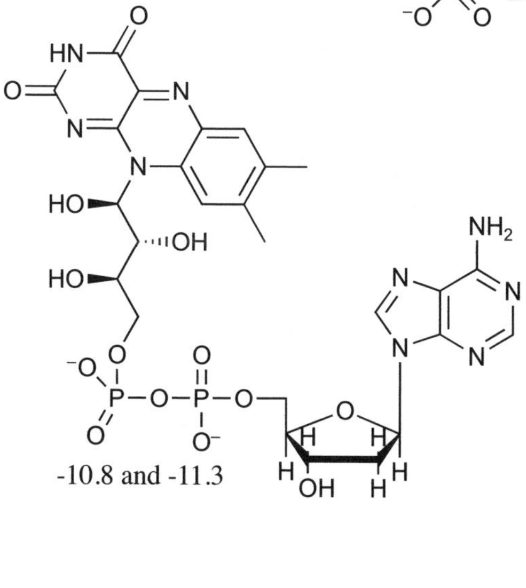
-10.8 and -11.3

7 IR Spectroscopy

7.1 Alkanes

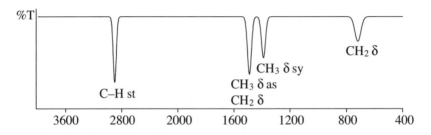

Typical Ranges ($\tilde{v}$ in cm^{-1})

Assignment	Range	Comments
C–H st	3000–2840	Intensity variable, often multiplet
	Beyond normal range:	
	2850–2815	CH_3–O, methyl ethers
	2880–2830	CH_2–O, ethers
	2880–2835, 2780–2750	O–CH_2–O, methylenedioxy
	≈2820	O–CH–O, acetals: weak
	3050–3000	⊳O, ⊳N
	2900–2800, 2780–2750	CH=O, aldehydes: Fermi resonance
	2820–2780	CH_3–N, CH_2–N; amines
	3100–3050, 3035–2995	▷
	2930–2915, 2900–2850	⬡ cyclohexanes: weak, comb at ≈2700
	3080–2900	CH–hal st

Assignment	Range	Comments
CH$_3$ δ as	1470–1430	Medium, coincides with CH$_2$ δ
	Beyond normal range:	
	1440–1400	CH$_3$–C=O, methyl ketones, acetals, CH$_3$–C=C
CH$_2$ δ	1475–1450	Medium, coincides with CH$_3$ δ as
	Beyond normal range:	
	≈1440	CH$_2$–C=C
		CH$_2$–C≡C
	≈1425	CH$_2$–C=O, CH$_2$– C≡N,
		CH$_2$–X (X: hal, NO$_2$, S, P)
CH$_3$ δ sy	1395–1365	Medium. Doublet in compounds with geminal methyl groups:
	≈1385, ≈1370	CH(CH$_3$)$_2$, of equal intensity (γ: 1175–1140, d)
	≈1385, ≈1365	C(CH$_3$)$_2$, 1385 weaker than 1365 (γ: 1220–1190, often d)
	≈1390, ≈1365	C(CH$_3$)$_3$, of equal intensity, sometimes triplet (γ: 1250–1200, d)
		N(CH$_3$)$_2$, no doublet
		Solid-state spectra: sometimes doublet also in the absence of geminal methyl groups
	Beyond normal range:	
	1325–1310	SO$_2$–CH$_3$
	1330–1290	S–CH$_3$, sulfides
	1310–1280	P–CH$_3$
	1275–1260	Si–CH$_3$, strong, sharp
CH$_3$ γ	1250–800	Intensity variable, of no practical significance. Strong band in compounds with geminal methyl groups:
	1175–1140	CH(CH$_3$)$_2$, doublet
	1220–1190	C(CH$_3$)$_2$, generally doublet
	1250–1200	C(CH$_3$)$_3$, doublet, often not resolved
	Beyond normal range:	
	≈765	SiCH$_3$
	≈855, ≈800	Si(CH$_3$)$_2$
	≈840, ≈765	Si(CH$_3$)$_3$

Assignment	Range	Comments
CH₂ γ	770–720	Medium, sometimes doublet
		C–(CH₂)ₙ–C for n > 4 at ≈720; for n < 4 at higher wavenumbers; in cyclohexanes at ≈890, weaker
	Beyond normal range:	
	1060–800	Cycloalkanes, numerous bands, unreliable
C–D st	2200–2080	In general, substitution of L by isotope L':

$$\tilde{\nu}_{X-L'} = \tilde{\nu}_{X-L} \sqrt{\frac{1/m_x + 1/m_{L'}}{1/m_x + 1/m_L}}$$

7.2 Alkenes

7.2.1 Monoenes

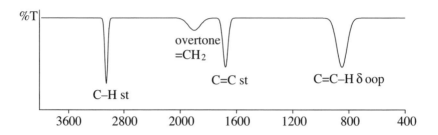

Typical Ranges ($\tilde{\nu}$ in cm^{-1})

Assignment	Range	Comments
=CH$_2$ st	3095–3075	Medium, often multiple bands
=CH st	3040–3010	Medium, often multiple bands
		CH st in aromatic hydrocarbons and three-membered rings fall into the same range
	In cyclic compounds:	
	≈3075	▷
	≈3060	☐
	≈3045	⬠
	≈3020	⬡
=CH δ ip	1420–1290	Of no practical significance
=CH δ oop	1005–675	A number of bands
	In the same range: ar CH δ oop, C–O–C γ, and C–N–C γ in saturated heterocyclics, OH δ oop in carboxylic acids, NH γ, NO st, SO st, CH$_2$ γ, CF st, CCl st	

Assignment	Range	Comments		
Subranges: C=C		C=C–C=O	C=C–OR	C=C–O–C=O
CH=CH$_2$	1005–985 920–900 (overtone at 1850–1800)	≈980 ≈960 ≈810	≈960 ≈815	≈950 ≈870
C=CH$_2$	900–880 (overtone at 1850–1780)	≈940 ≈810	≈795	
(cis-type, both H on same side)	990–960	≈975	≈960	≈950
(H, H cis disubstituted)	725–675	≈820		
(trans-type)	840–800	≈820		

C=C *(margin tab)*

Assignment	Range	Comments
C=C st	1690–1635	Of variable intensity, weak for highly symmetric compounds, strong for N–C=C and O–C=C
	Subranges:	
	1650–1635	CH=CH$_2$
	1660–1640	C=CH$_2$
	1690–1665	*(structure)* Weak
	1665–1635	*(structure)*
	1690–1660	*(structure)* Weak, often absent
	1690–1650	*(structure)* Weak, often absent
	Beyond normal range:	
	down to ≈1590	C=C–X with X: O, N, S; of higher intensity; in vinyl ethers often doublet due to rotational isomers

C = C

At lower frequency if conjugated with:

C=C	≈1650 ≈1600	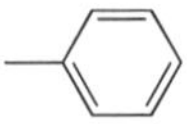 ⬡—	≈1630
C≡C	≈1600	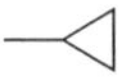 —◁	≈1640
C≡N	≈1620	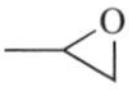 —◁^{O}	≈1640
C=O	≈1630		

Examples (ṽ in cm^{-1})

structure	values	structure	values	structure	values
	 1645 994 912		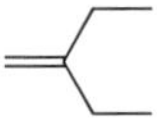 1647 889 669		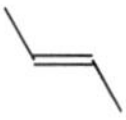 1682 972 963
	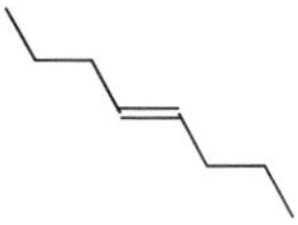 1670 968		 1650 709		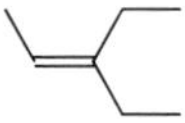 1667 825
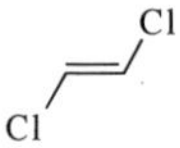	1575 826 761	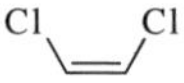	1595 848 714	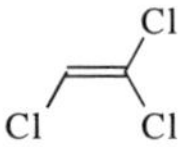	1587 929 835 780
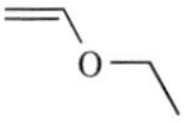	neat: 1610 987 810	CCl$_4$: 1634 1608 964 943	1655 1592 958 793	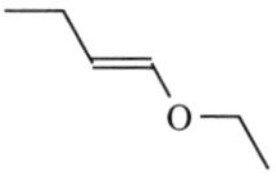	1670 1652 937 925
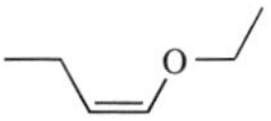	1663	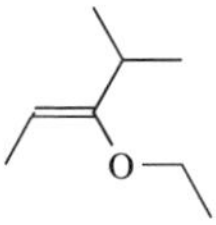	1660	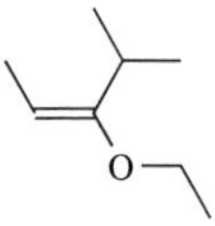	1673
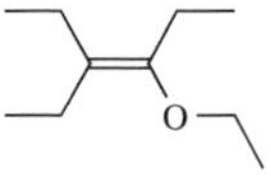	1663	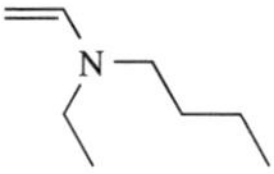	1628	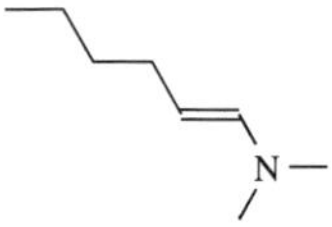	1650
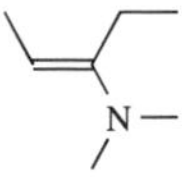	1640	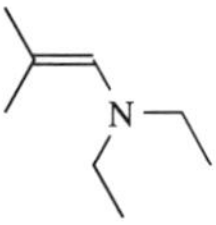	1662		

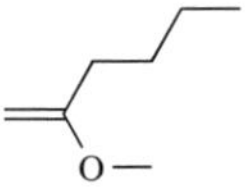

(isoprene structure)	1652 1612	(diene structure)	1830 1621 987 818	(diene structure)	1800 1621 941 899	
(enyne structure)	1607 (2270)	(styrene structure)	1636	(acrylonitrile structure) CN	1645 1612	**C = C**
CHO	1618 (1704)	(vinyl ketone) O	1618 (1684)	COOH	1635 1615	
COOCH₃	1637 (1735)				(1730) (1706)	

7.2.2 Allenes

Typical Ranges ($\tilde{v}$ in cm⁻¹)

Assignment	Range	Comments
(C=C)=C–H st	3050–2950	
C=C=C st as	1950–1930	Strong, doublet in X–C=C=CH$_2$ if X other than alkyl Ring strain increases frequency: (cyclopropylidene structure) =C=CH$_2$ ≈2020
C=C=C st sy	1075–1060	Weak, absent with highly symmetric substitution. In Raman, strong
(C=C)=CH$_2$ δ oop	≈850	Strong; overtone at ≈1700 (weak)

7.3 Alkynes

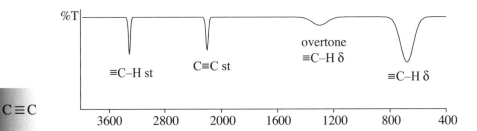

C≡C

Typical Ranges (ṽ in cm⁻¹)

Assignment	Range	Comments
≡C–H st	3340–3250	Strong, sharp; in the same region also OH st, NH st
C≡C st	2260–2100	Weak, sharp. In Raman, strong
	Beyond normal range:	
		R–C≡C–H; at the lower end of the cited range
		R–C≡C–R; usually 2 bands (Fermi resonance), often missing if symmetrical, strong in Raman
	Subranges:	
	≈2120	C–C≡C–H
	≈2220	C–C≡C–C
	≈2240	C–C≡C–CN
	≈2240	C–C≡C–COOH
	≈2240, ≈2140	C–C≡C–COOCH₃
	In the same range: C≡Z st, X=Y=Z st, Si–H st	
≡C–H δ	700–600	Strong, broad; overtone at 1370–1220 (broad, weak)

7.4 Alicyclics

Cyclic Alkanes

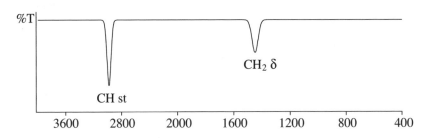

Cyclic Alkenes

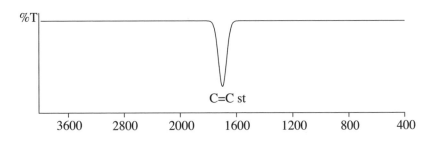

The other vibrations are similar to those in noncyclic alkenes and cyclic alkanes.

Typical Ranges ($\tilde{v}$ in cm^{-1})

Assignment	Range	Comments
C–H st	3090–2860	Strong
H–C–H δ	1470–1430	Weak
C=C st	1780–1610	Varies with ring size and substitution

Twisting and wagging CH_2 as well as C–C st do not significantly differ from the corresponding vibrations in noncyclic compounds and are of limited diagnostic value.

Examples ($\tilde{v}$ in cm^{-1})

▷	3090 3019 2933 1434	☐	2974 2896 1450	⬠	2951 2871 1455
⬡	2920 2860 1447	⬡	2933 2865 1462	⯃	2941 1471 1451

▷	≈1640	▷=	≈1780	O▷=	≈1650
☐	≈1570	☐⟋	≈1640	☐⟍	≈1680
	≈1690		≈1610		≈1660
	≈1660		≈1670		≈1690
	≈1570		≈1650		≈1675
	≈1650		≈1665		≈1670
	≈1615				

7.5 Aromatic Hydrocarbons

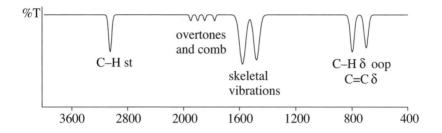

Typical Ranges ($\tilde{\nu}$ in cm^{-1})

Assignment	Range	Comments
ar C–H st	3080–3030	Often numerous bands; in the same range also CH st of alkenes and small rings
ar C–C	1625–1575	Medium, often doublet; generally weak in benzene derivatives having a center of symmetry in the ring

In the same range: C=C st, C=N st, C=O st, N=O st, C–C in heterocyclics, NH δ

	1525–1450	Medium, often doublet: Weak in:

In the same range: C=O st, N=O st, C–C in heterocyclics, B–N st, CH$_3$ δ, CH$_2$ δ, NH δ

| **comb** | 2000–1650 | Very weak; useful for determining substitution patterns in 6-membered aromatic rings |

In the same range: C=O st, B–H$\cdots$B st, N$^+$–H st, H$_2$O δ

| **ar C–H δ ip** | 1250–950 | Numerous bands of variable intensity; of no practical significance. May be very strong in Raman and, thereby, indicative of substitution type |

Assignment	Range	Comments
ar C–H δ oop	900–650	One or more strong bands; useful for determining substitution patterns in 6-membered aromatic rings. In Raman, generally weak
	In the same range:	=C–H δ oop, C–O–C γ and C–N–C γ in saturated heterocyclics, OH δ oop in carboxylic acids, NH δ, N–O st, S–O st, CH$_2$ γ, C–F δ, C–Cl st

Determination of Substitution Patterns in 6-Membered Aromatic Rings: Position and Shape of Bands Related to the Number of Adjacent H Atoms ($\tilde{\nu}$ in cm^{-1})

Not to be used for ring systems with strongly conjugated substituents such as C=O, NO$_2$, C≡N.

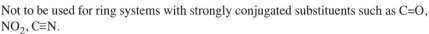

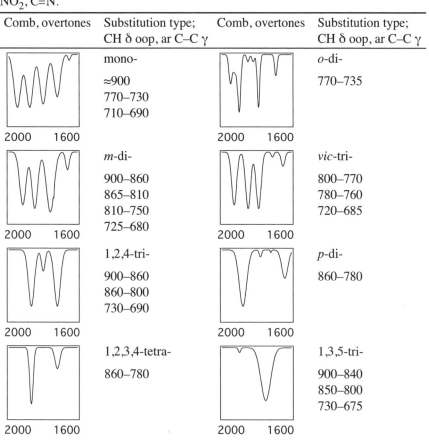

Comb, overtones	Substitution type; CH δ oop, ar C–C γ	Comb, overtones	Substitution type; CH δ oop, ar C–C γ
	mono- ≈900 770–730 710–690		o-di- 770–735
	m-di- 900–860 865–810 810–750 725–680		vic-tri- 800–770 780–760 720–685
	1,2,4-tri- 900–860 860–800 730–690		p-di- 860–780
	1,2,3,4-tetra- 860–780		1,3,5-tri- 900–840 850–800 730–675

Comb, overtones	Substitution type; CH δ oop, ar C–C γ	Comb, overtones	Substitution type; CH δ oop, ar C–C γ
(spectrum) 2000 1600	1,2,3,5-tetra- 900–840	(spectrum) 2000 1600	1,2,4,5-tetra- 900–840
(spectrum) 2000 1600	penta- 900–840	(spectrum) 2000 1600	hexa- –

Examples (ṽ in cm⁻¹)

(benzene)	3080 3040 1968 1818	(toluene)	3021 1945 1862 1808 1739	(styrene)	3086
(chlorobenzene) Cl	3080	(xylene)	3040 1915 1845 1775	(catechol) OH OH	1927 1887 1764

7.6 **Heteroaromatic Compounds**

Pyridines

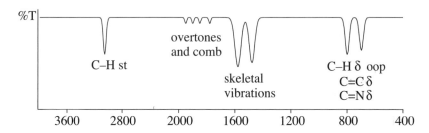

Furans

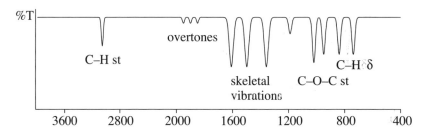

Pyrroles

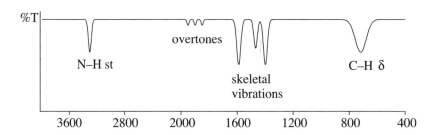

Typical Ranges ($\tilde{v}$ in cm^{-1})

Assignment	Range	Comments
N–H st	3450–3200	Medium, narrow; shifted by formation of hydrogen bonds
Overtones	2100–1800	Weak, characteristic
Ring skeleton	1610–1360	Strong, sharp bands
C–H δ	1000–700	Strong, broad; difficult to identify
C–H st	3100–3000	Medium, sharp
CO–C st	1190–990	Medium or strong; of variable intensity

Pyridines:

The frequencies of pyridines are very similar to those observed in benzenes. The nitrogen atom behaves like a substituted carbon atom in benzenes.

5-Ring Heteroaromatics

NH st free		3500–3400	
NH st H-bonded		3400–2800	
CH st	≈3100	≈3100	≈3100
Ring skeleton: intensity variable, generally multiplets	1610–1560 1510–1475	1590–1560 1540–1500	1535–1515 1455–1410
CH δ oop: generally strong	990–725	770–710	935–700

7.7 Halogen Compounds

7.7.1 Fluoro Compounds

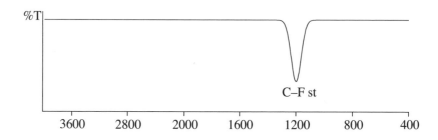

C–F st

3600   2800   2000   1600   1200   800   400

Hal

Typical Ranges ($\tilde{\nu}$ in cm^{-1})

Assignment	Range	Comments
C–F st	1400–1000	Strong, often more than one band (rotational isomers), often not resolved. In Raman, weak to medium
	Subranges:	
	1100–1000	al CF$_2$ (FC–H st: 3080–2990)
	1150–1000	al CF$_2$
	1350–1100	al CF$_3$
	1350–1150	C=CF
	≈1745	C=CF$_2$ st
	1250–1100	ar CF
	In the same range: strong bands for C–O st, NO$_2$ st sym, C=S st, S=O st	
CF$_2$	780–680	Medium or weak, assignment uncertain
CF$_3$	780–680	(C–F δ?)
S–F st	815–755	Strong
P–F st	1110–760	
Si–F st	980–820	
B–F st	1500–800	

7.7.2 Chloro Compounds

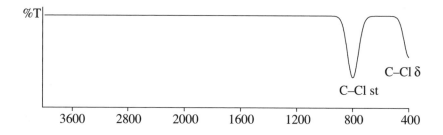

Typical Ranges (ṽ in cm⁻¹)

Assignment	Range	Comments
C–Cl st	830– <600	Strong, often broad (rotational isomers), absent in chloroaromatics
C–Cl δ	400–280	Of medium strength and width
Other	1100–1020	Strong, narrow or of medium width; chloroaromatics
P–Cl st	<600	
Si–Cl st	<625	
B–Cl st	1100–650	

In disubstituted halobenzenes, characteristic skeletal vibrations:

	X	*ortho*	*meta*	*para*
	Cl	1055–1035	1080–1075	1095–1090
	Br	1045–1030	1075–1065	1075–1070
	I			1060–1055

Hal

7.7.3 Bromo Compounds

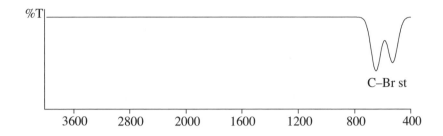

Typical Ranges ($\tilde{v}$ in cm^{-1})

Assignment	Range	Comments
C–Br st	700–500	Strong, of medium width; absent in bromoaromatics
C–Br δ	350–250	Of medium strength and width
Other	1080–1000	Strong, narrow or of medium width; bromoaromatics

Hal

7.7.4 Iodo Compounds

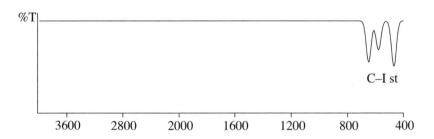

Typical Ranges ($\tilde{v}$ in cm^{-1})

Assignment	Range	Comments
C–I st	650–450	Strong, two or more bands
C–I δ	300–50	Of medium strength and width

7.8 Alcohols, Ethers, and Related Compounds

7.8.1 Alcohols and Phenols

Alcohols

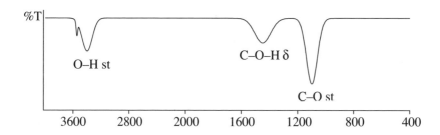

Phenols

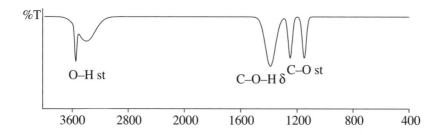

Typical Ranges ($\tilde{v}$ in cm^{-1})

Assignment	Range	Comments
O–H st	3650–3200	Of variable intensity. In Raman, generally weak
	Subranges:	
	3650–3590	Free OH; sharp
	3550–3450	H-bonded OH; broad
	3500–3200	Polymer OH; broad, often numerous bands
	Beyond normal range:	
	3200–2500	Enols, chelates; often very broad
	In the same range: NH st, ≡CH st (≈3300, sharp), H$_2$O	
O–H δ ip	1450–1200	Medium, of no practical significance

Assignment	Range	Comments
C–O st	1260–970	Strong, often doublet
	Subranges:	
	1075–1000	CH_2–OH
	1125–1000	CH–OH
	1210–1100	C–OH
	1275–1150	ar C–OH
	In the same range: C–F st, C–N st, N–O st, P–O st, C=S st,	
		S=O st, P=O st, Si–O st, Si–H δ
O–H δ oop	<700	Medium, of no practical significance

Examples ($\tilde{v}$ in cm^{-1})

$\diagdown\!\diagup\!\diagdown$OH	3250 1430 1075 1050	OH	3335 1350	OH	3290 1430 1020
OH	3215 1368 1220	OH OH	3450 1370 1260 1195	OH	3460 1315 1237 1210

7.8.2 Ethers, Acetals, and Ketals

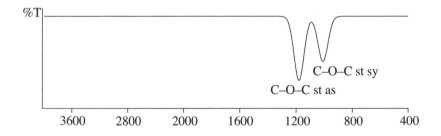

In acetals and ketals, the C–O stretching vibrations are split into 3, sometimes even 4 to 5 bands.

Acetals have an additional band due to a special C–H δ vibration.

The C–H st vibration frequency is especially low for OCH_3 st (2850–2815) and OCH_2 st (2880–2835).

Typical Ranges ($\tilde{v}$ in cm^{-1})

Assignment	Range	Comments
C–O–C st as	1310–1000	Strong, sometimes split
	Subranges for noncyclic ethers:	
	1150–1085	CH$_2$–O–CH$_2$
	1170–1115	CH–O–CH, often split
	1225–1180	C=C–O–al C
	1275–1200	ar C–O–al C
	Subranges for cyclic ethers:	
	1280 sy 870 as	
	≈1030 sy ≈980 as	
	≈1070 sy ≈915 as	
	≈1235	
	≈1100 as ≈815 sy	
	≈950	ketals, acetals: 4 to 5 bands
	≈925	
	1024, 1086 as ≈880 sy	
	≈800	in acetals: C–H st, ≈2820, weak
C–O–C st sy	1055–870	Strong, sometimes multiple bands
	Subranges for noncyclic ethers:	
	1125–1080	C=C–O–al C, medium
	1075–1020	ar C–O–al C, medium
	In the same range: strong bands for C–O st, C–F st, C–N st, N–O st, P–O st, C=S st, S=O st, P=O st, Si–O st, Si–H δ	

O

Examples (ṽ in cm⁻¹)

	1136		1225			1250
	935		1218			1040
	917		1211			
			1003			
	1188		1172			
	1138		1132			
	1111		1077			
	1046		1057			
			1038			

7.8.3 Epoxides

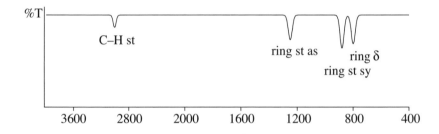

O

Typical Ranges (ṽ in cm⁻¹)

Assignment	Range	Comments
C–H st	3050–2990	Frequency higher than normally found in alkanes
ring st as	1280–1230	Variable intensity
ring st sy	950–815	Variable intensity
ring δ	880–750	Variable intensity

Examples (ṽ in cm⁻¹)

	1280		1230 sy		1260 sy
	870		885 as		890 as
			845 δ		780 δ

7.8.4 Peroxides and Hydroperoxides

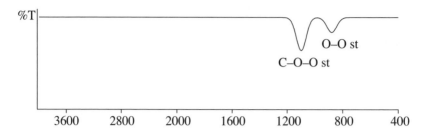

Typical Ranges ($\tilde{v}$ in cm^{-1})

Assignment	Range	Comments
O–O–H st	3450–3200	Of variable intensity
	Subranges:	
	≈3450	Free OOH; H-bonded: ≈30 cm^{-1} higher than in corresponding alcohols
	In the same range: OH st, NH st, ≡CH st, H_2O	
C–O–O st	1200–1000	Strong, ≈20 cm^{-1} lower than in corresponding alcohols
	In the same range: strong bands for C–O st, C–F st, C–N st, N–O st, P–O st, C=S st, S=O st, P=O st, Si–O st, Si–H δ	
O–O st	1000–800	Medium or weak, often doublet, assignment uncertain
Also:	1760–1745	C=O st in peracids
	1820–1770	C=O st in diacylperoxides (two bands)

Examples ($\tilde{v}$ in cm^{-1})

	1017		1070		1100
	880		1060		852
			943		

7.9 Nitrogen Compounds

7.9.1 Amines and Related Compounds

Primary Amines

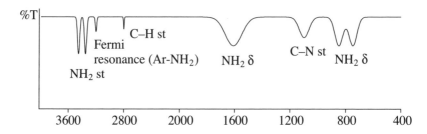

Secondary Amines

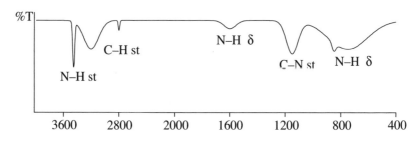

N   *Ammonium*

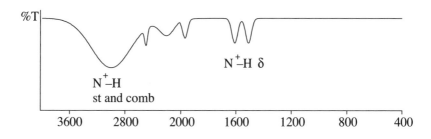

Typical Ranges ($\tilde{v}$ in cm^{-1})

Assignment	Range	Comments
NH$_2$ st	3500–3300	Of variable intensity, generally 2 sharp bands, $\Delta\tilde{v} = 65$–75
		At lower wavenumbers (<3200) and broader if H-bonded. Free and H-bonded forms often simultaneously observed
		In primary aromatic amines, additional combination band at ≈3200
		In the same range: OH st, ≡CH st
NH st	3450–3300	Of variable intensity, only one band
		At lower wavenumbers (<3200) and broader if H-bonded. Free and H-bonded forms often simultaneously observed
		In the same range: OH st, ≡CH st, H$_2$O
NH$_3^+$ st	3000–2000	Medium, broad, highly structured
	3000–2700	Major maximum, comb: ≈2000
NH$_2^+$ st	3000–2000	Medium, broad, highly structured
	3000–2700	Major maximum
NH$^+$ st	3000–2000	Medium, broad, highly structured
	2700–2250	Major maximum
		In the same range: OH st, NH st, CH st, SH st, PH st, SiH st, BH st, X=Y=Z st, X≡Y st
NH$_2$ δ	1650–1590	Medium or weak
NH δ	1650–1550	Weak
NH$_3^+$ δ	1600–1460	Medium, often more than one band; weak in aliphatic amines
NH$_2^+$ δ	1600–1460	Medium, often more than one band; weak in aliphatic amines
NH$^+$ δ	1600–1460	Medium, often more than one band; weak in aliphatic amines
C–N st	1400–1000	Medium, of no practical significance
NH$_2$ δ	850–700	Medium or weak; 2 bands in primary amines
NH δ	850–700	Medium or weak
P–N–C st	1110–930	
	770–680	

N

Examples ($\tilde{v}$ in cm^{-1})

CH$_3$—NH$_2$	3470
	3360
	1622

NH$_2$ (cyclohexylamine) 3357
3278
3200 sh
1605

H$_2$N   NH$_2$ 3356
3274
3175
1650

H N   N H   3279

NH$_2$ (p-toluidine) 3487
3405

NH (N-methylaniline) 3416
3386
1322
1266

7.9.2 Nitro and Nitroso Compounds

Nitro Compounds

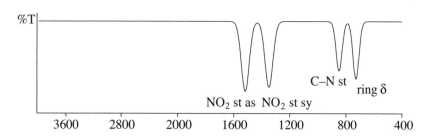

NO$_2$ st as NO$_2$ st sy   C–N st   ring δ

Nitroso Compounds

N

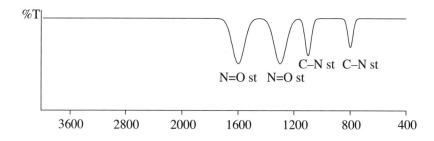

N=O st N=O st   C–N st C–N st

Typical Ranges ($\tilde{\nu}$ in cm^{-1})

Assignment	Range	Comments
NO$_2$ st as	1660–1490	Very strong, of medium width. In Raman, of weak to medium intensity
	Subranges:	
	1660–1625	O–NO$_2$, nitrates; missing in Raman
	1570–1540	C–NO$_2$, aliphatic nitro compounds
	1560–1490	C–NO$_2$, aromatic nitro compounds
	1630–1530	N–NO$_2$, nitramines
NO$_2$ st sy	1390–1260	Strong, of medium width
	Subranges:	
	1285–1270	O–NO$_2$, nitrates
	1390–1340	C–NO$_2$, aliphatic nitro compounds
	1360–1310	C–NO$_2$, aromatic nitro compounds; often 2 bands
	1315–1260	N–NO$_2$, nitramines
	In nitrates also:	
	≈870	N–O st, strong
	≈760	NO$_2$ γ
	≈700	NO$_2$ δ
Ring δ	760–705	Strong; modified deformation of aromatic ring
N=O st	1680–1450	Very strong, in monomers
	1420–1250	Very strong, in dimers
	Subranges:	
	1680–1650	O–NO (nitrites) *trans*; 1625–1610: *cis*
	1585–1540	C–NO, aliphatic *C*-nitroso compounds
	1510–1490	C–NO, aromatic *C*-nitroso compounds
	≈1450	N–NO, *N*-nitroso compounds
	In nitrites also:	
	3300–3200, ≈2500, 2300–2250	comb
	≈800	N–O st *trans*; *cis*: very weak
	≈600	O–NO δ *trans*; *cis*: ≈650
C–N st	≈850	C–NO, aliphatic *C*-nitroso compounds; coupled with other vibrations
	≈1100	C–NO, aromatic *C*-nitroso compounds
N–N st	≈1040	*N*-Nitroso compounds

N

Examples ($\tilde{v}$ in cm^{-1})

CH$_3$−NO   1564   NO   1506   NO   1497
            842   1110   1112
                               810   858

CH$_3$−NO	1564	NO (phenyl)	1506	NO (4-methoxyphenyl)	1497
	842		1110		1112
			810		858
(CH$_3$)$_2$CH−NO$_2$	1524	NO$_2$ (4-methylphenyl)	1527	NO$_2$ (1,2-dinitrophenyl)	1506
	1359		1351		1351
	851		853		1261
			720		873
					748

7.9.3 Imines and Oximes

Imines

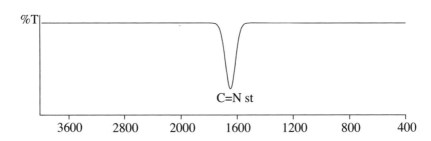

Oximes

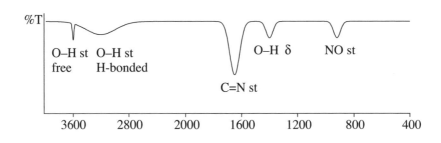

Typical Ranges ($\tilde{v}$ in cm^{-1})

Assignment	Range	Comments
C=N st	1690–1520	Generally strong. In Raman, generally strong
	Subranges:	
	≈1670	R–CH=N–R' R, R': al
	≈1645	R–CH=N–R' R or R': conjugated
	≈1630	R–CH=N–R' R, R': conjugated
	≈1655	R, R', R": al
	≈1645	R''\C=N–R' R: conjugated
	≈1635	R/ R, R': conjugated
	≈1555	O···\C=N–phenyl Additional band: ≈1655 C=O st
	≈1645	R, R': al
	≈1625	R\C=NH R, R': conjugated
	1685–1580	H₂N\C=N–R Additional band at 1540–1515 in: RHN\C=N–R
	1670–1600	CH=N–N=CH
	1690–1645	RO\C=NH (RO/) Additional bands: NH st: ≈3300, C–O st: ≈1325, ≈1100
	1680–1635	RO\C=NH₂⁺ (RO/) Additional bands: NH₂⁺ st: ≈3000, NH₂⁺ δ: 1590–1540
	2050–2000	C=C=N; ketimines, very strong, sometimes doublet
	1580–1520	Quinone oximes: C=O st 1680–1620
	1685–1650	Aliphatic oximes
	1650–1615	Aromatic oximes
	1690–1645	O–C=N
	1640–1605	S–C=N
	1640–1580	S–S–C=N
OH st	3600–2700	Strong
	Subranges:	
	≈3600	Free
	3300–3100	H-bonded, broad
	≥ ≈2700	Quinone oximes, more than one band
OH δ	1475–1315	Of no practical significance
N–O st	1050–400	Of no practical significance

N

Examples (ṽ in cm⁻¹)

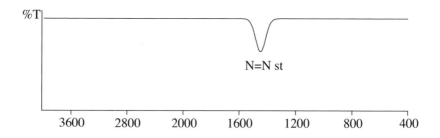

| | 1667 | | 1603 | | 1637 |
| | 1675 | | 1672 (solid)
1662 (gas) | | 1684 |

7.9.4 Azo, Azoxy, and Azothio Compounds

%T

N=N st

3600   2800   2000   1600   1200   800   400

Typical Ranges (ṽ in cm⁻¹)

Assignment	Range	Comments
N=N st	1580–1400	Very weak, missing in compounds of high symmetry. In Raman, generally strong
	1480–1450 1335–1315	st as (mainly N=N st) st sy (mainly N–O st)
	≈1450 ≈1060	st as (mainly N=N st) st sy (mainly N–S st)
	1410–1175	Dimers of *C*-nitroso compounds
	Subranges:	
	1290–1175	Aliphatic *trans*
	1425–1385, 1345–1320	Aliphatic *cis*
	1300–1250	Aromatic *trans*
	≈1410, ≈1395	Aromatic *cis*

N

7.9.5 Nitriles and Isonitriles

Nitriles

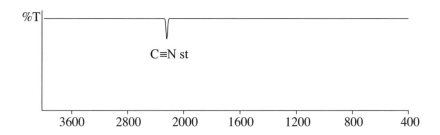

Isonitriles

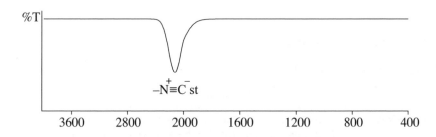

Typical Ranges ($\tilde{v}$ in cm^{-1})

Assignment	Range	Comments
C≡N st	2260–2240	Medium to strong, sharp; for O–CH$_2$–C≡N, N–CH$_2$–C≡N: of low intensity or absent. In Raman, of medium to high intensity
	Beyond normal range:	
	2240–2215	C=C–C≡N
	2240–2215	⬡–C≡N
	2240–2230	XC–C≡N, X: Cl, Br, I
	≈2275	–CF$_2$–C≡N
	2225–2175	\N–C≡N ⟷ \N$^+$=C=N$^-$
	2210–2185	>N–C=C–C≡N
	2200–2070	C≡N$^-$
–N$^+$≡C$^-$	2150–2110	Strong

N

Examples ($\tilde{v}$ in cm^{-1})

$\bigvee$–CN 2222	$\diagdown\!\diagup\!\diagdown$CN 2235	(benzyl)CN 2252
NC$\diagdown\!\diagup$CN 2273	NC$\diagdown\!\diagup\!\diagdown$CN 2235	NC$\diagup\!\diagdown$CN 2252
NC$\diagdown$C=C$\diagup$CN NC$\diagup$ $\diagdown$CN 2257 2222	(aryl)CN 2245	(aryl)CN, OH 2220

NaCN, KCN   2080–2070   AgCN   2178   NH$_2$–CN   2268

7.9.6 Diazo Compounds

Typical Ranges ($\tilde{v}$ in cm^{-1})

Assignment	Range	Comments
–N$^+$≡N st	2310–2130	Medium, frequency depends on anion
		In the same range: C≡C st, X=Y=Z st as, NH$^+$ st, PH st, POH st, SiH st, BH st
C=N$^+$=N$^-$	2050–2010	Very strong
	Subranges:	
	2050–2035	R–CH=N$^+$=N$^-$, R: al or ar
	2035–2010	R$_2$–C=N$^+$=N$^-$, R: al or ar
	Beyond normal range:	
	2100–2050	R–CO–C=N$^+$=N$^-$ C=O st ≈1645 (R: al) C=O st ≈1615 (R: ar) C=N$^+$=N$^-$ st sy: ≈1350, strong
	2180–2010	O=⟨ring⟩=N$^+$=N$^-$ C=O st 1655–1560

7.9.7 Cyanates and Isocyanates

Cyanates

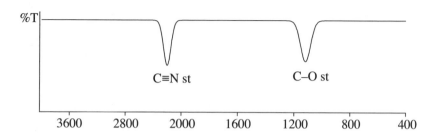

Isocyanates

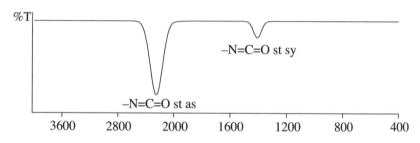

Typical Ranges ($\tilde{v}$ in cm^{-1})

Assignment	Range	Comments
OC≡N st	2260–2130	Medium to strong
	2220–2130	(OC≡N)⁻ st as
	1335–1290	(OC≡N)⁻ st sy
C–O st	1200–1080	Strong
N=C=O st as	2280–2230	Strong, sharp. In Raman, weak or absent
	≈2300	–CF₂–NCO
N=C=O st sy	1450–1380	Weak
	Beyond normal range:	
	2220–2130	(N=C=O)⁻

N

Examples ($\tilde{v}$ in cm^{-1})

CH$_3$–OCN   2248

$\diagup$OCN   2248
2282

OCN   2235
2261
2282

CH$_3$–NCO   2265

$\diagup$NCO   2280

NCO   2270

$\diagdown$NCO   2256
(1629 C=C)

NCO   2267

$\overset{O}{\diagdown}$NCO   2246

7.9.8 Thiocyanates and Isothiocyanates

Thiocyanates

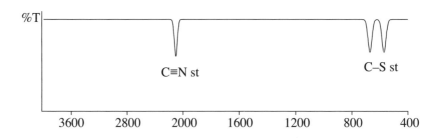

Isothiocyanates

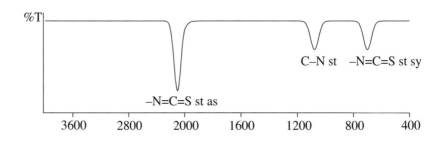

Typical Ranges ($\tilde{\nu}$ in cm^{-1})

Assignment	Range	Comments
SC≡N st	2170–2130	Medium, sharp
	2090–2020	(SC≡N)⁻
C–S st	750-550	Often doublet
N=C=S st as	2200–2050	Very strong, generally doublet, Fermi resonance
N=C=S st sy	950–650	
	≈950	aliphatic –N=C=S
	700–650	aromatic –N=C=S
	Beyond normal range:	
	2090–2020	(N=C=S)⁻
C–N st	1090–1075	

Examples ($\tilde{\nu}$ in cm^{-1})

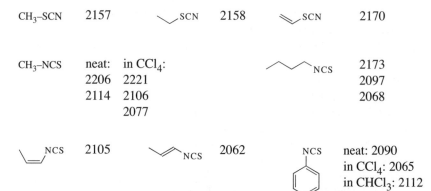

CH₃–SCN   2157   SCN   2158   SCN   2170

CH₃–NCS   neat: in CCl₄: NCS   2173
          2206   2221   2097
          2114   2106   2068
                  2077

NCS   2105   NCS   2062   NCS   neat: 2090
                                          in CCl₄: 2065
                                          in CHCl₃: 2112

N

7.10　Sulfur Compounds

7.10.1　Thiols and Sulfides

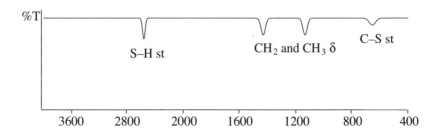

Typical Ranges ($\tilde{v}$ in cm^{-1})

Assignment	Range	Comments
S–H st	2600–2540	Often weak, narrow. In Raman, strong
S–H δ	915–800	Weak, of no practical significance
C–S st	710–570	Weak, broad, of no practical significance. In Raman, strong
S–S st	≈500	Weak, of no practical significance
Also:	≈2880	(S–)CH$_3$ st as
	≈2860	(S–)CH$_2$ st as
	≈1430	(S–)CH$_3$ δ as
	1330–1290	(S–)CH$_3$ δ sy
	≈1425	(S–)CH$_2$ δ
	815–755	S–F st, strong
	≈630	S–N st in S–N=O
	725–550	S–C in S– C≡N, often doublet

Examples ($\tilde{v}$ in cm^{-1})

SH　2950　　HS⌒⌒SH　2525　　2585

698
668　566　662
641

7.10.2 Sulfoxides and Sulfones

Sulfoxides

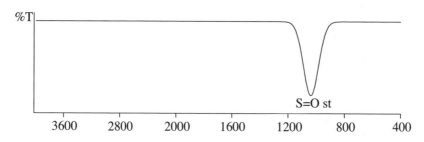

Sulfones

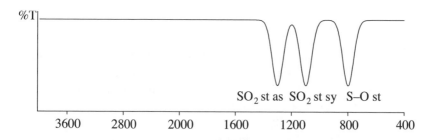

Typical Ranges ($\tilde{v}$ in cm^{-1})

Assignment	Range	Comments	
S=O st	1225–980	Strong, sometimes multiple bands. In Raman, weak to medium	
	Subranges:		
	1060–1015	R–SO–R	
	≈1100	R–SO–OH	S–O st 870–810 OH st free ≈3700, H-bonded ≈2900, ≈2500
	≈1135	R–SO–OR	S–O st 740–720, 710–690
	1225–1195	RO–SO–OR	
	≈1135	R–SO–Cl	
	≈1030, ≈980	R–SO$_2^-$	
	≈1100, ≈1050	R=SO	N=SO: ≈1250, ≈1135

S

Assignment	Range	Comments	
$\diagdown\!\!\underset{\diagup}{S}\!\!\diagup^{O}$ st as	1420–1300	Very strong; in Raman, often missing	
$\underset{O}{S}{<}_{O}$ st sy	1200–1000	Very strong; in Raman, strong	
	Subranges:		
	1370–1290, 1170–1110	R–SO$_2$–R	
	1375–1350, 1185–1165	R–SO$_2$–OR	
	≈1340, ≈1150	R–SO$_2$–SR	
	1415–1390, 1200–1185	RO–SO$_2$–OR	
	1365–1315, 1180–1150	R–SO$_2$–N	N–H st: 3330–3250; N–H δ: ≈1570; S–N st: 910–900
	1410–1375, 1205–1170	R–SO$_2$–hal	
	1355–1340, 1165–1150	R–SO$_2$–OH	O–H st, H-bonded: ≈2900, ≈2400 hydrated: 2800–1650, broad
	1250–1140, 1070–1030	R–SO$_3^-$	
	1315–1220, 1140–1050	RO–SO$_3^-$	
S–O st	870–690	Of variable intensity, weak in sulfites	

S

7.10.3   Thiocarbonyl Derivatives

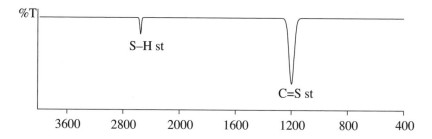

Typical Ranges ($\tilde{v}$ in cm^{-1})

Assignment	Range	Comments	
C=S st	1275–1030	Strong, narrow. In Raman, strong	
	Subranges:		
	1075–1030	Thioketones	
	1210–1080	Thioesters	
	≈1215	Dithiocarboxylic acids	SH st: ≈2550
			SH δ: ≈860
	1125–1075	Thiocarboxylic acid fluoride	perfluorinated: 1130–1105
	1100–1065	Thiocarboxylic acid chloride	perchlorinated: 1100–1075
	1140–1090	Thioamides and thiolactams	C–N st: 1535–1520 NH δ: 1380–1300
Also:	750–580	P=S st	

S

7.10.4   Thiocarbonic Acid Derivatives

Trithiocarbonates

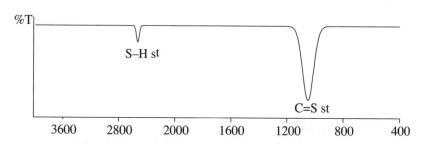

Xanthates

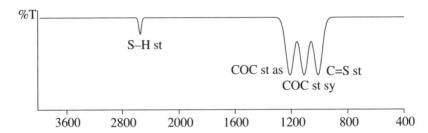

Thiocarbonates

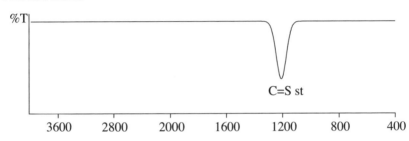

Thioureas

S

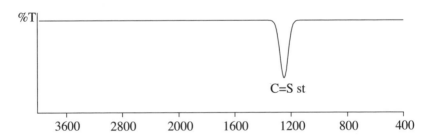

Typical Ranges ($\tilde{v}$ in cm^{-1})

Assignment	Range	Comments	
S–H st	2560–2510	Weak, narrow	trithiocarbonates
	2600–2500	Weak, narrow	xanthates
C=S st	1100–1020	Very strong	trithiocarbonates
	1070–1000	Strong	xanthates
	1250–1180	Strong	thiocarbonates
	1400–1100	Strong	thioureas
COC st as	1260–1140	Strong	xanthates
COC st sy	1150–1090	Strong to medium	xanthates

Examples ($\tilde{v}$ in cm^{-1})

Structure	Value	Structure	Value	Structure	Value
(methyl thiocarbonate ester)	in CCl$_4$: 1719	(dithiocarbonate)	in CCl$_4$: 1653	(1,3-oxathiolan-2-one)	in CCl$_4$: 1757
(1,3-dithiolan-2-one)	in CCl$_4$: 1718 1677 1640	(trithiocarbonate)	neat: 1076	(1,3-dithiolane-2-thione)	solid: 1058 in CCl$_4$: 1083 1079
(N-methyl thiocarbamate)	in CCl$_4$: 1662			(HS–C(=S)–O–)	gas: 2593 2548 neat: 2470
(HS–C(=S)–O–ethyl)	in CS$_2$: 2562 2522	(dithiocarbonate diester)	solid: 1212		solid: 1234
(thiourea H$_2$N–C(=S)–NH$_2$)	solid: 1400	(N,N'-dimethyl thiourea)	solid: 1130	(N,N'-diphenyl thiourea)	solid: 1131

S

7.11 Carbonyl Compounds

7.11.1 Aldehydes

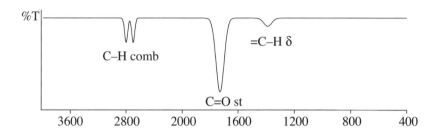

Typical Ranges ($\tilde{v}$ in cm^{-1})

Assignment	Range	Comments
C–H comb	2900–2800	Weak, Fermi resonance with C–H δ at ≈1390
	2780–2680	(for extreme position of C–H δ only one band)
	Subranges:	
	2830–2810,	Aliphatic
	2720–2690	
	2830–2810,	Aromatic, with *o*-substitution often higher
	2750–2720	
	In the same range: cyclohexanes at ≈2700, weak	
C=O st	1765–1645	Strong; in Raman, weak to medium
	Subranges:	
	1740–1720	Aliphatic
	1765–1730	α-Halogenated aliphatic
	1710–1685	Aromatic
	1695–1660	α,β-Unsaturated aromatic
	1670–1645	With intramolecular H bonds
C–H δ	1390	Weak, of no practical significance

C = X

Examples (ṽ in cm⁻¹)

CH_3–CHO	1748		1725		1742
CCl_3–CHO	1760		1687		1700
	1670		1696		in CCl_4: 1717 in $CHCl_3$: 1710

7.11.2 Ketones

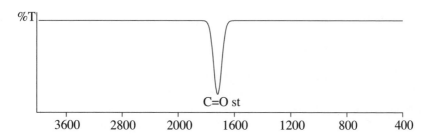

%T

C=O st

3600   2800   2000   1600   1200   800   400

Typical Ranges (ṽ in cm⁻¹)

Assignment	Range	Comments
C=O st	1775–1650	Strong; in Raman, weak to medium
	Subranges:	
	≈1715	Aliphatic, branching at α position causes shift to lower wavenumbers: ≈1695 ≈1685
	≈1775–1705	Cyclic, ṽ decreases with increasing ring size

C = X

[contd.]

Assignment	Range	Comments

≈1775   ≈1750

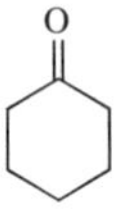

≈1715   ≈1705

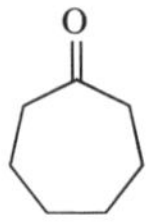

Conjugated:	≈1675	α,β-Unsaturated, often 2 bands (rotational isomers)
	1650–1600	C=C st
	≈1695	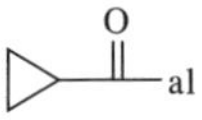
	≈1665	$\alpha,\beta,\gamma,\delta$-Diunsaturated; $\alpha,\beta;\alpha',\beta'$-diunsaturated
	≈1670	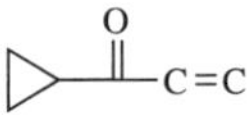
	≈1690	Aryl ketones
	≈1675	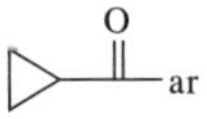
	≈1665	Diaryl ketones, with N or O in *p*-position: down to ≈1600
α-Halogenated ketones:	Shifted toward higher wavenumbers depending on dihedral angle ϕ between C=O and C-hal; largest effect for $\phi = 0°$, no effect for $\phi = 90°$	

Maximal shifts: α-chloro   ≈25 α-bromo   ≈20
  α,α-dichloro   ≈45 α-iodo   ≈0
  α,α'-dichloro   ≈45 α,α-difluoro   ≈60
    perfluoro   ≈90

C = X

α-Diketones:	≈1720	Aliphatic
	≈1775, ≈1760	Aliphatic 5-ring
	≈1760, ≈1730	Aliphatic 6-ring
	≈1675	Aliphatic enolized, C=C st: ≈1650
	≈1680	Aromatic
	≈1675	*o*-Quinones, with *peri*-OH: ≈1675, ≈1630
β-Diketones:	≈1720	Keto form, sometimes doublet
	≈1650	Enol form

Assignment	Range	Comments
	≈1615	Enol with intramolecular H bonds, C=C st: ≈1600, strong
γ-Diketones:		As monoketones
	≈1675	*p*-Quinones; with *peri*-OH: ≈1675, ≈1630 C=C st: ≈1600
C=C=O st as	2155–2130	Very strong

Examples ($\tilde{v}$ in cm^{-1})

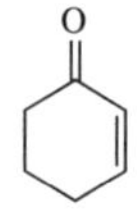

1691

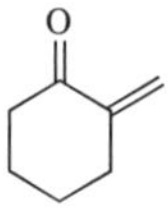

1697

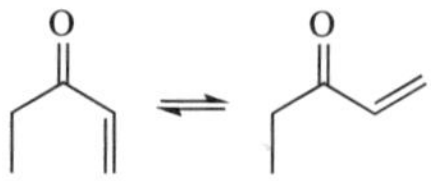

s-*trans* 1690
s-*cis*   1707

s-*trans*   s-*cis*

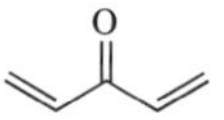

1672
1660

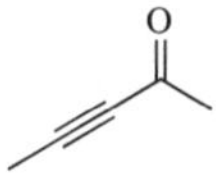

1678
(2222)

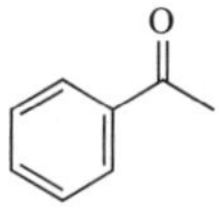

1692

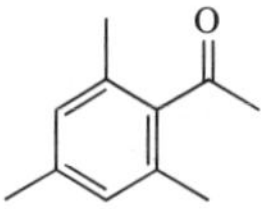

1701

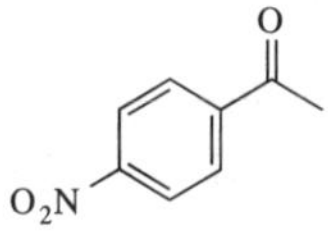

1702

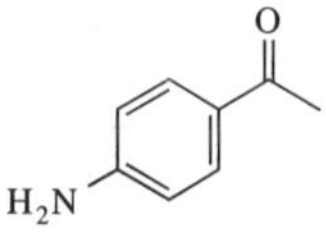

1676

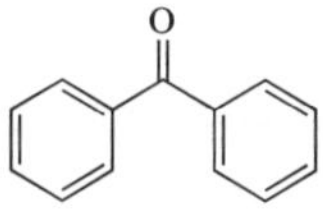

1664

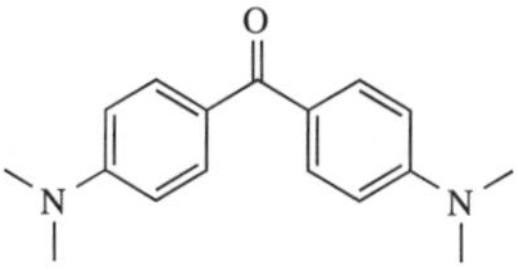

1639

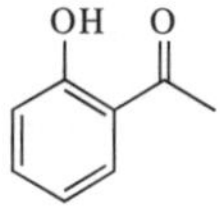

1648

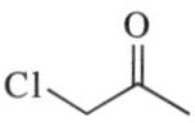

1752
1726
(rotamers)

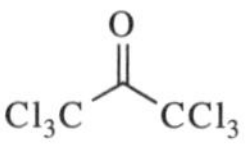

1780
1751

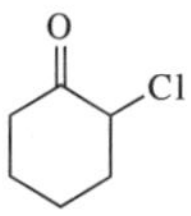

1722   C = X

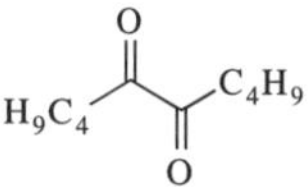

1710

1700
1655

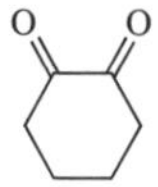

1735

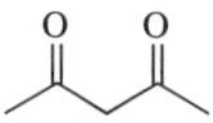

1724 (keto form)
1608 (enol form)

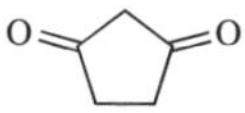

1755
1725
1635
1590

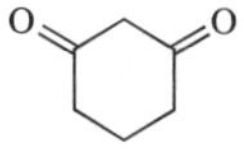

1630
1607

1669

1669

1675

1623

1662

1678

7.11.3 Carboxylic Acids

Carboxylic Acids

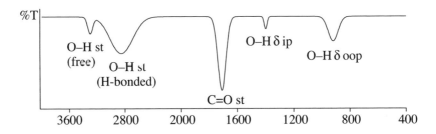

Carboxylate Anions

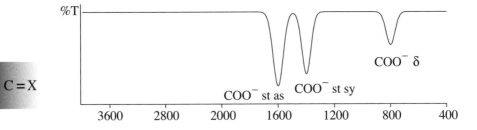

C = X

Typical Ranges ($\tilde{v}$ in cm^{-1})

Assignment	Range	Comments
COO–H st	3550–2500	Intensity variable
	Subranges:	
	3550–3500	Free, sharp, only in highly diluted solutions
	3300–2500	H-bonded, broad, often more than one band
	In the same range: OH st, NH st, CH st, SiH st, SH st, PH st	
C=O st	1800–1650	Strong; in Raman, weak to medium
	1800–1740	Free (also in dicarboxylic acids)
	1740–1650	H-bonded (dimer, also in dicarboxylic acids)
	Subranges for H-bonded C=O:	
	1725–1700	al–COOH
	1715–1690	C=C–COOH
	1700–1680	ar–COOH
	1740–1720	hal–C–COOH
	1670–1650	Intramolecular H bond
OC–OH st, C–OH δ	1440–1210	Of no practical significance
OC–OH δ oop	960–880	Medium, generally broad (only in dimers); in the same range: =CH δ, ar CH δ, NH δ
(COO)$^-$ st as	1610–1550	Very strong; in α-halogen carboxylates near the higher value, with more than one α-hal beyond the normal range; in polypeptides at ≈1575
(COO)$^-$ st sy	1450–1400	Strong, of no practical significance, in polypeptides at ≈1470
(COO)$^-$ δ	≈775	Formates, weak
	≈925	Acetates
	≈680	Benzoates
	≈600	CF$_3$COO$^-$

C = X

Examples ($\tilde{v}$ in cm^{-1})

	neat:	1727
	in CCl$_4$:	1756
		1724

	neat:	1759
		1718
	in CCl$_4$:	1768
		1717

	in CCl$_4$:	1704
	solid:	1686

(structure)	neat: 1700 in CCl$_4$: 1694	(structure)	neat: 1725	(structure)	neat: 1730

(structure)	solid: 1605	(structure)	solid:
			    1740
				(structure)	in CCl$_4$:
					    1788
					    1725

(structure)	solid: 1650	(structure)	solid: 1735 1703	(structure)	solid: 1724

(structure)	solid: 1690
	in CCl$_4$:
	    1730
	in CHCl$_3$:
	    1706
		(structure)	solid:
			1690
				(structure)	solid:
					    1690

(structure)	solid: 1667
	in CCl$_4$:
	    1696
	in CHCl$_3$:
	    1661
		(structure)	in CCl$_4$:
			1750
				(structure)	solid:
					    1693
					in CCl$_4$:
					    1696

7.11.4 Esters and Lactones

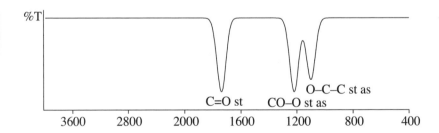

C = X

Typical Ranges ($\tilde{v}$ in cm^{-1})

Assignment	Range	Comments
C=O st	1790–1650	Strong. In Raman, weak to medium
	Subranges:	
	1750–1735	Aliphatic esters
Conjugated	1730–1710	α,β-Unsaturated esters
esters:	1730–1715	Aromatic esters
	1690–1670	With intramolecular H bonds
	1790–1740	α-Halogenated esters
	≈1760	Vinyl esters, C=C st: 1690–1650, strong
	≈1760	Phenol esters
	≈1735	Phenol esters of aromatic acids
Diesters:		As the corresponding monoesters
Keto esters:	1755–1725	α-Keto esters, generally one band
	≈1750 (ketone)	β-Keto esters, keto form
	≈1735 (ester)	
	≈1650	β-Keto esters, enol form, C=C st: ≈1630, strong
	≈1740, ≈1715	γ-Keto esters, pseudoesters: ≈1770
Lactones:		

β-lactone ≈1840   γ-butyrolactone ≈1770   α,β-unsaturated γ-lactone ≈1800

≈1750 (additional band at ≈1780 if α position is not substituted)   δ-valerolactone ≈1735

≈1760   ≈1720   ≈1730 (often doublet)

Assignment	Range	Comments
C–O st	1330–1050	2 bands: st as, very strong, at higher wavenumbers; st sy, strong, at lower wavenumbers
C–O st as:	*Subranges:*	
	≈1185	Formates, propionates, higher aliphatic esters
	≈1240	Acetates
	≈1210	Vinyl esters, phenol esters
	≈1180	γ-Lactones, δ-lactones
	≈1165	Methyl esters of aliphatic carboxylic acids
	In the same range:	Strong bands for C–F st, C–N st, N–O st, P–O st, C=S st, S=O st, P=O st, Si–O st, Si–H δ

C = X

Examples (ṽ in cm⁻¹)

H–CO–O–butyl　1730

CH₃–CO–O–butyl　1743

F₃C–CO–O–butyl　1787

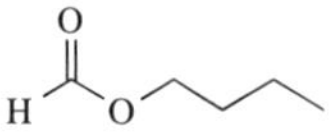

CH₃CH₂–CO–O–CH(Br)CH₃　1747

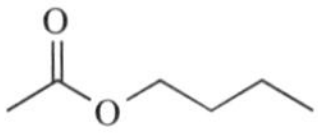

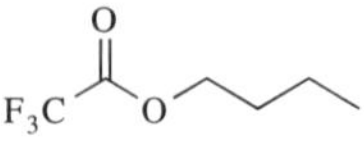

CH₃CH(Br)–CO–O–ethyl　1743

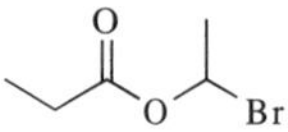

(cis) CH=CH–CO–O–CH₃　1724

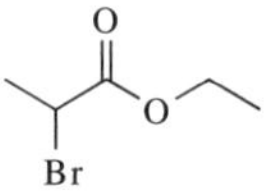

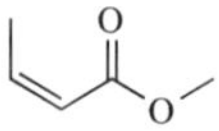

(trans) CH=CH–CO–O–CH₃　1726

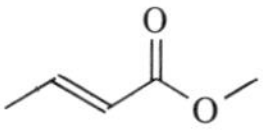

CH₃–CO–O–CH=CH (cis)　1758 (1690)

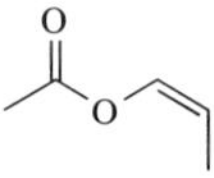

CH₃–CO–O–CH=CH (trans)　1752 (1675)

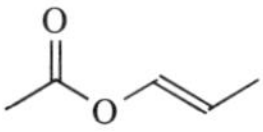

(trans) CH=CH–CO–O–CH=CH₂　1730 (1658) (1638)

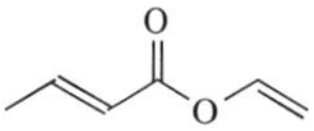

CH₃–CO–O–Si(CH₃)₃　1725

CH₃–CO–CO–O–CH₃　1725

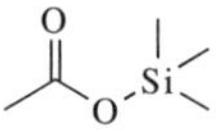

CH₃–CO–CH₂–CO–O–CH₃　ester: 1704 ketone: 1690 enol: 1645

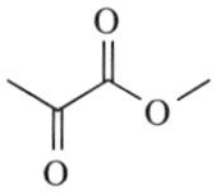

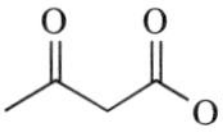

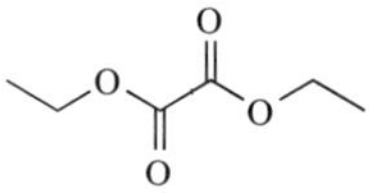

ethyl–O–CO–CO–O–ethyl　1774 1754

CH₂(CO–O–ethyl)₂ (malonate)　1760 1742

ethyl–O–CO–CH₂CH₂–CO–O–ethyl　1740

(cis) ethyl–O–CO–CH=CH–CO–O–ethyl　1734

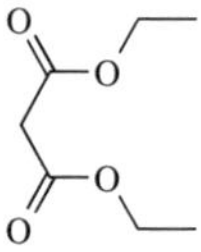

(trans) ethyl–O–CO–CH=CH–CO–O–ethyl　1727

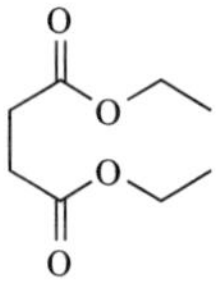

C₆H₅–CO–O–CH₃　1727

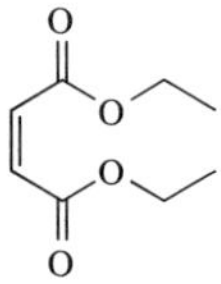

2-HO–C₆H₄–CO–O–CH₃　1684

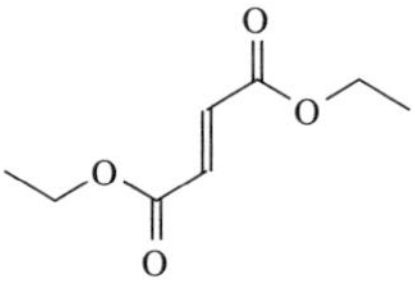

4-(CH₃)₂N–C₆H₄–CO–O–CH₃　1715

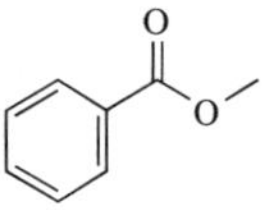

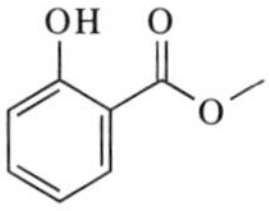

C = X

4-O₂N–C₆H₄–CO–O–CH₃　1737

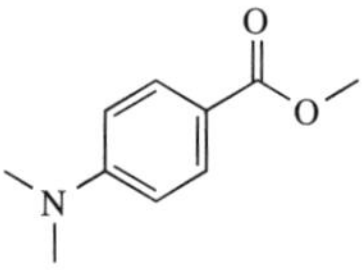

CH₃–CO–O–C₆H₅　1766

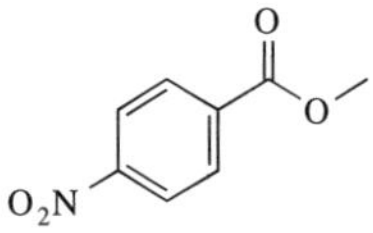

C₆H₅–CO–O–C₆H₅　1743

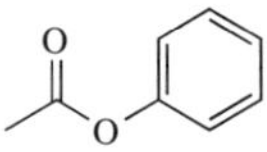

phthalate (1,2-C₆H₄(CO–O–ethyl)₂)　1746

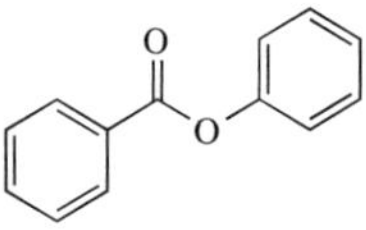

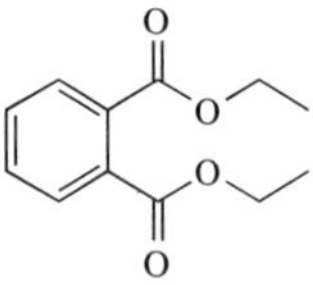

7.11.5 Amides and Lactams

Primary Amides

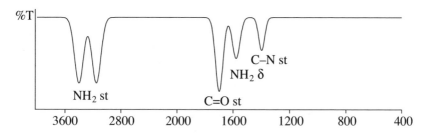

Secondary Amides

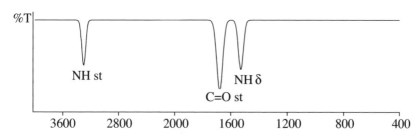

Tertiary Amides

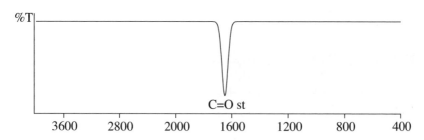

C = X

Typical Ranges ($\tilde{\nu}$ in cm^{-1})

Assignment	Range	Comments
N–H st	3500–3100	Medium, in primary amides two bands, in proteins multiplet
	Subranges:	
	3500–3400	Free
	3350–3100	H-bonded
	≈3350, ≈3180	In primary amides generally two bands
	≈3200, ≈3100	In lactams generally two bands
	≈3200	Monohydrazides

Assignment	Range	Comments
	≈3100	Dihydrazides
	≈3250	Imides
	In the same range: OH st, ≡CH st (≈3300, sharp), H_2O	
C=O st	1740–1630	Generally strong. In Raman, weak to medium
(amide I)	*Subranges:*	
	≈1690	$NH_2C=O$ free amides, H-bonded: ≈1650
	≈1685	$NHC=O$ free amides, H-bonded: ≈1660
	≈1650	$NC=O$ free amides, H-bonded: ≈1650
	≈1745	4-Ring lactams
	≈1700	5-Ring lactams
	≈1650	6-, 7-Ring lactams
	≈1670	Monohydrazides
	≈1600	Dihydrazides
	1740–1670	Imides
	≈1750, 1700	5-Ring imides, 2 bands
	1655–1630	Polypeptides
	≈1690	Isocyanurates; with aromatic substitution: ≈1770
	≈1720, 1755 sh	Trifluoroacetamides
NH δ and	1630–1510	Generally strong, absent in lactams
N–C=O st sy	*Subranges:*	
(amide II)	≈1610	$NH_2C=O$ free, H-bonded: ≈1630
	≈1530	$NHC=O$ free, H-bonded: ≈1540
	1560–1510	Polypeptides
	≈1555	Trifluoroacetamides
C–N st (?)	≈1400	$NH_2C=O$
	≈1250	$NHC=O$
	≈1330	Lactams
NH δ ip	≈1150	$NH_2C=O$
	≈1465	Lactams
NH δ oop	750–600	$NH_2C=O$
	≈700	$NHC=O$
	≈800	Lactams

C = X

Examples ($\tilde{\nu}$ in cm^{-1})

Structure	Values
H–C(=O)–NH–H (formamide)	neat: 1672; in CHCl$_3$: 1709
H–C(=O)–N(H)(–)	neat: 1672
H–C(=O)–N(–)(–)	neat: 1670; in CHCl$_3$: 1673
butanoyl–NH–H	solid: 1631; in CHCl$_3$: 1679
acetyl–NH–CH$_2$CH$_2$CH$_3$	in CCl$_4$: 1690
acetyl–N(propyl)$_2$	in CCl$_4$: 1647
crotonyl–NH–H	solid: 1677
acetyl–N(CH=CH–)(propyl)	in CS$_2$: 1675, 1650
benzamide (Ph–C(=O)–NH$_2$)	solid: 1656; in CHCl$_3$: 1678
acetanilide	solid: 1658; in CHCl$_3$: 1691; in CCl$_4$: 1705
N-ethyl acetanilide	in CCl$_4$: 1667
4-amino-3-penten-2-one	neat: 1700, 1625, 1540
enaminone (NH–CH$_3$)	solid: 1628, 1595
enaminone (N(–))	solid: 1631, 1584
diacetamide (N–H)	solid: 1734, 1505
diacetamide (N–)	solid: 1736, 1706, 1689
succinimide (NH)	solid: 1771, 1698; in CCl$_4$: 1753, 1727
N-methyl succinimide	solid: 1760, 1690; in CCl$_4$: 1721, 1705
N-bromo succinimide	in CHCl$_3$: 1783, 1733
glutarimide (NH)	in CCl$_4$: 1742, 1730, 1718
glutarimide (N–)	solid: 1718, 1670; in CCl$_4$: 1729, 1686
phthalimide (NH)	solid: 1774, 1749, 1724; in CHCl$_3$: 1778, 1735
N-methyl phthalimide	in CHCl$_3$: 1772, 1712
N-hydroxy phthalimide (N–O–)	solid: 1790, 1735

C = X

7.11.6 Acid Anhydrides

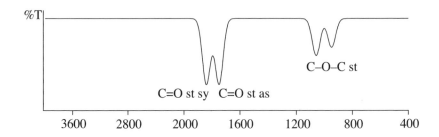

C–O–C st

C=O st sy   C=O st as

| | 3600 | 2800 | 2000 | 1600 | 1200 | 800 | 400 |

Typical Ranges (ṽ in cm⁻¹)

Assignment	Range	Comments
C=O st sy	1870–1770	Strong. In Raman, weak to medium
C=O st as	1800–1720	Strong. In Raman, weak to medium
	Subranges:	
	≈1820, ≈1760	Linear anhydrides, higher band stronger
	≈1850, ≈1775	5-Ring, lower band stronger
	≈1800, ≈1760	6-Ring, lower band stronger
C–O–C st	1300–900	Strong, several bands
	≈1040	Linear anhydrides
	≈920	Cyclic anhydrides

Examples (ṽ in cm⁻¹)

C = X

1825
1748

1810
1740
1045
1040

1803
1743

1780
1725

1790
1727
1035
1015
995

1865
1782
920

Structure	Values	Structure	Values	Structure	Values
	1859 1789		1845 1780		1850 1800 900
	1840 1810 1760 912		1802 1761		1802 1761

7.11.7 Acid Halides

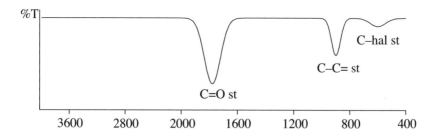

Typical Ranges (ṽ in cm⁻¹) — *Typical Ranges ($\tilde{v}$ in cm^{-1})*

Assignment	Range	Comments
C=O st	1820–1750	Chlorides, strong; in Raman, weak to medium. Of narrow or medium width, for bromides and iodides at lower wavenumber
	1900–1870	Fluorides, strong, of narrow or medium width, additional band at ≈1725 in aromatic acid chlorides and bromides
C–CO st	1000–800	1000–900 aliphatic, assignment uncertain 900–800 aromatic, assignment uncertain
C–hal st	1200–500	1200–800 F 750–550 Cl 700–500 Br 600–500 I

C = X

7.11.8 Carbonic Acid Derivatives

Carbonic Acid Esters

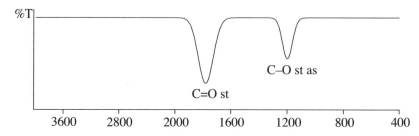

Carbamates

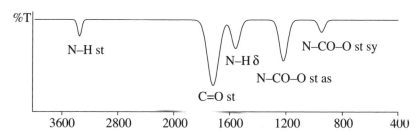

Ureas

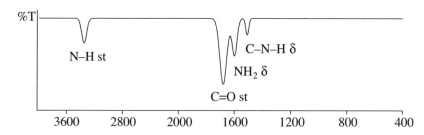

$C = X$

Typical Ranges ($\tilde{v}$ in cm⁻¹)

Assignment	Range	Comments	
C=O st	1820–1740	Strong. In Raman, weak to medium	Carbonic acid esters
	1750–1680	Strong. In Raman, weak to medium	Carbamates
	1690–1620	Strong. In Raman, weak to medium	Ureas
C–O st as	1260–1150	Strong	Carbonic acid esters

Assignment	Range	Comments	
N–H st	3500–3250	Medium, two bands for NH$_2$, one for NH	Carbamates
	3500–3200	Medium, two bands for NH$_2$	Ureas
N–H δ	1650–1500	Medium	Carbamates
NH$_2$ δ	1650–1600	Medium	Ureas
N–CO–O st as	1270–1210	Medium	Carbamates
N–CO–O st sy	1050–850	Weak	Carbamates
C–N–H δ	1600–1500	Weak	Ureas

Examples ($\tilde{v}$ in cm^{-1})

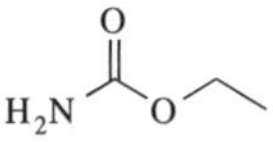

in CHCl$_3$: 1725

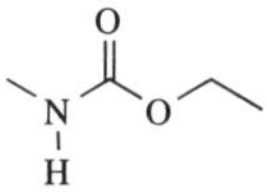

in CCl$_4$: 1727

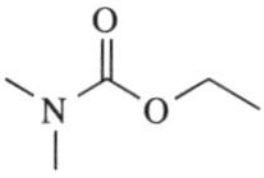

in CHCl$_3$: 1684

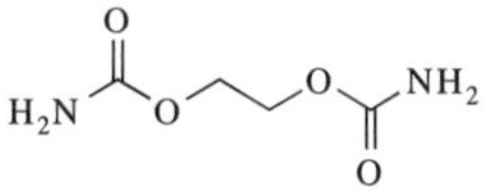

solid: 1690

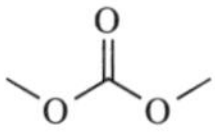

in CCl$_4$: 1758
in CHCl$_3$: 1751

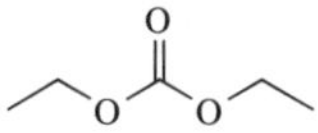

in CCl$_4$: 1748

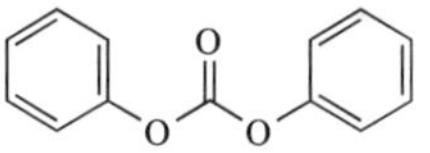

in CCl$_4$: 1786

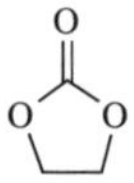

in CCl$_4$: 1822
1748

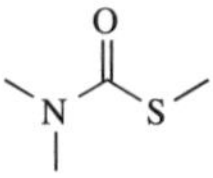

in CCl$_4$: 1662

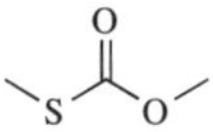

in CCl$_4$: 1719

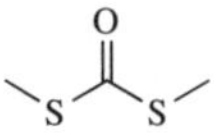

in CCl$_4$: 1653

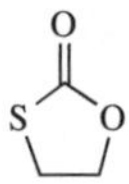

in CCl$_4$: 1757

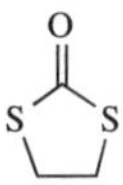

in CCl$_4$: 1718
1677
1640

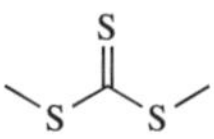

neat: 1076

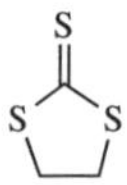

solid: 1058
in CCl$_4$: 1083
1079

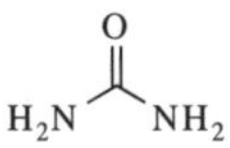

C = X

solid: 1679
1627

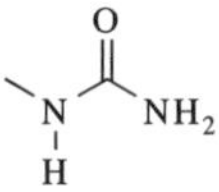

solid: 1645
1567
1418

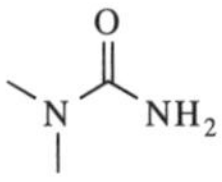

solid: 1656
1610
1511

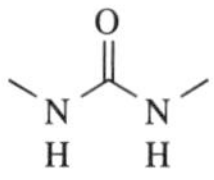

solid: 1622
1580
1530
in CHCl$_3$:
1663
1548

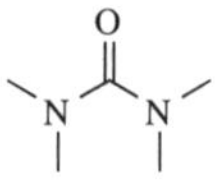

solid: 1645
1560
1497
in CHCl$_3$:
1675

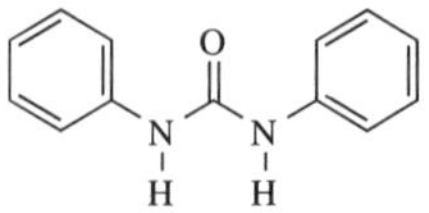

solid:
1650

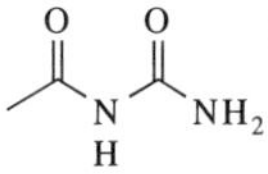

solid:
1667
1634

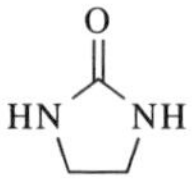

in CCl$_4$: 1735
1718

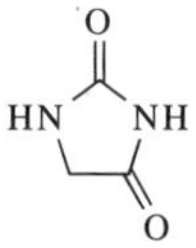

solid: 1776
1697

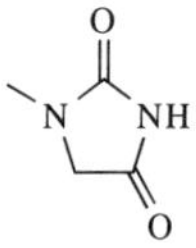

solid:
1712
1676

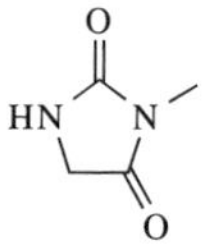

solid: 1748
1706

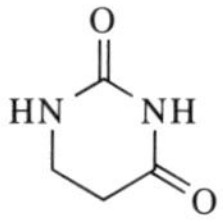

solid: 1767
1695

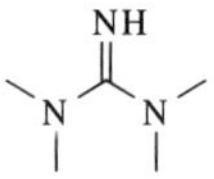

neat:
1600

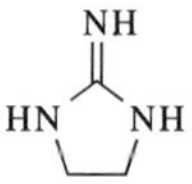

solid: 1767
1681
1621

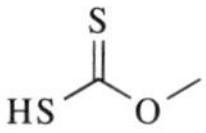

gas: 2593
2548
neat: 2470

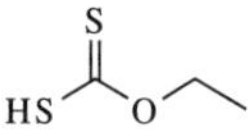

in CS$_2$:
2562
2522

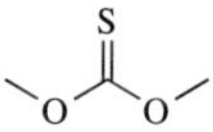

solid: 1212

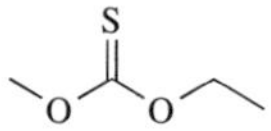

solid: 1234

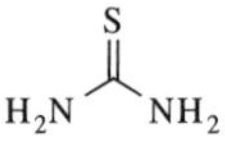

solid:
1400

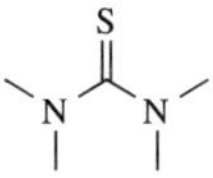

solid: 1130

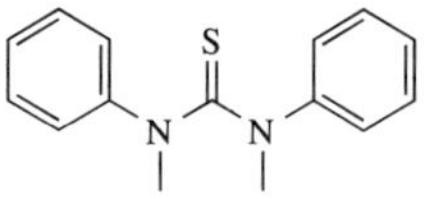

solid:
1131

C = X

7.12 Miscellaneous Compounds

7.12.1 Silicon Compounds

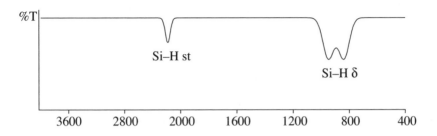

Typical Ranges ($\tilde{\nu}$ in cm^{-1})

Assignment	Range	Comments
Si–H st	2250–2090	Medium. In Raman, medium to strong
	Subranges:	
	2160–2090	R_3Si–H; also for R as H; for SiH$_3$: 2 bands
	≈2250	hal–Si–H
	2220–2120	(Si–O)Si–H
Si–H δ	1010–700	Strong, broad, generally 2 bands
(Si–)CH$_3$ δ as	≈1410	Weak
(Si–)CH$_3$ δ sy	1275–1260	Very strong, sharp, typical for SiCH$_3$, not split for Si(CH$_3$)$_2$
(Si–)CH$_3$ γ	860–760	
	≈765	SiCH$_3$
	≈855, ≈800	Si(CH$_3$)$_2$
	≈840, ≈765	Si(CH$_3$)$_3$
Si–O st	1110–1000, 900– <600	
	1110–1000, 850–800	Si–O–C
	1090–1030, < 650	Si–O–Si
	900–800	Si–OH
	3700–3200	Si–OH st
	≈1030	Si–OH δ
Si–C st	850–650	
Si–N st	1250–830	
	Subranges:	
	950–830	Si–N–Si
	≈3400	Si$_2$NH st

P Si

Assignment	Range	Comments
	950–830	N–Si–N
	1250–1100	Si–NH$_2$
	≈3570, ≈3390	SiN–H$_2$ st
	≈1540	Si–NH$_2$ δ
Si–F st	980–820	
	Subranges:	
	920–820	Si–F
	945–870	SiF$_2$, 2 bands
	980–860	SiF$_3$, 2 bands
Si–Cl st	< 625	

7.12.2 Phosphorus Compounds

Phosphorus Compounds

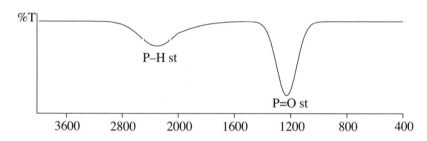

Phosphines

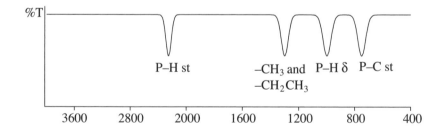

P Si

Typical Ranges ($\tilde{v}$ in cm^{-1})

Assignment	Range	Comments
P–H st	2440–2275	Weak to medium, generally one band, in R_3PH^+ very broad. In Raman, weak to medium
PO–H st	2700–2650	Weak, very broad
POH comb	2300–2250	Weak, very broad
	1740–1600	Additional band in O=P–OH (dimer?)
P–O st	1260–855	
	Subranges:	
	1050–970, 830–740	P–O–C al st; strong for upper band, often weak for lower band
	1260–1160	P–O–C ar st
	995–915	P(V)
	875–855	P(III)
	1100–940	P–OH st, broad, for $P(OH)_2$ often two bands
	980–900	P–O–P st
P=O st	1300–960	Strong. In Raman, weak to medium
	Subranges:	
	1190–1150	$R_3P=O$, also for R: H
	1265–1200	$R_2(R'O)P=O$, also for R: H
	1280–1240	$R(R'O)_2P=O$, also for R: H
	1300–1260	$(RO)_3P=O$
	1220–1150	$R(HO)_2P=O$
	1250–990	$R(HO)PO_2^-$, more than one band
	1125–970, 1000–960	RPO_3^{2-}
	1205–1090	$R_2(HO)P=O$
	1200–1090, 1090–995	$R_2PO_2^-$
	≈1250	$RO(HO)_2P=O$
	1230–1210, 1030–1020	$RO(HO)PO_2^-$
	1140–1050, 1010–970	$ROPO_3^{2-}$
	1250–1210	$(RO)_2(HO)P=O$
	1285–1120, 1120–1050	$(RO)_2PO_2^-$
	1220–1170	$R(RO)(HO)P=O$
	1245–1150, 1110–1050	$R(RO)PO_2^-$

P Si

Assignment	Range	Comments
	1240–1205	$R_2P(=O)\text{-}O\text{-}P(=O)R_2$
	1310–1260	$(RO)_2P(=O)\text{-}O\text{-}P(=O)(OR)_2$
	≈1195	$(HO)(R)P(=O)\text{-}O\text{-}P(=O)(OR)(OH)$
	≈1275	$(RO)(R_2N)P(=O)\text{-}O\text{-}P(=O)(OR)(NR_2)$
	1265–1250	$(R)(RO)P(=O)\text{-}O\text{-}P(=O)(R)(OR)$
	≈1300, ≈1240	$(RO)_2P(=O)\text{-}O\text{-}P(=O)(NR_2)_2$
	≈1250	$(RO)(HO)P(=O)\text{-}O\text{-}P(=O)(OR)(OH)$
	≈1235	$(R_2N)_2P(=O)\text{-}O\text{-}P(=O)(NR_2)_2$
	1265–1240	$R_2(X)P=O$, X: F, Cl, Br
	1365–1260	$R(X)_2P=O$, X: F, Cl, Br
	1330–1280	$(RO)_2(X)P=O$, X: F, Cl, Br
	1365–1260	$RO(X)_2P=O$, X: F, Cl, Br
P=N	1500–1170	
P–OH δ	≈1280	Weak, of no practical significance
P–C st	800–700	Intensity varies widely, of no practical significance
P–H δ	1090–910	Strong, for $(RO)_2HP=O$ very strong
P–N–C st	1110–930, 770–680	
P=N–al st	1500–1230	
P=N–ar st	1390–1300	
P=N–C=O st	1370–1310	
P=N–PR$_2$ st	1295–1170	
P=S st	750–580	Intensity varies widely
P–S st	<600	
(P–)CH$_3$ δ sy	1310–1280	
P–F st	905–760	

Assignment	Range	Comments
PF$_2$	1110–800	More than one band
P–Cl st	<600	

7.12.3 Boron Compounds

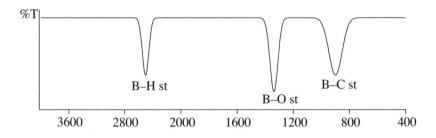

Typical Ranges ($\tilde{v}$ in cm^{-1})

Assignment	Range	Comments
B–H st	2640–2200	Strong, in Raman weak to medium
	2200–1540	B–H···B, more than one band
B–O st	1380–1310	Very strong
	≈1500	Haloboroxines
BO–H st	3300–3200	Very broad
B–N st	1550–1330	Very strong
B–C st	1240–620	Strong, 2 bands if substitution highly asymmetric
B–F st	1500–800	
B–Cl st	1100–650	

P Si

7.13 Amino Acids

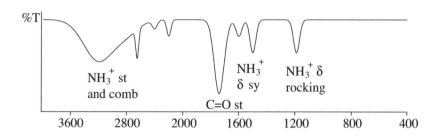

Typical Ranges ($\tilde{v}$ in cm^{-1})

Assignment	Range	Comments
N–H st **O–H st**	3400–2000	Generally strong, broad, very structured
	Subranges:	
	3100–2000	Zwitterions, distinct side band at 2200–2000
	3350–2000	Hydrochlorides
	3400–3200	Na$^+$ salts
NH$_3^+$ δ as	1660–1590	Weak, for hydrochlorides near the lower limit
NH$_3^+$ δ sy	1550–1480	Medium
COO$^-$ st as	1760–1595	Strong
	Subranges:	
	≈1595	Zwitterions
	1755–1700	Hydrochlorides; in α-amino acids: 1760–1730
	≈1595	Na$^+$ salts

7.14   Solvents, Suspension Media, and Interferences

7.14.1   Infrared Spectra of Common Solvents

The low transmission in regions where the solvent absorbs may lead to artifacts. For the interpretation of spectra, these regions should be disregarded. In the following, they are indicated by bars.

Chloroform: 0.2 mm cell

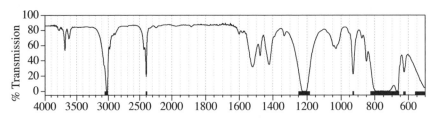

Chloroform: 1 mm cell

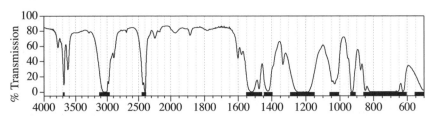

Carbon tetrachloride: 0.2 mm cell

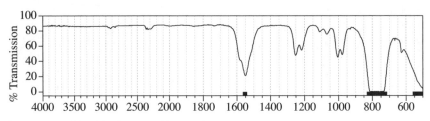

Carbon tetrachloride: 1 mm cell

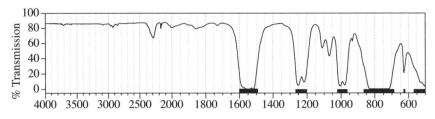

Solvents

Carbon disulfide: 0.2 mm cell

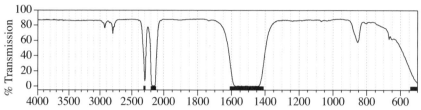

Carbon disulfide: 1 mm cell

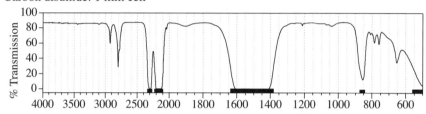

7.14.2 Infrared Spectra of Suspension Media

As it is difficult to prepare pellets and thin mineral oil films of reproducible thickness, the bands of these suspension matrixes are always found superimposed on the sample spectra.

Mineral oil (nujol): 10 μm thickness

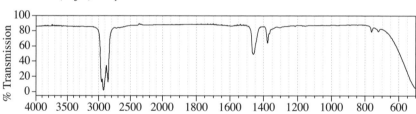

Potassium bromide: pellet

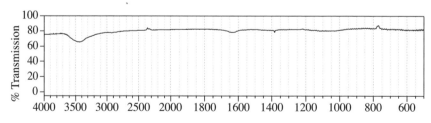

Solvents

7.14.3   Interferences in Infrared Spectra

Traces of water in carbon tetrachloride or chloroform may give rise to two bands in the vicinity of 3700 and 3600 cm^{-1} as well as one around 1600 cm^{-1}. At higher concentrations, a broad band at 3450 cm^{-1} is found. Water in the vapor phase exhibits many sharp bands between 2000 and 1280 cm^{-1}. If present in high concentration, they may temporarily block the detector and appear as shoulders when occurring at a steep side of a strong signal.

Dissolved carbon dioxide shows an absorption band at 2325 cm^{-1}. In solutions that contain amines and traces of water, CO_2 can form carbonates, which lead to the appearance of unexpected bands of protonated N-containing groups. In improperly balanced double beam instruments, gaseous CO_2 can give rise to two signals at approximately 2360 and 2335 cm^{-1} as well as a signal at 667 cm^{-1}.

Chloroform, saturated with water: 0.2 mm cell

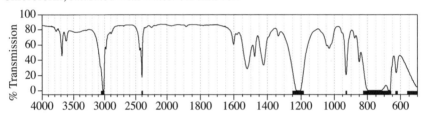

Water vapor with carbon dioxide

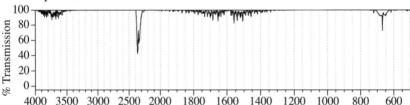

Commercially available polymers often contain phthalates as plasticizers, which can be found in apparently pure samples and give rise to a band at 1725 cm^{-1}. The presence of such phthalates can be confirmed by MS (m/z 149). In the course of chemical reactions, phthalates may be transformed into phthalic anhydride, which shows a band at 1755 cm^{-1}.

Other frequently encountered contaminants are silicones, which generally exhibit a band at 1625 cm^{-1}, together with a broad signal in the region from 1100 to 1000 cm^{-1}.

Solvents

8 Mass Spectrometry

8.1 Alkanes [1]

Unbranched Alkanes [2,3]

Fragmentation: Larger alkyl fragments (with $C_{n>4}$) are chiefly formed by direct cleavage. They dehydrogenate and undergo substantial H and skeleton rearrangements. Smaller alkyl fragments (C_2 to C_4) are mainly formed by secondary decomposition of higher alkyl fragments. Eliminations of groups from within the chain (and recombination of its ends) also occur.

Ion series: Consecutive peaks corresponding to C_nH_{2n+1} (m/z 29, 43, 57, 71, ...), accompanied by C_nH_{2n-1} (m/z 27, 41, 55, 69, ...) and C_nH_{2n} (m/z 28, 42, 56, 70, ...) of lower intensity.

Intensities: Maximum intensity at m/z 43 or 57; with increasing masses, intensity of local maxima smoothly decreasing to a minimum at [M-15]⁺.

Molecular ion: Medium intensity.

Branched Alkanes [2,3]

Fragmentation: In most cases, apparently simple bond cleavages, preferably at branched C atoms. The positive charge remains mainly on the branched C atom. Mechanistically, many H and skeleton rearrangements take place. This is reflected by the fact that no specific localization of heavy isotopes is possible.

$$\left[\begin{array}{c} R^1 \\ {}^{\diagdown}C{=}CHR^2 \\ H \end{array} \right]^{+\cdot} \xleftarrow[-R^3H]{} \left[\begin{array}{c} R^1 \\ {}^{\diagdown}CH{-}CH_2R^2 \\ R^3 \end{array} \right]^{+\cdot} \xrightarrow[-R^3]{} \left[R_1{-}CH{-}CH_2R^2 \right]^{+}$$

Ion series: Consecutive peaks corresponding to C_nH_{2n+1} (m/z 29, 43, 57, 71, ...), accompanied by C_nH_{2n-1} (m/z 27, 41, 55, 69, ...) and C_nH_{2n} (m/z 28, 42, 56, 70, ...) of lower intensity.

Intensities: Local intensity maxima at those masses that result from cleavage at branched C atoms if the charge is localized there. Both C_nH_{2n+1} and (often more characteristically) C_nH_{2n} show this tendency.

Molecular ion: Intensity decreasing with increasing degree of branching. No M⁺˙ is observed in highly branched systems.

References

[1] J.T. Bursey, M.M. Bursey, D.G. Kingston, Intramolecular hydrogen transfer in mass spectra. 1. Rearrangements in aliphatic hydrocarbons and aromatic compounds, *Chem. Rev.* **1973**, *73*, 191.

[2] K. Levsen, H. Heimbach, G.J. Shaw, G.W.A. Milne, Isomerization of hydrocarbon ions. VIII. The electron impact induced decomposition of *n*-dodecane, *Org. Mass Spectrom.* **1977**, *12*, 663.

[3] A. Lavanchy, R. Houriet, T. Gäumann, The mass spectrometric fragmentation of *n*-alkanes, *Org. Mass Spectrom.* **1979**, *14*, 79.

8.2 Alkenes [1–4]

Unbranched Alkenes

Fragmentation: Dominant loss of alkyl residues and neutral alkenes. The position of highly substituted double bonds can be localized because in this case alkene eliminations are specific McLafferty-type reactions. Otherwise, double bonds can be localized in derivatives, such as epoxides and glycols, or by means of low energy ionization techniques. Branching effects are less characteristic than in isoalkanes. Alicyclic compounds exhibit very similar spectra.

$C = C$

Ion series: Consecutive peaks corresponding to C_nH_{2n-1} (m/z 41, 55, 69, 83, ...), accompanied by alkyl and alkene ions, C_nH_{2n+1} (m/z 43, 57, 71, 85, ...) and C_nH_{2n} (m/z 42, 56, 70, 84, ...), mostly of lower intensity.
Intensities: Dominant maxima in the lower mass range, peaking around C_4. Local even-mass maxima due to alkene eliminations if the double bond is highly substituted.
Molecular ion: Significant, but not necessarily strong.

Branched Alkenes

Fragmentation: Highly substituted double bonds are less easily displaced than the unsubstituted ones and give rise to specific alkene eliminations of the McLafferty type, resulting in significant local maxima corresponding to C_nH_{2n} (see scheme). The latter may allow to localize the double bond. With unsubstituted double bonds, no reliable localization is possible and only moderately useful branching effects can be observed. The branching position is more easily determined after reduction to an alkane (in situ in GC/MS with H_2 as carrier gas and heated Pt wool as catalyst).

Ion series: Maxima of the alkene type (C_nH_{2n-1}; m/z 41, 55, 69, 83, ...), accompanied by weaker alkyl fragments, C_nH_{2n+1} (m/z 43, 57, 71, 85, ...), in the low mass range and more significant alkene ions, C_nH_{2n} (m/z 42, 56, 70, 84, ...).
Intensities: Intensive peaks in the lower mass range. Diagnostically important local maxima of even mass, frequently also in the higher mass range.
Molecular ion: Usually significant.

Polyenes and Polyynes

Fragmentation: The spectra of aliphatic compounds with several triple and/or double bonds are similar to those of aromatic hydrocarbons. A characteristic difference in the case of polyenes and polyynes is the presence of a signal at m/z 27, which is absent from spectra of purely aromatic compounds.
Ion series: Very similar to those of aromatic hydrocarbons, but fragments with higher hydrogen contents than in aromatics (m/z 54, 55; 66, 67; 79, 80) are usually found in polyenes and polyynes.

Intensities: Very similar distribution of peak intensities as for aromatic hydrocarbons.

Molecular ion: Usually strong, as in aromatic hydrocarbons.

References

C = C

[1] A.G. Loudon, A. Maccoll, The mass spectrometry of the double bond. In: *The Chemistry of Alkenes, Vol. 2*; J. Zabicky, Ed.; Interscience: London, 1970; p 327.

[2] J.T. Bursey, M.M. Bursey, D.G. Kingston, Intramolecular hydrogen transfer in mass spectra. 1. Rearrangements in aliphatic hydrocarbons and aromatic compounds, *Chem. Rev.* **1973**, *73*, 191.

[3] N.J. Jensen, M.L. Gross, Localization of double bonds. *Mass Spectrom. Rev.* **1987**, *6*, 497.

[4] C. Dass, Ion–molecule reactions of [ketene]$^{+\cdot}$ as a diagnostic probe for distinguishing isomeric alkenes, alkynes, and dienes: A study of the C_4H_8 and C_5H_8 isomeric hydrocarbons, *Org. Mass Spectrom.* **1993**, *28*, 940.

8.3 Alkynes [1]

Aliphatic Alkynes

Fragmentation: Tendency to lose a non-acetylenic H· from M$^{+·}$. Extensive rearrangements (including consecutive McLafferty rearrangements to the triple bond) result in uncharacteristic degradation:

C≡C

(base peak for 5-decyne)

In nonbranched alkynes with C$_{n>8}$, the rearrangement products at m/z 82 and 96 are dominant. Consecutive loss of methyl radical occurs. In general, no reliable localization of the triple bond is possible except in derivatives (as in ethylene glycol adducts [1], see scheme).

Ion series: Prominent peaks for C$_n$H$_{2n-3}$ (m/z 25, 39, 53, 67, 81, ...), accompanied by C$_n$H$_{2n-1}$ (m/z 41, 55, 69, 83, ...) and alkyl ions C$_n$H$_{2n+1}$ (m/z 43, 57, 71, 85, ...). Occasionally, even-mass maxima for C$_n$H$_{2n-2}$ (m/z 26, 40, 54, 68, 82, ...).
Intensities: Intensive peaks mainly in the lower mass range.
Molecular ion: Weak or missing in spectra of smaller molecules, significant in those of larger ones. Generally, [M-1]$^+$ is present. In terminal acetylenes, it is normally more abundant than M$^{+·}$.

References

[1] C. Lifshitz, A. Mandelbaum, Mass spectrometry of acetylenes. In: *The Chemistry of the Carbon-Carbon Triple Bond, Part 1*; S. Patai, Ed.; Wiley: Chichester, 1978; p 157.

8.4 Alicyclics [1]

Cyclopropanes [2,3]

Fragmentation: Generally, spectra of cyclopropanes and alkenes are very similar because at 70 eV ionization, the ring readily isomerizes to the corresponding alkene radical cations.

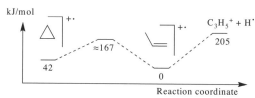

Preferred primary fragmentation by bond cleavage at branched C atoms. Loss of alkyl residues and of neutral alkenes dominates. The ring of monosubstituted cyclopropanes is opened exclusively at the 1,2- and not at the 2,3-bond. The primarily formed double bond is predominantly (for R: OCH_3) or exclusively (for R: H, alk, $COOCH_3$) found in the β,γ-position (even for $COOCH_3$, where the α,β-unsaturation is thermodynamically more stable).

Molecular ions of cyclopropyl cyanide, allyl cyanide, methacrylonitrile, and pyrrole rearrange to one common radical cation, most likely that of pyrrole [4].
Ion series: Consecutive maxima corresponding to C_nH_{2n-1} (m/z 41, 55, 69, 83, ...), accompanied by alkyl and alkenyl ions of the type C_nH_{2n+1} (m/z 43, 57, 71, 85, ...) and C_nH_{2n} (m/z 42, 56, 70, 84, ...), mostly of lower intensity.
Intensities: Dominant peaks in the low mass range, peaking around C_4. Local even-mass maxima due to alkene eliminations if the resulting double bond is highly substituted.
Molecular ion: Significant, but not necessarily strong.

Saturated Monocyclic Alicyclics [5]

Fragmentation: Preferred primary fragmentation by bond cleavage at branched C atoms, followed by loss of alkyl residues and alkenes.
Ion series: Consecutive maxima corresponding to C_nH_{2n-1} (m/z 41, 55, 69, 83, ...), accompanied by C_nH_{2n+1} (m/z 43, 57, 71, 85, ...) and C_nH_{2n} (m/z 42, 56, 70, 84, ...) of lower intensities. In general, the maxima are so similar to those of alkenes that no clear distinction is possible.
Intensities: Overall distribution of peaks maximizing in the lower mass range, around C_4 or C_5. Local maxima can result from branching effects.
Molecular ion: Significant, mostly of medium intensity.

Polycyclic Alicyclics

Fragmentation: Most important primary cleavage at highly branched carbon atoms, followed by H rearrangements and complex fragmentations.

Ion series: With increasing number of rings, the position of unsaturated hydrocarbon fragments in the upper m/z range shifts from C_nH_{2n-1} (m/z 41, 55, 69, 83, ...) to C_nH_{2n-3} (m/z 39, 53, 67, 81, ...) and to C_nH_{2n-5} (m/z 51, 65, 79, 93, ...). Typically, maxima in the lower m/z range have a lower degree of unsaturation than those in the upper m/z range.

Intensities: Major maxima evenly distributed, somewhat more intensive in the high mass or $M^{+\cdot}$ range.

Molecular ion: Strong.

Cyclohexenes

Fragmentation: Loss of larger ring substituents as well as retro-Diels–Alder reaction, yielding fragments of even-mass maxima with one or two double-bond equivalents, C_nH_{2n} (m/z 42, 56, 70, 84, ...) and C_nH_{2n-2} (m/z 40, 54, 68, 82, ...), unless the retro-Diels–Alder product corresponds to ethylene. Somewhat unexpectedly, the base peak of cyclohexene is at $[M-15]^+$.

The retro-Diels–Alder reaction often accounts for prominent fragments of cyclohexenes and 1,4-cyclohexadienes:

$$\left[\begin{array}{c} R^2 \\ R^1 \quad R^3 \end{array} \right]^{+\cdot} \longrightarrow \left[\begin{array}{c} \\ R^1 \end{array} + \begin{array}{c} R^2 \\ R^3 \end{array} \right]^{+\cdot}$$

However, double-bond migration may or may not occur beforehand. Also, other fragmentation pathways may dominate. Therefore, a reliable localization of the double bond in cyclohexene derivatives of unknown structure is not necessarily possible. For example, the base peak of 1,2-dimethylcyclohexene is at m/z 68 rather than at the expected m/z 82.

Ion series: Unsaturated hydrocarbon fragments in the upper m/z range are shifted, relative to cyclohexane fragments, by two mass units to C_nH_{2n-3} (m/z 39, 53, 67, 81, ...). Typically, maxima in the lower m/z range correspond to a lower degree of unsaturation than those in the upper m/z range.

Intensities: Intensive peaks evenly distributed over whole mass range.

Molecular ion: Medium intensity (ca. 40% in cyclohexene).

References

[1] J.T. Bursey, M.M. Bursey, D.G. Kingston, Intramolecular hydrogen transfer in mass spectra. 1. Rearrangements in aliphatic hydrocarbons and aromatic compounds, *Chem. Rev.* **1973**, *73*, 191.

[2] H. Schwarz, The chemistry of ionized cyclopropanes in the gas phase. In: *The Chemistry of the Cyclopropyl Group, Part 1*; Z. Rappoport, Ed.; Wiley: Chichester, 1987; p 173.

[3] J.R. Collins, G.A. Gallup, Energy surfaces in the cyclopropane radical ion and the photoelectron spectrum of cyclopropane, *J. Am. Chem. Soc.* **1982**, 104, 1530.

[4] G.D. Willet, T. Baer, Thermochemistry and dissociation dynamics of state-selected C_4H_4X ions. 3. $C_4H_5N^+$, *J. Am. Chem. Soc.* **1980**, *102*, 6774.

[5] E.F.H. Brittain, C.H.J. Wells, H.M. Paisley, Mass spectra of cyclobutanes and cyclohexanes of molecular formula $C_{10}H_{16}$, *J. Chem. Soc. B* **1968**, 304.

8.5 Aromatic Hydrocarbons [1–4]

Aromatic Hydrocarbons

Fragmentation: Weak tendency of fragmentation. Elimination of H˙ and successive H_2 eliminations, yielding $[M-1]^+$, $[M-3]^+$, and $[M-5]^+$ of decreasing intensities. In condensed aromatics, $[M-2]^{+\cdot}$ can be a dominating fragment. Further typical fragmentation reactions are the eliminations of acetylene (Δm 26) and C_3H_3 (Δm 39). Some CH_3 elimination frequently occurs in pure aromatic compounds. In the case of diphenyl compounds, biphenylene (m/z 152) and, if a CH_2 group is available, fluorene (m/z 165) ions are typically observed.

m/z 152   m/z 165

Ion series: C_nH_n and $C_nH_{n\pm1}$ (m/z 39, 51–53, 63–65, 75–77, …), for polycyclic aromatics gradually changing to more highly unsaturated ions. Doubly charged ions occur frequently, in particular as the size of the π-electron system increases.
Intensities: Weak fragments. The intensity pattern of doubly charged ions does not follow that of the corresponding singly charged ions.
Molecular ion: Strong.

Alkylsubstituted Aromatic Hydrocarbons

Fragmentation: Dominant loss of alkyl residues by benzylic cleavage, followed by elimination of alkenes.

At low resolution, methylbenzyl and β-phenylethyl have the same mass as benzoyl (m/z 105). In contrast to benzoyl, dehydrogenation products (m/z 104, 103) as well as protonated benzene (m/z 79) are also present if m/z 105 is a hydrocarbon rest.
Ion series: Aromatic hydrocarbon fragments, C_nH_n and $C_nH_{n\pm1}$ (m/z 39, 51–53, 63–65, 75–77, …), in the lower mass range.
Intensities: Intensive peaks mainly in the higher mass range. Maxima by benzylic cleavage.
Molecular ion: Strong or medium.

References

[1] J.T. Bursey, M.M. Bursey, D.G. Kingston, Intramolecular hydrogen transfer in mass spectra. 1. Rearrangements in aliphatic hydrocarbons and aromatic compounds, *Chem. Rev.* **1973**, *73*, 191.
[2] W. Schönfeld, Fragmentation diagrams for elucidation of decomposition reactions of organic compounds. 1. Aromatic hydrocarbons (in German), *Org. Mass Spectrom.* **1975**, *10*, 321.
[3] C. Lifshitz, Tropylium ion formation from toluene: Solution of an old problem in organic mass spectrometry. *Acc. Chem. Res.* **1994**, *27*, 138.
[4] M.V. Buchanan, B. Olerich, Differentiation of polycyclic aromatic hydrocarbons using electron-capture negative chemical ionization, *Org. Mass Spectrom.* **1984**, *19*, 486.

8.6 Heteroaromatic Compounds [1,2]

General Characteristics

Fragmentation: Mostly fragments of aromatic character with specific eliminations including heteroatoms, e.g., elimination of HCN, CO, CHO, CS, and CHS from $M^{+\cdot}$, and of HCN, CO, and CS from fragments. In the case of alkyl-substituted heteroaromatics, occurrence of benzylic-type cleavage and McLafferty rearrangements of substituents with $C_{n>1}$ as well as specific rearrangements including heteroatoms, especially in N aromatics.
Ion series: Aromatic fragments, C_nH_n and $C_nH_{n\pm1}$ (m/z 39, 51–53, 63–65, …), in the lower mass range if the necessary number of C atoms is present (no such fragments, e.g., in pyrazine). Ions including heteroatoms like $HCN^{+\cdot}$ (m/z 27), CH_3CNH^+ (m/z 42), and $CS^{+\cdot}$ (m/z 44).
Intensities: Intensive peaks mainly in the higher mass range.
Molecular ion: Generally strong. $[M-1]^+$ is often relevant in alkyl-substituted heteroaromatics.

Furans [3]

Fragmentation: Oxygen can be lost from $M^{+\cdot}$ together with the neighboring C as CHO (Δm 29). In 2- or 6-methylfurans, CH_3CO^+ (m/z 43) can be seen (base peak in 2,5-dimethylfuran). As in aromatic methyl ethers, $[M-43]^+$ is a product of a two-step reaction: ($M^{+\cdot}-CH_3^\cdot-CO$). Furans substituted with an alkyl group ($C_{n>1}$): benzylic-type cleavage (to pyrylium ion, $C_5H_5O^+$, m/z 81), followed by loss of CO.
Ion series: Mainly aromatic hydrocarbon fragments, C_nH_n and $C_nH_{n\pm1}$ (m/z 39, 51–53, 63–65, …).
Intensities: Intensive peaks mainly in the higher mass range. The fragments are usually more important than in purely aromatic hydrocarbons.
Molecular ion: Strong. No pronounced tendency to protonate. Usually, $[M-1]^+$ is very strong in methylfurans.

Thiophenes [4]

Fragmentation: Sulfur can be lost from $M^{+\cdot}$ together with the neighboring C as CHS (Δm 45) or CS (Δm 44). Typical for thiophenes substituted with an alkyl group ($C_{n>1}$) is benzylic-type cleavage followed by loss of CS (Δm 44). Protonated thiophene (m/z 85) is a characteristic product of monoalkylated thiophenes.
Ion series: Aromatic hydrocarbon fragments, C_nH_n and $C_nH_{n\pm1}$ (m/z 39, 51–53, 63–65, …). Besides the isotope peak at $[M+2]^{+\cdot}$, the signals at m/z 44 and 45 ($CS^{+\cdot}$ and CHS^+) are indicators of sulfur.
Intensities: Dominant peaks for $M^{+\cdot}$ and products of benzylic-type cleavage.
Molecular ion: Strong. Characteristic S isotope signal ($[M+2]^{+\cdot}$ corresponds to 4.5% of $M^{+\cdot}$). No pronounced tendency of protonation. Usually, $[M-1]^+$ is very strong in methylthiophenes.

Pyrroles [5]

Fragmentation: HCN elimination (Δm 27) from M$^{+\cdot}$ and from fragments. In methylpyrroles, [M-1]$^+$ is dominant. Benzylic-type cleavage in *C*- and *N*-alkyl-pyrroles with or without (nonspecific) H rearrangements.
Ion series: Aromatic hydrocarbon fragments, C_nH_n and $C_nH_{n\pm1}$ (m/z 39, 51–53, 63–65, ...).
Intensities: Dominant peaks for M$^{+\cdot}$ and products of benzylic-type cleavage.
Molecular ion: Strong (odd mass for odd number of N in the molecule). No tendency to protonate. In methyl-substituted pyrroles, [M-1]$^+$ is dominant.

Pyridines

Fragmentation: HCN elimination (Δm 27) from fragments and the ion H_2CN^+ (m/z 28) are characteristic. Additional reactions in 2- or 6-methylpyridines are CH_3CN elimination (Δm 41) and the formation of CH_3CNH^+ (m/z 42). Benzylic cleavage is dominant for 3-alkyl-, strong for 4-alkyl-, and weak for 2-alkylpyridines. Typical rearrangements with participation of the N atom in 2- and 6-alkylpyridine derivatives.

Intramolecular *N*-alkylation in 2-alkyl derivatives:

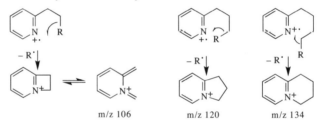

m/z 106   m/z 120   m/z 134

McLafferty rearrangements are important in 2- and 4-alkylpyridines:

Ion series: Aromatic hydrocarbon fragments, C_nH_n, $C_nH_{n\pm1}$ and $C_nH_{n\pm1}N$ (m/z 39–41, 51–54, 63–67, 75–80, ...).
Intensities: Dominant peaks for M$^{+\cdot}$ or, if possible, for products of benzylic-type cleavage.
Molecular ion: Strong, except when benzylic-type cleavage is possible. Odd mass for an odd number of N in the molecule. No tendency to protonate. [M-1]$^+$ is usually present and is strong in alkyl-substituted pyridines.

N-*Oxides of Pyridines and Quinolines*

Fragmentation: The [M-O]$^{+\cdot}$ radical ion, of variable intensity, is probably due to thermal decomposition. The fragments [M-CO]$^{+\cdot}$ and, if an alkyl group is present on the neighboring C atom, [M-OH]$^+$ are relevant for quinoline *N*-oxides. Rearrangements with ring formation including the N−O moiety if alkyl or aryl groups are present in the neighboring positions.

Ion series: As for the corresponding heteroaromatics, aromatic hydrocarbon fragments, C_nH_n, $C_nH_{n\pm1}$ and $C_nH_{n\pm1}N$ (m/z 39–41, 51–54, 63–67, 75–80, ...), are observed.

Intensities: Dominant peaks for $M^{+\cdot}$ and products of benzylic-type cleavage.

Molecular ion: Strong, except when $[M-O]^{+\cdot}$ dominates due to experimental conditions or when benzylic-type cleavage is possible. Odd mass for odd number of N atoms in the molecule. No tendency to protonate.

Pyridazines and Pyrimidines

Fragmentation: Loss of N_2 or $CH_2N^\cdot$ (Δm 28) from pyridazines. Also, loss of $N_2H^\cdot$ (especially important in methylpyridazines) to give $[M-29]^+$. In pyridazine *N*-oxides, consecutive loss of $NO^\cdot$ and HCN. Consecutive losses of two HCN (2 × Δm 27) molecules from pyrimidines. From 2-, 4-, and 6-methylpyrimidines, CH_3CN (Δm 41) is eliminated and the ion CH_3CNH^+ (m/z 42) occurs.

Ion series: Aromatic hydrocarbon fragments (C_nH_n, $C_nH_{n\pm1}$) and, for pyrimidines, $C_nH_{n\pm1}N$, at low masses (m/z 39, 51–53).

Intensities: Dominant peak for $M^{+\cdot}$.

Molecular ion: Strong. No tendency to protonate. For pyrimidines, $[M-1]^+$ is usually observable.

Pyrazines

Fragmentation: Consecutive losses of two HCN (2 × Δm 27) molecules. For methylpyrazines, elimination of CH_3CN (Δm 41) and formation of CH_3CNH^+ (m/z 42).

Ion series: No aromatic character of the spectra.

Intensities: Dominant peak for $M^{+\cdot}$.

Molecular ion: Strong. No tendency to protonate. Usually, $[M-1]^+$ is observable; it can be stronger than $M^{+\cdot}$ in alkyl-substituted ($C_{n>1}$) pyrazines.

Indoles

Fragmentation: Analogous to pyrrole; HCN elimination (Δm 27) from $M^{+\cdot}$ and from fragments. From $M^{+\cdot}$ also $CH_2N^\cdot$ (Δm 28) elimination (in one or two steps). In methyl-substituted indoles, $[M-1]^+$ is dominant. In *N*-methylindoles, $[M-15]^+$ is significant. Benzylic-type cleavage in *C*- and *N*-alkylindoles with or without (non-specific) H rearrangements.

Ion series: Aromatic ion series.

Intensities: Dominant maxima in the higher mass range.

Molecular ion: Strong. No tendency to protonate. In methyl-substituted indoles, strong signal for $[M-1]^+$.

Quinolines and Isoquinolines

Fragmentation: Similar to pyridine: HCN elimination (Δm 27) from $M^{+\cdot}$, $[M-1]^+$, and fragments. In methylquinolines and methylisoquinolines also CH_3CN elimination (Δm 41). In alkyl-substituted ($C_{n>1}$) quinolines, benzylic cleavage dominates except when neighboring effects of N play a role. For 2- and 8-alkylquinolines as well as 1- and 3-alkylisoquinolines, see rearrangements in pyridines.

Ion series: Aromatic hydrocarbon fragments, C_nH_n, $C_nH_{n\pm1}$, and $C_nH_{n\pm1}N$ (m/z 39–41, 51–54, 63–67, 75–80, ...).

Intensities: Dominant peak for $M^{+\cdot}$ or, if possible, for products of benzylic-type cleavage.

Molecular ion: Strong, except when benzylic-type cleavage is possible. Odd mass for odd number of N atoms in the molecule. No tendency to protonate. $[M-1]^+$ is usually present and is strong in alkyl-substituted quinolines.

Rearrangements in 8-alkylquinolines:

Cinnoline, Phthalazine, Quinazoline, Quinoxaline

Fragmentation: Same as for the corresponding monocyclic heteroaromatics pyridazine, pyrimidine, and pyrazine. Characteristic for pyridazine, cinnoline, and phthalazine is the elimination of N_2 (Δm 28) and $N_2H^\cdot$ (Δm 29) from their alkyl derivatives. Phthalazine loses HCN (Δm 27) twice.

Ion series: Aromatic hydrocarbon fragments, (C_nH_n, $C_nH_{n\pm1}$) and $C_nH_{n\pm1}N$ (m/z 39–41, 51–54, 63–67, 75–80, ...).

Intensities: Dominant maximum for $M^{+\cdot}$ or, if possible, for products of benzylic-type cleavage.

Molecular ion: Strong, except when benzylic-type cleavage is possible. Odd mass for odd number of N atoms in the molecule. No tendency to protonate. $[M-1]^+$ is usually present and is strong in alkyl-substituted compounds.

References

[1] Q.N. Porter, *Mass Spectrometry of Heterocyclic Compounds*, 2nd ed.; Wiley: New York, 1985.
[2] D.G.I. Kingston, B.W. Hobrock, M.M. Bursey, J.T. Bursey, Intramolecular hydrogen transfer in mass spectra. III. Rearrangements involving the loss of small neutral molecules, *Chem. Rev.* **1975**, *75*, 693.

[3] R. Spilker, H.-F. Grützmacher, Isomerization and fragmentation of methylfuran ions and pyran ions in the gas phase, *Org. Mass Spectrom.* **1986**, *21*, 459.

[4] W. Riepe, M. Zander, Mass-spectrometric fragmentation behavior of thiophene benzologs. *Org. Mass Spectrom.* **1979**, *14*, 455.

[5] H. Budzikiewicz, C. Djerassi, A.H. Jackson, G.W. Kenner, D.J. Newmann, J.M. Wilson, Mass spectra of monocyclic derivatives of pyrrole, *J. Chem. Soc.* **1964**, 1949.

8.7 Halogen Compounds [1–3]

Saturated Aliphatic Halides

Fragmentation: Loss of halogen radical (I > Br > Cl > F) followed by elimination of alkenes. Loss of alkyl radical followed by elimination of acid HX. Loss of acid HX to give an alkene radical cation.

Ion series: The dominant hydrocarbon fragments are mainly alkenyl fragments (C_nH_{2n-1}) for F and Cl, mixed alkyl (C_nH_{2n+1}) and alkenyl fragments (C_nH_{2n-1}) for Br, and mainly alkyl fragments (C_nH_{2n+1}) for I.

Intensities: Intensive peaks mainly in the lower mass range. Characteristic maxima for Cl and Br at $C_4H_8X^+$ (m/z 91, 93 and 135, 137, respectively), which has a cyclic structure:

Alkyl substituents on the chain reduce the intensity of this fragment. If it is strong, $[M-X]^+$ is weak. In the case of iodoalkanes, some I^+ and $HI^{+\cdot}$ at m/z 127, 128 is usually detectable.

Molecular ion: Strong for the smallest alkanes, with increasing intensity in the sequence F, Cl, Br, I. Decreases rapidly with increasing mass and with increasing branching. It is negligible for F and Cl if the *n*-alkyl chains are longer than pentyl, and for Br and I if they are longer than heptyl and nonyl, respectively. Low tendency to protonate. Characteristic isotope patterns for Cl and Br. Iodine can be detected because of its high mass; the ^{13}C signals of $M^{+\cdot}$ and its fragments are conspicuously weak.

Polyhaloalkanes

Fragmentation: Preferred fragmentation of the C–C bond if several halogen atoms are bonded to one of these carbon atoms. CF_3 (m/z 69) is often the base peak in terminally perfluorinated alkanes, and so is $CHCl_2$ (m/z 83, 85, 87) in terminally dichlorinated compounds. Often, X_2 is eliminated besides the usual fragmentation of $X^\cdot$ and HX. Interchange of halogens may occur. For example, m/z 85 (CF_2Cl) is a dominant signal (ca. 60%) for CF_3CFCl_2.

Ion series: Most fragments are halogenated alkyl and alkenyl groups, easily detectable on the basis of the isotope signals in the cases of Cl and Br.

Intensities: Intensive peaks mainly in the lower mass range.
Molecular ion: Weak, decreasing with increasing number of halogen atoms. Absent from the spectra of many polyhalogenated compounds.

Aromatic Halides

Fragmentation: Consecutive losses of halogen radicals and/or acid HX. In perhalogenated aromatics, decomposition down to C_x^+, with x from 1 to 6 (m/z 12, 24, 36, 48, 60, 72). If alkyl-substituted ($C_{n>1}$), the base peak is mostly the result of benzylic cleavage. In an otherwise aromatic environment, m/z 57 is a F indicator ($C_3H_2F^+$). Elimination of CF_2 (Δm 50) from CF_3 groups attached to the aromatic ring (from $M^{+\cdot}$ or fragments).
Ion series: Aromatic fragments, C_nH_n, C_nH_{n-1}, and C_nH_{n-2} (m/z 39, 51–53, 63–65, 75–77, ...). In the higher mass range: $C_n(H,X)_n$.
Intensities: Dominant peaks in the $M^{+\cdot}$ region.
Molecular ion: Usually very strong. Characteristic isotope signals for Cl and Br.

References

[1] A.G. Loudon, Mass spectrometry and the carbon-halogen bond. In: *The Chemistry of the Carbon-Halogen Bond, Part 1*; S. Patai, Ed.; Wiley: London, 1973; p 223.
[2] D.G.I. Kingston, B.W. Hobrock, M.M. Bursey, J.T. Bursey, Intramolecular hydrogen transfer in mass spectra. III. Rearrangements involving the loss of small neutral molecules, *Chem. Rev.* **1975**, *75*, 693.
[3] J.M. Miller, T.R.B. Jones, The mass spectra of azides and halides. In: *Suppl. D: The Chemistry of Halides, Pseudo-Halides, and Azides, Part 1*; S. Patai, Z. Rappoport, Eds.; Wiley: Chichester, 1983; p 75.

Hal

8.8 Alcohols, Ethers, and Related Compounds [1,2]

8.8.1 Alcohols and Phenols

Aliphatic Alcohols [3]

Fragmentation: Elimination of water from $M^{+\cdot}$ and from fragments. Strong for primary alcohols. If an aliphatic H atom can be transferred in a 6-ring process, it is involved in the water elimination in 90% of the investigated cases. If a CH_2CH_2 group is attached to the O-bearing C atom, water elimination is often followed by loss of ethylene. Water elimination is dominant for long-chain alcohols, rendering their spectra similar to those of alkenes.

Cleavage of bonds next to the OH-bearing C atom to form oxonium ions, then elimination of water and of alkenes. The α-cleavage is often dominant. Usually, its importance increases with increasing branching at the α-carbon atom. The larger substituent is lost most readily.

m/z 31 for primary alcohols (R^1, R^2: H)
m/z 30 + R^1 (45, 59, 73, ...) for secondary alcohols ($R^?$, II)
m/z 29 + R^1 + R^2 (59, 73, 87, ...) for tertiary alcohols

Consecutive H_2O and alkene eliminations in longer-chain primary alcohols lead to $[M-46]^{+\cdot}$, $[M-74]^{+\cdot}$, $[M-102]^{+\cdot}$, ... In particular, branched alcohols frequently show a typical series of fragments at $[M-15]^+$, $[M-18]^+$, and $[M-33]^+$.
Ion series: Dominant alkene ions corresponding to C_nH_{2n-1} (m/z 41, 55, 69, ...), C_nH_{2n} (m/z 42, 56, 70, ...), accompanied by weaker fragments, $C_nH_{2n+1}O$ (m/z 31, 45, 59, ...), with one or more local maxima in the latter series (m/z 31 dominates in primary alcohols).
Intensities: Intensive peaks in the lower mass range, local maxima among alkene-type fragment ions of the type $C_nH_{2n+1}O^+$.
Molecular ion: Mostly weak, often missing, especially in tertiary and long-chain alcohols. Indirect determination of $M^{+\cdot}$ is often possible from the fragments at $[M-15]^+$, $[M-18]^{+\cdot}$ and $[M-33]^+$. $[M+1]^+$ is often significant. In primary and secondary alcohols also $[M-1]^+$ can usually be seen. Sometimes, $[M-2]^{+\cdot}$ is formed because of oxidation to carbonyl compounds during sample introduction.

Alicyclic Alcohols

Fragmentation: Elimination of water from $M^{+\cdot}$, followed by loss of alkyl or alkenyl residues. Ring cleavage at the O-bearing C atom, followed by loss of alkyl residues after H rearrangement (see scheme).
Ion series: Alkene hydrocarbon fragments C_nH_{2n-1} (m/z 41, 55, 69, ...), C_nH_{2n-3} (m/z 39, 53, 67, 81, ...), and unsaturated O fragments, $C_nH_{2n-1}O$ (m/z 43, 57, 71, ...), as well as acetaldehyde and its homologues (m/z 44, 58, 72, ...).

Intensities: Local maxima evenly distributed over the whole mass range.
Molecular ion: Usually weak but in contrast to aliphatic alcohols practically never missing. $[M+1]^+$ typically contains a significant amount of $[M+H]^+$.

Unsaturated Aliphatic Alcohols [3]

Allyl alcohols: The spectra are similar to those of the corresponding carbonyl compounds, which are (partly) formed by double H rearrangement of $M^{+\cdot}$.

γ,δ-Unsaturated alcohols: Aldehyde elimination through a McLafferty-type rearrangement.

Vicinal Glycols

Fragmentation: Cleavage of bonds next to the OH-bearing C atom (α-cleavage) dominates. Preferable fragmentation of the C–C bond between the two oxygens, the charge remaining predominantly on the larger fragment. Water elimination from these fragments, but scarcely from $M^{+\cdot}$.
Ion series: Saturated and unsaturated aliphatic ions (m/z 43, 57, 71, ... and 41, 55, 69, ...) and intensive peaks from O-containing saturated rests (m/z 45, 59, 73, ...).
Intensities: Dominant peaks for the products of α-cleavages and their dehydrated derivatives.
Molecular ion: Weak.

Phenols

Fragmentation: Decarbonylation (Δm 28) and loss of CHO˙ (Δm 29) followed by elimination of acetylene. An important fragment of alkyl derivatives is $[M-1]^+$, as is $[M-15]^+$ if at least two alkyl carbons are present (dimethyl or ethyl). Elimination of CO from the primary fragments. $[M-18]^{+\cdot}$ mainly with *ortho*-alkylphenols. In derivatives with a longer alkyl chain, benzylic cleavage and alkene elimination (McLafferty rearrangement) are the dominant primary fragmentation processes. The fragments then lose CO (Δm 28).
Ion series: Aromatic hydrocarbon fragments, C_nH_n and $C_nH_{n\pm1}$ (m/z 39, 51–53, 63–65, 75–77, ...). The presence of some m/z 55 (C_3H_3O) is common. A peak at m/z 69 (O≡CCH=C=O) is characteristic of 1,3-dihydroxy substitution.
Intensities: Dominant peaks in the higher mass range.
Molecular ion: Dominant, no tendency to form $[M+H]^+$; $[M-1]^+$ is weak.

Benzyl Alcohols

Fragmentation: Loss of H˙ and consecutive elimination of CO (Δm 28) to give a protonated benzene molecule, which further loses H_2.

$$\xrightarrow{-\text{CO}} \quad C_6H_7^+ \xrightarrow{-H_2} C_6H_5^+$$

M⁺˙ (80%)   [M-1]⁺ (65%)   m/z 79   m/z 77
                                          (100%)   (65%)

Elimination of OH˙ (Δm 17) to yield the tropylium cation is the second important fragmentation path:

M⁺˙ (80%)   [M-17]⁺, $C_7H_7^+$, m/z 91 (25%)

Ion series: Aromatic fragments corresponding to C_nH_n and $C_nH_{n\pm1}$ (m/z 39, 51–53, 63–65, 75–77, …).
Intensities: Dominant peaks for the products described under *Fragmentation*. For benzyl alcohol decreasing in the sequence of [M-29]⁺, M⁺˙, [M-1]⁺, [M-31]⁺, [M-17]⁺.
Molecular ion: Strong.

8.8.2 Hydroperoxides

Aliphatic Hydroperoxides [4]

Fragmentation: Most pronounced is the loss of the hydroperoxy radical HO_2˙ (Δm 33), especially when a tertiary alkyl cation is formed. Important, in decreasing order, is loss of H_2O_2 (Δm 34), H_2O (Δm 18), HO˙ (Δm 17), and O (Δm 16).
Ion series: Mainly saturated and unsaturated alkyl fragments, C_nH_{2n+1} (m/z 43, 57, 71, …) and C_nH_{2n-1} (m/z 41, 55, 69, …). The oxygen-indicating fragment at m/z 31 and its homologues are always present.
Intensities: Intensive peaks mainly in the lower mass range.
Molecular ion: Weak.

8.8.3 Ethers

Aliphatic Ethers [5,6]

Fragmentation: Homolysis of the C–C bond next to the O atom to yield oxygen-containing fragments. Preferably, the bond at the highest substituted C atom breaks and the larger alkyl group is lost.

$$R^1 \quad O=CH-R^2 \longleftrightarrow R^1 \quad O-CH-R^2$$

$C_nH_{2n+1}O^+$, m/z 31, 45, 59, ...

This homolysis is followed by the elimination of alkenes, aldehydes, or, less importantly, of water.

$$- R^1CH=CH_2$$

$$\overset{+}{HO}=CHR^2$$
$$[30 + R^2]^+$$

$$- R^2CH=O$$

$$R^1CH_2CH_2^+$$
$$m/z\ 29, 43, 57, ...$$

As a competing process, especially with increasing molecular weight, heterolysis at the O atom takes place to yield strong alkyl ion signals. The larger as well as the branched alkyl rests are fragmented preferably. The base peak often arises from heterolysis of the C–O bond.

$$- R^1CH_2CH_2O^\bullet$$

$$m/z\ 29, 43, 57, ...$$

In contrast to the H_2O elimination from alcohols, the H transfer involved in the elimination of RCH_2CH_2OH from ethers is nonspecific.

$$- R^1CH_2CH_2OH$$

$$R^3CH=R^2$$

$$m/z\ 28, 42, 56, ...$$

Ion series: Alkyl fragments, C_nH_{2n+1} (m/z 29, 43, 57, …), with maxima due to cleavage of the C–O bond. Alkene ion series, C_nH_{2n} (m/z 28, 42, 56, …), due to elimination of alcohol. Oxygen-containing fragments, $C_nH_{2n+1}O$ (m/z 31, 45, 59, …), with maxima due to cleavage of the C–C bond next to the oxygen.
Intensities: Intensive peaks mainly in the lower mass range.
Molecular ion: Significant or weak. Decreasing with increasing chain length and branching.

Unsaturated Ethers [7]

Fragmentation of vinylic and acetylenic alkyl ethers: Dominant homolysis of the alkyl C–C bond next to the O atom on the saturated side, leading to $C_3H_5O^+$ (m/z 57) for vinylic and $C_3H_3O^+$ (m/z 55) for acetylenic ethers of primary aliphatic alcohols. For alkyl ($C_{n>5}$) vinyl ethers, ethanol elimination after triple H transfer. [M-15]$^+$ in vinyl ethers predominantly by elimination of the vinyl CH_2 after H rearrangement.

$$- \overset{\bullet}{C}H_3$$

$$[84 + alk]^+$$

Fragmentation of allylic ethers: Heterolysis of both C–O bonds, leading to strong $C_3H_5^+$ (m/z 41) and alkyl (m/z 29, 43, 57, …) cations. Formation of ionized allyl alcohol ($C_3H_6O^{+\bullet}$, m/z 58) by nonspecific H transfer from the alkyl rest. In allylic and propargylic ethers, no cleavage of the C–C bond next to the O atom of the alkenyl group occurs. Hence, loss of vinyl or acetylenyl cannot be observed.

Ion series: $C_nH_{2n}O$ (m/z 44, 58, 72, ...) for alkenyl alkyl ethers and $C_nH_{2n-2}O$ (m/z 42, 56, 70, ...) for dialkenyl ethers. Unsaturated aliphatic (C_nH_{2n-1}; m/z 41, 55, 69, ...) as well as saturated aliphatic and unsaturated oxygen-containing fragments (C_nH_{2n+1} and $C_nH_{2n-1}O$; m/z 43, 57, 71, ...).

Intensities: Intensive peaks mainly in the lower mass range.

Molecular ion: Weak to medium, very weak for acetylenic ethers.

Alkyl Cycloalkyl Ethers

Fragmentation of methyl ethers of cycloalkanols with > 3 C atoms: After primary cleavage of the ring C–C bond next to the O atom, the prominent fragments formed are $CH_3OCH=CH_2^{+\cdot}$ (m/z 58) and, for alicyclics with > 4 C atoms, $CH_3O=CHCH=CH_2^+$ (m/z 71, rearrangement in analogy to that observed for cycloalkanols). Loss of methanol to give hydrocarbon fragments, C_nH_{2n-2} (m/z 54, 68, 82, ...).

Fragmentation of ethyl and higher alkyl ethers of cycloalkanols with > 3 C atoms: Alkene elimination to yield the protonated cycloalkanol (m/z 72, 86, 100, ...) and heterolytic cleavage of the C–O bond to give dominating cycloalkyl ions (m/z 69, 83, ...).

Ion series: Besides the fragments already mentioned, mainly unsaturated hydrocarbon fragments (C_nH_{2n-1}, m/z 27, 41, 55, 69, ...).

Intensities: The above mentioned fragments dominate the spectrum.

Molecular ion: Weak or intermediate.

Cyclic Ethers

Fragmentation: Primary ring cleavage at C–C bonds next to the O atom, followed by loss of CH_2O (Δm 30), H_2O (Δm 18), or alkyl (Δm 15, 29, ...). Elimination of H˙ to give $[M-1]^+$, followed by CO elimination (Δm 28) to $[M-29]^+$. When α-substituted, dominant loss of substituents, followed by water elimination. Formation of acyl cation if two α-substituents are present.

Ion series: Mainly ions of the alkene type. Weak saturated, oxygen-containing fragments (m/z 31, 45, ...).

Intensities: Intensive peaks evenly distributed over the whole mass range.

Molecular ion: Often significant but sometimes weak, especially when α-substituted. Intensity of $[M-1]^+$ usually comparable to that of $M^{+\cdot}$ if no α substituent is present.

Methoxybenzenes

Fragmentation: Loss of methyl radical, followed by decarbonylation to $[M-43]^+$; elimination of formaldehyde (Δm 30) from $M^{+\cdot}$ or from primary fragments.
Ion series: Aromatic fragments corresponding to C_nH_n and $C_nH_{n\pm1}$ (m/z 39, 51–53, 63–65, 75–77, ...).
Intensities: Intensive peaks in the $M^{+\cdot}$ region.
Molecular ion: Strong.

Alkyl Aryl Ethers [8]

Fragmentation: Commonly dominating alkene elimination to give the corresponding phenol ion (nonspecific hydrogen migration), followed by decarbonylation. In the case of aryl methyl ethers, loss of CH_2O from $M^{+\cdot}$ or from primary fragments as well as $CH_3^\cdot$ elimination followed by decarbonylation.
Ion series: Mostly aromatic fragments, C_nH_n and $C_nH_{n\pm1}$ (m/z 39, 51–53, 63–65, 75–77, ...).
Intensities: Usually maximum at the mass of the corresponding phenol. Otherwise, intensive peaks mainly concentrated in the high and medium mass range.
Molecular ion: Strong.

Aromatic Ethers

Fragmentation: Loss of $H^\cdot$ (Δm 1), CO (Δm 28), and $CHO^\cdot$ (Δm 29) from $M^{+\cdot}$. Cleavage at the C–O bond and decarbonylation of the resulting product, followed by dehydrogenation.
Ion series: Aromatic fragments corresponding to C_nH_n and $C_nH_{n\pm1}$ (m/z 39, 51–53, 63–65, 75–77, ...).
Intensities: Intensive peaks mainly in the $M^{+\cdot}$ region.
Molecular ion: Strong.

O

8.8.4 Aliphatic Epoxides [9]

Fragmentation: The most important primary fragmentation is the cleavage of C–C bonds next to the O atom (α-cleavage), resulting in complex degradation due to the related multiple choice and extensive secondary rearrangements. The products allow mass-spectrometric localization of double bonds after epoxidation.

Due to ring opening prior to fragmentation, β-cleavage is as relevant as the α-cleavage.

m/z 57

γ-Cleavage is the most important fragmentation mechanism, especially in terminal epoxides:

m/z 71

Mainly in terminal epoxides, rearrangement with alkene elimination, formally leading to alkene-OH$^{+\cdot}$ ($C_nH_{2n}O$, m/z 44, 58, 72, ...) and alkene$^{+\cdot}$ (C_nH_{2n}, m/z 28, 42, 56, ...):

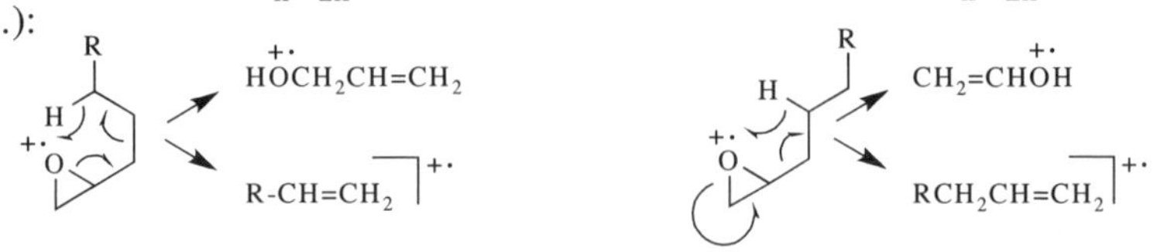

Mainly in nonterminal epoxides, transannular cleavage with H transfer and elimination of an alkenyl radical, leading to $C_nH_{2n+1}O$ fragments (m/z 45, 59, 73, ...):

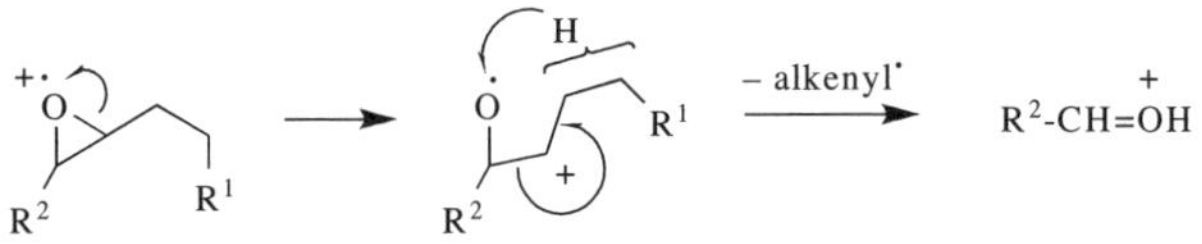

Ion series: Mixed, not characteristic.
Intensities: Intensive peaks mainly in the lower mass range.
Molecular ion: Usually weak.

8.8.5 Aliphatic Peroxides [4]

Fragmentation: Alkene elimination to give hydroperoxide radical cations and hydroperoxide elimination to yield alkene radical cations (dominating if larger alkyl groups are present). Alkene elimination can be followed by loss of OH$^{\cdot}$, resulting in products that formally correspond to those obtained by O–O cleavage, which probably is not a one-step process:

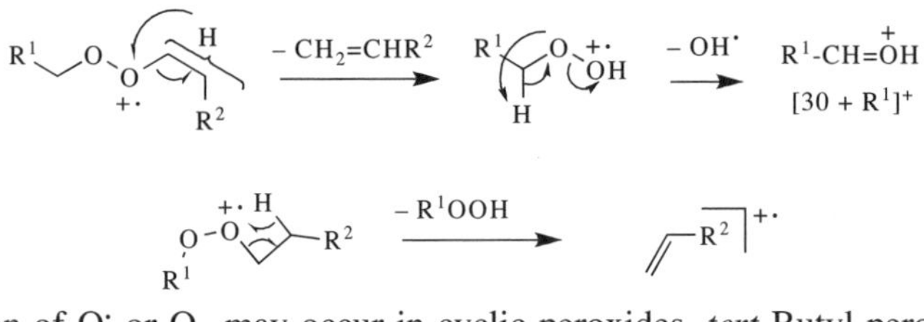

Elimination of O$^{\cdot}$ or O_2 may occur in cyclic peroxides. *tert*-Butyl peroxides predominantly eliminate *tert*-butyl-OO$^{\cdot}$ to give [M-89]$^+$.
Ion series: Saturated or unsaturated alkyl groups (C_nH_{2n+1}, m/z 29, 43, 57, ...; C_nH_{2n-1}, m/z 27, 41, 55, ...) and alkenyl ions (C_nH_{2n}, m/z 28, 42, 56, ...) dominate. The fragment at m/z 31 and sometimes its homologues indicate the presence of oxygen.
Intensities: Intensive peaks mainly in the lower mass range.
Molecular ion: Weak to moderate.

8.8.6 References

[1] D.G.I. Kingston, J.T. Bursey, M.M. Bursey, Intramolecular hydrogen transfer in mass spectra. II. The McLafferty rearrangement and related reactions, *Chem. Rev.* **1974**, *74*, 215.

[2] D.G.I. Kingston, B.W. Hobrock, M.M.Bursey, J.T. Bursey, Intramolecular hydrogen transfer in mass spectra. III. Rearrangements involving the loss of small neutral molecules, *Chem. Rev.* **1975**, *75*, 693.

[3] R.G. Cooks, The mass spectra of hydroxyl compounds. In: *The Chemistry of the Hydroxyl Group, Part 2*; S. Patai, Ed.; Interscience: London, 1971; p 1045.

[4] H. Schwarz, H.-M. Schiebel, Mass spectrometry of organic peroxides. In: *The Chemistry of Peroxides*; S. Patai, Ed.; Wiley: Chichester, 1983; p 105.

[5] C.C. van de Sande, The mass spectra of ethers and sulphides. In: *The Chemistry of Ethers, Crown Ethers, Hydroxyl Groups and Their Sulphur Analogues, Suppl. E, Part 1*; S. Patai, Ed.; Wiley: Chichester, 1980; p 299.

[6] S.L. Bernasek, R.G. Cooks, The β-cleavage reaction in ethers, *Org. Mass Spectrom.* **1970**, *3*, 127.

[7] J.P. Morizur, C. Djerassi, Mass spectrometric fragmentation of unsaturated ethers, *Org. Mass Spectrom.* **1971**, *5*, 895.

[8] G. Sozzi, H.E. Audier, P. Morgues, A. Millet, Alkyl phenyl ether radical cations in the gas phase: A reaction model, *Org. Mass Spectrom.* **1987**, *22*, 746.

[9] Q.N. Porter, *Mass Spectrometry of Heterocyclic Compounds*, 2nd ed.; Wiley: New York, 1985.

O

8.9 Nitrogen Compounds [1,2]

8.9.1 Amines

Saturated Aliphatic Amines [3]

Fragmentation: Dominating loss of alkyl residues by cleavage of the C–C bond next to the N atom ("N-cleavage"). Larger substituents are eliminated preferably. When a γ-H is available, subsequent elimination of alkenes by McLafferty-type reactions:

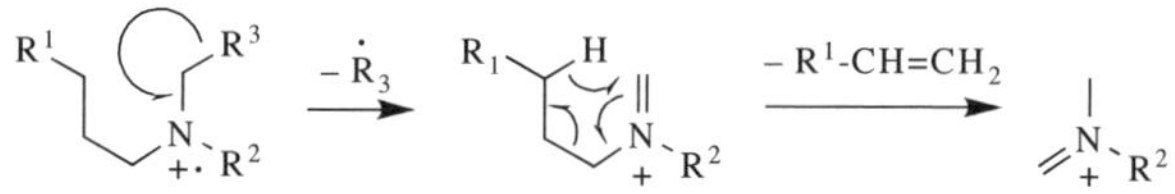

Otherwise, unspecific H transfer onto the N atom:

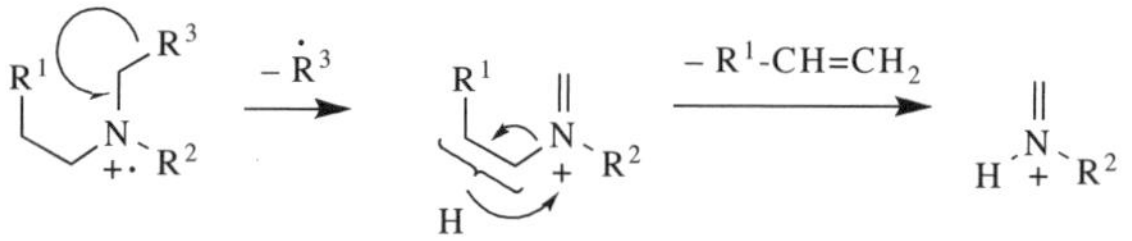

NH_3, RNH_2, and $RR'NH$ eliminations from primary, secondary, and tertiary amines, respectively, are negligible except from some multifunctional compounds (e.g., diamines and phenyl-phenoxy-substituted amines).

Ion series: Even-mass fragments of the type $C_nH_{2n+2}N$ (m/z 30, 44, 58, 72, 86, …).

Intensities: Mainly peaks in the low mass range. Dominating base peak from "N-cleavage" at $[28 + m(R^1) + m(R^2) + m(R^4) + m(R^5)]^+$ for $R^1R^2R^3CNR^4R^5$ (e.g., m/z 30 for RCH_2NH_2, m/z 44 for RCH_2NHCH_3, m/z 58 for $RCH_2N(CH_3)_2$, and m/z 86 for $RCH_2N(CH_2CH_3)_2$). Local maximum at m/z 86 ($C_5H_{12}N^+$) for *n*-alk–NH_2 (protonated piperidine, 6-membered ring).

Molecular ion: Usually weak or absent, especially if the α-C atom is substituted. Decreasing intensity with increasing molecular weight. Tendency to protonate to $[M+H]^+$. Odd mass for odd number of N atoms in the molecule.

Cycloalkylamines

Fragmentation: The most important primary reaction is the ring cleavage next to the N atom, followed by H rearrangement and loss of an alkyl residue. Some elimination of amine, R^1R^2NH.

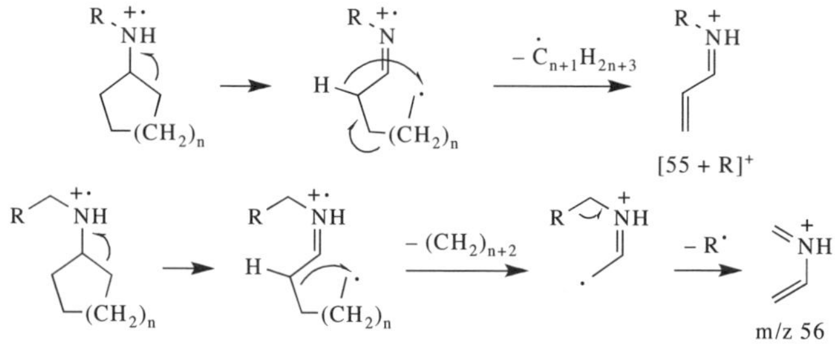

Ion series: Even-mass fragments of the type $C_nH_{2n}N$ (m/z 42, 56, 70, 84, ...).
Intensities: Intensive local maxima evenly distributed over the whole mass range.
Molecular ion: Usually significant. Odd mass for odd number of N atoms in the molecule.

Cyclic Amines

Fragmentation: Dominating primary reaction is the cleavage of C–C bonds next to N, resulting in the loss of substituents next to N or in primary ring cleavage. Primary ring cleavage is followed by H rearrangement and loss of alkenes or alkyl groups. The most important primary fragmentation for substituted cyclic amines is the loss of substituents at C atoms next to N.
Piperidine:

Ion series: Even-mass fragments of the type $C_nH_{2n}N$ (m/z 42, 56, 70, 84, ...) and $C_nH_{2n+2}N$ (m/z 30, 44, 58, ...) as well as odd-mass fragments of the type $C_nH_{2n+1}N$ (m/z 43, 57, 71, 85, ...).
Intensities: Intensive local maxima evenly distributed over the whole mass range if no substituent is bonded to the C atom next to N. Otherwise, dominating maxima by loss of such substituents.
Molecular ion: Significant or strong if no substituent is bonded to the C atom next to N; otherwise, weak. Tendency to form [M-H]+. Odd mass for odd number of N atoms in the molecule.

Piperazines

Fragmentation: As for cyclic amines, enhanced primary ring cleavage at C–C bonds next to the N atom.
Ion series: Even-mass fragments of the type $C_nH_{2n}N$ (m/z 42, 56, 70, 84, ...) and $C_nH_{2n+2}N$ (m/z 30, 44, 58, ...) as well as odd-mass series of the type $C_nH_{2n+1}N$ (m/z 43, 57, 71, 85, ...).
Intensities: Intensive local maxima evenly distributed over the whole mass range if no substituent is bonded to the C atom next to N. Otherwise, dominating maxima by loss of such substituents.
Molecular ion: Significant or strong if no substituent is bonded to the C atom next to N; otherwise, weak. Tendency to form [M-H]+. Odd mass for odd number of N atoms in the molecule.

Aromatic Amines

Fragmentation: Dominating cleavage of alkyl bond at N-bearing C atom ("N-cleavage") followed by alkene elimination if aliphatic substituents with $C_{n \geq 2}$ are present. Otherwise, loss of H· from primary and secondary anilines and benzylic amines. Loss of HCN from $M^{+\cdot}$ or from fragments. A local maximum at m/z 42 is typical of an aromatically bonded dimethylamino group.
Ion series: Aromatic hydrocarbon fragments (C_nH_n and $C_nH_{n\pm1}$; m/z 39, 51–53, 63–65, 75–77, …).
Intensities: Dominating maxima by "N-cleavage" and following alkene loss if aliphatic substituents with $C_{n>1}$ are present.
Molecular ion: Abundant if no aliphatic substituents with more than one C atom are present, otherwise, medium or weak. No tendency to protonate. In primary and secondary aromatic and benzylic amines, $[M-H]^+$ is important. Odd mass for odd number of N atoms in the molecule.

8.9.2 Nitro Compounds

Aliphatic Nitro Compounds

Fragmentation: Loss of NO· (Δm 30), NO_2· (Δm 46), and HNO_2 (Δm 47) as well as the formation of some m/z 30 as N indicator. Spectra with only few characteristic features.
Ion series: Mixed alkyl and alkenyl fragments, C_nH_{2n+1} (m/z 43, 57, 71, …) and C_nH_{2n-1} (m/z 41, 55, 69, …).
Intensities: Dominant peaks in the lower mass range.
Molecular ion: Weak or missing. Odd mass for odd number of N atoms in the molecule.

Aromatic Nitro Compounds

Fragmentation: Loss of O (Δm 16), NO· (Δm 30, followed by elimination of CO, Δm 28), and NO_2· (Δm 46) from $M^{+\cdot}$ or from a major primary cleavage product. Extensive rearrangement of the functional group to a nitroso ester.
Ion series: Aromatic fragments corresponding to C_nH_n and $C_nH_{n\pm1}$ (m/z 39, 51–53, 63–65, 75–77, …).
Intensities: Intensive peaks mainly in the upper mass range.
Molecular ion: Strong. Odd mass for odd number of N atoms in the molecule.

8.9.3 Diazo Compounds and Azobenzenes

Diazo Compounds [4,5]

Diazonium: Because of the low volatility of diazo compounds, their electron impact mass spectra show thermal decomposition products. These are formed by loss of N_2

(e.g., a diazonium chloride gives rise to the corresponding aromatic chloro compound). From a phenyl diazonium *ortho*-carboxylate zwitterion, biphenylene is formed as dimerization product.

Diazomethane and derivatives: $M^{+\cdot}$ is strong except when catalytic decomposition occurs on metal surfaces of the inlet system. Loss of N_2 is a dominant reaction of diazomethane and diazoketones.

Azobenzenes

Fragmentation: Cleavage at the azo group followed by loss of N_2, giving rise to the dominant base peak.
Ion series: Aromatic fragments (C_nH_n, $C_nH_{n\pm1}$; m/z 39, 51–53, 63–65, 75–77, …).
Intensities: Dominant $M^{+\cdot}$ and azo cleavage products.
Molecular ion: Strong. Odd mass for odd number of N atoms in the molecule.

8.9.4 Azides

Aliphatic Azides [6]

Fragmentation: $[M-42]^+$ ($N_3^{\cdot}$ elimination) or $[M-28]^{+\cdot}$ (N_2 elimination) dominant in most cases. The spectra are similar to those of the corresponding aliphatic compounds.
Ion series: Aliphatic hydrocarbon series.
Intensities: Dominant peaks in the lower mass range, as in aliphatic compounds.
Molecular ion: Absent or weak. Odd mass for odd number of N atoms in the molecule.

Aromatic Azides [7]

N

Fragmentation: In most cases, $[M-28]^{+\cdot}$ (N_2 elimination) is the base peak. The next step is the elimination of HCN (Δm 27) or acetylene (Δm 26), or, if there is a substituent X on the ring, of $X^{\cdot}$ or HX.

Ion series: Aromatic hydrocarbon fragments (C_nH_n and $C_nH_{n\pm1}$; m/z 39, 51–53, 63–65, 75–77, …).

Intensities: Dominant peaks in the higher mass range; $[M-28]^{+\cdot}$ (N_2 elimination) and $[M-55]^{+\cdot}$ (N_2 and HCN elimination) are the most intensive peaks.

Molecular ion: Weak. Odd mass for odd number of N atoms in the molecule.

8.9.5 Nitriles and Isonitriles

Aliphatic Nitriles (R–CN) [4]

Fragmentation: Elimination of alkyl radicals to give $(CH_2)_nCN^+$ (m/z 40, 54, 68, …). McLafferty rearrangement yielding $CR_2=C=NH^{+\cdot}$ (m/z 41 for R: H). In most cases, C–CN cleavage and HCN elimination are not significant reactions. Complex rearrangements in unsaturated nitriles if other functional groups are present.

Ion series: Saturated and unsaturated alkyl ions mainly in the lower mass range (C_nH_{2n+1} and C_nH_{2n-1}; m/z 29, 43, 57, … and 27, 41, 55, …). Rearrangement products corresponding to $C_nH_{2n-1}N$ contribute, to a significant extent, to the ion series m/z 41, 55, 69, …. For alkyl chains with $C_{n>5}$, dominating $(CH_2)_nCN^+$ (i.e., $C_nH_{2n-2}N$, m/z 82, 96, 110, …, probably with a cyclic structure).

Intensities: Intensive peaks due to the above mentioned ions.

Molecular ion: Weak or missing. Both $[M+H]^+$ and $[M-H]^+$ are usually more intensive than $M^{+\cdot}$. In some aliphatic nitriles, $[M+2H]^{+\cdot}$ is as intensive as $M^{+\cdot}$. Odd mass for odd number of N atoms in the molecule.

Aromatic Nitriles (R–CN)

Fragmentation: Consecutive elimination of HCN and acetylene.

Ion series: Aromatic fragments corresponding to C_nH_n and $C_nH_{n\pm1}$ (m/z 39, 51–53, 63–65, 75–77, …).

Intensities: Intensive peaks in the $M^{+\cdot}$ region.

Molecular ion: Dominant intensity, often base peak. In contrast to aliphatic and benzylic nitriles, $[M-1]^+$ is usually not important. Odd mass for odd number of N atoms in the molecule.

Aliphatic Isonitriles (R–NC)

Fragmentation: In general, the spectra are similar to those of the corresponding nitriles. The most important difference lies in the loss of $CN^{\cdot}$ (Δm 26) and the higher probability of losing HCN (Δm 27). Further important fragmentations are the elimination of alkyl radicals to give $(CH_2)_nCN^+$ ions and the McLafferty rearrangement to yield $CR_2=N=CH^{+\cdot}$ (m/z 41 for R: H).

Ion series: Saturated and unsaturated alkyl ions mainly in the lower mass range (C_nH_{2n+1}, m/z 29, 43, 57, … and C_nH_{2n-1}, m/z 27, 41, 55, …). Rearrangement products corresponding to $C_nH_{2n-1}N$ contribute, to a significant extent, to the ion series of m/z 41, 55, 69, ….

Intensities: Intensive peaks in the lower mass range.

Molecular ion: Weak, decreasing with increasing chain length and degree of branching. Both [M+H]$^+$ and [M-H]$^+$ can be stronger than M$^{+\cdot}$. Odd mass for odd number of N atoms in the molecule.

Aromatic Isonitriles (R–NC) [4]

Fragmentation: Dominant loss of HCN ([M-27]$^{+\cdot}$). In methylphenyl and benzyl isocyanides also formation of isocyanotropylium ion, [M-1]$^+$, followed by loss of HCN to [M-28]$^+$.
Ion series: Aromatic (C$_n$H$_n$ and C$_n$H$_{n\pm1}$; m/z 39, 51–53, 63–65, 75–77, …).
Intensities: Intensive peaks in the higher mass range.
Molecular ion: Dominant; base peak for phenyl isocyanide. Odd mass for odd number of N atoms in the molecule.

8.9.6 Cyanates, Isocyanates, Thiocyanates, and Isothiocyanates

Aliphatic Cyanates (R–OCN) [8]

Fragmentation: Spectra often very similar to those of the corresponding isocyanates (see below). Cleavage of the C–C bond next to O, with the charge remaining on CH$_2$OCN (m/z 56) for short-chain cyanates and preferably on the alkyl substituent if it has a C$_{n>2}$ chain (m/z 29, 43, 57, …). Cleavage of the C–O bond with H rearrangement to give HCNO$^{+\cdot}$ (m/z 43) or alkene$^{+\cdot}$ (m/z 42, 56, 70, …). For cyanates with C$_{n>5}$ substituents, alkene elimination yields m/z 99.
Ion series: Saturated and unsaturated alkyl cations (C$_n$H$_{2n+1}$, m/z 29, 43, 57, … and C$_n$H$_{2n-1}$, m/z 27, 41, 55, …). Alkene radical cations (C$_n$H$_{2n}$, m/z 42, 56, 70, …) together with isobaric ions of the composition C$_n$H$_{2n}$NCO.
Intensities: Intensive peaks mainly in the lower mass range.
Molecular ion: Usually weak or absent. [M-H]$^+$ is often more intensive. Odd mass for odd number of N atoms in the molecule.

N

Aromatic Cyanates (R–OCN) [8]

Fragmentation: Loss of OCN$^\cdot$ (Δm 42) or, to a lesser extent, of CO (Δm 28), with subsequent HCN elimination (Δm 27).
Ion series: Aromatic fragments corresponding to C$_n$H$_n$ and C$_n$H$_{n\pm1}$ (m/z 39, 51–53, 63–65, 75–77, …).
Intensities: Intensive peaks in the higher mass range.
Molecular ion: Strong. Odd mass for odd number of N atoms in the molecule.

Aliphatic Isocyanates (R–NCO) [8]

Fragmentation: Spectra often very similar to those of the corresponding cyanates. Cleavage of the C–C bond next to N, the charge remaining on the CH$_2$NCO (m/z 56) for short-chain isocyanates and preferably on the alkyl substituent for compounds with a C$_{n>2}$ chain (m/z 29, 43, 57, …). Cleavage of the C–N bond with H rearrangement to give HNCO$^{+\cdot}$ (m/z 43) or alkene$^{+\cdot}$ (m/z 42, 56, 70, ...) ions. For isocyanates with $C_{n>5}$ alkyl chains, alkene elimination, yielding m/z 99.

Ion series: Saturated and unsaturated alkyl cations (C_nH_{2n+1}, m/z 29, 43, 57, ... and C_nH_{2n-1}, m/z 27, 41, 55, ...). Alkene radical cations (C_nH_{2n}, m/z 42, 56, 70, ...) together with isobaric ions of the composition $C_nH_{2n}OCN$.
Intensities: Intensive peaks mainly in the lower mass range.
Molecular ion: Usually weak or absent. [M-H]$^+$ is often more intensive. Odd mass for odd number of N atoms in the molecule.

Aromatic Isocyanates (R–NCO) [8]

Fragmentation: Consecutive elimination of CO (Δm 28) and HCN (Δm 27). In contrast to aromatic cyanates, practically no elimination of NCO$^\cdot$ (Δm 42).
Ion series: Aromatic fragments corresponding to C_nH_n and $C_nH_{n\pm1}$ (m/z 39, 51–53, 63–65, 75–77, ...).
Intensities: Intensive peaks in the higher mass range.
Molecular ion: Dominating; base peak for phenyl isocyanate. Odd mass for odd number of N atoms in the molecule.

Aliphatic Thiocyanates (R–SCN) [8]

Fragmentation: Elimination of HCN (Δm 27) followed by loss of an alkyl group. The cleavage of the C–C bond next to SCN is unimportant except in short-chain thiocyanates.
Ion series: Saturated and unsaturated alkyl cations (C_nH_{2n+1}, m/z 29, 43, 57, ... and C_nH_{2n-1}, m/z 27, 41, 55, ...).
Intensities: Intensive peaks in the lower mass range.
Molecular ion: Weak. Decreasing with increasing chain length and degree of branching; absent from the spectrum of hexyl thiocyanate. Odd mass for odd number of N atoms in the molecule. Both [M+H]$^+$ and [M-H]$^+$ are detectable. Characteristic ^{34}S isotope peak at [M+2]$^{+\cdot}$ and [frag+2] for S-containing fragments (4.5% per S atom).

Aromatic Thiocyanates (R–SCN) [8]

Fragmentation: The most important fragmentation is the elimination of SCN$^\cdot$ (Δm 58). Further elimination reactions are loss of CN$^\cdot$ (Δm 26), HCN (Δm 27),

and CS (Δm 44).
Ion series: Aromatic fragments corresponding to C_nH_n and $C_nH_{n\pm1}$ (m/z 39, 51–53, 63–65, 75–77, ...). Weak signal at m/z 45 (CHS$^+$) indicates sulfur.
Intensities: Intensive peaks in the higher mass range.
Molecular ion: Dominant; base peak in phenyl thiocyanate. Odd mass for odd number of N atoms in the molecule. Characteristic ^{34}S isotope peak at [M+2]$^{+\cdot}$ and [frag+2] for S-containing fragments (4.5% per S atom).

Aliphatic Isothiocyanates (R–NCS) [8]

Fragmentation: Cleavage of the C–C bond next to NCS, leading to m/z 72 (CH$_2$NCS) or to its homologues if the α-C atom is substituted. Loss of the alkyl residue with concomitant double hydrogen rearrangement to yield H$_2$NCS$^+$ (m/z 60). With a $C_{n>4}$ alkyl chain, loss of SH$^\cdot$ (Δm 33). With $C_{n>5}$ alkyl chains, loss of alkene leading to m/z 115, probably according to the mechanism shown for aliphatic isocyanates.
Ion series: Mainly saturated and unsaturated alkyl cations (C_nH_{2n+1}, m/z 29, 43, 57, ... and C_nH_{2n-1}, m/z 27, 41, 55, ...). Signal for CH$_2$NCS$^+$ (m/z 72) or its homologues (m/z 86, 100, 114, ...) if the α-C atom is substituted.
Intensities: Intensive peaks mainly in the lower mass range.
Molecular ion: Medium to weak, decreasing with increasing chain length and degree of branching. More intensive than in the corresponding thiocyanates; 1% for hexadecyl isothiocyanate. Both [M+H]$^+$ and [M-H]$^+$ are relevant. Odd mass for odd number of N atoms in the molecule. Characteristic ^{34}S isotope peak at [M+2]$^{+\cdot}$ and [frag+2] for S-containing fragments (4.5% per S atom).

Aromatic Isothiocyanates (Ar-NCS) [8]

Fragmentation: Dominant loss of NCS$^\cdot$ (Δm 58). In contrast to aromatic thiocyanates, the loss of HCN (Δm 27) or CS (Δm 44) leads to very weak fragments only.
Ion series: Aromatic fragments corresponding to C_nH_n and $C_nH_{n\pm1}$ (m/z 39, 51–53, 63–65, 75–77, ...). Weak signal at m/z 45 (CHS$^+$) indicates sulfur.
Intensities: Intensive peaks in the higher mass range.
Molecular ion: Dominant; base peak in phenyl isothiocyanate. Odd mass for odd number of N atoms in the molecule. Characteristic ^{34}S isotope peak at [M+2]$^{+\cdot}$ and [frag+2] for S-containing fragments (4.5% per S atom).

N

8.9.7 References

[1] H. Schwarz, K. Levsen, The chemistry of ionized amino, nitroso and nitro compounds in the gas phase. In: *Suppl. F, The Chemistry of the Amino, Nitroso and Nitro Compounds and Their Derivatives, Part 1*; S. Patai, Ed.; Wiley: Chichester, 1982; p 85.
[2] D.G.I. Kingston, B.W. Hobrock, M.M. Bursey, J.T. Bursey, Intramolecular hydrogen transfer in mass spectra. III. Rearrangements involving the loss of small neutral molecules, *Chem. Rev.* **1975**, *75*, 693.

[3] R.D. Bowen, The chemistry of $C_nH_{2n+2}N^+$ ions. *Mass Spectrom. Rev.* **1991**, *10*, 225.

[4] K.-P. Zeller, Mass spectra of cyano, isocyano and diazo compounds. In: *Suppl. C, The Chemistry of Triple-Bonded Functional Groups, Part 1*; S. Patai, Z. Rappoport, Eds.; Wiley: Chichester, 1983; p 57.

[5] C.W. Thomas, L.L. Levsen, Electron-impact spectra of 2-diazoacetophenones, *Org. Mass Spectrom.* **1978**, *13*, 39.

[6] J.M. Miller, T.R.B. Jones, The mass spectra of azides and halides. In: *Suppl. D, The Chemistry of Halides, Pseudo-Halides and Azides, Part 1*; S. Patai, Z. Rappoport, Eds.; Wiley: Chichester, 1983; p 75.

[7] R.A. Abramovitch, E.P. Kyba, E.F. Scriven, Mass spectrometry of aryl azides, *J. Org. Chem.* **1971**, *36*, 3796.

[8] K.A. Jensen, G. Schroll, Mass spectra of cyanates, isocyanates, and related compounds. In: *The Chemistry of Cyanates and Their Thio Derivatives, Part 1*; S. Patai, Ed.; Wiley: Chichester, 1977, p 273.

N

8.10 Sulfur Compounds [1]

8.10.1 Thiols

Aliphatic Thiols [2]

Fragmentation: Elimination of H_2S (Δm 34; or SH, Δm 33, from secondary thiols) followed by loss of alkenes; consecutive losses of ethylene from unbranched thiols. Cleavage of the α,β-C–C bond (next to the SH group) leads to CH_2SH^+ (m/z 47). Note that this fragment also occurs in secondary and tertiary thiols. The S atom is poorer than N, but better than O, at stabilizing such a fragment. Cleavage at the next C–C bonds leads to signals at m/z 61, 75, and 89. In secondary and tertiary thiols, prominent fragments are formed by loss of the largest α-alkyl group.

Ion series: Dominant alkenyl fragments (C_nH_{2n-1}, m/z 41, 55, 69, ...) and smaller aliphatic fragments (C_nH_{2n+1}, m/z 43, 57, 71, ...). Sulfur-containing aliphatic fragments: $C_nH_{2n+1}S$ (m/z 47, 61, 75, 89, ...). Often significant sulfur-indicating fragments: HS^+, $H_2S^{+\cdot}$, H_3S^+, and CHS^+ (m/z 33, 34, 35, and 45).

Intensities: More intensive peaks in the lower mass range, mostly of the alkene type. Characteristic local maxima from S-containing fragments, $C_nH_{2n+1}S$ (m/z 47, 61, 75, 89, ...). In *n*-alkyl thiols, the intensity of the signal at m/z 61 is roughly half that of m/z 47; the signal at m/z 89 is more intensive than that at m/z 75, presumably because it is stabilized by cyclization.

Molecular ion: Relatively strong except for higher tertiary thiols. Characteristic ^{34}S isotope peak at $[M+2]^{+\cdot}$ and [frag+2] for S-containing fragments (4.5% per S atom).

Aromatic Thiols [2]

Fragmentation: CS elimination from $M^{+\cdot}$ and $[M-1]^+$, yielding $[M-44]^{+\cdot}$ and $[M-45]^+$. HS$^\cdot$ elimination from $M^{+\cdot}$ to give $[M-33]^+$.

Ion series: HCS^+ (m/z 45) is characteristic besides the aromatic fragments, C_nH_n and $C_nH_{n\pm1}$ (m/z 39, 51–53, 63–65, 75–77, ...).

Intensities: Intensive peaks in the higher mass range.

Molecular ion: Usually dominating; base peak in thiophenol. $[M-1]^+$ is usually strong. Characteristic ^{34}S isotope peak at $[M+2]^{+\cdot}$ and [frag+2] for S-containing fragments (4.5% per S atom).

S

8.10.2 Sulfides and Disulfides

Aliphatic Sulfides [1]

Fragmentation: Loss of alkyl radicals by cleavage of the C–C bond next to S (the largest group being lost preferably) and of the C–S bond, followed by alkene and H_2S elimination. Alkene elimination from $M^{+\cdot}$ to form the corresponding thiol ions. In contrast to thiols and cyclic sulfides, no H_2S or HS$^\cdot$ elimination from $M^{+\cdot}$.

$$\overset{+\cdot}{alk-SH} \xleftarrow{\ -\ alkene\ } \overset{+\cdot}{\underset{R}{\overset{R}{S-C-R}}} \xrightarrow{\ -\ R^\cdot\ } \overset{+}{alk-S=CR_2} \xrightarrow{\ -\ alkene\ } \overset{+}{HS=CR_2}$$

$$\overset{R}{\underset{R}{\overset{+}{C-R}}} \xleftarrow{\ -\ alk\text{-}S^\cdot\ } \quad \overset{-\ ^\cdot CR_3}{\xrightarrow{\hspace{1.5cm}}} alk-S^+$$

$$\overset{+\cdot}{alk^{\diagdown}S^{\diagup\diagdown}R} \xrightarrow{\ -\ R^\cdot\ } alk-\overset{+}{S}{\diagup} \xrightarrow{\ -\ alkene\ } H-\overset{+}{S}{\diagup} \quad m/z\ 61$$

In general, the H rearrangements are nonspecific. The transfer of secondary H predominates over that of primary H.

Ion series: Sulfur-containing aliphatic fragments, $C_nH_{2n+1}S$ (m/z 47, 61, 75, 89, ...). The hydrocarbon fragments may dominate in long-chain sulfides.

Intensities: Intensive peaks in the lower mass range. Characteristic local maxima from S-containing fragments, $C_nH_{2n+1}S$ (m/z 47, 61, 75, 89, ...).

Molecular ion: Usually strong. Characteristic ^{34}S isotope peak at $[M+2]^{+\cdot}$ and [frag+2] for S-containing fragments (4.5% per S atom).

Alkyl Vinyl Sulfides

Fragmentation: Loss of alkyl radicals (Δm 15, 29, 43, ...). Elimination of thioethanol (Δm 62) after triple H rearrangement. Dominant m/z 60 ($CH_3CH=S^{+\cdot}$) accompanied by m/z 61 ($CH_3CH_2S^+$).

Ion series: Sulfur-containing unsaturated aliphatic fragments, $C_nH_{2n-1}S$ (m/z 45, 59, 73, ...). Unsaturated hydrocarbon ions, C_nH_{2n} (m/z 42, 56, 70, ...) and C_nH_{2n-2} (m/z 40, 54, 68, ...)

Intensities: Intensive peaks evenly distributed over the whole mass range.

Molecular ion: Of medium intensity. Characteristic ^{34}S isotope peak at $[M+2]^{+\cdot}$ and [frag+2] for S-containing fragments (4.5% per S atom).

Cyclic Sulfides

Fragmentation: Primary cleavage of the C–C bond next to S, followed by rearrangements and elimination of $CH_3^\cdot$ (base peak for tetrahydrothiapyrane) and $C_2H_5^\cdot$. In tetrahydrothiophene, $[M-1]^+$ is also significant. $HS^\cdot$, H_2S, and C_2H_4 elimination from $M^{+\cdot}$.

Ion series: Sulfur-containing aliphatic fragments with one degree of unsaturation, $C_nH_{2n-1}S$ (m/z 45, 59, 73, 87, 101, ...), m/z 87 being of special dominance.

Intensities: Overall distribution of peaks maximizing in the low mass range due to S-containing fragments, $C_nH_{2n-1}S$ (m/z 45, 59, 73, 87, ...).

Molecular ion: Very strong. Characteristic ^{34}S isotope peak at $[M+2]^{+\cdot}$ and [frag+2] for S-containing fragments (4.5% per S atom).

Aromatic Sulfides [2]

Fragmentation: Loss of CS (Δm 44) and of $HS^\cdot$ (Δm 33) from $M^{+\cdot}$.

Ion series: HCS^+ (m/z 45) is characteristic besides the aromatic fragments, C_nH_n and $C_nH_{n\pm1}$ (m/z 39, 51–53, 63–65, 75–77, ...).

Intensities: Intensive peaks mainly in the higher mass range.
Molecular ion: Strong. Characteristic ^{34}S isotope peak at $[M+2]^{+\cdot}$ (4.5% relative to $M^{+\cdot}$ per S atom) and [frag+2] for S-containing fragments.

Disulfides

Fragmentation: Loss of RSS$^{\cdot}$, leading to alkyl cations and alkene elimination to give RSSH$^{+\cdot}$. Cleavage of the S–S bond with or without H rearrangements, leading to RS^{+}, $[RS\text{-}H]^{+\cdot}$, and $[RS\text{-}2H]^{+}$. Loss of one or two S with or without H atoms is a common process in cyclic, unsaturated, and aromatic disulfides.
Ion series: In saturated aliphatic disulfides, H_2S_2 and its alkyl homologues are characteristic (m/z 66, 80, 94, ...).
Intensities: Variable.
Molecular ion: Usually strong. Characteristic ^{34}S isotope peak at $[M+2]^{+\cdot}$ and [frag+2] for S-containing fragments (4.5% per S atom).

8.10.3 Sulfoxides and Sulfones

Aliphatic Sulfoxides [4,5]

Fragmentation: Most fragments are produced after rearrangement with non-specific H transfer to the O atom and subsequent OH$^{\cdot}$ elimination to yield $[M\text{-}17]^{+}$ or alkene elimination to $[M\text{-}alkene]^{+\cdot}$, followed by OH$^{\cdot}$, SOH$^{\cdot}$ (giving alk^{+} ions), or alk$^{\cdot}$ elimination (yielding $CH_2=S\text{-}OH^{+}$, m/z 63).

Ion series: Characteristic ion at m/z 63 ($CH_2=S\text{-}OH^{+}$) as well as alkyl and alkenyl fragments, C_nH_{2n+1} (29, 43, 57, 71, ...) and C_nH_{2n-1} (27, 41, 55, 69, ...).
Intensities: Intensive peaks evenly distributed over the whole mass range.
Molecular ion: Of medium intensity. Characteristic ^{34}S isotope peak at $[M+2]^{+\cdot}$ and [frag+2] for S-containing fragments (4.5% per S atom).

Alkyl Aryl and Diaryl Sulfoxides [4,5]

Fragmentation: Most fragments of methyl aryl sulfoxides are produced, after rearrangement to $CH_3S–O–ar^{+\cdot}$, by elimination of CH_2S (yielding $[M-46]^{+\cdot}$, a phenol), of CO (to $[M-28]^{+\cdot}$), and of $CH_3\cdot$ (to $[M-15]^+$). The latter ion loses CO to give the thiapyranyl cation (m/z 97 if ar is phenyl).

The skeletal rearrangement is not relevant for the fragmentation of higher alkyl aryl sulfoxides. Here, direct cleavage of the C–S bonds and McLafferty rearrangements dominate.

For diaryl sulfoxides, elimination of SO (to give $[M-48]^{+\cdot}$) as well as of O, $OH\cdot$, and $CHO\cdot$ (yielding $[M-16]^{+\cdot}$, $[M-17]^+$, and $[M-29]^+$, respectively). After rearrangement to sulfenates, cleavage of the S–O bond to produce $ar–S^+$ and $ar–O^+$ ions, which further lose CS and CO, respectively, to give $C_5H_5^+$ (m/z 65).

Ion series: Besides the ions described under *Fragmentation*, mainly fragments of the aromatic type, i.e., C_nH_n and $C_nH_{n\pm1}$ (m/z 39, 51–53, 63–65, 75–77, ...), as well as O- and S-containing ions.

Intensities: Intensive peaks mainly in the high mass range.

Molecular ion: Very strong. Characteristic ^{34}S isotope peak at $[M+2]^{+\cdot}$ and [frag+2] for S-containing fragments (4.5% per S atom).

Aliphatic Sulfones [4,5]

Fragmentation: Fragmentation of the S–C bond with the charge remaining on either side. Single and double H rearrangements to give $RS(O)OH^{+\cdot}$ and $RS(OH)_2^+$. The probability of the double H rearrangement increases with increasing chain length. If one of the substituents is unsaturated, rearrangement to $RS(O)O–alkene$ followed by cleavage of the S–O bond yields the ion RSO^+.

Ion series: Dominating aliphatic fragments, C_nH_{2n+1} (m/z 29, 43, 57, ...) and C_nH_{2n-1} (m/z 27, 41, 55, ...). Usually, one significant fragment corresponding to alk–S(O)OH$^{+\cdot}$ (from the series of m/z 80, 94, 108, ...) or alk–S(OH)$_2{}^+$ (from the series of m/z 81, 95, 109, ...) can be observed.

Intensities: Intensive peaks mainly of aliphatic fragments in the lower mass range.

Molecular ion: Weak. Characteristic ^{34}S isotope peak at [M+2]$^{+\cdot}$ and [frag+2] for S-containing fragments (4.5% per S atom).

Cyclic Sulfones [4]

Fragmentation: Dominant elimination of SO_2 (Δm 64, followed by loss of $CH_3{}^{\cdot}$), $HSO_2{}^{\cdot}$ (Δm 65, followed by loss of C_2H_4), or CH_2SO_2 (Δm 78). Weak signal at [M-17]$^+$ due to OH$^{\cdot}$ elimination.

Ion series: Mainly unsaturated hydrocarbon fragments, C_nH_{2n-1} (m/z 27, 41, 55, ...).

Intensities: Intensive peaks in the lower mass range.

Molecular ion: Moderate. Characteristic ^{34}S isotope peak at [M+2]$^{+\cdot}$ and [frag+2] for S-containing fragments (4.5% per S atom).

Alkyl Aryl Sulfones [4]

Fragmentation: Isomerization of M$^{+\cdot}$ to ar–OS(=O)alk and formation of the phenoxy ion or the phenol radical cation with H rearrangement. The migration of the aryl group depends on the type of substituents. It is facilitated by electron donors and hindered by acceptors. Mainly in substituted or unsaturated alkyl derivatives also isomerization to ar–S(=O)O–alk(ene) and formation of ar–S=O$^+$ (m/z 125 if ar is phenyl). Single and double H rearrangements to give ar–S(=O)OH$^{+\cdot}$ and ar–S(OH)$_2{}^+$. The probability of the double H rearrangement increases with increasing chain length. In some derivatives, SO_2 elimination from M$^{+\cdot}$ dominates. Substituents X of the alkyl group may migrate to the aryl group to yield X–ar–S=O$^+$ ions.

Ion series: Aromatic hydrocarbon fragments, C_nH_n and $C_nH_{n\pm1}$ (m/z 39, 51–53, 63–65, 75–77, ...), as well as S- and O-containing aromatic fragments at higher masses.

Intensities: Intensive peaks mainly in the higher mass range.

Molecular ion: Strong. Characteristic ^{34}S isotope peak at [M+2]$^{+\cdot}$ and [frag+2] for S-containing fragments (4.5% per S atom).

S

Diaryl Sulfones [4,5]

Fragmentation: Predominant aromatic fragments of the type ar–O$^+$ and ar–SO$^+$ (m/z 125 if ar is phenyl), formed after migration of one of the aryl groups. The ar–SO$_2$$^+$ ion is unimportant; ar$^+$ is intensive. Small signals due to SO$_2$, SO$_2$H$^.$, and SO$_2$H$_2$ eliminations (Δm 64, 65, and 66, respectively). With alkyl substituents in *ortho* position, [M-OH]$^+$ and [M-H$_2$O]$^{+.}$ are formed, upon which SO elimination follows.

Ion series: Aromatic fragments, C$_n$H$_n$ and C$_n$H$_{n\pm1}$ (m/z 39, 51–53, 63–65, 75–77, ...) and the S- and O-containing aromatic fragments at higher masses. Usually, ar–SO$^+$ (m/z 125 if ar is phenyl) is very strong.

Intensities: Intensive peaks mainly in the higher mass range.

Molecular ion: Strong. Characteristic ^{34}S isotope peak at [M+2]$^{+.}$ and [frag+2] for S-containing fragments (4.5% per S atom).

8.10.4   Sulfonic Acids and Their Esters and Amides

Aromatic Sulfonic Acids [6]

Fragmentation: The most prominent fragment, [M-HSO$_3$]$^+$ (Δm 81), is formed in a two-step process. In the first step, OH$^.$ elimination leads to a weak fragment ion [M-OH]$^+$ (Δm 17). If an alkyl group is present in *ortho* position, [M-H$_2$SO$_3$]$^{+.}$ (Δm 82) is formed instead of [M-81]$^+$. Other important fragments are [M-SO$_2$]$^{+.}$ (Δm 64), [M-HSO$_2$]$^+$ (Δm 65), and [M-SO$_3$]$^{+.}$ (Δm 80).

Ion series: Aromatic hydrocarbon fragments, C$_n$H$_n$ and C$_n$H$_{n\pm1}$ (m/z 39, 51–53, 63–65, 75–77, ...), and O-containing aromatic fragments at higher masses.

Intensities: Intensive peaks mainly in the higher mass range.

Molecular ion: Very strong. Characteristic ^{34}S isotope peak at [M+2]$^{+.}$ and [frag+2] for S-containing fragments (4.5% per S atom).

Alkylsulfonic Acid Esters [6]

Fragmentation: Loss of alkyl by fragmentation of the C–O bond with concomitant double H rearrangement to form the protonated sulfonic acid ion (m/z 97 for methanesulfonates), which then loses water. Loss of the alkoxyl residue (fragmentation of the S–O bond). Formation of an alkene ion from the alkyl ester group by a McLafferty-type rearrangement. In aryl esters, the phenoxy ion and the phenol radical cations dominate the spectrum.

Ion series: Besides RSO$_3$H$_2$$^+$ and RSO$_2$$^+$ (m/z 97 and 79 for methanesulfonates), for aliphatic esters mainly alkene fragments. In aryl esters, aromatic fragments, C$_n$H$_n$ and C$_n$H$_{n\pm1}$ (m/z 39, 51–53, 63–65, 75–77, ...), as well as O-containing aromatic fragments at higher masses.

Intensities: Intensive peaks in the lower mass range.

Molecular ion: Small or negligible signal for alkyl esters; intensive for aryl esters. Characteristic ^{34}S isotope peak at [M+2]$^{+.}$ and [frag+2] for S-containing fragments (4.5% per S atom).

Arylsulfonic Acid Esters [6]

Fragmentation: Dominating fragments resulting from cleavage of the S—O bond (leading to the ar–SO$_2^+$ ion), which loses SO$_2$ (m/z 155 and 91 for *p*-toluenesulfonates). In arylsulfonates with longer chains, double H rearrangement to give the protonated acid (m/z 173 for *p*-toluenesulfonates).
Ion series: Aromatic hydrocarbon fragments, C$_n$H$_n$ and C$_n$H$_{n\pm1}$ (m/z 39, 51–53, 63–65, 75–77, ...).
Intensities: Intensive peaks mainly in the higher mass range.
Molecular ion: Medium or weak. Characteristic ^{34}S isotope peak at [M+2]$^{+\cdot}$ and [frag+2] for S-containing fragments (4.5% per S atom).

Aromatic Sulfonamides [6]

Fragmentation: In *N*-alkylamides, the C—C bond next to N is split preferably. In *N*-arylsulfonamides, besides [M-SO$_2$]$^{+\cdot}$ and [M-HSO$_2$]$^+$, the ions ar–SO$_2^+$ and ar'–NH$^+$ are formed.

Ion series: Typical for the tosyl group are ions at m/z 155, 91, and 65.
Molecular ion: In arylsulfonamides, M$^{+\cdot}$ is dominant. Characteristic ^{34}S isotope peak at [M+2]$^{+\cdot}$ and [frag+2] for S-containing fragments (4.5% per S atom).

S

8.10.5 Thiocarboxylic Acid Esters [7]

In contrast to esters, the major fragmentation process is elimination of the alkyl radical from the thiol site. Ethylene sulfide is eliminated from thioesters with longer alkyl chains. Aromatic dithiocarboxylic acid esters usually fragment in two steps to the aryl cation.

8.10.6 References

[1] C.C. van de Sande, The mass spectra of ethers and sulphides. In: *Suppl. E, The Chemistry of Ethers, Crown Ethers, Hydroxyl Groups and Their Sulphur Analogues, Part 1*; S. Patai, Ed.; Wiley: Chichester, 1980; p 299.

[2] C. Lifshitz, Z.V. Zaretskii, The mass spectra of thiols. In: *The Chemistry of the Thiol Group, Part 1*; S. Patai, Ed.; Wiley: London, 1974; p 325.

[3] Q.N. Porter, Mass Spectrometry of Heterocyclic Compounds, 2nd ed.; Wiley: New York, 1985.

[4] K. Pihlaja, Mass spectra of sulfoxides and sulfones. In: The Chemistry of Sulphones and Sulphoxides; S. Patai, Z. Rappoport, C.G. Stirling, Eds.; Wiley: Chichester, 1988; p 125.

[5] R.A. Khmel'nitskii, Y.A. Efremov, Rearrangements in sulphoxides and sulphones induced by electron impact, *Russ. Chem. Rev.* **1977**, *46*, 46.

[6] S. Fornarini, Mass spectrometry of sulfonic acids and their derivatives; In: *The Chemistry of Sulphonic Acids, Esters, and their Derivatives*; S. Patai, Z. Rappoport, Eds.; Wiley: Chichester, 1991; p. 73.

[7] K.B. Tomer, C. Djerassi, Mass spectrometry in structural and stereochemical problems—CCXXV: Sulfur migration in the $[M{-}C_2H_4]^{+\cdot}$ ion of S-ethyl thiobenzoate, *Org. Mass Spectrom.* **1973**, *7*, 771.

S

8.11 Carbonyl Compounds [1–4]

8.11.1 Aldehydes

Aliphatic Aldehydes [5]

Fragmentation: Cleavage of the bond next to CO. The fragmentation of the hydrocarbon chain is similar to that in corresponding alkanes. McLafferty rearrangement with localization of the charge on either side, giving rise to $C_nH_{2n}^{+\cdot}$ (m/z 28, 42, 56, ...) and, often less important, to $C_nH_{2n}O^{+\cdot}$ ions (m/z 44, 58, 72, ...). At least one product (often both) is significant. Elimination of water from the molecular ion to give $[M-18]^{+\cdot}$, occasionally very pronounced.

Ion series: Dominating fragments of the series of C_nH_{2n+1} and $C_nH_{2n-1}O$ (in both cases: m/z 29, 43, 57, ...). Weaker signals of the series C_nH_{2n-1} (m/z 41, 55, 69, ...) and rearrangement products, C_nH_{2n} (m/z 28, 42, 56, ...).

Intensities: Intensive peaks concentrated in the lower mass range. Local even-mass maxima from McLafferty-type reactions ($[M-44]^{+\cdot}$ when the aldehyde is not substituted in α-position).

Molecular ion: Only strong for molecules of low molecular weight; very weak for $C_{n>9}$. $[M-1]^+$ may be more relevant than $M^{+\cdot}$.

Unsaturated Aliphatic Aldehydes

Fragmentation: Cleavage of the bond next to CO, leading to $[M-1]^+$ (more significant than in saturated aldehydes), $[M-29]^+$, and m/z 29. No McLafferty rearrangement occurs if the γ-hydrogen atom is attached to a double-bonded carbon or if there is a double bond in α,β-position.

Ion series: Fragments of the series of C_nH_{2n-1} and $C_nH_{2n-3}O$ (in both cases, m/z 41, 55, 69, ...).

Molecular ion: Stronger than in saturated aldehydes. Usually, $[M-1]^+$ is relevant.

Aromatic Aldehydes

Fragmentation: Characteristic H$^\cdot$ loss to yield the corresponding benzoyl ion, $[M-1]^+$, followed by decarbonylation to a phenyl ion, $[M-1-28]^+$, of lower intensity. To a small extent also decarbonylation of the molecular ion, leading to $[M-28]^{+\cdot}$. Weak signal at m/z 29 (CHO^+).

Ion series: Aromatic fragments corresponding to C_nH_n and $C_nH_{n\pm1}$ (m/z 39, 51–53, 63–65, 75–77, ...).

Intensities: Intensive peaks predominantly in the molecular ion region.

Molecular ion: Usually prominent. $[M-1]^+$ is strong.

$C = X$

8.11.2 Ketones

Aliphatic Ketones

Fragmentation: Cleavage of the bond next to CO is the most important primary fragmentation. The charge can remain on either side. The acyl ions then lose CO. McLafferty rearrangement giving rise to $C_nH_{2n}O^{+\cdot}$ ions (m/z 58, 72, 86, ...). Consecutive rearrangements occur if both alkyl chains contain a γ-H atom. Keto-enol tautomerism of the first rearrangement product is not a prerequisite for the second rearrangement to occur. Oxygen is sometimes indicated by weak signals at $[M-18]^{+\cdot}$ and m/z 31, 45, 59. Fragmentation of the hydrocarbon chain similar to that in the corresponding alkanes.
Ion series: Dominating fragments of the series C_nH_{2n+1} and $C_nH_{2n-1}O$ (in both cases m/z 29, 43, 57, ..., but often distinguishable by the intensity of the ^{13}C isotope signal), with maxima due to cleavage at the CO group to give acyl ions and their decarbonylation products. Weaker signals in the series C_nH_{2n-1} (m/z 41, 55, 69, ...). Even-mass maxima, $C_nH_{2n}O$ (m/z 58, 72, 86, ...), due to alkene elimination (McLafferty rearrangement). Usually, m/z 43 (CH_3CO^+) is strong if an unsubstituted α-CH_2 group is present.
Intensities: Intensive peaks mainly in the lower mass range.
Molecular ion: Relatively abundant, weak in long-chain and branched ketones.

Unsaturated Ketones

Fragmentation: Cleavage of the bond next to CO, more favorably on the saturated side, is the most important primary fragmentation. The acyl ion then loses CO. The McLafferty rearrangement occurs neither when the unsaturated substituents are in α,β-position nor when the only available γ-hydrogen atom is attached to a double-bonded carbon.
Molecular ion: Relatively abundant.

Cyclic Ketones

Fragmentation: Major primary fragmentation by bond cleavage next to carbonyl, followed by loss of alkyl residue.

C = X

Prominent McLafferty-type elimination of larger alkyl groups in position 2 or 6 as alkenes. This rearrangement is very favorable; even aromatically bonded H atoms can rearrange. For cyclohexanones, a consecutive retro-Diels–Alder reaction can occur:

Oxygen is sometimes indicated by a weak signal at $[M-18]^{+\cdot}$.

Ion series: Alkene fragments of the type of C_nH_{2n-1} or $C_nH_{2n-3}O$ (for both: m/z 41, 55, 69, ...) with maxima due to alkyl loss after ring opening next to the carbonyl group and H transfer. Prominent even-mass maxima by elimination of substituents at position 2 or 6 as alkenes via sterically favored McLafferty rearrangements.

Intensities: Overall more intensive peaks in the lower mass range or even distribution of major peaks over the whole mass range. Local maxima from major fragmentation pathway.

Molecular ion: Abundant.

Aromatic Ketones

Fragmentation: Dominant α-cleavage to give the benzoyl ion, followed by decarbonylation to a phenyl ion of lower intensity. α-Cleavage in acetophenone also produces the acetyl cation (m/z 43). Even-mass maxima due to alkene elimination via McLafferty rearrangement. CO elimination from diaryl ketones through skeletal rearrangements.

Ion series: Aromatic fragments corresponding to C_nH_n and $C_nH_{n\pm1}$ (m/z 39, 51–53, 63–65, 75–77, ...).

Intensities: Intensive peaks predominantly in the molecular ion region.

Molecular ion: Strong.

8.11.3 Carboxylic Acids

Aliphatic Carboxylic Acids

Fragmentation: Cleavage of the C–CO bond leading to m/z 45 and to $[M-45]^+$. Loss of OH· leading to $[M-17]^+$; may be followed by decarbonylation. Cleavage of the γ-bond (relative to CO) leads to $^+CH_2CH_2COOH$ (m/z 73) if there is no branching on the α- and β-C atoms. Loss of H· (not the carboxylic one) gives $[M-1]^+$. Water elimination to give $[M-18]^{+\cdot}$ if the alkyl group consists of at least 4 C atoms; may be followed by decarbonylation. McLafferty rearrangement to m/z 60 (acetic acid) if there is no α-substituent.

Ion series: Saturated and unsaturated alkyl ions mainly in the lower mass range $C = X$ (C_nH_{2n+1} and C_nH_{2n-1}, m/z 29, 43, 57, ... and 27, 41, 55, ...). With long-chain aliphatic acids, $C_nH_{2n-1}O_2$ series (m/z 59, 73, 87, ...), exhibiting maxima for n = 3, 7, 11, 15, ... (m/z 73, 129, 185, 241, ...). Even-mass maxima, $C_nH_{2n}O_2$ (m/z 60, 74, 88, ...), due to McLafferty rearrangements.

Intensities: Intensive peaks due to the above mentioned ions.

Molecular ion: Generally detectable. Easily protonated to $[M+H]^+$.

Aromatic Carboxylic Acids

Fragmentation: Pronounced loss of OH·, leading to $[M-17]^+$ and followed by decarbonylation (Δm 28) to a phenyl ion of lower intensity. Water elimination to

$[M-18]^{+\cdot}$ if a H-bearing *ortho*-substituent is present. Some acids decarboxylate (Δm 44). Loss of CO (Δm 28) from $M^{+\cdot}$.

X: CH_2, m/z 118
X: NH, m/z 119
X: O, m/z 120

Ion series: Aromatic hydrocarbon fragments, C_nH_n and $C_nH_{n\pm1}$ (m/z 39, 51–53, 63–65, 75–77, …).
Intensities: Intensive peaks predominantly in the molecular ion region.
Molecular ion: Strong.

8.11.4 Carboxylic Acid Anhydrides

Fragmentation: In the case of linear anhydrides, abundant acyl ions due to cleavage next to carbonyl group. For cyclic anhydrides, maxima due to decarboxylation (Δm 44), followed by decarbonylation.
Molecular ion: Weak or absent (especially in linear aliphatic anhydrides), easily protonated to $[M+H]^+$. Relatively strong for phthalic anhydrides.

8.11.5 Esters and Lactones

Esters of Aliphatic Carboxylic Acids

Fragmentation: Dominant fragmentation of the bonds next to the carbonyl C, leading to alk-CO^+ (m/z 43, 57, 71, …; decreasing intensity with increasing length of the alkyl chain) and followed by decarbonylation, as well as fragmentation to $COOR^+$ (m/z 59, 73, 87, …) and to alk^+ (m/z 15, 29, 43, …). Alcohol elimination to give $C_nH_{2n-2}O$ (m/z 42, 56, 70, …), followed by decarbonylation (Δm 28) or ketene elimination (Δm 42). Alkene elimination from the acid side via McLafferty rearrangements, leading to $C_nH_{2n}O_2$ (m/z 60, 74, 88, …). The larger alkyl group participates in the rearrangement if several γ-H atoms are available. In the following example, the alternative process leading to $[M-C_2H_4]^{+\cdot}$ is negligible:

$- C_5H_{10}$ $[M-70]^{+\cdot}$ $- {}^\cdot CH_3$

Nonspecific H rearrangements on the alcohol side (from $M^{+\cdot}$ or the McLafferty product) lead to $C_nH_{2n}O_2$ and to the corresponding alkene, C_nH_{2n} (m/z 28, 42, 56, …). In methyl esters of long chain acids, the ions $[(CH_2)_{2+4n}COOCH_3]^+$ (m/z 87, 143, 199, …) correspond to maxima. For esters of higher alcohols ($C_{n\geq3}$), double H rearrangement to the protonated acid, $C_nH_{2n+1}CO_2H_2^+$ (m/z 61, 75, 89, …). α-Substituted esters may lose the substituent and then CO (Δm 28) via alkoxyl rearrangement. Analogously, β-substituted esters may eliminate ketene (Δm 42).

Besides usual ester reactions, specific rearrangements can be observed in formates.

$$R^1 \quad (m/z\ 31\ for\ R^1\!:\ H)$$

Ion series: C_nH_{2n+1} (m/z 29, 43, 57, ...) for the alkyl groups at the ester oxygen (except for methyl esters). C_nH_{2n-1} (m/z 27, 41, 55, ...). $C_nH_{2n-1}O_2$ (m/z 59, 73, 87, ...), exhibiting maxima for n = 4, 8, 12, ... (m/z 87, 143, 199, ...) in the case of methyl esters of long-chain acids. Even-mass maxima for $C_nH_{2n}O_2$ (m/z 60, 74, 88, ...) due to alkene elimination via McLafferty rearrangements on both sides of the carboxyl group. C_nH_{2n} (m/z 28, 42, 56, ...) as H rearrangement product from the alcohol side.

Intensities: Intensive peaks due to the above mentioned ions in the lower mass range.

Molecular ion: Often of low abundance. Easily protonated to $[M+H]^+$.

Esters of Unsaturated Carboxylic Acids

α,β-Unsaturated esters: Loss of alk–O˙ followed by CO elimination is the dominant fragmentation path. Also, loss of the δ-substituent yields a 6-membered oxonium ring:

$$m/z\ 113$$

Significant difference between Z and E isomers of long-chain α,β-unsaturated esters: Single H rearrangement occurs with Z esters, and double H rearrangements (leading to protonated acids) have been found for E esters.

β,γ-Unsaturated esters: Only slight qualitative, but significant quantitative differences have been observed as compared to α,β-unsaturated esters (e.g., less intensive signals for $M^{+\cdot}$ of β,γ- than of α,β-unsaturated esters).

γ,δ-Unsaturated esters: Loss of the alcohol chain as a radical, R˙, followed by ketene elimination.

Aliphatic enol esters and aryl esters: Formation of alk–CO^+ (m/z 43, 57, 71, ...). Elimination of a ketene to give the enol or phenol radical cation. The rearrangement occurs prodominantly, but not exclusively, through a 4-membered transition state:

$C = X$

$$[M–42]^{+\cdot}\ for\ R\!:\ H$$

Esters of Aromatic Acids

Fragmentation: Dominant loss of RO˙ to form the benzoyl ion, followed by decarbonylation (Δm 28) and further loss of acetylene (Δm 26). Ethyl esters also eliminate C_2H_4 (Δm 28) to give the acid radical cation, which then loses OH˙ to yield the benzoyl ion. In higher alkyl esters, besides the acid, the protonated acid is formed (double H rearrangement). In *ortho*-substituted aryl esters with an α-hydrogen atom on the substituent, an alcohol is eliminated from $M^{+\cdot}$. In the case of alkyl phthalates (other than dimethyl phthalate), alkenyl elimination from one ester group to give the protonated ester acid, followed by alkene elimination from the other ester group, and subsequent water elimination to the protonated anhydride ion, which forms the base peak at m/z 149.

Ion series: Aromatic hydrocarbon fragments, C_nH_n and $C_nH_{n\pm1}$ (m/z 39, 51–53, 63–65, 75–77, …).

Intensities: Prominent maximum at the mass of the related benzoyl ion and its decarbonylation product.

Molecular ion: Usually strong.

Lactones

Fragmentation: The most prominent reaction is the loss of substituents (or H·) at the O-bearing C atom, followed by decarbonylation (Δm 28), decarboxylation (Δm 44, mainly in smaller molecules), and ketene elimination (Δm 42). Decarboxylation of $M^{+\cdot}$ is rarely significant. Competing reactions are several kinds of primary ring cleavages. Aromatic lactones show maxima due to two consecutive decarbonylations.

Ion series: No specific ion series. The acetyl ion (m/z 43) is often an important fragment.

Intensities: Maxima at the mass resulting from loss of substituents at the C atom next to oxygen. Otherwise, intensive peaks evenly distributed over the whole mass range.

Molecular ion: Usually of low intensity and easily protonated to $[M+H]^+$ in aliphatic lactones; abundant in the case of aromatic lactones.

8.11.6 Amides and Lactams

Amides of Aliphatic Carboxylic Acids

Fragmentation: Alkene elimination on the acid side via McLafferty reaction to yield the corresponding acetamide radical cation. Loss of alkenes on the amine side to give the ion of the desalkyl amide, often via double H rearrangement to the protonated desalkyl amide ion. Bond cleavage on both sides of the carbonyl group. Cleavage of the C–C bond attached to N, and the β,γ-C–C bond (relative to N):

Cleavage of the bonds to the β-C (see scheme) and to the γ-C on the acid side.

Ion series: Even-mass fragments corresponding to $C_nH_{2n}NO$ (m/z 44, 58, 72, …)

produced by cleavage of the bond next to CO on the acid side. Odd-mass fragments (in secondary and tertiary amides), $C_nH_{2n-1}O$ (m/z 43, 57, 71, ...), produced by cleavage of the bond next to CO on the amine side.

Intensities: Overall peak distribution maximizing in the low mass range. Local maxima from McLafferty and from γ-cleavage products.

Molecular ion: Significant. Strong tendency to protonate to $[M+H]^+$.

Amides of Aromatic Carboxylic Acids

Fragmentation: Maxima due to amide bond cleavage yielding the benzoyl ion, followed by decarbonylation (Δm 28).

Ion series: Aromatic fragments corresponding to C_nH_n and $C_nH_{n\pm1}$ (m/z 39, 51–53, 63–65, 75–77, ...).

Intensities: Intensive peaks predominantly in the molecular ion region.

Molecular ion: Abundant. $[M-H]^+$ is significant in *N,N*-disubstituted anilides, weaker in monosubstituted derivatives, and absent from the spectrum of benzamide. It is formed exclusively by loss of *ortho*-hydrogens of the aromatic ring.

Anilides

Formanilides: Loss of CO (Δm 28) to give the aniline radical cation and consecutive HCN elimination (Δm 27).

Acetanilides: Ketene elimination gives the aniline radical cation (often base peak), which can eliminate HCN (Δm 27), and formation of the acetyl cation (m/z 43).

Trichloroacetanilides: Dominant loss of $CCl_3^{\cdot}$ (Δm 117).

Pivalanilides: Besides reactions analogous to those of acetanilides (Δm 84, formation of the aniline radical cation), also formation of the *tert*-butylbenzene radical cation through elimination of HNCO (Δm 43).

Lactams

Fragmentation: Cleavage of the C–C bond at the N-bearing C atom. Cleavage of the CO–N bond, followed by loss of CO (Δm 28) or by further cleavage of the C–C bond next to N, giving an iminium ion. In 2-pyrrolidone and 2-piperidone, the signal at m/z 30 ($[CH_2=NH_2]^+$) is strong. The base peak of 2-pyridone is formed by CO elimination (Δm 28).

C = X

2-Pyrrolidone:

2-Piperidone:

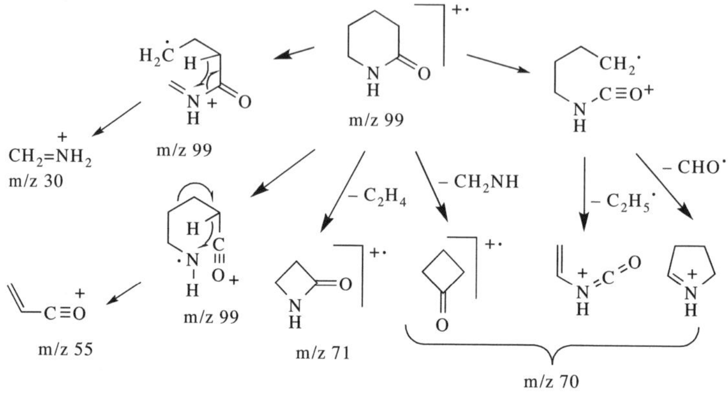

Molecular ion: Often observable; more abundant than for the corresponding lactones.

8.11.7 Imides

Saturated acyclic imides: Consecutive CO (Δm 28) and alkoxy elimination:

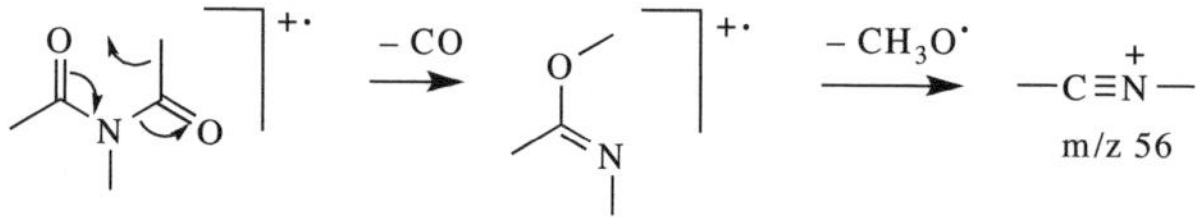

Ketene elimination:

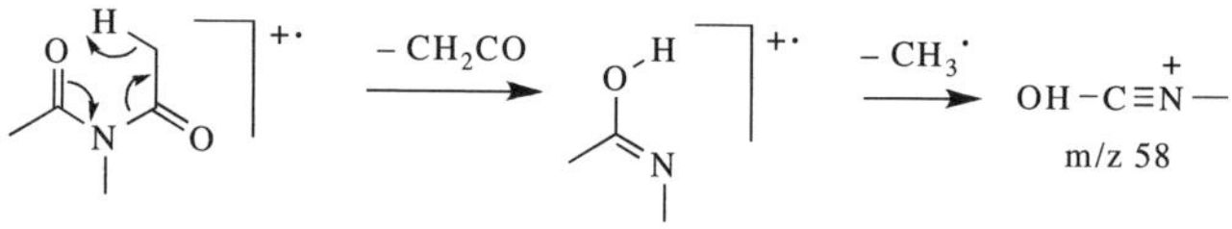

If the *N*-substituent chain is sufficiently long, cleavage of the C–C bonds next to N, with or without H rearrangement.

C = X

Dibenzoylamine: Loss of CO to *N*-phenylbenzamide:

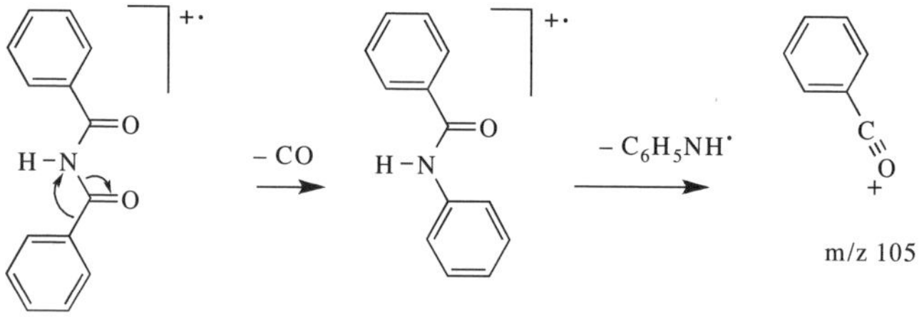

Cyclic imides: The spectra of saturated cyclic imides are almost identical to those of the corresponding diketones. Loss of HNCO (Δm 43) from succinimide, followed by CO elimination (Δm 28). Aroyl migration and loss of CO_2 from aromatic cyclic imides.

8.11.8 References

[1] J.H. Bowie, Mass spectrometry of carbonyl compounds. In: *The Chemistry of the Carbonyl Group, Vol. 2*; J. Zabicky, Ed.; Interscience: London, 1970; p 277.

[2] S.W. Tam, Mass spectra of acid derivatives. In: *Suppl. B, The Chemistry of Acid Derivatives, Part 1*; S. Patai, Ed.; Wiley: Chichester, 1979; p 121.

[3] D.G.I. Kingston, J.T. Bursey, M.M. Bursey, Intramolecular hydrogen transfer in mass spectra. II. The McLafferty rearrangement and related reactions, *Chem. Rev.* **1974**, *74*, 215.

[4] D.G.I. Kingston, B.W. Hobrock, M.M.Bursey, J.T. Bursey, Intramolecular hydrogen transfer in mass spectra. III. Rearrangements involving the loss of small neutral molecules, *Chem. Rev.* **1975**, *75*, 693.

[5] A.G. Harrison, High-resolution mass spectra of aliphatic aldehydes, *Org. Mass Spectrom.* **1970**, *3*, 549.

C = X

8.12 Miscellaneous Compounds

8.12.1 Trialkylsilyl Ethers [1,2]

Fragmentation: Loss of alkyl attached to Si (preferential loss of larger groups). Cleavage of the C–C bond adjacent to O, followed by alkene elimination. Loss of alkoxyl, followed by alkene eliminations. Elimination of trialkylsilanol. The R_2Si–OR' cation has the tendency to attack, in an electrophilic manner and even over long distances, free electron pairs and π-electron centers, causing the expulsion of neutral fragments from the interior of the molecule via a rearrangement:

$$Br-(CH_2)_{10}-O-\overset{|+\cdot}{\underset{|}{Si}}\!\!\!-\!\!\!<\quad\xrightarrow[\Delta m\ 57]{-\overset{\cdot}{C}(CH_3)_3}\quad\xrightarrow[\Delta m\ 156]{-(CH_2)_{10}O}\quad Br-\overset{/}{\underset{\backslash}{Si}}+$$

Ion series: $[C_nH_{2n+3}OSi]^+$ (m/z 75, 89, 103, 117, …). $[C_nH_{2n+3}Si]^+$ (m/z 45, 59, 73, 87, …). Occasionally, maxima at even mass due to elimination of trialkylsilanol.
Molecular ion: $M^{+\cdot}$ often of low abundance or absent, easily protonated to $[M+H]^+$. Typical isotope patterns owing to ^{28}Si, ^{29}Si, and ^{30}Si (see Chapter 2.5.5).

8.12.2 Phosphorus Compounds

Alkyl Phosphates [3]

Fragmentation: Maxima due to alkenyl loss from $M^{+\cdot}$ via double H rearrangement, followed by successive alkene eliminations down to protonated phosphoric acid (m/z 99).
Ion series: PO^+ (m/z 47), $H_2PO_2^+$ (m/z 65), $H_2PO_3^+$ (m/z 81), often as nonspecific P indicators.
Molecular ion: $M^{+\cdot}$ observable.

Aliphatic Phosphines

Ion series: Maxima of the ion series of $[C_nH_{2n+3}P]^+$ (m/z 48, 62, 76, 90, …) due to alkene eliminations.
Molecular ion: $M^{+\cdot}$ observable.

P Si

Aromatic Phosphines and Phosphine Oxides

Fragmentation: Maxima due to loss of an aryl group, followed by H_2 elimination to yield the 9-phosphafluorenyl ion (m/z 183).
Molecular ion: $M^{+\cdot}$ abundant, easily losing $H^\cdot$ to give $[M-1]^+$.

m/z 183

8.12.3 References

[1] D.G.I. Kingston, B.W. Hobrock, M.M. Bursey, J.T. Bursey, Intramolecular hydrogen transfer in mass spectra. III. Rearrangements involving the loss of small neutral molecules, *Chem. Rev.* **1975**, *75*, 693.

[2] H. Schwarz, Positive and negative ion chemistry of silicon-containing molecules in the gas phase. In: *The Chemistry of Organic Silicon Compounds, Part 1*; S. Patai, Z. Rappoport, Eds.; Wiley: Chichester, 1989; p 445.

[3] D.G.I. Kingston, J.T. Bursey, M.M. Bursey, Intramolecular hydrogen transfer in mass spectra. II. The McLafferty rearrangement and related reactions, *Chem. Rev.* **1974**, *74*, 215.

P Si

8.13   Mass Spectra of Common Solvents and Matrix Compounds

8.13.1   Electron Impact Ionization Mass Spectra of Common Solvents

The label {50} indicates that the intensity scale ends at 50% relative intensity and is subdivided in 10% steps. In these cases, the height of the base peak has to be doubled to bring it to 100%. All spectra represent positive ions only.

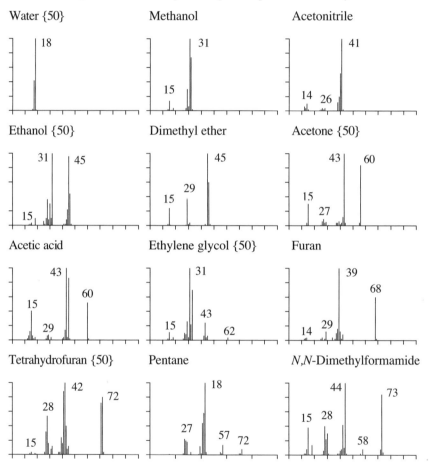

Water {50}   Methanol   Acetonitrile

Ethanol {50}   Dimethyl ether   Acetone {50}

Acetic acid   Ethylene glycol {50}   Furan

Tetrahydrofuran {50}   Pentane   N,N-Dimethylformamide

Solvents

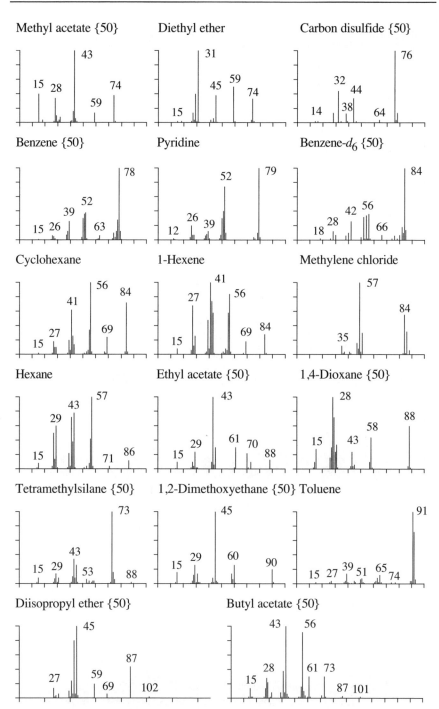

Methyl acetate {50}

Diethyl ether

Carbon disulfide {50}

Benzene {50}

Pyridine

Benzene-d_6 {50}

Cyclohexane

1-Hexene

Methylene chloride

Hexane

Ethyl acetate {50}

1,4-Dioxane {50}

Tetramethylsilane {50}

1,2-Dimethoxyethane {50}

Toluene

Diisopropyl ether {50}

Butyl acetate {50}

Solvents

Chloroform

Chloroform-*d*

Trichloroethylene

Carbon tetrachloride

Tetrachloroethylene

Dibutyl phthalate [25] (frequent impurity due to its use as polymer plasticizer)

Dioctyl phthalate (frequent impurity due to its use as polymer plasticizer)

Heptacosafluorotributylamine (calibration reagent)

Solvents

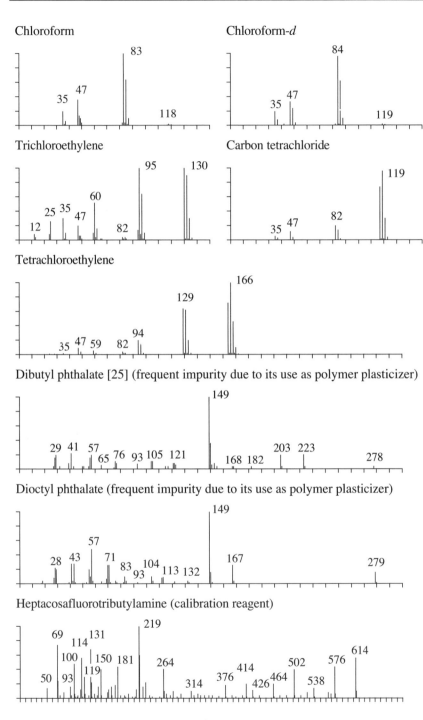

8.13.2 Spectra of Common FAB MS Matrix and Calibration Compounds

Fast atom bombardment (FAB) mass spectra (MS) usually exhibit signals for the protonated or deprotonated molecular ions, $[M\pm H]^{\pm}$, and protonated clusters, $[M_n+X_m\pm H]^{\pm}$ $(n, m = 0, 1, 2, \ldots)$, of the sample and matrix molecules, X. Even traces of metal salts in the sample give rise to clusters of the type $[M_n+X_m+\text{metal cation}]^+$. Na^+ (23 u) and K^+ (39 u) adducts are often found. The nature of the clusters is often revealed by the regular intervals at which their peaks occur in the spectra.

Calibration Compounds in Positive Ionization FAB Mass Spectra

Ultramark 1621 (erroneously also referred to as perfluoroalkyl phosphazine)

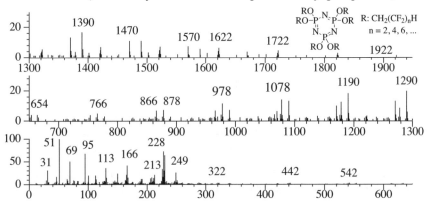

Polyethylene glycol 600 (often used as internal reference for high resolution m/z determinations)

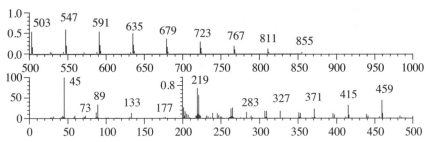

CsI (Cs^+, 132.9; I^-, 126.9) in glycerol (formation of $[\text{glycerol}_m\text{-}H_n+Cs_p+I_q]^+$)

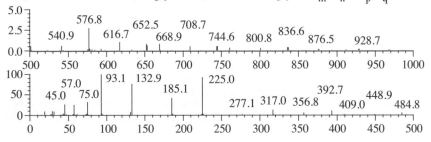

Solvents

Matrix Compounds in Positive Ionization FAB Mass Spectra

3-Nitrobenzyl alcohol (M_r 153.1)

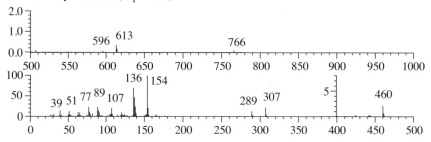

Glycerol (M_r 92.1)

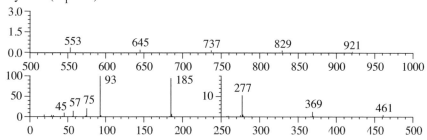

1-Thioglycerol (M_r 108.2. Note m/z 23, Na$^+$; 131, [M+Na]$^+$; 239, [2M+Na]$^+$. Similarly, small K$^+$ impurities give signals at m/z 39, 147, 255)

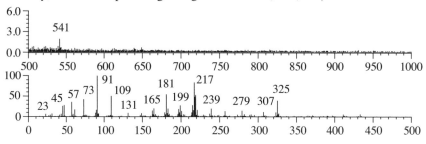

1,4,7,10,13,16-Hexaoxacyclooctadecane (18-crown-6, M_r 264.3. Also used as an additive; binds metal ions and reduces [M+metal ion]$^+$ in favor of [M+H]$^+$, which can be important for samples with exchangeable H$^+$, such as for peptides [1])

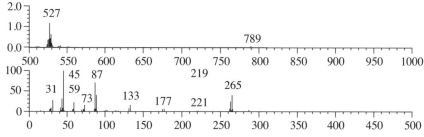

Solvents

2-Nitrophenyl octyl ether (M$_r$ 251.3)

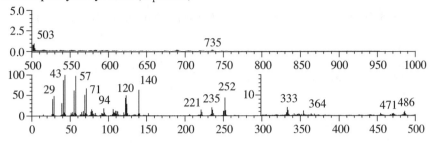

Triethanolamine (M$_r$ 149.2)

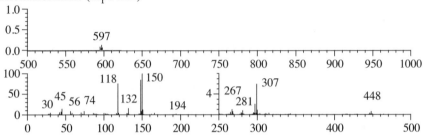

Sulfolane (M$_r$ 120.2) [2]

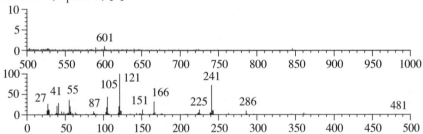

Hexadecylpyridinium bromide (M$_r$ 384.4; for [hexadecylpyridinium]$^+$ m/z 304.3) in 2-nitrobenzyl alcohol

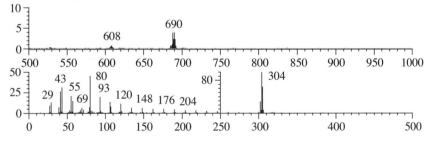

Solvents

Calibration Compounds in Negative Ionization FAB Mass Spectra

Ultramark 1621 (erroneously also referred to as perfluoroalkyl phosphazine)

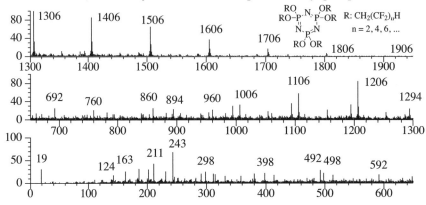

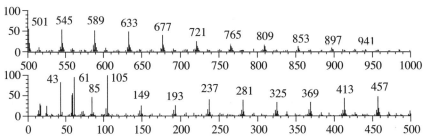

Polyethylene glycol 600 (often used as internal reference for high resolution m/z determinations)

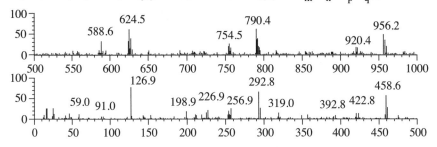

KI (K^+, 39.1; I^-, 126.9) in glycerol (formation of $[glycerol_m-H_n+K_p+I_q]^-$)

Matrix Compounds in Negative Ionization FAB Mass Spectra

3-Nitrobenzyl alcohol (M_r 153.1)

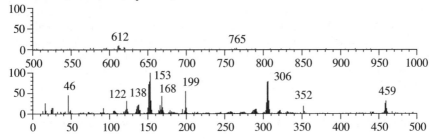

Glycerol (M_r 92.1)

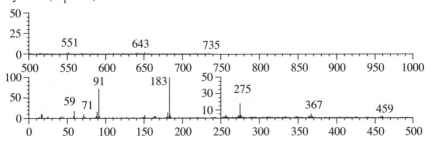

2-Nitrophenyl octyl ether (M_r 251.3)

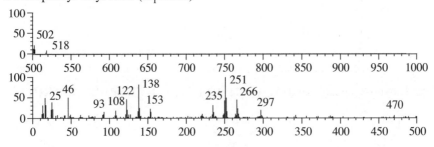

2-Nitrobenzyl alcohol solution of hexadecylpyridinium bromide (M_r 384.4; enhances detectability and reduces metal ion adducts of sample [3])

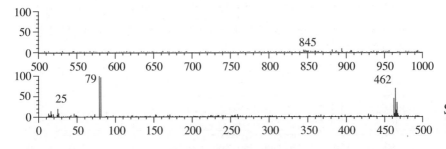

Solvents

8.13.3 Spectra of Common MALDI MS Matrix Compounds

Matrix-assisted laser desorption ionization (MALDI) mass spectra (MS) usually show signals for protonated or deprotonated molecular ions, $[M\pm H]^{\pm}$, and protonated clusters, $[M_n+X_m\pm H]^{\pm}$ (n, m = 0, 1, 2, ...), of the sample and matrix molecules, X. In positive ionization mass spectra, clusters of the type $[M_n+X_m+\text{metal cation}]^+$ occur even if there are only traces of metal salts in the sample. Sodium (23 u) and potassium (39 u) ion adducts are often encountered. The nature of the clusters is revealed by the regular intervals at which their signals occur in the spectra [4].

Matrix Compounds in Positive Ionization MALDI Mass Spectra

3-Aminoquinoline (M_r 144.2)

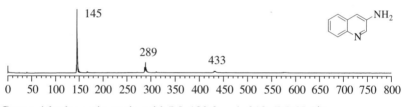

α-Cyano-4-hydroxycinnamic acid (M_r 189.2; m/z 212, $[M+Na]^+$)

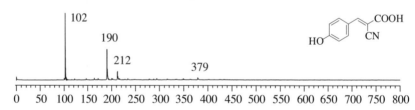

2,5-Dihydroxybenzoic acid (M_r 154.1; m/z 177, $[M+Na]^+$; m/z 193, $[M+K]^+$)

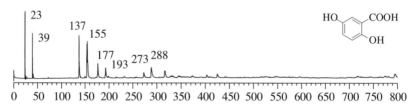

2,6-Dihydroxyacetophenone (M_r 152.1; m/z 175, $[M+Na]^+$; m/z 191, $[M+K]^+$; m/z 365, $[2M+Na+K-H]^+$?)

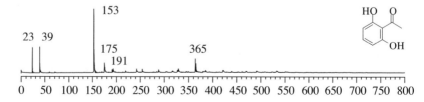

Dithranol (M_r 226.2)

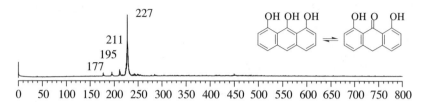

Ferulic acid (4-hydroxy-4-methoxycinnamic acid; M_r 194.2)

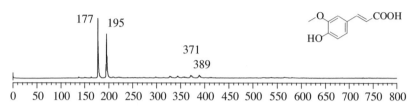

Sinapinic acid (3,5-dimethoxy-4-hydroxycinnamic acid; M_r 224.2; m/z 471, $[2M+Na]^+$)

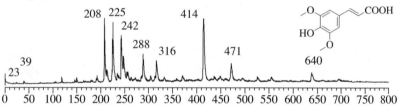

Matrix Compounds in Negative Ionization MALDI Mass Spectra

3-Aminoquinoline (M_r 144.2)

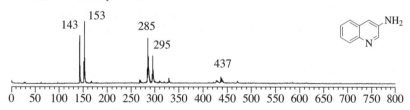

α-Cyano-4-hydroxycinnamic acid (M_r 189.2; m/z 399, $[2M+Na-2H]^-$)

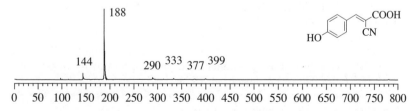

Solvents

2,5-Dihydroxybenzoic acid (M_r 154.1)

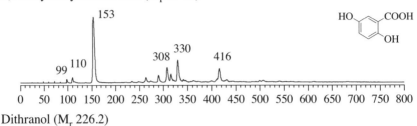

Dithranol (M_r 226.2)

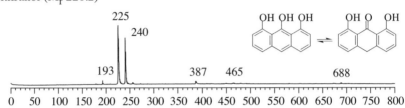

Ferulic acid (4-hydroxy-4-methoxycinnamic acid; M_r 194.2)

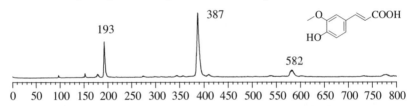

Sinapinic acid (3,5-dimethoxy-4-hydroxycinnamic acid; M_r 224.2)

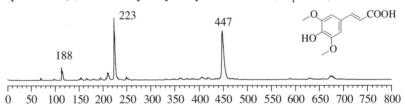

8.13.4 References

[1] R. Orlando, Analysis of peptides contaminated with alkali-metal salts by fast atom bombardment mass spectrometry using crown ethers, *Anal. Chem.* **1992**, *64*, 332.

[2] P.K. Singh, L. Field, B.J. Sweetman, Organic disulfides and related substances, *J. Org. Chem.* **1988**, *53*, 2608.

[3] Z.-H. Huang, B.-J. Shyong, D.A. Gage, K.R. Noon, J. Allison, N Alkylnicotinium halides: A class of cationic matrix additives for enhancing the sensitivity in negative ion fast-atom bombardment mass spectrometry of polyanionic analytes, *J. Am. Soc. Mass Spectrom.* **1994**, *5*, 935.

[4] A.E. Ashcroft, *Ionization Methods in Organic Mass Spectrometry*, The Royal Society of Chemistry: Cambridge, 1997.

Solvents

9 UV/Vis Spectroscopy

9.1 Correlation between Wavelength of Absorbed Radiation and Observed Color

Absorbed light		Observed (transmitted) color
Wavelength [nm]	Corresponding color	
400	violet	yellow-green
425	indigo blue	yellow
450	blue	orange
490	blue-green	red
510	green	purple
530	yellow-green	violet
550	yellow	indigo blue
590	orange	blue
640	red	blue-green
730	purple	green

9.2 Simple Chromophores

Chromophore	Compound	Transition	λ_{max} [nm]	ε_{max}	Solvent
C–H	CH_4	$\sigma \rightarrow \sigma^*$	122	strong	gas
C–C	CH_3–CH_3	$\sigma \rightarrow \sigma^*$	135	strong	gas
C=C	CH_2=CH_2	$\pi \rightarrow \pi^*$	162	15000	heptane
	$(CH_3)_2C$=$C(CH_3)_2$	$\pi \rightarrow \pi^*$	196	11500	heptane
C=C=C	CH_2=C=CH_2		170	4000	
			227	630	
C≡C	HC≡CH		173	6000	gas
	n-C_5H_{11}–C≡C–CH_3		178	10000	hexane
			196	2000	
			222	160	
C–Cl	CH_3Cl	$n \rightarrow \sigma^*$	173	200	hexane
C–Br	n-C_3H_7Br	$n \rightarrow \sigma^*$	208	300	hexane

Chromophore	Compound	Transition	λ_{max} [nm]	ε_{max}	Solvent
C–I	CH_3I	$n \rightarrow \sigma^*$	259	400	hexane
C–O	CH_3OH	$n \rightarrow \sigma^*$	177	200	hexane
	CH_3OCH_3	$n \rightarrow \sigma^*$	184	2500	gas
C–N	$(C_2H_5)_2NH$	$n \rightarrow \sigma^*$	193	2500	hexane
	$(CH_3)_3N$	$n \rightarrow \sigma^*$	199	4000	hexane
C=N	$H_2N-C(=NH)-NH_2 \cdot HCl$		265	15	water
	$(CH_3)_2C=NOH$		193	2000	ethanol
	$(CH_3)_2C=NONa$		265	200	ethanol
N=N	$CH_3-N=N-CH_3$		340	16	ethanol
N=O	$(CH_3)_3C-NO$		300	100	ether
			665	20	
	$(CH_3)_3C-NO_2$		276	27	ethanol
	$n\text{-}C_4H_9-O-NO$		218	1050	ethanol
			313–384	20–40	ethanol
	$C_2H_5-O-NO_2$		260	15	ethanol
C≡N	$CH_3C \equiv N$		<190		
X=Y=Z	$C_2H_5-N=C=S$		250	1200	hexane
	$C_2H_5-N=C=N-C_2H_5$		230	4000	
			270	25	
C–S	CH_3SH	$n \rightarrow \sigma^*$	195	1800	gas
		$n \rightarrow \sigma^*$	235	180	
	$C_2H_5-S-C_2H_5$	$n \rightarrow \sigma^*$	194	4500	gas
		$n \rightarrow \sigma^*$	225	1800	
	$C_2H_5-S-S-C_2H_5$	$n \rightarrow \sigma^*$	194	5500	hexane
		$n \rightarrow \sigma^*$	250	380	
C=S	$(CH_3)_2C=S$		460	weak	
	cyclohexanethione (=S)		495	weak	ethanol
C=O	$(CH_3)_2C=O$	$n \rightarrow \sigma^*$	166	16000	gas
		$\pi \rightarrow \pi^*$	189	900	hexane
		$n \rightarrow \pi^*$	279	15	hexane
	CH_3COOH	$n \rightarrow \pi^*$	200	50	gas
	CH_3COONa	$n \rightarrow \pi^*$	210	150	water
	$CH_3COOC_2H_5$	$n \rightarrow \pi^*$	210	50	gas
	CH_3CONH_2	$n \rightarrow \pi^*$	220	63	water
	succinimide (O=, N–H, =O)		191	15200	CH_3CN
C=C=O	$(C_2H_5)_2C=C=O$		227	360	
			375	20	

9.3 Conjugated Alkenes

9.3.1 Dienes and Polyenes

The $\pi \rightarrow \pi^*$ transition of conjugated double bonds is above ≈ 200 nm with typical intensities of the order of $\log \varepsilon \approx 4$. Its position can be estimated with the Woodward–Fieser rule. For cross-conjugated systems, the value for the chromophore absorbing at the longest wavelength has to be calculated.

Woodward–Fieser rule for estimating the position of the $\pi \rightarrow \pi^*$ transition (λ_{max} in nm)

Parent system		acyclic	217
		heteroannular	214
		homoannular	253
Increments	for each additional conjugated double bond		+30
	for each exocyclic double bond		+5
	for each substituent	C-substituent	+5
		Cl	+5
		Br	+5
		O–alkyl	+6
		OCOCH$_3$	0
		N(alkyl)$_2$	+60
		S–alkyl	+30
Solvent correction			≈ 0

Example: Estimation of the absorption maximum for

base value (homoannular)	253
1 additional conjugated double bond	30
1 exocyclic double bond	5
3 C-substituents	15
1 OCOCH$_3$	0
estimated	303
experimental	306

9.3.2 α,β-Unsaturated Carbonyl Compounds

The $\pi \rightarrow \pi^*$ transition of α,β-unsaturated carbonyl compounds is above ≈200 nm with typical intensities of the order of log ε ≈ 4. Its position can be estimated with the extended Woodward rule. For cross-conjugated systems, the value for the chromophore absorbing at the longest wavelength must be calculated.

Extended Woodward rule for estimating the position of the $\pi \rightarrow \pi^*$ transition (λ_{max} in nm)

Parent system			
		X: alkyl	215
		X: H	207
		X: OH	193
		X: O–alkyl	193
			215
			202

Increments	for each additional conjugated double bond	+30
	for each exocyclic double bond C=C	+5
	for each homoannular diene system	+39

| For each substituent on double bond system | Increment | | | |
	α	β	γ	δ and beyond
C-substituent	10	12	18	18
Cl	15	12		
Br	25	30		
OH	35	30		50
O–alkyl	35	30	17	31
O–COCH$_3$	6	6	6	6
S–alkyl		85		
N(alkyl)$_2$		95		

Solvent corrections	Solvent	Correction term
	water	-8
	hexane	11
	cyclohexane	11
	chloroform	1
	methanol	0
	ethanol	0
	diethyl ether	7
	dioxane	5

Example: Estimation of the absorption maximum in ethanol for

base value	215
2 additional conjugated double bonds	60
exocyclic double bond	5
homoannular diene system	39
1 β-C-substituent	12
3 additional C-substituents	54
solvent correction	0
estimated	385
experimental	388

9.4 Aromatic Hydrocarbons

9.4.1 Monosubstituted Benzenes

Typical Ranges for Monosubstituted Benzenes (λ_{max} in nm)

Transition	λ_{max}	ε
$\pi \rightarrow \pi^*$ (allowed)	180–230	2000–10000
$\pi \rightarrow \pi^*$ (forbidden)	250–290	100–2000
$\pi \rightarrow \pi^*$ (substituent delocalized by aryl; K band)	220–250	10000–30000
$n \rightarrow \pi^*$ (substituent with lone pair; R band)	275–350	10–100

Specific Examples of Monosubstituted Benzenes (λ_{max} in nm)

Substituent R (solvent)	$\pi \rightarrow \pi^*$ (allowed) λ_{max}	ε	$\pi \rightarrow \pi^*$ (forbidden) λ_{max}	ε	$\pi \rightarrow \pi^*$ (K band) λ_{max}	ε	$n \rightarrow \pi^*$ (R band) λ_{max}	ε
–H (cyclohexane)	198	8000	255	230				
–CH_3 (hexane)	208	7900	262	230				
–$CH=CH_2$ (ethanol)			282	450	244	12000		
–$C\equiv CH$ (hexane)			278	650	236	12500		
–Cl (ethanol)	210	7500	257	170				
–OH (water)	211	6200	270	1450				
–O^- (water)	235	9400	287	2600				
–NH_2 (water)	230	8600	280	1430				
–NH_3^+ (water)	203	7500	254	160				
–NO_2 (Hexan)	208	9800	270	800	251	9000	322	150
	213	8100						
–$C\equiv N$ (water)			271	1000	224	13000		
–CHO (hexane)			280	1400	242	14000	≈330	≈60
–$COCH_3$ (ethanol)			278	1100	243	13000	319	50
–COOH (water)	202	8000	270	800	230	10000		

9.4.2 Polysubstituted Benzenes

Estimation of the position of the allowed $\pi \rightarrow \pi^*$ transition in multiply substituted benzenes (λ_{max} in nm, log $\varepsilon \approx 4$)

Base value: 203.5

Substituent	Increment [nm]
–CH$_3$	3.0
–Cl	6.0
–Br	6.5
–OH	7.0
–O$^-$	31.5
–OCH$_3$	13.5
–NH$_2$	26.5
–NHCOCH$_3$	38.5
–NO$_2$	65.0
–C≡N	20.5
–CHO	46.0
–COCH$_3$	42.0
–COOH	25.5

9.4.3 Aromatic Carbonyl Compounds

Scott rules for estimating the position of the K band (solvent: ethanol; λ_{max} in nm, $\varepsilon = 10000–30000$)

Parent systems

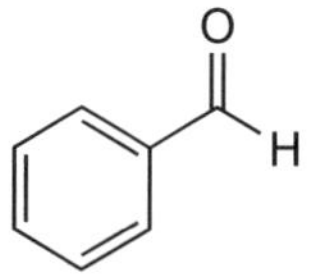

H   250

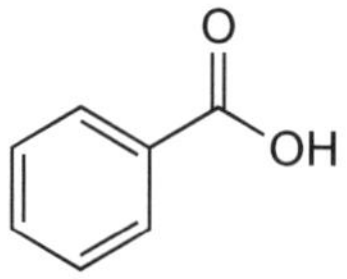

OH   230

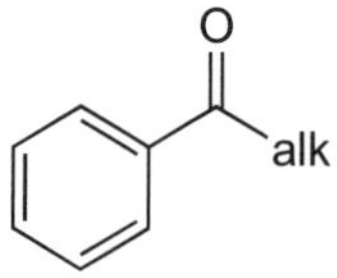

alk   246

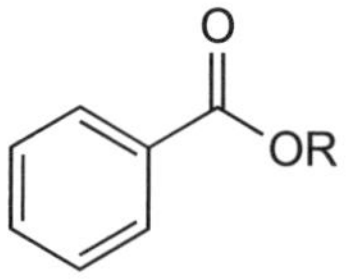

OR   230

Increments	Substituent	*ortho*	*meta*	*para*
	–alkyl	3	3	10
	–cycloalkyl	3	3	10
	–Cl	0	0	
	–Br	2	2	15
	–OH	7	7	25
	–O–alkyl	7	7	25
	–O⁻	11	20	78
	–NH$_2$	13	13	58
	–N(CH$_3$)$_2$	20	20	85
	–NHCOCH$_3$	20	20	45

Example: Estimation of the absorption maximum (K band) for

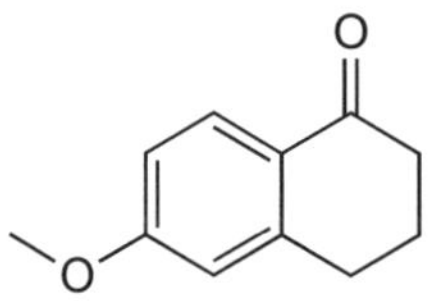

base value	246
ortho-cycloalkyl	3
para-O–alkyl	25
estimated	274
experimental	276

9.5 Reference Spectra

9.5.1 Alkenes and Alkynes

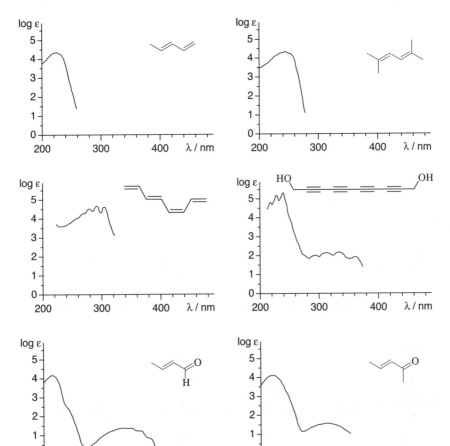

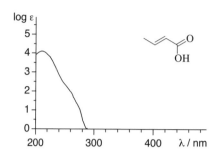

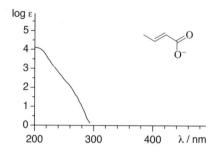

9.5.2 Aromatic Compounds

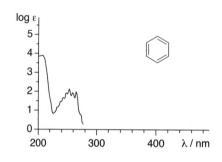

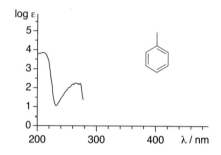

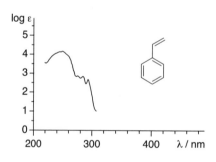

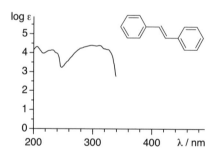

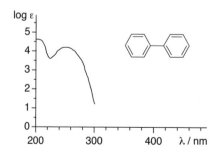

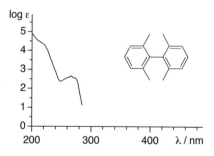

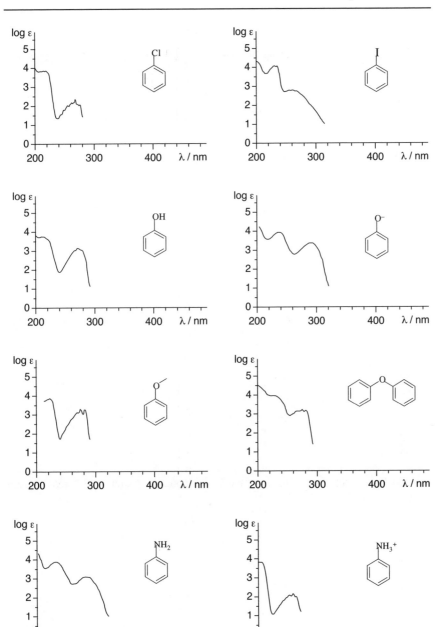

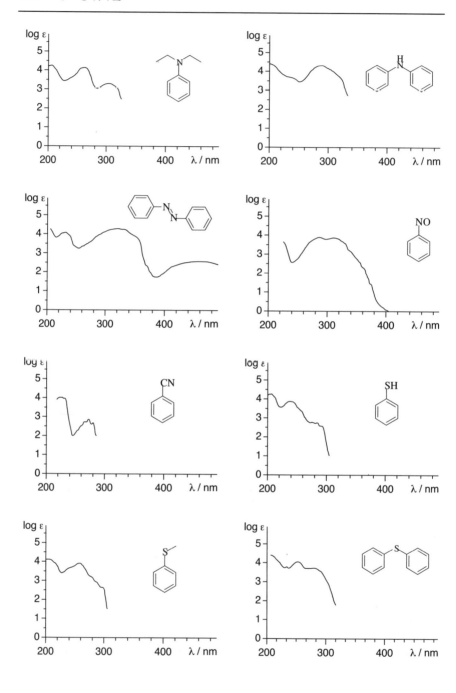

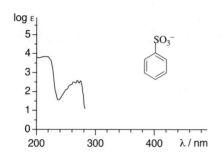

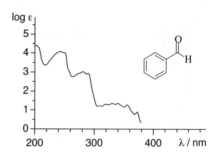

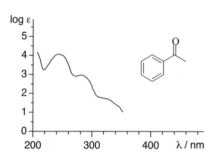

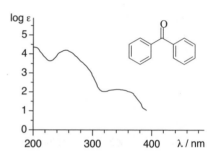

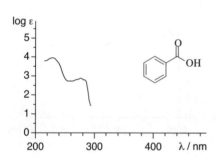

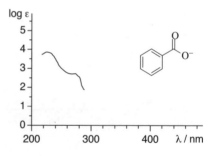

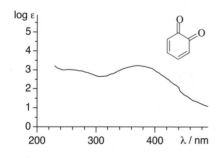

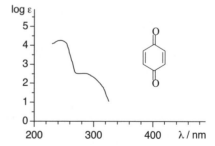

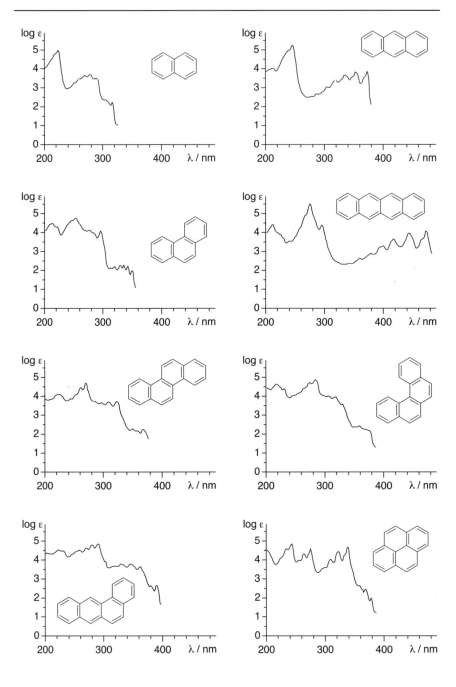

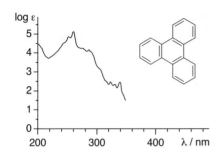

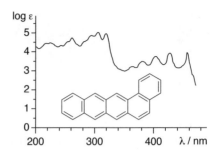

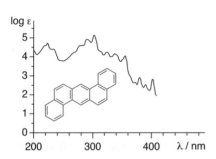

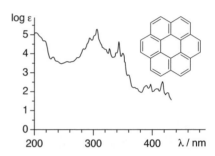

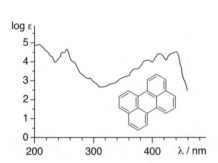

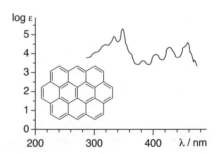

9.5.3 Heteroaromatic Compounds

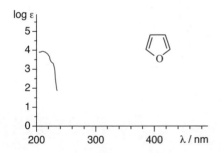

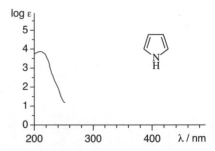

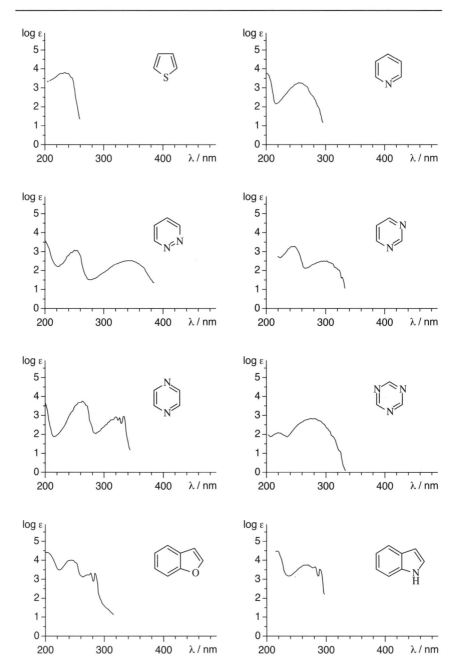

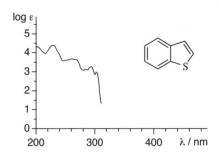

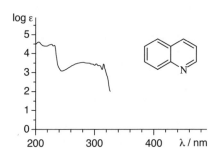

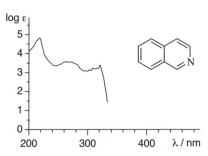

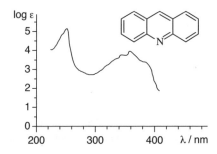

9.5.4 Miscellaneous Compounds

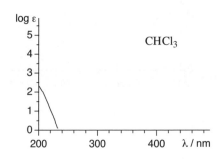

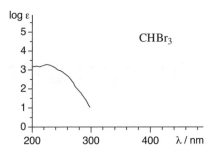

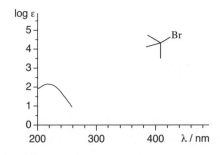

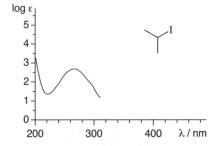

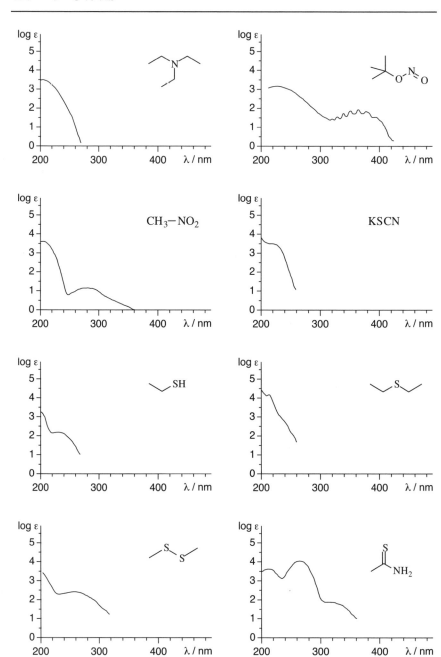

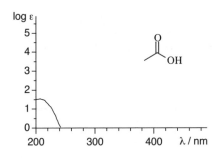

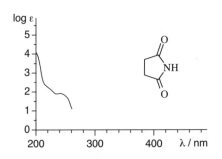

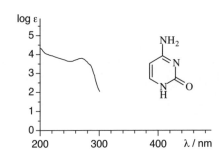

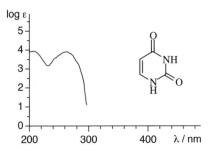

9.5.5 Nucleotides

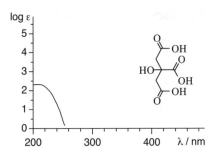

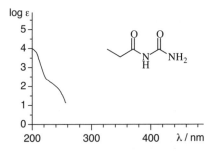

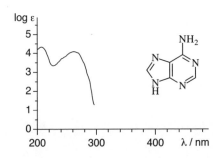

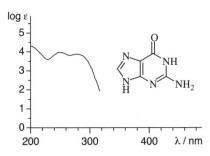

9.6 Common Solvents

The end absorption, λ_{end}, of several common solvents is given here as the wavelength at which the solvents absorb 80% of the irradiated light (λ_{end} in nm; cell length, 1 cm; reference, water).

Solvent	λ_{end}	Solvent	λ_{end}
acetone	335	ethyl acetate	205
acetonitrile	190	heptane	195
benzene	285	hexane	195
carbon disulfide	380	methanol	205
carbon tetrachloride	265	pentane	200
chloroform	245	2-propanol	205
cyclohexane	210	pyridine	305
dichloromethane	230	tetrahydrofuran	230
diethyl ether	210	toluene	285
1,4-dioxane	215	2,2,4-trimethylpentane	210
ethanol	205	xylene	290

Subject Index

Printing: Krips bv, Meppel, The Netherlands
Binding: Stürtz, Würzburg, Germany